BIOORGANIC CHEMISTRY

Volume IV Electron Transfer and Energy Conversion; Cofactors; Probes

CONTRIBUTORS

DANIEL I. ARNON

KATHRYN E. CARLSON

C. K. CHANG

BARRY S. COOPERMAN

DONALD J. CREIGHTON

MICHAEL A. CUSANOVICH

D. DOLPHIN

DAVID J. ECKERMAN

ANTHONY A. GALLO

PHILIP GREENWELL

JOSEPH HAJDU

RAYMOND J. HARRIS

SIDNEY M. HECHT

HOWARD J. JOHNSON, JR.

JOHN A. KATZENELLENBOGEN

ROBERT J. KEMPTON

R. J. KILL

RONALD KLUGER

EDWARD M. KOSOWER

D. MAUZERALL

JOHN J. MIEYAL

HARVEY N. MYERS

AKITSUGU NAKAHARA

YASUO NAKAO

HENRY Z. SABLE

DAVID S. SIGMAN

ROBERT H. SYMONS

T. G. TRAYLOR

ELIO F. VANIN

D. A. WIDDOWSON

OSAMU YAMAUCHI

BIOORGANIC CHEMISTRY

Edited by
E. E. van Tamelen
Department of Chemistry
Stanford University
Stanford, California

Volume IV
ELECTRON TRANSFER AND ENERGY CONVERSION; COFACTORS; PROBES

A treatise to supplement Bioorganic Chemistry:
An International Journal

Edited by
E. E. van Tamelen

EDITORIAL ASSOCIATES

D. H. R. Barton   A. Kornberg   Lord Todd

HONORARY BOARD OF ADVISORS

D. Arigoni   M. Calvin   D. E. Koshland, Jr.
A. R. Battersby   R. W. Holley   H. M. McConnell
K. Bloch   W. S. Johnson   A. S. Scott
R. Breslow   T. Kametani   L. L. M. van Deenen

ACADEMIC PRESS   New York   San Francisco   London   1978

A Subsidiary of Harcourt Brace Jovanovich, Publishers

COPYRIGHT © 1978, BY ACADEMIC PRESS, INC.
ALL RIGHTS RESERVED.
NO PART OF THIS PUBLICATION MAY BE REPRODUCED OR
TRANSMITTED IN ANY FORM OR BY ANY MEANS, ELECTRONIC
OR MECHANICAL, INCLUDING PHOTOCOPY, RECORDING, OR ANY
INFORMATION STORAGE AND RETRIEVAL SYSTEM, WITHOUT
PERMISSION IN WRITING FROM THE PUBLISHER.

ACADEMIC PRESS, INC.
111 Fifth Avenue, New York, New York 10003

United Kingdom Edition published by
ACADEMIC PRESS, INC. (LONDON) LTD.
24/28 Oval Road, London NW1 7DX

Library of Congress Cataloging in Publication Data

Main entry under title:

Bioorganic chemistry.

      Includes bibliographies and index.
      CONTENTS: v. 1. Enzyme   action.—v.2. Substrate
behavior.—v. 3. Macro- and multimolecular systems.
[etc.]
      1. Biological chemistry.   2. Chemistry, Organic.
I.   Van Tamelen, Eugene E., Date [DNLM: 1. Biochemistry.   2. Chemistry, Organic. QU4 B61597]
QP514.2.358   574.1'92   76-45994
ISBN 0-12-714304-1 (v. 4)

PRINTED IN THE UNITED STATES OF AMERICA

Contents

Chapter 1   Photosynthetic Phosphorylation: Conversion of Sunlight
             to Biochemical Energy
             DANIEL I. ARNON

1127

Chapter 2   Oxidation and Oxygen Activation by Heme Proteins
C. K. CHANG AND D. DOLPHIN

Chapter 3   Affinity Labeling Studies on *Escherichia coli* Ribosomes
BARRY S. COOPERMAN

Chapter 4   Mechanisms of Electron Transfer by High-Potential *c*-Type Cytochromes
MICHAEL A. CUSANOVICH

Chapter 5   Structural and Mechanistic Aspects of Catalysis by Thiamin
ANTHONY A. GALLO, JOHN J. MIEYAL, AND HENRY Z. SABLE

List of Contributors

Numbers in parentheses indicate the pages on which the authors' contributions begin.

DANIEL I. ARNON (1), Department of Cell Physiology, University of California, Berkeley, California

KATHRYN E. CARLSON (207), Department of Chemistry, University of Illinois, Urbana, Illinois

C. K. CHANG* (37), Department of Chemistry, The University of British Columbia, Vancouver, British Columbia, Canada

BARRY S. COOPERMAN (81), Department of Chemistry, University of Pennsylvania, Philadelphia, Pennsylvania

DONALD J. CREIGHTON (385), Department of Biological Chemistry and Molecular Biology Institute, UCLA School of Medicine, Los Angeles, California

MICHAEL A. CUSANOVICH (117), Department of Chemistry, University of Arizona, Tucson, Arizona

D. DOLPHIN (37), Department of Chemistry, The University of British Columbia, Vancouver, British Columbia, Canada

DAVID J. ECKERMAN (409), Department of Biochemistry, University of Adelaide, Adelaide, South Australia

ANTHONY A. GALLO (147), Department of Biochemistry, Case Western Reserve University, Cleveland, Ohio

PHILIP GREENWELL (409), Department of Biochemistry, National Institute for Medical Research, Mill Hill, London, England

JOSEPH HAJDU† (385), Department of Biological Chemistry and Molecular Biology Institute, UCLA School of Medicine, Los Angeles, California

*Present address: Department of Chemistry, Michigan State University, East Lansing, Michigan.

†Present address: Department of Chemistry, Boston College, Chestnut Hill, Massachusetts.

RAYMOND J. HARRIS (409), School of Pharmacy, South Australian Institute of Technology, Adelaide, South Australia

SIDNEY M. HECHT (179), Department of Chemistry, Massachusetts Institute of Technology, Cambridge, Massachusetts

HOWARD J. JOHNSON, JR. (207), Department of Biochemistry, University of Arkansas for Medical Sciences, Little Rock, Arkansas

JOHN A. KATZENELLENBOGEN (207), Department of Chemistry, University of Illinois, Urbana, Illinois

ROBERT J. KEMPTON (207), Department of Physical Sciences, Northern Kentucky University, Highland Heights, Kentucky

R. J. KILL (239), Department of Chemistry, Imperial College, London, England

RONALD KLUGER (277), Department of Chemistry, University of Toronto, Toronto, Ontario, Canada

EDWARD M. KOSOWER (293), Department of Chemistry, Tel Aviv University, Tel Aviv, Israel, and Department of Chemistry, State University of New York, Stony Brook, New York

D. MAUZERALL (303), The Rockefeller University, New York, New York

JOHN J. MIEYAL* (147, 315), Department of Biochemistry, Case Western Reserve University, Cleveland, Ohio

HARVEY N. MYERS (207), The Upjohn Company, Kalamazoo, Michigan

AKITSUGU NAKAHARA (349), Institute of Chemistry, College of General Education, Osaka University, Toyonaka, Osaka, Japan

YASUO NAKAO (349), Institute of Chemistry, College of General Education, Osaka University, Toyonaka, Osaka, Japan

HENRY Z. SABLE (147), Department of Biochemistry, Case Western Reserve University, Cleveland, Ohio

DAVID S. SIGMAN (385), Department of Biological Chemistry and Molecular Biology Institute, UCLA School of Medicine, Los Angeles, California

ROBERT H. SYMONS (409), Department of Biochemistry, University of Adelaide, Adelaide, South Australia

T. G. TRAYLOR (437), Department of Chemistry, University of California, San Diego, La Jolla, California

ELIO F. VANIN (409), Department of Biochemistry, University of Adelaide, Adelaide, South Australia

D. A. WIDDOWSON (239), Department of Chemistry, Imperial College, London, England

OSAMU YAMAUCHI (349), Institute of Chemistry, College of General Education, Osaka University, Toyonaka, Osaka, Japan

*Present address: Department of Pharmacology, Northwestern University Medical School, Chicago, Illinois.

Foreword

What is bioorganic chemistry? It is the field of research in which organic chemists interested in natural product chemistry interact with biochemistry. For many decades the natural product chemist has been concerned with the way in which Nature makes organic molecules. In the absence of any information other than that provided by structure, conclusions had necessarily to be derived from structural analysis. Broad groups of natural products could be recognized, such as alkaloids, isoprenoids, and polyketides (acetogenins), which clearly had elements of structure indicating a common biosynthetic origin. Indeed, for alkaloids and terpenoids, structural work was greatly helped by such biogenetic hypothesis. Similarly, after A. J. Birch had made an extensive analysis of polyketides, the repeating structural element postulated also helped in the determination of structure.

The alternative, and complement, to the above analysis is to consider the chemical mechanisms whereby the units of structure are assembled into the final natural product. For example, alkaloid structure can often be analyzed in terms of anion–carbonium ion combination. Also, the later stages of biosynthesis of many alkaloids can be analyzed by the concept of phenolate radical coupling. In polyisoprenoids the critical mechanism for carbon–carbon bond formation is the carbonium ion–olefin interaction to give a carbon–carbon bond and regenerate a further carbonium ion.

The analysis of natural product structures in terms of either structural units or mechanisms of bond formation has been subjected to rigorous tests since radioactively labeled compounds became generally available. It is gratifying that, on the whole, the theories developed from structural and mechanistic analysis have been fully confirmed by *in vivo* experiments.

Organic chemists have always been fascinated by the possibility of imitating in the laboratory, but without the use of enzymes, the precise steps of a biosynthetic pathway. Such work may be called biogenetic-type, or biomimetic, synthesis. This type of synthesis is a proper activity for the bioorganic chemist and undoubtedly deserves much attention. Nearly all such efforts are, however, much less successful than Nature's synthetic activities using enzymes. It is well appreciated that Nature has solved the outstanding problem of synthetic chemistry, viz., how to obtain 100% yield and complete stereospecificity in a chemical synthesis. It, therefore, remains a major task for bioorganic chemists to understand the mechanism of enzyme action and the precise reason why an enzyme is so efficient. We are still far from the day when we can construct an organic molecule which will be as efficient a catalyst as an enzyme but which will not be based on the conventional polypeptide chain.

Much of contemporary bioorganic chemistry is presented in these volumes. It will be seen that much progress has been made, especially in the last two decades, but that there are still many fundamental problems left of great intellectual challenge and practical importance.

The world community of natural product chemists and biochemists will be grateful to the editor and to all the authors for the effort that they have expended to make this work an outstanding success.

DEREK BARTON
Chemistry Department
Imperial College of Science and Technology
London, England

Preface

Although natural scientists have always been concerned with the development and behavior of living systems, only in the twentieth century have investigators been in a position to study on a molecular level the intimate behavior of organic entities in biological environments. By mid-century, the form and function of various natural products were being defined, and complex biosynthetic reactions were even being simulated in the nonenzymatic laboratory. As the cinematographic focus on biomolecules sharpened, one heard increasingly the adjective *bioorganic* applied to the interdisciplinary area into which such activity falls.

In 1971, publication of a new journal, *Bioorganic Chemistry*, was begun. As a follow-up, what could be more timely and useful than a well-planned, multivolume collection of bioorganic review articles, solicited from carefully chosen professionals, surveying the entire field from all possible vantage points? This four-volume work contains a collection, but it did not originate in this manner.

As the journal *Bioorganic Chemistry* developed, the number and quality of regular, original research articles were maintained at an acceptable level. However, comprehensive review articles appeared only sporadically, despite their intrinsic value at a time when general interest in bioorganic chemistry was burgeoning. In order to enhance this function of the journal, as well as to mark the fifth anniversary of its birth, we originally planned to publish in 1976 a special issue comprised entirely of reviews by active practitioners. After contact with a handful of stalwart bioorganic chemists, about two hundred written invitations for reviews were mailed during late 1975 to appropriate, diverse scientists throughout

the world. The response was overwhelming! More than seventy preliminary acceptances were received within a few months, and it soon became
evident that the volume could not be handled adequately through publication by journal means. After consultation with representatives of Academic
Press, we agreed to publish the manuscripts in book form.

Although the stringency of journal deadlines disappeared, the weightier
matter of editorial treatment had to be reconsidered. Should contributions
be published in the same, piecemeal, random fashion as received? Such
practice would be acceptable for journal dissemination, but for book
purposes, broader, more orderly, and inclusive treatment might be desir-
able and also expected. Partly because of editorial indolence, but mostly
because of a predilection for maintaining the candor and spontaneity
which might be lost with increased editorial control, we decided not to
attempt coverage of all identifiable areas of bioorganic chemistry, not to
seek out preferentially the recognized leaders in particular areas, and
even not to utilize outside referees. Consequently, we present reviews
composed by scientists who were not coerced or pressured, but who
wrote freely on subjects they wanted to write about and treated them as
they wanted to, at the cost perhaps of a certain amount of objectivity and
restraint as well as proper coverage of some important bioorganic areas.

We turn now to the results of this publication project. Because of the
inevitable attrition for the usual reasons, fewer than the promised number
of reviews materialized: fifty-seven manuscripts were received in good
time and accepted by this office. Eight countries are represented by the
entire collection, which emanates almost entirely from academia, as
would be expected. A great variety of topics congregated—greater than
we had foreseen. Inclusion of all papers in one volume was impractical,
and thus the problem arose of logically dividing the heterogeneous material into several unified subsections, each suitable for one volume, a
problem compounded by the fact that an occasional author elected to
treat, in one manuscript, several unconnected topics happening to fall in
his purview. Therefore, perfect classification without discarding or dis-
secting bodies of material as received was simply not possible.

After some reflection and a few misconceptions, we evolved a plan for
division into four more or less scientifically integral sections; these, hap-
pily, also constitute approximately equal volumes of written material, an
aspect of some importance to the publisher. The enzyme–substrate interaction was expected to be a well-represented subject, and, in fact, too
many manuscripts on this subject for one proportionally sized volume
were received. Although the separation of enzyme action and substrate
behavior is contrived and not basically justifiable, it turned out that, for
the most part, a group of authors heavily emphasized the former, while
another concentrated on the latter. Accordingly, Volume I was entitled

"Enzyme Action," and Volume II "Substrate Behavior." Admittedly, in a few cases, articles could be considered appropriate for either volume.

A gratifyingly significant number of contributions dealt with the behavior of biologically important polymers and related matters, sent in by authors having quite different investigational approaches. In addition, several discourses were concerned with molecular aggregates, e.g., micelles. All of these were incorporated into Volume III, "Macro- and Multimolecular Systems."

Whatever papers did not belong in Volumes I–III were combined and constitute Volume IV. Fortunately, in these remaining papers some elements of unity could be discerned; in fact, their entire content falls into the following categories: "Electron Transfer and Energy Conversion (photosynthesis, porphyrins, NAD^+, cytochromes); Cofactors (coenzymes, NAD^+, metal ions); Probes (cytokinin behavior, steroid hormone action, peptidyl transferase reactivity)."

Finally, early in this enterprise, we asked Derek Barton to compose a Foreword. Sir Derek complied graciously, and in every volume his personalized view on the nature of bioorganic chemistry appears.

E. E. VAN TAMELEN

Contents of Other Volumes

VOLUME III   Macro- and Multimolecular Systems

1

Photosynthetic Phosphorylation: Conversion of Sunlight to Biochemical Energy

Daniel I. Arnon

INTRODUCTION

All life forms, except a few chemosynthetic bacteria, obtain, directly or indirectly, their energy from sunlight through the process of photosynthesis. Until recently, it was a settled principle of biology that living cells obtain the energy needed for life by consuming and degrading foodstuffs of direct or indirect photosynthetic origin. In biochemical terms, cells were thought to depend on respiration or fermentation for the release of reducing power and free energy needed for the formation of ATP, the universal cellular energy carrier. ATP and reducing power, singly or jointly, drive the multitude of endergonic reactions essential for life.

This concept remains valid for nonphotosynthetic cells, but photosynthetic cells are now known to have the capacity for generating ATP and reducing power, independently of fermentation and respiration, by the process of photosynthetic phosphorylation (photophosphorylation). Photophosphorylation denotes the ability of the photosynthetic apparatus to use directly the electromagnetic energy of sunlight to generate ATP and reducing power for cellular use. Photophosphorylation thus constitutes the conversion of solar energy into the main forms of biochemical energy that first enter the biosphere via photosynthesis. Photosynthetic cells conserve and store the energy of the

1

photochemically generated ATP and reducing power through the biosynthesis of organic compounds from CO_2. When these are later degraded by fermentation and respiration (oxidative phosphorylation), ATP and reducing power are regenerated for general use in cellular metabolism.

The aim of this chapter is to trace the emergence and development of our knowledge of photophosphorylation. No attempt is made to deal with certain topics, such as coupling factors and proton gradients, that are applicable to both photophosphorylation and oxidative phosphorylation; these, along with other topics, were recently reviewed by Jagendorf [1].

ROLE OF LIGHT IN PHOTOSYNTHESIS

Diverse Hypotheses

The first hypothesis about the role of light in what we now call photosynthesis is appropriately linked with the discoverer of the photochemical nature of this process, Jan Ingenhousz, who wrote in 1796 that the green plant absorbs from "carbonic acid in the sunshine, the carbon, throwing out at that time the oxygen alone, and keeping the carbon to itself as nourishment" [2]. The idea that light liberates oxygen by photodecomposition of CO_2 had, with some modifications, dominated photosynthesis research for well over a century and had attracted the support of some of the most illustrious chemists and biochemists of the time, e.g., von Baeyer [3], Willstätter [4], and Warburg [5]. After it was shown that water is a reactant [6] and carbohydrates are the first products of photosynthesis [7,8], the CO_2 cleavage hypothesis readily accounted for what became the deceptively simple overall reaction of photosynthesis [Eq. (1), where (CH_2O) represents one-sixth of a glucose molecule].

$$CO_2 + H_2O \xrightarrow{h\nu} (CH_2O) + O_2 \tag{1}$$

The C:2H:O proportions of the carbohydrate products of photosynthesis were viewed as the obvious consequence of, and indisputable evidence for, the recombination of the carbon released by the photodecomposition of CO_2 with the elements of water.

In contemporary photosynthesis research, most investigators, with the notable exception of Warburg and his associates [5], abandoned the hypothesis of the photodecomposition of CO_2 in favor of an alternative hypothesis put forward by van Niel [9] on the basis of his studies of bacterial photosynthesis. Van Niel proposed that both bacterial and plant photosynthesis are special cases of a general process in which light energy is used to photo-decompose a hydrogen donor (H_2A) into hydrogen, used for the reduction of CO_2, and a residue ("A") that becomes the excreted by-product of photosynthesis. In bacterial photosynthesis H_2A may be a compound such as H_2S that gives rise to the liberation of elemental sulfur. Plant photosynthesis is unique in that it uses water as the hydrogen donor with the important consequence that all of the oxygen produced in plant photosynthesis comes from water and not from CO_2 [Eq. (2)].

$$CO_2 + 2H_2O^* \xrightarrow{h\nu} (CH_2O) + H_2O + O_2^* \qquad (2)$$

In later formulations [10,11], van Niel no longer considered the photo-decomposition of water as being unique to plant photosynthesis but postulated that "the photochemical reaction in the photosynthetic process of green bacteria, purple bacteria, and green plants represents, in all cases, a photodecomposition of water" [11]. According to this revised concept, the distinction between plant and bacterial photosynthesis turned on the events that followed the photodecomposition of water into [H] and [OH]. The [H] was used for CO_2 reduction, and [OH] formed a complex with an appropriate acceptor. In plant photosynthesis, the acceptor was regenerated when the complex was decomposed by liberation of molecular oxygen. In bacterial photosynthesis, oxygen was not liberated and the acceptor could be regenerated only when the [OH]–acceptor complex was reduced by a special hydrogen donor (other than water) that is always required in bacterial photosynthesis.

Neither the concept of photodecomposition of CO_2 nor that of photo-decomposition of water envisaged a light-induced formation of ATP. The first comprehensive hypothesis that the role of light in photosynthesis may be to generate ATP was formulated by Ruben [12], who suggested that photosynthetic CO_2 assimilation is a dark process dependent on photochemically generated ATP and reduced pyridine nucleotides. Although brilliant in conception, Ruben's hypothesis, lacking any supportive experimental evidence, could not be effectively defended against the theoretical objections that were soon raised by Rabinowitch [13], who argued that the use of light quanta (containing about 43 kcal/Einstein) for the synthesis of ATP molecules (storing about 10 kcal/mole) constituted "*dissipation* rather than . . . accumulation of energy" (original emphasis).

These theoretical objections were directed not only against the hypothesis of Ruben [12] but even more so against the more extreme hypothesis of Emerson, Stauffer, and Umbreit [14], who proposed, on the basis of experiments with intact *Chlorella* cells, that the "*sole function* of light energy in photosynthesis is the formation of energy-rich phosphate bonds" (emphasis added). No strong evidence was presented in support of this far-reaching hypothesis [13]. It is not surprising, therefore, that it, as well as Ruben's hypothesis, had little influence on subsequent research in photosynthesis.

Energy Requirements for Carbon Assimilation

In marked contrast to the controversy over speculative or insufficiently documented earlier proposals for a role of ATP in photosynthesis was new evidence that began to emerge from research on photosynthetic CO_2 assimilation after the introduction of powerful new research tools, such as radioisotopes (particularly that of ^{14}C [15]), paper chromatography [16], and radioautography [17]. With their use, Calvin, Benson, and their associates discovered phosphorylated compounds—pointing to an involvement of ATP—among the early products of $^{14}CO_2$ assimilation; the products included phosphoglyceric acid (the first stable product) and phosphate esters of ribulose, sedoheptulose, and other sugars [18–20]. These findings culminated in the formulation of a photosynthetic carbon cycle (reductive pentose phosphate cycle) which became accepted as the major pathway of photosynthetic carbon assimilation in green plants [21,22].

All of the component steps of the carbon cycle proved to be enzymatic reactions that in a direct sense are independent of light. Indeed, the entire reductive pentose phosphate cycle has also been identified in CO_2 assimilation by autotrophic, nonphotosynthetic bacteria (see, e.g., Trudinger [23] and Aubert *et al.* [24]). Insofar as the light reactions of photosynthesis were concerned, the reductive pentose phosphate cycle clarified for the first time the energy demands for photosynthetic CO_2 assimilation. The cycle revealed which of its component enzymatic reactions require an input of energy-rich chemical intermediates that must be formed at the expense of light energy. The cycle showed that the conversion of 1 mole of CO_2 to the level of hexose phosphate requires 3 moles of ATP and 2 moles of reduced NAD(P), i.e., an ATP:NAD(P)H ratio of 1.5. The need for light energy in photosynthesis could thus be traced to those photochemical reactions that generate ATP and reduced pyridine nucleotide, two energy-rich products that jointly constitute the assimilatory power required for CO_2 assimilation.

PHOTOCHEMICAL REACTIONS OF
PHOTOSYNTHESIS

Experiments with Whole Cells

The elucidation of the photosynthetic carbon cycle made it clear that the required ATP and reduced pyridine nucleotides must be formed by photochemical reactions prior to CO_2 assimilation. It was supposed at first that these early photochemical reactions might be demonstrated by kinetic experiments with whole cells, using tracers and short exposures to light. Similar kinetic experiments had played a crucial role in the identification of the early products of CO_2 assimilation. But kinetic experiments with whole cells provided no direct evidence for the photoproduction of either ATP or reduced pyridine nucleotides. For example, an investigation of the photo-assimilation of phosphorus by *Scenedesmus* cells with the aid of carrier-free ^{32}P showed that the shortest exposure to light gave the lowest incorporation of ^{32}P into ATP; the highest incorporation of ^{32}P into ATP occurred not in the light but after a brief exposure of the cells to $KH_2{}^{32}PO_4$ in the dark [25]. The first compound to be labeled in the light was not the expected ATP but 3-phosphoglycerate [25].

Other investigations of light-induced ATP formation in intact cells prior to CO_2 assimilation gave results that were at best suggestive (see Arnon [26]). In short, experiments with whole cells proved, for reasons discussed below, incapable of yielding evidence for a special, light-induced phosphorylation or for the photoreduction of pyridine nucleotides. The occurrence of such light reactions in photosynthesis was discovered not in whole cells but in isolated chloroplasts.

Experiments with Isolated Chloroplasts

Chloroplasts were once widely believed to be the site of complete photosynthesis, but this view was not supported by critical evidence [27,28] and was largely abandoned after Hill [29–31], who initiated the modern period of experimentation with isolated chloroplasts, demonstrated that chloroplasts could evolve oxygen but could not assimilate CO_2. (The failure of isolated chloroplasts to fix CO_2 was also reported with the sensitive $^{14}CO_2$ technique [32].) In the oxygen-producing reaction, isolated chloroplasts evolved oxygen only in the presence of artificial oxidants with distinctly positive oxidation–reduction potentials, e.g., ferric oxalate, ferricyanide, and benzoquinone [31,33].

Oxygen production by isolated chloroplasts, now appropriately known as the Hill reaction, established that oxygen evolution by chloroplasts was basically independent of CO_2 assimilation. This evidence provided strong support for van Niel's concept that the source of oxygen in plant photosynthesis is water. The importance of the Hill reaction was that it provided compelling evidence for this conclusion, whereas (contrary to a widely held belief) other evidence on this point from experiments with ^{18}O was not decisive [34,35].

Left in doubt was the role of chloroplasts in the energy-storing reactions needed for CO_2 assimilation. The photochemical generation by isolated chloroplasts of a strong reductant capable of reducing CO_2 was deemed unlikely on experimental and theoretical grounds [31,33]. Likewise, the first experiments with the sensitive ^{32}P technique to test the ability of isolated chloroplasts to form ATP on illumination gave negative results [36].

A different perspective on the photosynthetic capacity of isolated chloroplasts began to emerge in 1951, when three laboratories [37–39] independently and simultaneously found that isolated chloroplasts could photoreduce pyridine nucleotides despite their strongly electronegative oxidation–reduction potential ($E_m = -320$ mV at pH 7). Vishniac and Ochoa [37] and Tolmach [38] observed that chloroplasts could photoreduce either NAD^+ or $NADP^+$; Arnon [39] observed only the photoreduction of $NADP^+$. These findings were followed by several other developments that drastically altered the then-prevalent ideas about the photosynthetic capacity of isolated chloroplasts.

In 1954, a reinvestigation of phorosynthesis in isolated chloroplasts by different methods yielded evidence for a light-dependent assimilation of CO_2 [40]. Chloroplasts isolated from spinach leaves assimilated [14]CO_2 to the level of carbohydrates, including starch, with a simultaneous evolution of oxygen [41,42]. When the conversion of [14]CO_2 by isolated chloroplasts to sugars and starch was confirmed and extended in other laboratories [43–47], the capacity of chloroplasts to carry on complete extracellular photosynthesis was no longer open to question.

Because of the earlier negative results, special experimental safeguards were used to establish that chloroplasts alone, without other organelles or enzyme systems and with light as the only energy source, were capable of a total synthesis of carbohydrates from CO_2. The chloroplasts were washed, and, to eliminate possible sources of chemical energy and intermediates, their isolation was performed in isotonic sodium chloride [40,42]. In comparison with the parent leaves, chloroplasts so treated gave low rates of CO_2 assimilation [42], a situation reminiscent of the first reconstruction of other cellular processes *in vitro*, e.g., fermentation [48,49], protein synthesis [50], and polymerization of DNA [51].

Crucial for the documentation of complete photosynthesis in isolated chloroplasts were not high rates but the fact that their newly found CO_2 assimilation was reproducible and yielded the same intermediate and final products as did photosynthesis by intact cells. More recently, when CO_2 assimilation by chloroplasts ceased to be a matter of dispute and the special experimental safeguards were no longer needed, chloroplasts that were isolated by modified procedures gave much higher rates of CO_2 assimilation that more closely resembled those of intact leaves [52–55].

The similarity of products indicated that the energy requirements for the light-dependent CO_2 assimilation by isolated chloroplasts were no different from those in intact cells. Since neither ATP nor reduced pyridine nucleotides were added to isolated chloroplasts that fixed CO_2, it was apparent that these energy-rich compounds were being photochemically generated from their respective precursors within the chloroplasts. As discussed above, chloroplasts had been observed to photoreduce added NAD^+ or $NADP^+$, but nothing was known about their ability (or that of any other photosynthetic structures) to form ATP at the expense of light energy. A renewed attack on this problem in isolated chloroplasts was therefore undertaken.

DISCOVERY OF PHOTOSYNTHETIC PHOSPHORYLATION

The likelihood of detecting a direct role of light in ATP formation was much greater in isolated chloroplasts than in intact cells. Chloroplasts cannot respire, they lack even the terminal respiration enzyme, cytochrome oxidase [56,57]. This feature insured that the photochemically generated ATP would not be confused with the ATP formed by respiration, a possibility that could not be excluded with certainty in intact cells. Moreover, intact cells contain only catalytic amounts of the precursors AMP and ADP, and these, because of permeability barriers, could not be increased by external additions. By contrast, in experiments with isolated and fragmented chloroplasts, it was possible to supply these normally catalytic substances in substrate amounts and to determine chemically their light-induced conversion to ATP. These features of isolated chloroplasts proved to be of great experimental advantage in the discovery and documentation of light-induced formation of ATP.

Chloroplasts at first were assigned an indirect role in light-induced ATP formation. The only cytoplasmic organelles known to form ATP were mitochondria, and the most plausible model for ATP formation in photosynthesis became one that envisaged a collaboration between chloroplasts and mitochondria [58]. In that collaboration, chloroplasts were thought to reduce NAD^+ photochemically, and mitochondria were thought to reoxidize NADH with oxygen and to form, concurrently, ATP by oxidative phosphorylation [58].

This model posed serious difficulties. First, isolated chloroplasts, without the collaboration of mitochondria, were found to be capable of CO_2 assimilation, a process that requires ATP. Second, photosynthesis in saturating light can proceed at a rate almost 30 times greater than the rate of respiration or oxidative phosphorylation. It was difficult to see, therefore, how oxidative

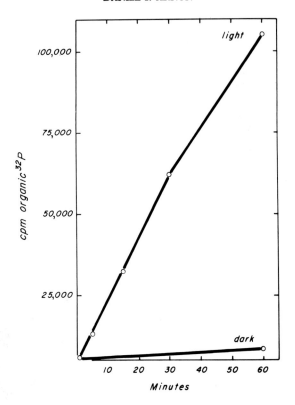

Fig. 1 Early evidence for light dependence of photophosphorylation [61]. (Reprinted with permission from Arnon *et al.*, *J. Am. Chem. Soc.* **76**, 6324–6329 (1954). Copyright by the American Chemical Society.)

phosphorylation by mitochondria could cope with the ATP requirements of photosynthesis.

In 1954, work with the same spinach chloroplast preparations that fixed CO_2 led to the discovery that they were able to use light energy for the formation of ATP, without the collaboration of mitochondria [59–61]. Several unique features distinguished photophosphorylation by chloroplasts (Fig. 1) from substrate-level phosphorylation in fermentation and oxidative phosphorylation in respiration: (a) ATP formation occurred in the chlorophyll-containing lamellae and was independent of other enzyme systems or organelles; (b) no energy-rich substrate, other than absorbed photons, served as a source of energy; (c) no oxygen was produced or consumed; and (d) ATP formation was not accompanied by a measurable oxidation or reduction of any external electron donor or acceptor [59–61]. The light-induced ATP formation could be expressed by Eq. (3).

$$ATP + P_1 \xrightarrow[\text{chloroplasts}]{h\nu} ATP \tag{3}$$

When the discovery of photophosphorylation in chloroplasts was followed by evidence of a similar phenomenon in cell-free preparations of diverse types of photosynthetic organisms, i.e., photosynthetic bacteria [62] and algae [63,64], it became evident that photophosphorylation is a major ATP-forming process in nature that supplies ATP for the biosynthetic reactions in all types of photosynthetic cells, including those that do not contain chloroplasts.

A question arose initially whether one fundamental property of photophosphorylation in chloroplasts, its independence from chemical substrates also applied to bacterial photophosphorylation. Frenkel [62] reported that photophosphorylation in a cell-free preparation from *Rhodospirillum rubrum* became dependent on a substrate (α-ketoglutarate) when the chlorophyll-containing particles were washed. However, in later experiments, Frenkel [65] and other investigators [66–68] found that the role of α-ketoglutarate and other organic acids in the bacterial system was regulatory and not that of substrate. When this basic point was clarified, the fundamental similarity of photophosphorylation in chloroplasts and in bacterial membrane preparations was no longer in doubt.

Photophosphorylation was discovered in isolated, whole chloroplasts to which no cofactors were added to increase the rate of ATP formation. Without the addition of cofactors, the integrity of the chloroplast structure was essential for ATP formation; chloroplast fragments, which gave good rates of Hill reaction, had only feeble photophosphorylation activity [60,69]. The rates of photophosphorylation in whole chloroplasts, even though much higher than in chloroplast fragments, were, like those of CO_2 assimilation, still low enough to give rise to questions about the quantitative importance of photophosphorylation in photosynthesis [70]. With further improvements in experimental methods (which included the addition of such cofactors as menadione and the use of broken chloroplasts with lowered permeability barriers) rates of photophosphorylation increased 170 times [71] and with phenazine methosulfate increased even more [72] over those originally described [60,61].

The improved rates of photophosphorylation were equal to or greater than the maximum known rates of carbon assimilation in intact leaves. It appeared, therefore, that isolated chloroplasts retain, without substantial loss, the enzymatic apparatus for photophosphorylation, a conclusion in harmony with evidence that the phosphorylating system was tightly bound in the water-insoluble lamellar portion of the chloroplasts.

CYCLIC PHOTOPHOSPHORYLATION AND
THE CONCEPT OF A LIGHT-INDUCED
ELECTRON FLOW

Once the main features of photophosphorylation were firmly established, the next objective was to explain its mechanism, particularly its absolute dependence on illumination (Fig. 1). On the one hand, photophosphorylation was independent of such classical manifestations of photosynthesis as oxygen evolution and CO_2 assimilation; on the other hand, it seemed unlikely that light was involved in the formation of ATP itself, a reaction universally occurring in all cells independently of photosynthesis. Light energy, therefore, had to be used in photophosphorylation before ATP synthesis and in a manner unrelated to CO_2 assimilation or oxygen evolution. The most probable mechanism for such a role seemed to be a light-induced electron flow [73,74].

It is often difficult for the student of photosynthesis today to realize that before the discovery of photophosphorylation the concept of a light-induced electron transport had no substantial basis in photosynthesis research. The idea that photon energy is used in photosynthesis to transfer electrons rather than cumbersome atoms had a few proponents at various times, for example, Katz [75] and Levitt [76,77], but, as the literature before the late 1950's shows, it did not become a viable concept in photosynthesis, it was merely one of several speculative ideas based on model systems. The situation changed with the discovery of light-induced ATP formation. ATP is formed in nonphotosynthetic cells at the expense of energy released by electron transport. The idea that ATP may also be formed in photosynthesis through a special, light-induced electron flow mechanism in chloroplasts now had a high probability that could be tested experimentally.

After the abandonment of early ideas that linked photophosphorylation to a photolysis of water and recombination of the resultant [H] and [OH] moieties [41,69], a hypothesis was put forward that a chloroplast molecule, on absorbing a quantum of light, becomes excited and promotes an electron to an outer orbital with a higher energy level [73,74]. This high-energy electron is then transferred to an adjacent electron-acceptor molecule, a catalyst (A) with a strongly electronegative oxidation–reduction potential. The transfer of an electron from excited chlorophyll to this first acceptor is the energy conversion step proper and terminates the photochemical phase of the process. By transforming a flow of photons into a flow of electrons, it constitutes a mechanism for the generation of a strongly electronegative reductant at the expense of the excitation energy of chlorophyll.

Once the strongly electronegative reductant molecule is formed, no further input of energy is needed. Subsequent electron transfers within the chloroplast liberate energy because they provide an electron flow from the electronegative reductant to electron acceptors (thought then to include chloroplast cytochromes) with more electropositive oxidation–reduction potentials [73,74]. Several of the exergonic electron transfer steps, particularly those involving cytochromes, were thought to be coupled with phosphorylation. At the end of one complete cycle, the emitted electron is returned to the electron-deficient chlorophyll molecule and the quantum absorption process is repeated.

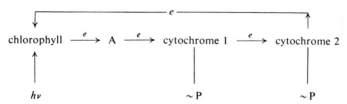

A mechanism of this kind explained why photophosphorylation was not accompanied by any net oxidation–reduction change and did not require any external electron donor or acceptor. Because of the envisaged cyclic pathway traversed by the emitted electron, the process was named *cyclic photophosphorylation* [73,74].

A cyclic electron flow that is driven by light and that liberates chemical energy, used for the synthesis of the pyrophosphate bonds of ATP, is unique to photosynthetic cells. As discussed elsewhere [73,74], cyclic photophosphorylation may be a primitive manifestation of photosynthetic activity that remains as a common denominator of plant and bacterial photosynthesis.

NONCYCLIC PHOTOPHOSPHORYLATION

At first there was no experimental evidence to link photophosphorylation to the generation of reducing power that, jointly with ATP, was required for CO_2 assimilation. In fact, the photoproduction of a reductant and photophosphorylation by chloroplasts were once thought to be antagonistic [69]. It was therefore wholly unexpected when a second type of photophosphorylation was discovered that provided direct evidence for a coupling between photoreduction of $NADP^+$ and the synthesis of ATP by chloroplasts [78]. Here, in contrast to cyclic photophosphorylation, the formation of 1ATP was stoichiometrically coupled with a light-driven transfer of a pair of electrons from water to $NADP^+$ (or to a nonphysiological electron acceptor, such as

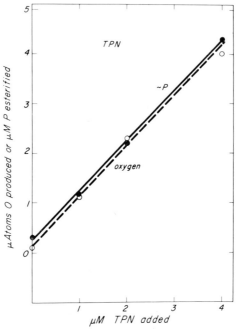

Fig. 2 Stoichiometry of noncyclic photophosphorylation; relation between moles of ATP formed ($\sim$P), oxygen evolved (μatoms), and NADP$^+$ (TPN) reduced [78]. [From Arnon *et. al.*, *Science* **127**, 1026–1034 (1958).]

ferricyanide) and a concomitant evolution of oxygen, i.e., $P/e_2 = P/O = 1$ [Fig. 2 and Eq. (4)].

$$NADP^+ + H_2O + ADP + P_i \xrightarrow[\text{chloroplasts}]{hv} NADPH + \tfrac{1}{2}O_2 + ATP + H^+ \quad (4)$$

A striking feature of this new chloroplast reaction was that the rate of the light-induced, thermodynamically "uphill" electron transport from water to NADP$^+$ was greatly increased by the addition of ADP (in the presence of P$_i$); that is, the rate of electron transport was at its maximum when it was coupled (as it would be under physiological conditions) to the synthesis of ATP [78,79]. Thus, photosynthetic electron transport in chloroplasts, like respiratory electron transport in mitochondria, was found to be under *phosphorylation control*. The phosphorylation control of photosynthetic electron transport was soon confirmed [80,81] and extended by the discovery that ammonium ions effectively uncouple electron transport from photophosphorylation in chloroplasts [82].

The electron flow concept postulated for cyclic photophosphorylation was also applicable to the new type of photophosphorylation [73,74]. It was envisaged that a chlorophyll molecule excited by a captured photon transfers an electron to $NADP^+$ and that electrons thus removed from chlorophyll are replaced by electrons from water (OH^- at pH 7) with a resultant evolution of oxygen. In this manner, light would induce an electron flow from OH^- to $NADP^+$ and a coupled phosphorylation. Because of the unidirectional, or noncyclic, nature of this electron flow, this process was named *noncyclic photophosphorylation* [74,83].

PHYSICAL SEPARATION OF LIGHT AND DARK PHASES OF PHOTOSYNTHESIS IN CHLOROPLASTS

It is clear from the foregoing that in the late 1950's work with isolated chloroplasts laid the basis for a concept that the light phase of photosynthesis consists of the photoproduction of ATP and NADPH, which are then used for CO_2 assimilation during a "dark" phase consisting of enzymatic reactions independent of light. The validity of this concept was directly substantiated through a physical separation of the light and dark phases of photosynthesis by fractionation of isolated chloroplasts [84]. The light phase was completed first by the complete chloroplast system in the absence of CO_2 and resulted in an evolution of oxygen accompanied by an accumulation of substrate amounts of NADPH and ATP in the reaction mixture. The green lamellar portion of the chloroplasts (grana) was then removed, and CO_2 was supplied to the chlorophyll-free extract (stroma) in the dark.

In the presence of substrate amounts of NADPH and ATP, the chlorophyll-free extract was able to fix $^{14}CO_2$ in the dark. Only feeble $^{14}CO_2$ fixation occurred in the dark without ATP and NADPH. The dark fixation of CO_2 by the chlorophyll-free extract supplemented with these two components of assimilatory power was comparable with fixation in the light by the complete system, to which $^{14}CO_2$ had been added at the beginning of the illumination period. In the complete chloroplast system, prior accumulation of assimilatory power was not needed because it was formed continuously in the light and used at once for CO_2 fixation. Very little CO_2 fixation occurred in the complete chloroplast system in the dark.

The products of CO_2 assimilation were found to be the same whether CO_2 assimilation occurred during continuous illumination or in the dark at the expense of assimilatory power generated during a preceding light period. The products included hexose and pentose monophosphates and diphosphates, phosphoglyceric acid, dihydroxyacetone phosphate, and small amounts of phosphoenolpyruvate and malate.

ROLE OF CYCLIC PHOTOPHOSPHORYLATION:
EARLY VIEWS

Prior to the discovery of noncyclic photophosphorylation, cyclic photophosphorylation was regarded as the only source of ATP for CO_2 assimilation [69]. This view could no longer be sustained when noncyclic photophosphorylation was found to account for all three products of the light phase of photosynthesis: ATP, NADPH, and O_2. Three possibilities had to be considered with respect to cyclic photophosphorylation: (a) that it produced ATP for processes other than CO_2 assimilation, (b) that it contributed ATP for CO_2 assimilation to supplement the insufficient ATP generated by noncyclic photophosphorylation, and (c) that it was an experimental artifact, i.e., that noncyclic photophosphorylation was the only physiological source of ATP in photosynthesis.

Arnon, Whatley, and Allen [78] favored possibilities (a) and (b). They suggested a physiological role for cyclic photophosphorylation as a supplementary source of ATP for CO_2 assimilation and also as a source of photochemically generated ATP when CO_2 assimilation was, for one reason or another, diminished or stopped altogether. In higher plants, this situation might arise during the well-known midday closure of stomata [85,86]. Thus, cyclic photophosphorylation would be a source of ATP for diverse metabolic purposes that might or might not include CO_2 assimilation.

When the observed stoichiometry of noncyclic photophosphorylation (ATP:NADPH = 1) was compared with the ATP:NADPH ratio of 1.5 required for the conversion of CO_2 to carbohydrates, the need for extra ATP in CO_2 assimilation became apparent. Early evidence that the extra ATP could be supplied by cyclic photophosphorylation came from experiments on CO_2 assimilation with reconstituted chloroplast systems that were supplied with catalytic amounts of ADP and $NADP^+$ under three conditions: (a) when the photochemical phase was limited to noncyclic photophosphorylation, (b) when the photochemical phase was limited to cyclic photophosphorylation, and (c) when the photochemical phase included both cyclic and noncyclic photophosphorylation [87].

Under condition (a) or (b), CO_2 assimilation was limited almost entirely to the formation of phosphoglycerate. Sugar phosphates were the predominant products of CO_2 assimilation only under condition (c). These results were interpreted as having been caused by a shortage of ATP in condition (a) and of NADPH in condition (b). Only in condition (c), when cyclic and noncyclic photophosphorylation operated concurrently, was a balance maintained between ATP and NADPH that made possible the formation of sugar phosphates, i.e., complete CO_2 assimilation [87].

Direct as this evidence was, it did not prevent later doubts about the role of cyclic photophosphorylation in chloroplasts. In the experiments on CO_2 assimilation, the balance between noncyclic and cyclic photophosphorylation that was needed to bring about the formation of sugar phosphates was achieved by adding to the reconstituted chloroplast system one of the then-known catalysts of cyclic photophosphorylation: FMN [88], menadione (vitamin K_3) [89], or a substance foreign to chloroplasts, phenazine methosulfate [72]. Each of these three substances was found to be an effective catalyst that greatly increased the rate of cyclic photophosphorylation under anaerobic conditions. Since FMN and menadione (as a component of vitamin K_1) could be regarded as natural components of chloroplasts, they were tentatively assumed to be physiological catalysts of cyclic photophosphorylation [83]. This assumption, however, was open to question in view of the even greater catalytic effectiveness of phenazine methosulfate, a substance foreign to chloroplasts.

To recapitulate, the capacity of isolated chloroplasts to carry on cyclic and noncyclic photophosphorylation became well established, and there was experimental evidence that both types might be needed for CO_2 assimilation. However, the identity of the physiological catalyst of cyclic photophosphorylation was unknown, and this uncertainty raised doubts about the physiological role of the process. The situation changed drastically with the discovery of ferredoxin.

FERREDOXINS IN CHLOROPLASTS AND BACTERIA

A major advance in modern photosynthesis research was the new knowledge of the key role of ferredoxins in that process. The name "ferredoxin" was introduced by Mortenson, Valentine, and Carnahan [90] to describe an iron-containing protein that they isolated from *Clostridium pasteurianum*. In *C. pasteurianum* and in other nonphotosynthetic anaerobic bacteria in which it was later found, ferredoxin appeared to function as an electron carrier either between molecular hydrogen (activated by hydrogenase) and various electron acceptors or in the breakdown of compounds that, like pyruvate, generate strong reducing power.

The isolation of ferredoxin from *C. pasteurianum*, an anaerobic bacterium devoid of chlorophyll and normally living in the soil at a depth to which sunlight does not penetrate, at first had no bearing on photosynthesis or any other photochemical process. A connection between ferredoxin and photosynthesis was established when Tagawa and Arnon [91] crystallized *C. pasteurianum* ferredoxin and found that it mediated the photoreduction of

NADP$^+$ by spinach chloroplasts. In this reaction, clostridial ferredoxin replaced a native chloroplast protein that had an unusual history, which can be discussed here only briefly and is given in detail elsewhere [92,93].

Beginning in 1952, three proteins were successively and independently isolated from chloroplasts, each under a different name to denote a different function. Davenport, Hill, and Whatley [94] isolated a "methemoglobin-reducing factor"; Arnon, Whatley, and Allen [95] isolated from an aqueous extract of spinach chloroplasts a "TPN-reducing factor," required for the photoreduction of NADP$^+$; and San Pietro and Lang [96] isolated a "photosynthetic pyridine nucleotide reductase," which they characterized as the enzyme required for the photoreduction of NADP$^+$ and NAD$^+$ by chloroplasts.

By 1960, it became clear that "methemoglobin-reducing factor," "TPN-reducing factor," and "photosynthetic pyridine nucleotide reductase" were different names for the same protein. Because the protein under its various names was isolated from, and was effective in, chloroplasts, it was thought at first to be a unique chloroplast protein. However, in 1961 the association of the protein with chloroplasts ceased to be unique when K. Tagawa and M. Nozaki (unpublished data from this laboratory) and Losada, Whatley, and Arnon [97] isolated a protein with similar properties from the photosynthetic bacterium *Chromatium*. Although *Chromatium* cells do not have chloroplasts, do not photoreduce NADP$^+$, and do not evolve oxygen, the "pyridine nucleotide reductase" isolated from these bacterial cells was able to replace the native protein of spinach chloroplasts in mediating the photoreduction of NADP$^+$ and the concomitant evolution of oxygen. The full implications of this finding, namely, that the *Chromatium* protein, the chloroplast protein, and the clostridial ferredoxin constituted a new group of electron carriers with sufficiently similar chemical similarities to warrant the common designation of ferredoxins, became clear only a year later, when the new nomenclature was introduced [91].

CHEMICAL AND PHYSICAL PROPERTIES
OF FERREDOXINS

The similarities initially found between the clostridial and chloroplast (spinach) ferredoxins were that the two proteins contained iron, were free of heme and flavin groups, had low molecular weights, underwent reversible reduction that could be measured by a decrease in their respective absorption peaks, and, what was then considered especially important, had a reducing power about equal to that of hydrogen gas, i.e., their oxidation–reduction potentials (at pH 7) were close to -420 mV [91].

Thus, ferredoxins were recognized as a group of electron-carrier proteins that transfer to appropriate enzyme systems some of the most reducing electrons in cellular metabolism. In recent years, the intensive study of the properties of ferredoxins and related proteins had added much to our knowledge of their chemical and physical properties. These will now be briefly recapitulated. They are discussed in detail elsewhere [92,93,98,99].

Ferredoxins are now recognized as a subdivision of a broad category of iron–sulfur proteins present in all living cells. Photosynthetic cells contain ferredoxins whose properties include an equal number of iron and acid-labile sulfur atoms, a low molecular weight, a strongly negative oxidation–reduction potential, and (in the reduced state and at low temperature) a characteristic electron paramagnetic resonanace (epr) spectrum with the principal signal near $g = 1.94$. Photosynthetic cells contain two types of soluble ferredoxins, distinguishable by different absorption spectra (which undergo reversible changes on oxidation–reduction): a plant or chloroplast type and a bacterial type [92,100].

In the oxidized state, all ferredoxins are colored proteins characterized either by a bacterial or by a chloroplast type of absorption spectrum. The bacterial type, found in different types of photosynthetic bacteria (*Rhodospirillum rubrum*, *Chromatium*, and *Chlorobium*) and the nonphotosynthetic anaerobes (e.g., *Clostridium pasteurianum*), has a single absorption peak in the visible region (around 390 nm) and a peak in the ultraviolet region at about 280 nm with a shoulder at 300 nm. The chloroplast type of ferredoxin has the usual protein peak in the ultraviolet region at about 280 nm and peaks at 463, 420, and 330 nm in the near-ultraviolet region. On reduction, the absorbance peaks in the visible region disappear; there is little change in the spectra in the ultraviolet region. In the reduced state, the absorption spectra of all types of ferredoxins are similar to one another.

X-Ray diffraction studies of the molecular structure of an eight iron–eight sulfur ferredoxin and a four iron–four sulfur iron–sulfur protein have shown that the iron and acid-labile sulfur (S^*) atoms are arranged in cubelike tetrameric clusters ($Fe_4 \cdot S_4^*$) linked to the protein through a crysteinyl sulfur bond to each of the iron atoms [101,102]. It is now generally accepted that the active site of eight iron–eight sulfur bacterial ferredoxins is accounted for by two such clusters ($Fe_4 \cdot S_4^*)_2$, whereas only one cluster accounts for the active sites of the four iron–sulfur bacterial ferredoxins ($4Fe \cdot 4S^*$) and the two iron–two sulfur chloroplast-type ferredoxins ($2Fe \cdot 2S^*$).

The number of electrons transferred by plant and bacterial ferredoxins has been a point of considerable interest. It now appears that a ferredoxin containing a single iron–sulfur cluster transfer one electron and that ferredoxins containing two iron–sulfur clusters transfer one electron at a time from each of the two clusters.

TABLE 1

Some Properties of Soluble Ferredoxins from Photosynthetic Organisms

| Property | Photosynthetic bacteria | | | | Chloroplast |
	Chlorobium ferredoxin	*Chromatium* ferredoxin	*Rhodospirillum rubrum* Ferredoxin I	Ferredoxin II	Spinach ferredoxin
Color	Brown	Brown	Brown	Brown	Red
Molecular weight	7000	10,000	8800	14,500	12,000
Oxidation–reduction potential (mV at pH 7)	—[a]	−490	—[a]	−430	−420
Iron (atoms/molecule)	8	8	8	4	2
Labile sulfide (atoms/molecule)	8	8	8	4	2
Absorption maximum ultra-violet (nm)	280	280	280	280	280,330
Absorption maximum, visible (nm)	385	385	385	385	463,420
Principal epr signals in reduced state (g value)	1.94	1.94	1.94	1.94	، 1.96

[a] No determination has as yet been made.

The chloroplast type of ferredoxin is found in all oxygen-evolving cells, including blue-green algae that do not have chloroplasts [100,103]. Table 1 summarizes the iron and labile sulfur content, molecular weight, and certain other properties of plant and bacterial iron–sulfur proteins.

ROLE OF FERREDOXIN IN NONCYCLIC PHOTOPHOSPHORYLATION

The elucidation of the nature of photosynthetic ferredoxins was paralleled by evidence for their important role in cyclic and noncyclic photophosphory-lation. With regard to noncyclic photophosphorylation, the first major advance concerned the mechanism of $NADP^+$ reduction by chloroplasts. Here, progress was greatly aided by the finding that hydrogen gas, in the presence of added hydrogenase, could substitute for light in the reduction of NADP$^+$ by chloroplasts [91]. Using the hydrogen–hydrogenase system as the source of reducing power, NADP$^+$ was reduced by a reconstituted system that required only two chloroplast components, ferredoxin and a flavoprotein enzyme, which was isolated in a crystalline form [104]. These experiments demonstrated that the role of chloroplast ferredoxin was that of an electron-carrier protein (replaceable by ferredoxins from other sources) that could not react directly with NADP$^+$. Ferredoxin transferred electrons to the flavoprotein enzyme that served as the specific ferredoxin-NADP$^+$ reductase of chloroplasts [105]. The mechanism of NADP$^+$ reduction was resolved into (a) a photochemical reduction of ferredoxin followed by two "dark" steps, (b) reoxidation of ferredoxin by ferredoxin-NADP$^+$ reductase, and (c) reoxidation of the reduced ferredoxin-NADP$^+$ reductase by NADP$^+$ [105].

$$\text{Illuminated chloroplasts} \xrightarrow{e} \text{ferredoxin} \xrightarrow{e} \text{ferredoxin-NADP}^+ \text{ reductase}$$
$$\downarrow 2e$$
$$\text{NADP}^+$$

A further advance in the understanding of the mechanism of NADP$^+$ reduction by chloroplasts was the finding that the ferredoxin-NADP$^+$ reductase forms stoichiometric 1:1 complexes with ferredoxin and NADP$^+$ [106–108].

The affinity of ferredoxin-NADP$^+$ reductase for NADP$^+$ was much greater than that for NAD$^+$. The NADP$^+$ concentration required to give half-maximal velocity of the reaction was found to be about 400 times smaller than that for NAD$^+$ [105]. This great difference in affinities accounts for the apparent specificity of the purified enzyme toward NADP$^+$ and for preferential reduction of NADP$^+$ by chloroplasts.

The evidence for ferredoxin as the terminal electron acceptor in the photochemical events that lead to NADP$^+$ reduction was subjected to a rigid test. The evolution of oxygen by chloroplasts is uniquely dependent on light, and it occurs only in the presence of a suitable electron acceptor. Thus, ferredoxin supplied in substrate amounts should support the production of oxygen by illuminated chloroplasts in the absence of any other electron acceptor.

Such evidence was indeed obtained [109]. Of special interest was the ratio between the ferredoxin reduced and the oxygen produced. Four molecules of ferredoxin were reduced for each molecule of oxygen produced. These findings were in accord with earlier evidence that the oxidation–reduction of ferredoxin involved a transfer of one electron [110]. The coupling of oxygen evolution to photoreduction of ferredoxin provided strong support for the important role assigned to ferredoxin in the photochemical reactions of photosynthesis. The evidence for that became even stronger when ferredoxin was also found to catalyze noncyclic photophosphorylation. Formation of ATP was linked to photoreduction of ferredoxin in the expected theoretical ratio of two molecules of ferredoxin reduced for each molecule of ATP formed. Here, as in oxygen evolution, the oxidized form of ferredoxin was the terminal electron acceptor. $NADP^+$ was not essential; its omission did not affect ATP formation [110]. The new equation for noncyclic photophosphorylation could now be written as Eq. (5).

$$4 \text{ Ferredoxin}_{\text{oxidized}} + 2ADP + 2P_i + 2H_2O \xrightarrow{h\nu}$$
$$4 \text{ Ferredoxin}_{\text{reduced}} + 2ATP + O_2 + 4H^+ \quad (5)$$

Although Eq. (5) became the true equation for noncyclic photophosphorylation, it is still experimentally more convenient and economical to use ferredoxin in only catalytic amounts by coupling reaction (5) with a re-oxidation of reduced ferredoxin by $NADP^+$ [Eq. (6)]. The sum of reactions (5) and (6) gives the earlier equation for noncyclic photophosphorylation [Eq. (4)].

$$4 \text{ Ferredoxin}_{\text{reduced}} + 2NADP^+ + 2H^+ \xrightarrow{\text{ferredoxin-NADP}^+ \text{ reductase}}$$
$$4 \text{ ferredoxin}_{\text{oxidized}} + 2NADPH \quad (6)$$

Under aerobic conditions and in the absence of $NADP^+$, the photoreduction of ferredoxin [Eq. (5)] can be coupled with its chemical reoxidation by molecular oxygen [Eq. (7)] [109,111].

$$4 \text{ Ferredoxin}_{\text{reduced}} + O_2 + 4H^+ \longrightarrow 4 \text{ ferredoxin}_{\text{oxidized}} + 2H_2O \quad (7)$$

Reactions (5) and (7) were observed in the presence of catalase, which is normally found in the preparation of spinach chloroplasts used [109,111]. However, when catalase is removed, the photoreduction of substrate amounts of ferredoxin is accompanied by the accumulation of hydrogen peroxide [112].

The sum of Eqs. (5) and (7) represents an oxygen-linked noncyclic photophosphorylation [Eq. (8)] that is catalyzed by ferredoxin. Reaction (8) is

$$\text{Sum Eq. (7)} + \text{Eq. (9):} \quad 2ADP + 2P_i \xrightarrow[\text{ferredoxin, } O_2]{h\nu} 2ATP \quad (8)$$

superficially similar to cyclic photophosphorylation in that it yields only ATP, but it differs from true cyclic photophosphorylation in being dependent on a continuous production and consumption of oxygen. It is therefore known as oxygen-dependent noncyclic or *pseudocyclic photophosphorylation*, a name coined in an earlier period, prior to our knowledge of ferredoxin, when other catalysts were used in lieu of ferredoxin in reactions (5), (7), and (8) [113,114].

Reduced ferredoxin has a stronger affinity for ferredoxin-NADP$^+$ reductase and NADP$^+$ than for oxygen [115]. This property of ferredoxin ensures that NADP$^+$-linked noncyclic photophosphorylation will continue to generate ATP and NADPH in an aerobic environment.

FERREDOXIN AS THE PHYSIOLOGICAL CATALYST OF CYCLIC PHOTOPHOSPHORYLATION

Work with spinach ferredoxin has revealed that its role is not limited to noncyclic photophosphorylation. It will be recalled that under anaerobic conditions cyclic photophosphorylation in chloroplasts depended on catalysts whose physiological role was at best uncertain. Moreover, experiments with labeled oxygen revealed that the catalysts of cyclic photophosphorylation, FMN and menadione (but not phenazine methosulfate), also catalyzed an exchange between O_2 and H_2O [116,117]. Such an exchange could give rise to a pseudocyclic type of photophosphorylation.

There were also differences between cyclic photophosphorylation in chloroplasts and bacterial chromatophores that cast doubt on the physiological role of menadione or FMN as catalysts of this process. Neither was required for cyclic photophosphorylation in freshly prepared bacterial chromatophores [66,67]. Furthermore, with these exogenous catalysts, cyclic photophosphorylation in chloroplasts was not sensitive to such characteristic inhibitors of cyclic bacterial photophosphorylation and mitochondrial oxidative phosphorylation as antimycin A [118].

A possible explanation for the need of exogenous catalysts in cyclic photophosphorylation by chloroplasts was that during isolation these organelles lost an endogenous soluble constituent that they require for this process. This possibility was verified when evidence was obtained for an anaerobic cyclic photophosphorylation in chloroplasts that was dependent on catalytic amounts of ferredoxin and on no other catalyst of photophosphorylation [115]. When catalyzed by ferredoxin, cyclic photophosphorylation in chloroplasts for the first time became sensitive to inhibition by low concentrations of antimycin A and oligomycin and resembled in this respect cyclic photophosphorylation in bacteria and oxidative phosphorylation in mitochondria [92,115,119,120].

Thus, ferredoxin was found to catalyze two distinctly different types of photophosphorylation, cyclic and noncyclic. They could be readily distinguished by their opposite responses to monochromatic light and to inhibitors. Noncyclic photophosphorylation exhibited a sharp "red drop" at the longer wavelengths of monochromatic light, coming to an almost complete halt at 714 nm. By contrast, cyclic photophosphorylation showed a marked "red rise"; i.e., its rate increased greatly under far-red illumination [111,119].

Differential sensitivity to antimycin A and to other inhibitors provided another sharp distinction between ferredoxin-catalyzed cyclic and noncyclic photophosphorylation. Low concentrations of antimycin A, oligomycin, and other inhibitors that sharply inhibited cyclic photophosphorylation had no effect on noncyclic photophosphorylation. By contrast, low concentrations of 3-(3,4-dichlorophenyl)-1,1-dimethylurea (DCMU) and o-phenanthroline, which markedly inhibited noncyclic photophosphorylation, actually stimulated cyclic photophosphorylation [92,111].

Several other lines of evidence pointed to ferredoxin as the physiological catalyst of cyclic photophosphorylation in chloroplasts: (a) As expected of a true catalyst, ferredoxin stimulated cyclic photophosphorylation at low concentrations (100 μM), comparable on a molar basis to those of the other known catalysts of the process; (b) when light intensity was restricted, ferredoxin catalyzed ATP formation more effectively than any other catalyst; and (c) in adequate light, cyclic photophosphorylation by ferredoxin produced ATP at a rate comparable with the maximim rates of photosynthesis *in vivo* [92,111].

The optimal concentration of ferredoxin for cyclic photophosphorylation, although still within a catalytic range (ca. 100 μM), was about an order of magnitude higher than that, (ca. 10 μM) needed to catalyze NADP$^+$ reduction and its coupled noncyclic photophosphorylation. This difference in concentration was sometimes used as an argument against ferredoxin being the physiological catalyst of cyclic photophosphorylation (see, for example, Avron and Neumann [121]). However, in more recent work, discussed below, the optimum concentration for ferredoxin-catalyzed cyclic photophosphorylation was found to be about 10 μM, the same as that for NADP$^+$ reduction and the coupled noncyclic photophosphorylation.

CYCLIC PHOTOPHOSPHORYLATION IN THE PRESENCE OF OXYGEN

Soon after its discovery, cyclic photophosphorylation in chloroplasts began to be preferentially investigated under anaerobic conditions. Historically, this ATP formation in the absence of oxygen played an important part in overcoming the early resistance to the concept that light-induced ATP synthesis by chloroplasts was totally independent of oxidative photophosphorylation by mitochondria that consume oxygen [26,41]. Although the distinction between photophosphorylation and oxidative phosphorylation is no longer in dispute, anaerobic conditions have continued to be widely used in studies of cyclic photophosphorylation in chloroplasts for another reason: Anaerobicity helped to distinguish cyclic photophosphorylation from the pseudocyclic type. Reduced ferredoxin can be reoxidized by air and, as already discussed, may catalyze pseudocyclic photophosphorylation, which, like the true cyclic type, yields only ATP but in reality is a variety of noncyclic photophosphorylation that depends on continuous production and consumption of oxygen.

Putting these historical and experimental reasons aside, the compelling consideration remains that *in vivo* oxygen is never excluded from the immediate environment of the thylakoids in which cyclic photophosphorylation occurs. Indeed, early evidence indicated that lower concentrations of ferredoxin were needed to catalyze cyclic photophosphorylation (at 708 nm) in the presence of air than were needed under anaerobic conditions [119,122]. A ferredoxin-catalyzed cyclic photophosphorylation under aerobic conditions was also observed in a cell-free preparation of a blue-green alga [123]. Recently, therefore, an investigation was carried out to characterize more fully cyclic photophosphorylation in the presence of air [124].

Figure 3 shows that, at 715 nm, ferredoxin was a far more effective catalyst of cyclic photophosphorylation when the reaction mixture contained dissolved air rather than dissolved N_2. This superiority of air over N_2 was pronounced at high but not at low light intensity [125]. The optimum ferredoxin concentration in air was about 10 μM, i.e., an order of magnitude less than that used earlier under argon and about the same as that needed to catalyze noncyclic photophosphorylation [111]. Much higher concentrations were less effective,

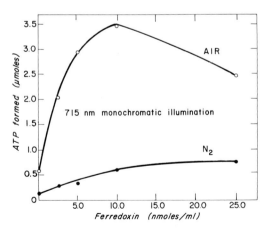

Fig. 3 Effect of ferredoxin concentration on cyclic photophosphorylation under aerobic and anaerobic conditions [124].

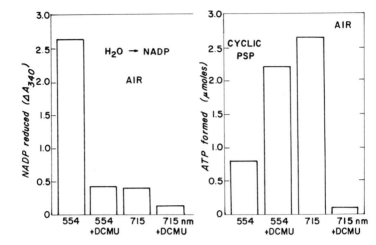

Fig. 4 Comparison of noncyclic electron transport (left) with ferredoxin-catalyzed cyclic photophosphorylation (right). Monochromatic illumination 554 or 715 nm [124].

probably because they facilitated a reoxidation of reduced ferredoxin by oxygen rather than by the cyclic electron transport chain of chloroplasts. Similar results were obtained in parallel experiments with green (554 nm) monochromatic illumination. Here again, low concentrations of ferredoxin were found to catalyze cyclic photophosphorylation more effectively under aerobic than anaerobic conditions but—and this was crucial—only when the light-induced electron flow from water was severely inhibited by 3-(3,4-dichlorophenyl)-1,1-dimethylurea. Thus, unlike pseudocyclic photophosphorylation, aerobic cyclic photophosphorylation at 554 nm was stimulated rather than inhibited by DCMU (Fig. 4).

In short, ferredoxin-catalyzed cyclic photophosphorylation functioned best when light-induced electron flow from water (and hence noncyclic photophosphorylation and its pseudocyclic variant) was materially restricted either by the use of 715 nm illumination or by combining 554 nm illumination with the addition of DCMU. The extent to which these conditions restricted electron flow from water is reflected in the parallel inhibition of the photoreduction of $NADP^+$ by water (Fig. 4).

Another observation was that under the aerobic conditions of these experiments a light-induced electron flow from water, although severely restricted, could not be dispensed with entirely. When this remaining "trickle" electron flow from water was totally suppressed, as by adding DCMU to chloroplasts illuminated with 715 nm light, cyclic photophosphorylation stopped (Fig. 4). This observation is contrary to the generally held view that DCMU does not inhibit cyclic photophosphorylation (see Avron and

Neumann [121]) but is in agreement with findings that in far-red light endogenous cyclic photophosphorylation in isolated chloroplasts is inhibited by DCMU [126].

In the presence of oxygen, the trickle of electrons from water appeared to maintain the proper oxidation–reduction balance or "poising" that is required for cyclic photophosphorylation. The need for poising of ferredoxin-catalyzed cyclic photophosphorylation in chloroplasts had become apparent in earlier experiments under conditions different from those used here [122,127]. The general importance of posing for cyclic photophosphorylation in both chloroplasts and bacterial chromatophores has been reviewed and stressed by Avron and Neumann [121].

A comparison of the effectiveness of ferredoxin, menadione, and phenazine methosulfate as catalysts of cyclic photophosphorylation in the presence of air is shown in Fig. 5. With equal 715 nm or 554 nm illumination (in the presence of DCMU) and at equal concentrations, ferredoxin was a decisively more effective catalyst of cyclic photophosphorylation than either menadione or phenazine methosulfate. The great superiority of low concentrations of ferredoxin over other catalysts in catalyzing the conversion of light energy into phosphate bond energy is consistent with the role assigned to ferredoxin as the physiological catalyst of cyclic photophosphorylation.

The conclusion that the aerobic ferredoxin-catalyzed cyclic photophosphorylation is physiological in nature is also supported by its sensitivity to low concentrations of dibromothymoquinone [128]. Dibromothymoquinone

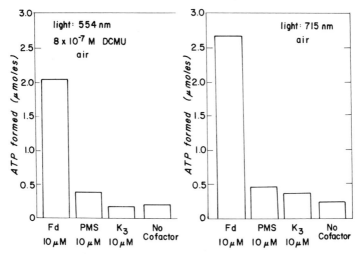

Fig. 5 Comparison of ferredoxin with other cofactors of cyclic photophosphorylation (Fd, ferredoxin; PMS, phenazine methosulfate; K_3, menadione) [124].

is an antagonist of plastoquinone and inhibits chloroplast thylakoid reactions in which plastoquinone is involved [128]. Recently, Hauska, Reimer, and Trebst [129] concluded from dibromothymoquinone inhibition experiments (under argon) that ferredoxin is the probable cofactor of cyclic photophosphorylation *in vivo*, in which plastoquinone serves as the natural energy-conserving site. A similar conclusion now seems applicable to ferredoxin-catalyzed cyclic photophosphorylation in the presence of air.

STOICHIOMETRY OF NONCYCLIC
AND ATP PHOTOPHOSPHORYLATION
REQUIREMENTS OF CO_2 ASSIMILATION

The new evidence for a ferredoxin-catalyzed, aerobic cyclic photophosphorylation helped to clarify a long-standing controversy over the stoichiometry of noncyclic photophosphorylation and the role of cyclic photophosphorylation in CO_2 assimilation. The basis of the controversy is as follows.

As stated earlier, the conversion of CO_2 to sugar phosphates requires, for each molecule of CO_2, three molecules of ATP and two molecules of NADPH, i.e., an ATP/NADPH (P/e_2) ratio of 1.5, which is appreciably greater than the P/e_2 ratio of 1.0 that was observed initially for noncyclic photophosphorylation [78] and confirmed in other studies [130,134].

To account for the extra ATP for which this stoichiometry does not provide, three explanations were put forward: (a) that, contrary to earlier findings, the P/e_2 ratio of noncyclic photophosphorylation is appreciably greater than 1.0, (b) that the P/e_2 ratio is close to 1.0 but the extra ATP comes from a concurrent pseudocyclic photophosphorylation, or (c) that the P/e_2 ratio is 1.0 but the extra ATP comes from a concurrent cyclic photophosphorylation.

Ratios of P/e_2 greater than 1.0 (up to 1.3) for noncyclic photophosphorylation were first reported by Winget, Izawa, and Good [135] with the use of buffers other than Tris–HCl, the buffer used in the earlier investigations. Using Tris–HCl buffer, Winget *et al.* [135] obtained a P/e_2 ratio of 1.0; with the same buffer Lynn and Brown [136] reported P/e_2 ratios approaching 4.0. Although these high P/e_2 ratios were not confirmed [134], a considerable literature has now accumulated in support of the view that the "corrected" P/e_2 ratio of noncyclic photophosphorylation is 2.0 rather than 1.0 [137–142]. In this view, a possible contribution of ATP from a concurrent cyclic or pseudocyclic photophosphorylation is negligible [137,141] and unnecessary: the ATP produced by noncyclic photophosphorylation alone exceeds the requirements of CO_2 assimilation.

A detailed critique of this view is beyond the scope of this chapter, but attention needs to be drawn to some of the underlying assumptions that enter into computations of the "corrected" high P/e_2 ratios. One assumption is that chloroplasts have two kinds of light-induced noncyclic electron transport: first a basal, nonphosphorylating kind that occurs without phosphorylation and, second, a phosphorylating kind that is coupled to ATP formation.

The rationale for an assumed basal electron transport in chloroplasts stems from comparisons with mitochondria. It was reasoned that since "chloroplast preparations, unlike the best mitochondrial preparations, transport electrons at appreciable rates in the absence of phosphorylation" [139] chloroplasts must have a separate nonphosphorylating electron transport pathway. However, the rate of nonphosphorylating electron transport, while low in intact mitochondria, may be appreciable in partly damaged mitochondrial preparations [143]. Similarly, light-induced, nonphosphorylating electron transport is usually measured not in intact chloroplasts but in partly damaged (envelope-free) preparations. Moreover, an appreciable rate of nonphosphorylating electron transport in chloroplasts may result from large energy perturbations that follow the absorption of light quanta (of over 40 kcal/Einstein) by chlorophyll pigments not found in mitochondria.

A much more likely analogy between chloroplasts and mitochondria is that in both organelles, the rate of electron transport is slowed unless an induced high-energy state in the membranes is discharged, either by a coupled phosphorylation or by the use of an uncoupling agent. Stated in other terms, phosphorylation control of electron transport in chloroplasts [78] is deemed to be analogous to respiratory control in mitochondria [143].

Another assumption, on which much of the argument in support of the $P/e_2 = 2$ stoichiometry of noncyclic photophosphorylation rests, comes from experiments with different classes of artificial electron acceptors [139]. These are assumed to mirror reliably physiological events in chloroplasts. However useful artificial acceptors or cofactors may be for specific and limited objectives, they often confer characteristics on a photophosphorylation reaction that are foreign to its physiological counterpart. For example, cyclic photophosphorylation catalyzed by the widely used phenazine methosulfate has extremely high light saturation requirements and is insensitive to inhibition by antimycin A and by dibromothymoquinone [128]—three features that are exact opposites of the characteristics of cyclic photophosphorylation when catalyzed by its physiological catalyst, ferredoxin.

It is concluded that work with artificial electron acceptors offers no compelling evidence that noncyclic photophosphorylation alone provides sufficient ATP for CO_2 assimilation and that therefore an additional supply of ATP must come either from cyclic or, as has lately been proposed [144,145], from pseudocyclic photophosphorylation. Recent work suggests that the extra ATP comes from an aerobic, ferredoxin-catalyzed cyclic photophosphorylation that operates concurrently with the noncyclic type. Noncyclic photophosphorylation with ferricyanide as the terminal electron acceptor (and in the absence of ferredoxin) gave a P/e_2 ratio of 1.0, but, with the physiological acceptor $NADP^+$ (and ferredoxin present), the P/e_2 ratio was about 1.5 [125]. Evidence that the extra ATP originated from cyclic rather than from pseudocyclic photophosphorylation came from inhibition studies with antimycin A. Antimycin A, within a concentration range that inhibited cyclic photophosphorylation and did not inhibit noncyclic (including pseudocyclic) photophosphorylation, suppressed the extra ATP formed by cyclic photophosphorylation, thereby restoring to noncyclic photophosphorylation its intrinsic P/e_2 ratio of 1.0 (Fig. 6).

In sum, the ATP and NADPH required for CO_2 assimilation can be adequately supplied only by a joint, concurrent operation of cyclic and noncyclic photophosphorylation, a conclusion that is in agreement with earlier evidence (see Role of Cyclic Photophosphorylation: Early Views) and with more recent experiments on CO_2 assimilation by isolated chloroplasts [146]; (compare however Klob *et al.* [147] and Miginiac–Maslow and Champigny [148].) This conclusion is buttressed by recent findings that a concurrent noncyclic electron flow from water to $NADP^+$ provides a physiological poising mechanism for the operation of ferredoxin-catalyzed cyclic photophosphorylation in chloroplasts illuminated by short-wavelength (or white) light [124].

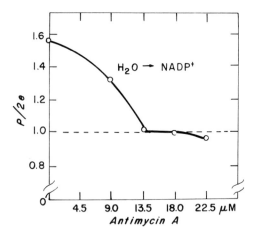

Fig. 6 Effect of antimycin A on concurrent cyclic and noncyclic photophosphorylation [125].

REGULATION OF
CYCLIC PHOTOPHOSPHORYLATION

As discussed earlier, ferredoxin-catalyzed cyclic photophosphorylation can operate effectively only in a poised system in which the light-induced electron pressure from water is greatly diminished. Experimentally, such poising was at first maintained by decreasing the electron flow from water by means foreign to photosynthesis in nature: far-red monochromatic illumination or shorter wavelength illumination combined with an addition of the inhibitor DCMU. In seeking a physiological regulation mechanism, Arnon and Chain [124] found that poising of cyclic photophosphorylation in chloroplasts is accomplished by its concurrent operation with noncyclic photophosphorylation. During CO_2 assimilation, electrons from water are used (via ferredoxin and $NADP^+$) for the reduction of 1,3-diphosphoglycerate. Thus, an overreduction of the thylakoid milieu is avoided, proper poising is maintained, and cyclic photophosphorylation operates concurrently with the noncyclic type. The result is that the P/NADPH ratio exceeds 1.0 [124].

When CO_2 assimilation is curtailed or stopped altogether, cyclic photophosphorylation may operate by itself as a general source of cellular ATP (see Role of Cyclic Photophosphorylation: Early Views). A curtailment or cessation of CO_2 assimilation would result in an accumulation of NADPH. Hence, the possibility was recently tested that NADPH might have a regulatory effect on the operation of ferredoxin-catalyzed cyclic photophosphorylation in isolated chloroplasts.

The addition of NADPH greatly stimulated ferredoxin-catalyzed cyclic photophosphorylation (sensitive to antimycin A) in 554 nm light [124]. Unexpectedly, no DCMU was required for poising. It appeared that NADPH, like DCMU, diminished the electron flow from water. Such a novel role of NADPH was tested further by measuring its effect on the photoreduction of C-550 [149], a primary indicator of electron flow from water [150]. Preincubation of chloroplasts with NADPH (in the presence of ferredoxin) in the dark diminished the subsequent photoreduction of C-550 (and photooxidation of cytochrome b_{559}), results that are consistent with a diminution of electron flow from water [124].

It appears, therefore, that when CO_2 assimilation is curtailed and NADPH accumulates, the backreaction of NADPH and ferredoxin with C-550 and possibly with cytochrome b_{559} provides a regulatory mechanism that, by diminishing electron pressure from water, maintains proper poising for the operation of cyclic photophosphorylation as a source of ATP for diverse cellular needs.

CYCLIC PHOTOPHOSPHORYLATION *IN VIVO*

Apart from photosynthetic CO_2 assimilation in which cyclic photophosphorylation acts as a source of additional ATP, cyclic photophosphorylation may, as discussed earlier, operate by itself and serve as a source of ATP for endergonic cellular processes, which, like protein synthesis, require ATP but are independent of reducing power generated (along with ATP) by noncyclic photophosphorylation. For example, evidence has been obtained that intact chloroplasts incorporate amino acids and synthesize proteins at the expense of ATP produced by endogenous cyclic photophosphorylation [151,152].

Evidence has also been accumulating for a physiological role of cyclic photophosphorylation in intact cells. Urbach and Simonis [153] and Simonis [154,155] observed a light-dependent incorporation of ^{32}P by intact cells in far-red light (708 nm) or in the presence of inhibitors that inhibit noncyclic but not cyclic photophosphorylation. Nultsch [156,157] found that light-induced changes in rates of movement (photokinesis) of blue-green algae depends on cyclic photophosphorylation as the main source of energy. Of special interest is the growing evidence that blue-green algae use for nitrogen fixation the ATP formed in these cells by cyclic photophosphorylation [158–160].

Other evidence for the operation of cyclic photophosphorylation in intact cells comes from investigations of the photoassimilation of organic compounds and light-dependent ion uptake. *Chlorella* cells photoassimilate glucose at 711 nm, a wavelength that can support cyclic photophosphorylation but not noncyclic photophosphorylation or CO_2 assimilation. Moreover, light-dependent glucose assimilation showed a striking sensitivity to inhibition by antimycin A [161,162]. Extensive evidence has been obtained for the role of cyclic photophosphorylation as an energy source of active potassium influx in *Hydrodictyon africanum* cells [163]. Other work in this and related areas of photophosphorylation *in vivo* has recently been reviewed by Simonis and Urbach [164].

CONCLUDING REMARKS

Photosynthetic phosphorylation (photophosphorylation) is viewed as the conversion of light energy into forms of biochemical energy, prior to, and independently from, carbon dioxide assimilation. Photophosphorylation is subdivided into two types, cyclic and noncyclic. In the cyclic type, light energy is used to form ATP from ADP and P_i, without a net change in the oxidation–reduction state of any external or internal electron donor or acceptor. In noncyclic photophosphorylation, light-induced ATP formation is coupled with the evolution of oxygen and the generation of reducing power.

Insofar as CO_2 assimilation is concerned, noncyclic photophosphorylation generates part of the needed ATP and all of the reducing power in the form of reduced ferredoxin, an iron–sulfur protein present in all photosynthetic cells. Reduced ferredoxin ($E_0' = -420$ mV) in turn serves as an electron donor for the reduction of $NADP^+$ by an enzymatic reaction that is independent of light. The remainder of the required ATP is supplied by cyclic photophosphorylation, a reaction that is also catalyzed by ferredoxin.

Recent evidence indicates that during normal photosynthesis ferredoxin-catalyzed cyclic and noncyclic photophosphorylation operate concurrently. Under special conditions, cyclic photophosphorylation may operate by itself and serve as a source of ATP for general cellular needs, other than CO_2 assimilation.

ACKNOWLEDGMENTS

The work from the author's laboratory described herein was initiated and pursued in its earlier stages with the support of grants from the National Institute of General Medical Sciences and in its later stages by grants from the National Science Foundation.

REFERENCES

1. A. T. Jagendorf, in "Bioenergetics of Photosynthesis" (Govindjee, ed.), pp. 413–492. Academic Press, New York, 1975.
2. J. Ingenhousz, "Essay on the Food of Plants and the Renovation of Soils," Appendix to Chapter 15 of General Report from Board of Agriculture, London, 1796.
3. A. von Baeyer, *Ber. Dtsch. Chem. Ges.* **3**, 63 (1864).
4. R. Willstätter and A. Stoll, "Untersuchungen über die Assimilation der Kohlensäure" Springer-Verlag, Berlin and New York, 1918.
5. O. Warburg, G. Krippahl, and A. Lehmann, *Am. J. Bot.* **56**, 961–971 (1969).
6. T. de Saussure, "Recherches chimiques sur la Végétation," V. Nyon, Paris, 1804.
7. J. Sachs, *Jahrb. Wiss. Bot.* **3**, 184 (1863).
8. J. Boehm, *Bot. Zg.* **41**, 33 and 49 (1883).
9. C. B. van Niel, *Arch. Mikrobiol.* **3**, 1–112 (1931).
10. C. B. van Niel, *Adv. Enzymol.* **1**, 263–328 (1941).
11. C. B. van Niel, in "Photosynthesis in Plants" (J. Franck and W. E. Loomis, eds.), pp. 437–495. Iowa State Coll. Press, Ames, 1949.
12. S. Ruben, *J. Am. Chem. Soc.* **65**, 279–282 (1943).

13. E. I. Rabinowitch, "Photosynthesis and Related Processes" p. 228. Wiley (Interscience), New York, 1945.
14. R. L. Emerson, J. F. Stauffer, and W. W. Umbreit, *Am. J. Bot.* **31**, 107–120 (1944).
15. S. Ruben and M. Kamen, *Phys. Rev.* **59**, 49 (1941).
16. A. J. P. Martin and R. L. M. Synge, "Les Prix Nobel en 1952." Norstedt, Stockholm, 1952.
17. R. M. Fink and K. Fink, *Science* **107**, 253 (1948).
18. A. A. Benson and M. Calvin, *Annu. Rev. Plant Physiol.* **1**, 25–42 (1950).
19. A. A. Benson, *J. Am. Chem. Soc.* **73**, 2971 (1951).
20. A. A. Benson, J. A. Bassham, M. Calvin, A. G. Hall, H. E. Hirsch, S. Kawaguchi, V. Lynch, and N. E. Tolbert, *J. Biol. Chem.* **196**, 703–716 (1952).
21. J. A. Bassham, A. A. Benson, L. D. Kay, A. Z. Harris, A. T. Wilson, and M. Calvin, *J. Am. Chem. Soc.* **76**, 1760 (1954).
22. J. A. Bassham and M. Calvin, "The Path of Carbon in Photosynthesis." Prentice-Hall, Englewood Cliffs, New Jersey, 1957.
23. P. A. Trudinger, *Biochem. J.* **64**, 273–286 (1956).
24. J. P. Aubert, G. Milhaud, and J. Millet, *Ann. Inst. Pasteur, Paris* **92**, 515–528 (1957).
25. M. Goodman, D. F. Bradley, and M. Calvin, *J. Am. Chem. Soc.* **75**, 1962 (1953).
26. D. I. Arnon, *Annu. Rev. Plant Physiol.* **7**, 325–354 (1956).
27. J. Sachs, "Lectures on the Physiology of Plants." Oxford Univ. Press (Clarendon), London and New York, 1887.
28. W. Pfeffer, "Physiology of Plants." Oxford Univ. Press (Clarendon), London and New York, 1900.
29. R. Hill, *Nature (London)* **139**, 881–882 (1937).
30. R. Hill, *Proc. R. Soc. London, Ser. B* **127**, 192–210 (1939).
31. R. Hill, *Symp. Soc. Exp. Biol.* **5**, 222–231 (1951).
32. A. H. Brown and J. Franck, *Arch. Biochem.* **16**, 55–60 (1948).
33. J. S. C. Wessels and E. Havinga, *Recl. Trav. Chim. Pays-Bas* **71**, 809–812 (1952).
34. A. H. Brown and A. W. Frenkel, *Annu. Rev. Plant Physiol.* **4**, 23–58 (1953).
35. H. Metzner, *J. Theor. Biol.* **51**, 201–231 (1975).
36. S. Aronoff and M. Calvin, *Plant Physiol.* **23**, 351–358 (1948).
37. W. Vishniac and S. Ochoa, *Nature (London)* **167**, 768–770 (1951).
38. L. J. Tolmach, *Nature (London)* **167**, 946–949 (1951).
39. D. I. Arnon, *Nature (London)* **167**, 1008–1010 (1951).
40. D. I. Arnon, M. B. Allen, and F. R. Whatley, *Nature (London)* **174**, 394–396 (1954).
41. D. I. Arnon, *Science* **122**, 9–16 (1955).
42. M. B. Allen, D. I. Arnon, J. B. Capindale, F. R. Whatley, and L. J. Durham, *J. Am. Chem. Soc.* **77**, 4149–4155 (1955).
43. M. Gibbs and M. A. Cykin, *Nature (London)* **182**, 1241–1242 (1958).
44. M. Gibbs and N. Calo, *Plant Physiol.* **34**, 318–323 (1959).
45. N. E. Tolbert, *Brookhaven Symp. Biol.* **11**, 271–275 (1958).
46. R. M. Smillie and R. C. Fuller, *Plant Physiol.* **34**, 651–656 (1959).
47. R. M. Smillie and G. Krotkov, *Can. J. Bot.* **37**, 1217–1225 (1959).
48. E. Buchner, *Ber. Dtsch. Chem. Ges.* **30**, 117–124 and 1110–1113 (1897).
49. M. Rubner, *in* "Die Ernährungsphysiologie de Hefezelle bei Alkoholischer Gärung," p. 39. Veit, Leipzig, 1931.
50. P. C. Zammecnik and E. B. Keller, *J. Biol. Chem.* **209**, 337–354 (1954).
51. A. Kornberg, "Les Prix Nobel en 1959." Norstedt, Stockholm, 1960.

52. D. A. Walker, *Plant Physiol.* **40**, 1157–1161 (1965).
53. D. A. Walker, *Biochem. J.* **92**, 22C–23C (1964).
54. R. G. Jensen and J. A. Bassham, *Proc. Natl. Acad. Sci. U.S.A.* **56**, 1096–1101 (1966).
55. P. P. Kalberer, B. B. Buchanan, and D. I. Arnon, *Proc. Natl. Acad. Sci. U.S.A.* **57** 1542–1549 (1967).
56. W. O. James and V. S. R. Das, *New Phytol.* **56**, 325–343 (1957).
57. H. Lundegårdh, *Nature (London)* **192**, 234–248 (1961).
58. W. Vishniac and S. Ochoa, *J. Biol. Chem.* **198**, 501–506 (1952).
59. D. I. Arnon, M. B. Allen, and F. R. Whatley, *Prog Int. Congr. Bot., 8th, 1954* Rep. Commun., Sects. 11 and 12, pp. 1–2 (1954).
60. D. I. Arnon, M. B. Allen, and F. R. Whatley, *Nature (London)* **174**, 394–396 (1954).
61. D. I. Arnon, F. R. Whatley, and M. B. Allen, *J. Am. Chem. Soc.* **76**, 6324–6329 (1954).
62. A. W. Frenkel, *J. Am. Chem. Soc.* **76**, 5568–5569 (1954).
63. J. B. Thomas and J. M. Haans, *Biochim. Biophys. Acta* **18**, 286–288 (1955).
64. B. Petrack and F. Lipmann, *in* "Light and Life" (W. D. McElroy and B. Glass, eds.), pp. 621–630. Johns Hopkins Press, Baltimore, Maryland, 1961.
65. A. W. Frenkel, *J. Biol. Chem.* **222**, 823–834 (1956).
66. J. W. Newton and M. Kamen, *Biochim. Biophys. Acta* **25**, 462–474 (1957).
67. I. C. Anderson and R. C. Fuller, *Arch. Biochem. Biophys.* **76**, 168–179 (1958).
68. D. Geller and F. Lipmann, *J. Biol. Chem.* **235**, 2478–2484 (1960).
69. D. I. Arnon, M. B. Allen, and F. R. Whatley, *Biochim. Biophys. Acta* **20**, 449–461 (1956).
70. E. Rabinowitch, *in* "Research in Photosynthesis" (H. Gaffron, ed.), p. 345. Wiley (Interscience), New York, 1957.
71. M. B. Allen, F. R. Whatley, and D. I. Arnon, *Biochim. Biophys. Acta* **27**, 16–23 (1958).
72. A. T. Jagendorf and M. Avron, *J. Biol. Chem.* **231**, 277–290 (1958).
73. D. I. Arnon, *Nature (London)* **184**, 10–21 (1959).
74. D. I. Arnon, *in* "Light and Life" (W. D. McElroy and B. Glass, eds.), pp. 489–566. Johns Hopkins Press, Baltimore, Maryland, 1961.
75. E. Katz, *in* "Photosynthesis in Plants" (J. Franck and W. E. Loomis, eds.), p. 287. Iowa State Coll. Press, Ames, 1949.
76. L. S. Levitt, *Science* **118**, 696 (1953).
77. L. S. Levitt, *Science* **120**, 33–35 (1954).
78. D. I. Arnon, F. R. Whatley, and M. B. Allen, *Science* **127**, 1026–1034 (1958).
79. D. I. Arnon, F. R. Whatley, and M. B. Allen, *Biochim, Biophys. Acta* **32**, 47–57 (1959).
80. M. Avron, D. W. Krogmann, and A. T. Jagendorf, *Biochim. Biophys. Acta* **30**, 114–153 (1958).
81. H. E. Davenport, *Nature (London)* **184**, 524–526 (1959).
82. D. W. Krogmann, A. T. Jagendorf, and M. Avron, *Plant Physiol.* **34**, 272–277 (1959).
83. D. I. Arnon, *in* "Handbuch der Planzenphysiologie" (W. Ruhland, ed.), Vol. 12, Part 1, pp. 773–829. Springer-Verlag, Berlin and New York, 1960.
84. A. V. Trebst, H. Y. Tsujimoto, and D. I. Arnon, *Nature (London)* **182**, 351–355 (1958).
85. M. G. Stålfelt, *Physiol. Plant.* **8**, 572–593 (1955).

86. O. V. S. Heath and B. Orchard, *Nature (London)* **180**, 180–181 (1957).
87. A. V. Trebst, M. Losada, and D. I. Arnon, *J. Biol. Chem.* **234**, 3055–3058 (1959).
88. F. R. Whatley, M. B. Allen, and D. I. Arnon, *Biochim. Biophys. Acta* **16**, 605–606 (1955).
89. D. I. Arnon, F. R. Whatley, and M. B. Allen, *Biochim. Biophys. Acta* **16**, 607–608 (1955).
90. L. E. Mortenson, R. C. Valentine, and J. E. Carnahan, *Biochem. Biophys. Res. Commun.* **7**, 448–452 (1962).
91. K. Tagawa and D. I. Arnon, *Nature (London)* **195**, 537–543 (1962).
92. D. I. Arnon, *Naturwissenschaften* **56**, 295–305 (1969).
93. B. B. Buchanan and D. I. Arnon, *Adv. Enzymol.* **33**, 119–176 (1970).
94. H. E. Davenport, R. Hill, and F. R. Whatley, *Proc. R. Soc. London, Ser. B* **139**, 346–358 (1952).
95. D. I. Arnon, F. R. Whatley, and M. B. Allen, *Nature (London)* **180**, 182–185 (1957); errata p. 1325.
96. A. San Pietro and H. M. Land, *J. Biol. Chem.* **231**, 211–229 (1958).
97. M. Losada, F. R. Whatley, and D. I. Arnon, *Nature (London)* **190**, 606–610 (1961).
98. R. Malkin, *in* "Iron-Sulfur Proteins" (W. Lovenberg, ed.), Vol. 2, pp. 1–26. Academic Press, New York, 1973.
99. W. H. Orme-Johnson, *Annu. Rev. Biochem.* **42**, 159–204 (1973).
100. D. I. Arnon, *Science* **149**, 1460–1469 (1965).
101. E. T. Adman, L. C. Sieker, and L. H. Jensen, *J. Biol. Chem.* **248**, 3987–3996 (1973).
102. C. W. Carter, Jr., J. Kraut, S. T. Freer, R. A. Alden, L. C. Sieker, E. Adman, and L. H. Jensen. *Proc. Natl. Acad. Sci. USA* **69**, 3526–3529 (1972).
103. A Mitsui and D. I. Arnon, *Physiol. Plant.* **25**, 135–140 (1971).
104. M. Shin, K. Tagawa, and D. I. Arnon, *Biochem. Z.* **338**, 84–96 (1963).
105. M. Shin and D. I. Arnon, *J. Biol. Chem.* **240**, 1405–1412 (1965).
106. M. Shin and A. San Pietro, *Biochem. Biophys. Res. Commun.* **33**, 38–42 (1968).
107. G. P. Foust, S. G. Mayhew, and V. Massey, *J. Biol. Chem.* **244**, 964–970 (1969).
108. W. Nelson and J. Neumann, *J. Biol. Chem.* **244**, 1932–1936 (1969).
109. D. I. Arnon, H. Y. Tsujimoto, and B. D. McSwain, *Proc. Natl. Acad. Sci. U.S.A.* **51**, 1274–1282 (1964).
110. F. R. Whatley, K. Tagawa, and D. I. Arnon, *Proc. Natl. Acad. Sci. U.S.A.* **49**, 266–270 (1963).
111. D. I. Arnon, H. Y. Tsujimoto, and B. D. McSwain, *Nature (London)* **214**, 562–566 (1967).
112. A. Telfer, R. Cammack, and M. C. W. Evans, *FEBS Lett.* **10**, 21–24 (1970).
113. D. I. Arnon, M. Losada, and F. R. Whatley, *Proc. Natl. Acad. Sci. U.S.A.* **47**, 1314–1344 (1961).
114. A. V. Trebst and H. Eck, *Z. Naturforsch., Teil B* **16**, 455–461 (1961).
115. K. Tagawa, H. Y. Tsujimoto, and D. I. Arnon, *Proc. Natl. Acad. Sci. U.S.A.* **49**, 567–572 (1963).
116. T. Nakamoto, D. W. Krogmann, and B. Mayne, *J. Biol. Chem.* **235**, 1843–1845 (1960).
117. A. B. Krall, N. E. Good, and B. C. Mayne, *Plant Physiol.* **36**, 44–47 (1961).
118. H. Baltscheffsky, *Sven. Kem. Tidskr.* **72**, 310–325 (1960).
119. K. Tagawa, H. Y. Tsujimoto, and D. I. Arnon, *Proc. Natl. Acad. Sci. U.S.A.* **50**, 544–549 (1963).
120. H. Böhme, S. Reimer, and A. Trebst, *Z. Naturforsch., Teil B* **26**, 341–352 (1971).

121. M. Avron and J. Neumann, *Annu. Rev. Plant Physiol.* **19**, 137–166 (1968).
122. K. Tagawa, H. Y. Tsujimoto, and D. I. Arnon, *Nature (London)* **199**, 1247–1252 (1963).
123. H. Bothe, *Z. Naturforsch, Teil B* **24**, 1574–1582 (1969).
124. D. I. Arnon and R. K. Chain, *Proc. Natl. Acad. Sci. U.S.A.* **72**, 4961–4965 (1975).
125. D. I. Arnon and R. K. Chain, *Plant Cell Physiol.* Special Issue No. 3, 129–147 (1977).
126. W. Kaiser and W. Urbach, *Ber. Dtsch. Bot. Ges.* **86**, 213–226 (1973).
127. B. R. Grant and F. R. Whatley, in "The Biochemistry of Chloroplasts" (T. W. Goodwin, ed.), vol. 2, pp. 505–521. Academic Press, New York, 1967.
128. A. Trebst, E. Harth, and W. Draber, *Z. Naturforsch., Teil B* **25**, 1157–1159 (1970).
129. G. Hauska, S. Reimer, and A. Trebst, *Biochim. Biophys. Acta* **357**, 1–12 (1974).
130. A. T. Jagendorf, *Brookhaven Symp. Biol.* **11**, 236–258 (1958).
131. M. Avron and A. T. Jagendorf, *J. Biol. Chem.* **234**, 1315–1320 (1959).
132. M. Stiller and B. Vennesland, *Biochim. Biophys. Acta* **60**, 562–579 (1962).
133. J. F. Turner, C. C. Black, and M. Gibbs, *J. Biol. Chem.* **237**, 577–579 (1962).
134. F. F. Del Campo, J. M. Ramirez, and D. I. Arnon, *J. Biol. Chem.* **243**, 2805–2809 (1968).
135. G. D. Winget, S. Izawa, and N. E. Good, *Biochem. Biophys. Res. Commun.* **21**, 438–443 (1965).
136. W. S. Lynn and R. H. Brown, *J. Biol. Chem.* **242**, 412–417 (1967).
137. S. Izawa and N. E. Good, *Biochim. Biophys. Acta* **162**, 380–391 (1968).
138. S. Saha and N. E. Good, *J. Biol. Chem.* **245**, 5017–5021 (1970).
139. S. Saha, R. Ouitrakul, S. Izawa, and N. E. Good, *J. Biol. Chem.* **246**, 3204–3209 (1971).
140. D. O. Hall, S. G. Reeves, and H. Baltscheffsky, *Biochem. Biophys. Res. Commun.* **43**, 359–366 (1971).
141. S. G. Reeves and D. O. Hall, *Biochim. Biophys. Acta* **314**, 66–78 (1973).
142. K. R. West and J. T. Wiskich, *Biochim. Biophys. Acta* **292**, 197–205 (1973).
143. A. L. Lehninger, "Biochemistry," p. 518. Worth Publ., New York, 1975.
144. J. F. Allen, *Nature (London)* **256**, 599–600 (1975).
145. H. Egneus, U. Heber, U. Matthiesen, and M. Kirk, *Biochim. Biophys. Acta* **408**, 252–268 (1975).
146. P. Schürmann, B. B. Buchanan, and D. I. Arnon, *Biochim. Biophys. Acta* **267**, 111–124 (1972).
147. W. Klob, O. Kandler, and W. Tanner, *Plant Physiol.* **51**, 825–827 (1973).
148. M. Miginiac-Maslow and M.-L. Champigny, *Plant Physiol.* **53**, 856–862 (1974).
149. D. B. Knaff and D. I. Arnon, *Proc. Natl. Acad. Sci. U.S.A.* **63**, 963–969 (1969).
150. W. L. Butler, *Acc. Chem. Res.* **6**, 177–184 (1973).
151. J. M. Ramirez, F. F. Del Campo, and D. I. Arnon, *Proc. Natl. Acad. Sci. U.S.A.* **59**, 606–612 (1968).
152. R. J. Ellis, in "Membrane Biogenesis: Mitochondrial, Chloroplast, and Bacterial" (A. Tzagoloff, ed.), pp. 247–278. Plenum, New York, 1975.
153. W. Urbach and W. Simonis, *Biochem. Biophys. Res. Commun.* **17**, 39–45 (1964).
154. W. Simonis, *Ber. Dtsch. Bot. Ges.* **77**, 5013 (1964).
155. W. Simonis, in "Currents in Photosynthesis" (J. B. Thomas and J. C. Goedheer, eds.), pp. 217–223. Donker Publ., Rotterdam, The Netherlands, 1966.
156. W. Nultsch, in "Currents in Photosynthesis" (J. B. Thomas and J. C. Goedheer, eds.), pp. 421–427. Donker Publ., Rotterdam, The Netherlands, 1966.
157. W. Nultsch, *Photochem. Photobiol.* **10**, 119–123 (1969).

158. P. Fay, *Biochim. Biophys. Acta* **216**, 353–356 (1970).
159. H. Bothe and E. Loos, *Arch. Mikrobiol.* **86**, 241–254 (1972).
160. R. L. Lyne and W. D. P. Stewart, *Planta* **109**, 27–38 (1973).
161. W. Tanner, E. Loos, and O. Kandler, *in* "Currents in Photosynthesis" (J. B. Thomas and J. C. Goedheer, eds.), pp. 243–250. Donker Publ., Rotterdam, The Netherlands, 1966.
162. W. Tanner, U. Zinecker, and O. Kandler, *Z. Naturforsch., Teil B* **22**, 358–359 (1967).
163. J. A. Raven, *J. Exp. Bot.* **22**, 420–433 (1971).
164. W. Simonis and W. Urbach, *Annu. Rev. Plant Physiol.* **24**, 89–114 (1973).

2

Oxidation and Oxygen Activation
by Heme Proteins

C. K. Chang and D. Dolphin

HISTORICAL PREVIEW

There is still considerable contention as to whether George III, by the grace of God king of Great Britain, France, and Ireland, defender of the faith, emperor of the British and Hanoverian Dominions and the American Colonies (mad George), suffered from porphyria or, as his detractors have suggested, cerebral syphilis. Nonetheless, there is an authoritative body of evidence [1] suggesting that the neurological disorders of George III were indeed the result of his pathological state of porphyria, characterized by abnormalities of porphyrin metabolism and excretion. Many forms of porphyria result in severe neurological disorders, and there is no doubt that George's attitudes toward the American Colonies two centuries ago were not those of a rational individual. Indeed, it is interesting to speculate on the fate of the colonies had George not been so grievously afflicted. For while there is no doubt that the "shot heard around the world" initiated the state of independence that was celebrated in 1976, it is also likely that the effects of porphyrins initiated a more subtle but just as devastating series of events leading to American independence.

It is with these thoughts in mind that we, a renegade Englishman and a new immigrant to North America, present our research on porphyrins.

INTRODUCTION

Heme proteins and the closely related chlorophylls and bacteriochlorophylls are ubiquitous in nature, and their primary biochemical roles are those of oxidation or oxygen transport and storage. Photosynthesis provides nature's principal mechanism for the utilization of solar energy through the interconversion of photonic energy to chemical energy. In green plants this process, which requires chlorophyll *a* (**1**) in the primary photochemical step, results in the reduction of CO_2 to carbohydrate with the concomitant oxidation of water to oxygen [2,3]. Photosynthetic bacteria, which utilize bacteriochlorophyll *a* (**2**), perform similar chemistry except that hydrogen donors other than water are used as the terminal reductants.

The oxygen produced during photosynthesis is the terminal electron acceptor for all aerobic organisms, and the processes by which oxygen is transported, reduced, and activated as an oxidant are often mediated by heme proteins, which with but a few exceptions contain protoporphyrin (**3**) as

their prosthetic group. In mammals, oxygen, no matter what its final fate, is transported by the cooperative tetrameric protein hemoglobin and stored by the monomeric protein myoglobin [4,5]. Both of these systems contain ferrous protoporphyrin and reversibly bind oxygen without suffering oxidation to the ferric state, a matter to which we address ourselves below. When oxygen is reduced to water, in the terminal step of respiration, four electrons are

transported via a series of cytochromes, which all contain iron protoporphyrin and function via reversible one-electron Fe(II) $\rightleftharpoons$ Fe(III) couples to
cytochrome oxidase. Cytochrome oxidase contains two atoms of copper
and two iron porphyrins with heme a (**4**), rather than protoporphyrin, as

4

their prosthetic group. The manner and mechanism by which cytochrome
oxidase reduces oxygen to 2 moles of water are still basically unknown [6].

In addition to a four-electron reduction to water, nature also brings about
both one- and two-electron reduction of oxygen to superoxide [7] and
peroxide [8], and the resultant decomposition and activation of hydrogen
peroxide are again mediated by the hemoproteins catalase and peroxidase [9].
As well as using oxygen as an electron acceptor, nature also brings about
oxidations of organic substrates whereby molecular oxygen is incorporated
into the substrate. The enzymes that catalyze such oxidation have been
classified [10] as monooxygenases or mixed-function oxidases when one atom
of molecular oxygen is incorporated into the substrate and the other is
reduced to water and as dioxygenases when both atoms of oxygen are
incorporated into the substrate. Hemoproteins are found among both the
mono- and dioxygenases. The biologically important monooxygenase
cytochrome *P*-450, to which we devote much of this review, is intriguing in
that both the electron transport properties of the cytochromes and the
oxygen binding of hemoglobin and myoglobin are combined in its functioning.

In this review we attempt to survey the electronic properties of metalloporphyrins and their interactions with molecular oxygen.

ELECTRON TRANSPORT PROPERTIES OF THE
PORPHYRIN MACROCYCLE

The apparent, albeit poorly understood, roles of the iron in hemoglobin,
myoglobin, and the respiratory chain cytochromes have for many years

focused attention on the metal in metalloporphyrins, but questions concerning the biochemical functions of the porphyrin macrocycle itself remain largely unanswered. We were curious as to why nature had constructed so complex a macrocyclic system as the porphyrin if its only function were that of modifying the redox potential of a coordinated metal or acting as a bridge between metal and protein. Surely some more definitive role could be found for the porphyrin, and it was this question to which we initially turned out attention.

Porphyrin π-Cation Radicals

Since metalloporphyrin are directly linked to so many biochemical redox systems, the redox properties of simple metalloporphyrins themselves were studied first. There is a wealth of knowledge concerning the redox chemistry of the metal coordinated to a porphyrin [11], and porphyrins, which can function as both electron sink and sources, can coordinate to essentially every element in the periodic table, as well as stabilize any particular element in a variety of oxidation states. Thus, iron porphyrins can exist in the $+1$, $+2$, $+3$, and $+4$ oxidation states, and oxidation states unusual for coordination complexes, such as Si^{2+}, Ag^{2+}, Os^{8+}, and P^{5+} [12] are known among the metalloporphyrins.

We chose, however, to examine the metalloporphyrins in which there was little chance of the metal undergoing redox processes. Zinc and magnesium, in their coordination chemistry, exhibit only the $+2$ oxidation state and form stable complexes when coordinated to dianionic tetradentate porphyrins. While the specific peripheral groups of the naturally occurring porphyrins are essential for proper biochemical function, their chemical reactivity and asymmetric distribution make the natural porphyrins poor substrates for *in vitro* model studies. Instead, we chose to use two synthetic porphyrins, *meso*-tetraphenylporphyrin (TPP) [13] (5) and octaethylporphyrin (OEP) [14] (6), which, due to their "high" solubility, chemical stability, and relative ease of preparation, have proved to be exceptionally useful compounds [9].

5   6

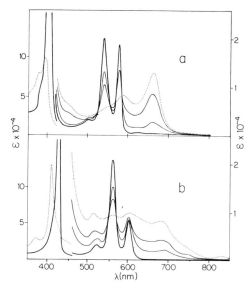

Fig. 1 (a) Oxidation of MgOEP in CH$_2$Cl$_2$: (——) Mg(II)OEP; (···) [Mg(II)OEP]$\overset{+}{\cdot}$ClO$_4$⁻. (b) Oxidation of MgTPP in CH$_2$Cl$_2$: (——) Mg(II)TPP; (···) [Mg(II)TPP]$\overset{+}{\cdot}$ClO$_4$⁻.

When magnesium OEP was oxidized either electrochemically or chemically, a clean, reversible one-electron oxidation was observed. Analysis of both the optical (Fig. 1a) and electron paramagnetic resonance (epr) spectra of the oxidized product and a deuterated analog enabled us to characterize it as a π-cation radical [15] in which the oxidation had not, as expected, changed the oxidation state of the metal but rather had removed an electron from the highest filled bonding π-molecular orbital. When the magnesium complex of TPP was oxidized, a similar reversible one-electron oxidation occurred to generate the corresponding π-cation radical. However, the optical spectra (Fig. 1b) of these two cation radicals as well as their epr spectra were very different. Indeed, an analysis of the epr spectra showed that the distribution

7

of the remaining unpaired electron in the free radicals was very different in the two cases [16].

Gouterman [17] has shown that in a metalloporphyrin the highest filled bonding orbitals, a_{1u} and a_{2u}, are essentially degenerate. Abstraction of an electron from one or other of these orbitals would generate a π-cation radical with either a $^2A_{1u}$ or a $^2A_{2u}$ ground state. Open-shell calculations support this hypothesis [15] and predict that the two ground states will differ in energy by about 6–8 kcal/mole whereby small perturbations such as changes in peripheral groups or centrally coordinated metal could determine the symmetry of the ground state. Moreover, the calculations showed that in the $^2A_{1u}$ state low spin density appears at both the *meso*-carbon and nitrogen atoms, while spin density appears on these atoms in the $^2A_{2u}$ state, and indeed the epr evidence confirms that the π-cation radical of MgOEP occupies a $^2A_{1u}$ and that of MgTPP a $^2A_{2u}$ ground state.

The Isocyclic Ring of Chlorophylls

When porphyrin cation radicals are further oxidized, the remaining unpaired electron can be reversibly removed to generate the corresponding π-dication [15]. These π-dications, unlike the π-cation radicals that are stable in solution for days, are powerful electrophiles. When the π-dication of ZnTPP was treated with methanol, attack at the *meso*-carbon occurred to give a stable isoporphyrin [18] (**7**); however, when the *meso*-carbon is substituted

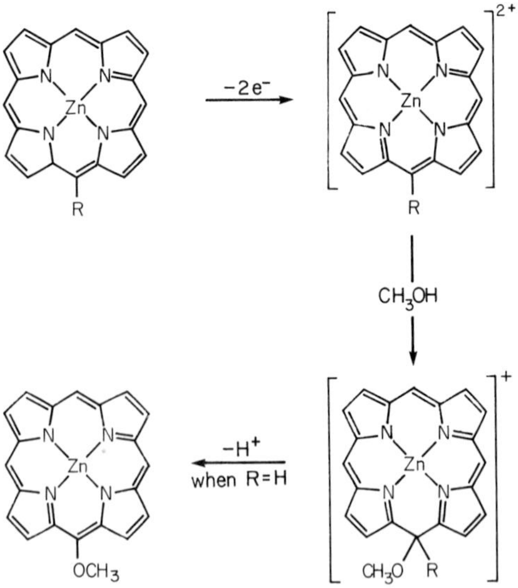

Scheme 1

by hydrogen rather than phenyl the intermediate isoporphyrin loses a proton to regenerate a *meso*-substituted porphyrin [19] (Scheme 1). At the time we observed these reactions it was reported [20] that porphyrins bearing a β-keto ester side chain could be cyclized in the presence of iodine in methanol containing K_2CO_3 to give the isocyclic ring of the chlorins and bacterio-chlorins. Both Woodward [21] and Kenner [22] had earlier suggested that the biosynthesis of this isocyclic ring could derive from the attack of an enolic β-keto ester onto the *meso* position of a metal-free diprotonated porphyrin. However, this scheme required nucleophilic attack at the porphyrin periphery, a reaction still unknown with porphyrins. Moreover, the overall sequence to the isocyclic ring bearing porphyrin would require a two-electron oxidation.

The iodine-catalyzed cyclization was thought [20] to proceed via the coupling of a diradical (a porphyrin and enolate radical) with subsequent loss of a proton from the *meso* position. We have found no examples, even among porphyrins bearing β-keto ester side chains, in which the removal of a second electron from a π-cation radical generated any species other than the π-dication. Indeed, oxidation of the zinc complex of a porphyrin bearing a β-keto ester side chain (**8**) gave, in our hands [19], a π-cation radical that showed no signs of cyclizing even over prolonged periods and gave on reduction a quantitative recovery of starting material. When, however, (**8**)

8

9

was oxidized at higher potentials an unstable π dication was produced which under basic conditions cyclized to the porphyrin (**9**) bearing an isocyclic ring [19]. We suggest, then, that the mechanism for this cyclization as well as that for the iodine-initiated reaction, proceeds via the attack of the nucleophillic β-keto ester onto the π-dication to give an isoporphyrin, which then collapses, by loss of a proton, to give the porphyrin bearing an isocyclic ring (**8 → 9**). Moreover, it is known that magnesium protoporphyrin, in which one of the propionic acid side chains has been oxidized to a β-keto acid (**10**), is the biosynthetic precursor of chlorophyll *a* [23]. We suggest that the biosynthesis of chlorophyll proceeds via the

10

two-electron oxidation of **10** followed by generation of the isocyclic ring. It is interesting that the potentials required to oxidize a variety of metallo-porphyrins to the π-cation radical stage are usually high enough to oxidize the corresponding *magnesium* porphyrin to the π-dication. Thus, if nature wishes to generate a metalloporphyrin π-dication, the metal that most favors this process is magnesium.

One might then ask: If the magnesium porphyrin **10** is required for the biosynthesis of chlorophyll why is the magnesium retained in chlorophyll (and bacteriochlorophyll) for the photosynthetic process? Again, we believe that the answer lies in the ease of oxidation of these macrocycles. In fact, magnesium chlorins and bacteriochlorins have even lower oxidation potentials than magnesium porphyrins [24,25].

Relevance of Porphyrin π-Cations to Photosynthesis

In green plants the majority of the chlorophyll *a* in a chloroplast has optical absorptions comparable to those of the cell-free pigment, and this chlorophyll functions primarily as an antenna to collect and transport photons to a specific chlorophyll *a* molecule, which absorbs at lower energy than (to the red of) the antenna [26]. This species, which absorbs at about 700 nm, is known as pigment 700 (*P*-700). The difference in the optical spectrum of *P*-700, which contains chlorophyll *a*, and that of the antenna chlorophyll *a* (677 nm) is a function of the local environment of *P*-700, which appears to involve two chlorophyll *a* molecules linked through their magnesium atoms by water [26]. A similar bathochromic shift is observed between the antenna bacteriochlorophyll and *P*-865. In both cases, these "special" pigments are involved in the primary photochemical step of photosynthesis when a photon causes the bleaching of the pigment coupled with the appearance of an epr-detectable free radical. It was suggested that this process involved the photooxidation of these species [27–32]. Having observed the ease of formation and stability of metalloporphyrin π-cation radicals, we anticipated that the *in vivo* photochemical oxidation of *P*-700, if indeed it was a photooxidation initiating the photosynthetic sequence, might involve the formation of a chlorophyll (or bacteriochlorophyll) π-cation radical.

Cell-free chlorophyll *a* undergoes a reversible one-electron electrochemical oxidation in methylene chloride. The resulting species is a radical with $g = 2.0026$ and behaves as a cation upon electrophoresis [25]. A comparison of the difference spectrum between neutral and electrochemically oxidized cell-free chlorophyll *a* and that of photosynthesizing chloroplasts suggested that the two oxidized species were similar; this was additionally supported by a comparison of the epr spectra of the two species. Unfortunately, in both cases no hyperfine structure could be detected and hence the electronic configuration of the radical could not be determined. The same series of experiments were performed with bacteriochlorophyll, which also underwent a one-electron oxidation. A comparison of the optical and epr spectra of the one-electron oxidation product of cell-free bacteriochlorophyll with that of photosynthesis chromophores showed that the two species were similar to each other and to those derived from chlorophyll, but, again, no hyperfine structure was observed in their epr spectra, and no conclusions concerning the electronic configuration of the oxidized species could be made [33]. We prepared the zinc complex of *meso*-tetraphenylchlorin (ZnTPC) (11) in the hope that this would be a suitable and more amenable model for chlorophyll *a*, but upon oxidation the optical and epr spectra were similar to those of both *P*-700 and photosynthesizing chloroplast but no hyperfine structure was observed.

11   12

At this stage we examined the only remaining analog in this series, namely, *meso*-tetraphenylbacteriochlorin (ZnTPBC) (12). The optical spectrum of 12 is similar to that of bacteriochlorophyll, suggesting that 12 is a good model for bacteriochlorophyll. Moreover, the difference spectra between the neutral and one-electron oxidized species (Fig. 2) and that for photosynthesizing chromophores suggested a close similarity between the structures of all three oxidized compounds. Moreover, the epr spectrum of oxidized 12 was rich

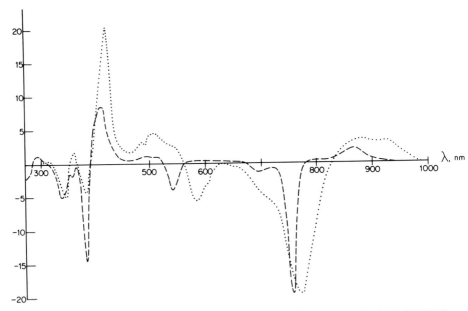

Fig. 2 The optical difference spectra of electrochemically oxidized Zn(II)TPBC (–––) and cell-free bacteriochlorophyll (···).

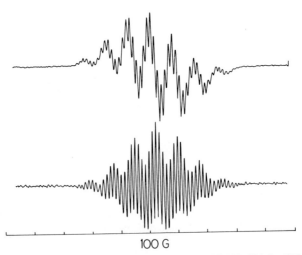

100 G

Fig. 3 First- and second-derivative epr spectra of $[Zn(II)TPBC]^{+}$ in CH_2Cl_2/Pr_4N^{+}-ClO_4^{-}.

with hyperfine structure (Fig. 3), showing that the product was a π-cation radical [33]. The close similarity between photobleached P-700 and P-865 and the electrochemically oxidized cell-free pigments and the metal complexes 11 and 12 allowed us to identify the oxidation product in every case as a π-cation radical and to demonstrate that the primary photochemical event in photosynthesis involves the photon-induced ejection of an electron from the π cloud of the "special" pigment and generation of the corresponding π cation radical.

The Cytochromes

What is the fate of this ejected electron? While the exact path that this electron follows and the sequence and identity of the various electron acceptors it encounters are still not fully understood, it is clear that cytochromes are involved. How do the cytochromes transport electrons? The Fe(II) $\rightleftharpoons$ Fe(III) couple clearly accounts for the overall redox process, but how is the electron transferred from and delivered to the iron atom coordinated to the porphyrin? The three-dimensional structures of both cytochromes c and b_5 have been determined by X-ray diffraction, and in both cases the iron atom is buried in the protein and well shielded from the exterior. How then, bearing in mind that the oxidants and reductants for these systems are themselves bulky macromolecules, does the iron atom interact with the electron? The occurrence of porphyrin π-cation radicals in other "porphyrin"-mediated biochemical processes prompted us to speculate [9] that the oxidation of a cytochrome could be envisaged as shown in Eq. (1). We have

$$\text{Fe(II) cytochrome} \xrightleftharpoons{-e^-}$$

$$\text{Fe(II) cytochrome porphyrin } \pi\text{-cation radical} \xrightleftharpoons[\substack{\text{electron} \\ \text{transfer}}]{\text{internal}} \text{Fe(III) cytochrome} \quad (1)$$

so far found no evidence for any intermediate during the oxidation of simple iron porphyrins. Nevertheless, an exact analogy for this peripheral oxidation coupled to an internal electron transfer is observed with another metalloporphyrin.

The one-electron oxidation of Ni(II)TPP gives the expected Ni(II)TPP π-cation radical, which occupies a $^2A_{1u}$ ground state, an observation that is quite unexceptional and that parallels the chemistry of most other metalloporphyrins. What is exceptional, however, is that when this green π-cation radical is cooled to liquid-nitrogen temperature it is converted to the red Ni(III)TPP [34]. Upon warming to room temperature the Ni(II)TPP π-cation radical is regenerated, and this internal electron transfer can be repeated numerous times. While the nature and cause of this process are not understood, the internal transfer occurs only in the presence of poorly coordinating counterions such as PF_6^- and ClO_4^-, suggesting that the coordination, or lack thereof, of ligands to the central metal influences the ground state that this species will occupy at different temperatures [35].

This analogy between the behavior of Ni(II)TPP $^+$ and our hypothesis concerning the mode of electron transfer in the cytochromes encourages us to look farther for an intermediate in the oxidation of Fe(II) to Fe(III) porphyrins. One might nevertheless ask, Is there really a difference in the electronic configuration between an Fe(II) porphyrin π-cation radical and an Fe(III) porphyrin, or are these really two resonance structures of the same delocalized species? It is our contention that two discrete species in fact exist. Iron in the $+2$ and $+3$ oxidation states has a different covalent radius, and in fact the iron in ferrous porphyrin frequently sits above the plane of the four porphyrin nitrogens, while iron in the corresponding ferric complex is invariably in plane [36]. If such a conformational change occurred between an Fe(II) porphyrin π-cation radical and an Fe(III) porphyrin, then these two species would be distinct entities with an energy barrier between.

So far we have shown only porphyrin π-cation radicals to have biochemical roles in photosynthesis; in retrospect it is not surprising that if the macrocycle is to lose an electron the site of electron abstraction should be from the macrocycle rather than from divalent magnesium. Of more importance is the question as to whether there is any definitive evidence for a π-cation radical in a hemoprotein, and indeed there is.

Catalases and Peroxidases

The catalases [37] and peroxidases [38] are two closely related series of enzymes in that both resting enzyme are ferrichemoproteins and both undergo a two-electron, hydrogen peroxide-mediated oxidation to the so-called primary compounds (Compounds I). At this stage the enzymatic activity of these two systems differs. Catalase is further reduced by hydrogen peroxide to the resting enzyme, and the hydrogen peroxide is oxidized to oxygen. The peroxidases, on the other hand, are stable toward excess hydrogen peroxide and instead react with a variety of hydrogen donors, generating an organic free radical and a one-electron reduction product of the enzyme, the secondary compound (Compound II), which can in turn bring about a further one-electron oxidation of substrate with regeneration of the resting enzyme. The primary compound of catalase can also suffer a nonenzymatic one-electron reduction to its secondary compound (Compound II).

A multitude of hypotheses concerning the electronic configurations of the primary and secondary compounds of these enzymes have been postulated. The original suggestion [39] that the primary complexes were coordination complexes between hydroperoxide and ferric iron, i.e., Fe(III)–OOH, was inconsistent with the observation that nonperoxidatic oxidants could also generate the primary compounds [40]. Addition of hydrogen peroxide to the periphery of the ferric protoporphyrin to give an isoporphyrin-like structure [41] was also invalidated when we prepared authentic isoporphyrins [18]. The only unifying theme among all the structural hypotheses for the primary complexes was the two-electron oxidation state above ferric, and indeed an Fe(V) complex was also suggested [42].

We had hoped that a simple two-electron oxidation of model ferric porphyrins might resolve the problem. This was not to be the case, for while a clean, reversible one-electron oxidation of ferric porphyrins to the Fe(IV) oxidation state was possible [43,44] the potentials required for higher oxidations brought about decomposition of solvent and substrate.

We showed earlier that the removal of an electron from a π orbital of a metalloporphyrin generates a π-cation radical and that the remaining unpaired electron occupies either an a_{1u} or a_{2u} orbital. More definitive evidence for these two ground states, and the electronic configurations for oxidized catalases and peroxidases, were found when we examined the redox chemistry of cobalt porphyrins. The electrochemical oxidation of Co(II)OEP in methylene chloride and tetrapropylammonium perchlorate gave, first, the cobaltic species $[\mathrm{Co(III)OEP}]^{+}\mathrm{ClO_4}^{-}$ and then the cobaltic π-cation radical $[\mathrm{Co(III)OEP}]^{2.+}$-$2\mathrm{ClO_4}^{-}$. The corresponding oxidation with elemental bromine also brought

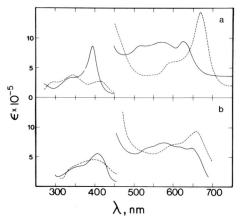

Fig. 4   (a) Comparison of the optical absorption spectra of $[\mathrm{Co(III)OEP}]^{2.+}2\mathrm{Br}^{-}$ (——) and $[\mathrm{Co(III)OEP}]^{2.+}2\mathrm{ClO_4}^{-}$ (– – –). (b) Comparison of the optical absorption spectra of CAT I (——) and HRP I (– – –).

about two separate one-electron oxidations to give the salts $[Co(III)OEP]^{\ddagger}$-Br^- and $[Co(III)OEP]^{2\cdot+}2Br^-$. When the optical spectra of these two cations were compared (Fig. 4a) it was apparent that the perchlorate salt occupied a $^2A_{2u}$ and the bromide salt a $^2A_{1u}$ ground state [45]. Moreover, addition of silver perchlorate to the dibromide gave the diperchlorate with a change in the ground state as evidenced by the change in the optical spectrum.

Of even greater significance was the observation that the optical spectra of the two ground states of the cobaltic π-cation radical parallel those of the primary compounds of catalase (CAT I) and peroxidase (typified by horse-radish peroxidase, HRP I) (Fig. 4b). This suggested that during the oxidation of the ferrihemoprotein to the primary compounds one, but only one, electron was removed from a π orbital. This leaves the metal as the other most likely site of electron abstraction, such that the overall redox sequence of these enzymes can be formulated as shown in Eq. (2). That the secondary compounds

$$
\underset{\substack{\text{resting} \\ \text{enzyme}}}{\text{Fe(III)}} \xrightleftharpoons{-e^-} \underset{\substack{\text{secondary} \\ \text{compound}}}{\text{Fe(IV)}} \xrightleftharpoons{-e^-} \underset{\text{primary compound}}{\text{Fe(IV) } \pi\text{-cation radical}} \tag{2}
$$

are derivatives of Fe(IV) is substantiated by their optical spectra [37], which are typical of metalloporphyrins and indicate oxidation of the metal rather than the ligand, by Mössbauer spectroscopy [46,47], which demonstrates a different electronic structure for the iron in the secondary compound compared to that in the resting enzyme, and by our electrochemical generation of Fe(IV) porphyrin complexes [43,44]. It is interesting to speculate that the different enzymatic functions of the catalases and peroxidases, which are in many ways similar enzymes, derive from the different ground states of the porphyrin π-cation in the enzymatically active primary compounds. As yet, however, the chemistry of the π-cation radicals, even in model systems, has hardly been explored.

THE NATURE OF OXYGEN AND OXYGEN BINDING

The electron transport properties of the cytochromes and the oxygen binding of hemoglobin and myoglobin are combined in the functioning of cytochrome P-450, and in order to obtain a clearer picture of how this enzyme might function we shall spend some time on the nature of oxygen binding to metalloporphyrins and the more general problem of oxygen activation.

Oxygen is one of the few stable molecules that contains an unfilled orbital. In the ground state molecular oxygen approximates a closed-shell nitrogen molecule with two extra electrons in the π_g orbital $[\pi_x^*(\uparrow)\pi_z^*(\uparrow)]$ when the molecule is in a triplet state. This triplet nature of O_2 prevents it from reacting readily with other singlet molecules [48], since angular momentum must be conserved. Hence, the reaction of triplet O_2 with a singlet organic compound must give an initial triplet intermediate, which may not live long enough to undergo spin inversion to a stable singlet product. One should bear in mind, however, that once this spin forbiddenness of the reaction is overcome, organic compounds undergo exceedingly rapid oxidation, and the concomitant reduction of oxygen releases considerable free energy, which is exceeded only by that of the reduction of molecular fluorine [49]. It is these two unique characteristics of oxygen, i.e., inertness on the one hand and the high oxidation potential on the other, that made aerobic life processes on earth possible.

To cope with the fact that O_2 is a triplet and still reacts readily with organic compounds under physiological conditions, nature appears to have employed three general methods to circumvent this problem [50]:

1. *Radical mechanisms.* When a triplet reacts with a singlet to give two separate molecules, each contains one unpaired electron; the reaction is a spin-allowed process. Energetically it is more favorable to have one unpaired electron on a single molecule than two electrons on one molecule. However, the initiation of such radical reactions [Eq. (3)] is normally so endothermic

$$R\!-\!H + O_2 \longrightarrow R\cdot + HO_2\cdot \qquad (3)$$

that this route is rendered unattractive unless the radical product $R\cdot$ can be sufficiently stabilized by electron delocalization. In biological systems the reaction of reduced flavin monooxygenase with O_2 has been speculated to involve a free-radical process [Eq. (4)] [51]. It should be apparent that the ease of this reaction must be due to the extensive delocalization available in the isoalloxazine ring structure.

$$+ H^+ + O_2^- \qquad (4)$$

2. *Singlet oxygen.* There are two low-energy singlet excited states of oxygen. The higher, $^1\Sigma g^+$, has an energy of 37 kcal/mole and is short-lived ($\tau \cong 10^{-11}$ sec). The lower-energy state, $^1\Delta g$ (22 kcal/mole above the ground), is longer-lived (10^{-5} sec) and can be easily obtained photochemically

in the presence of a sensitizer. The chemistry of singlet oxygen and photo-sensitized oxygenation has been studied extensively [52,53], and it is now apparent that $^1\Delta g$ oxygen behaves as an electrophile that is capable of reacting with double bonds or electron-rich chromophores much like the alkene in a Diels–Adler reaction, and free-radical mechanisms have been completely ruled out. The biochemical relevance of photosensitized oxygena-tion lies in its damaging effects on certain amino acids, enzymes, nucleosides, lipids, and other cell constituents [53], and the association between photo-sensitizing ability and carcinogenicity of polynuclear aromatic hydrocarbons has also been noted. An interesting case in which singlet oxygen may play a benevolent role is in the medical treatment of neonatal jaundice by irradiation. Singlet oxygen, formed by bilirubin as sensitizer, has been shown to be involved in the destruction and elimination of excess bilirubin in the skin of the patient [54,55].

　　　3. *Oxygenation mediated by transition metals.* Ground-state triplet oxygen can be ligated to a transition metal which itself has unpaired electrons (for example, a cobalt(II) porphyrin has one unpaired electron). The result is that the oxygen–metal complex can be readily formed without violating the spin-conservation rule; moreover, it is a spin-allowed process for such a complex to react with singlet organic compounds as long as the number of unpaired electrons on the overall metal ion complex remains unchanged throughout the reaction [56]. Furthermore, the spin-conservation rule has no restrictions on the mechanism by which complexed O_2 may react with substrates. However, the high specificity of enzymatic reactions suggests specific ways by which the oxygen molecules are activated and transferred from the oxygen–metal complex to the substrate. The precise molecular events that enable the oxygen–metal complex to undertake such excursions have so far eluded definition, and before discussing the reactions of oxygen–metal complexes we shall examine the chemistry of oxygen binding by metalloporphyrins.

REVERSIBLE OXYGEN BINDING TO METALLOPORPHYRINS

　　　Reversible oxygen complexes of transition metals can be regarded as reaction intermediates of transient stability during oxygen fixation processes. Under suitable conditions the oxygenated species may be stable enough to be isolated. The oxygen carriers hemoglobin and myoglobin represent examples of this extreme in that the reversibility of heme–O_2 binding is optimized to suit their biological purposes. This requires a subtle balance between the thermodynamic stability of the oxygen complexes and the kinetic reversibility of oxygenation. It was not until recently that the mechanism of heme oxygenation became clearer from model studies, but since this topic has been reviewed thoroughly [57–62] it is only highlighted here.

Electronic Structure of the O_2 Complex

There are three possible bonding modes that are consistent with the diamagnetism of oxyheme complexes:

1. Oxygen reacts in the singlet state:

$$Fe\ (d^6, S = 0) + O_2 \rightleftharpoons FeO_2\ (S = 0)$$

The influence of the electrostatic field of a transition-metal ion might remove the degeneracy of the ground-state O_2 π_g orbital level, such that the two unpaired electrons of O_2 might be paired in the lower π_g orbital to give a singlet [63]. The oxygen molecule in this situation would have an electronic configuration comparable to that of ethylene. One may therefore visualize the electron distribution in terms of trigonal sp^2 hybridization at the 2 oxygen atoms. This molecule may interact with hemes in two possible modes as shown in Fig. 5.

According to ligand field theory, the five d orbitals of the metal ion are split into two high-lying e_g levels ($d_{x^2-y^2}, d_{z^2}$) and three lower-energy t_{2g} levels (d_{xy}, d_{xz}, d_{yz}) in an octahedral hemochrome. The t_{2g} orbitals play no part in σ bonding with O_2, whereas the e_g orbitals combine with O_2 ligand orbitals to give bonding and antibonding orbitals. When the ligand molecule has empty orbitals of the appropriate symmetry, i.e., the antibonding e_g orbital, t_{2g} orbitals of the metal can further interact with them to give π bonding. If such interaction takes place this will result in a net increase of the bonding stability. Both the σ and π bonding between oxygen and metal

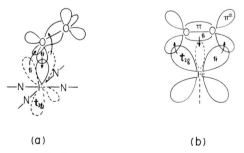

(a)   (b)

Fig. 5 Heme–oxygen binding mode. (a) Pauling's model; (b) Griffith's model.

can take place with either the "end-on" model of Pauling [64] or the "side-on" model of Griffith [63]. Mingos [65] has discussed the preferred geometries by consideration of the symmetries of the appropriate metal d and oxygen orbitals. Metals susceptible to a two-electron oxidation (d^8, d^{10} configurations) favor the side-on O_2 binding, while metals capable of undergoing a one-electron oxidation appear to favor the end-on binding. Recent X-ray studies on model compounds revealing the $Co-O_2$ angle of $126°$ [66] and the $Fe-O_2$ angle of $136°$ [67] appear to be consistent with the sp^2 hybridization, end-on binding of oxygen. However, theoretical arguments based on extended Hückel MO calculations [68] suggest that the most stable bonding mode of oxygen, in oxyhemes, is in fact between that of the Pauling and Griffith models such that the center of gravity of the $O-O$ bond is moved sideways with respect to the z axis and tilted at an angle of $20°$ from the plane of the porphyrin. Moreover, this model gives the best fit for the Mössbauer hyperfine splitting parameters of heme–O_2 complexes [69].

2. Oxygen reacts in the triplet state [70]:

$$Fe\ (d^6,\ S = 1) + O_2\ (S = 1) \rightleftharpoons FeO_2\ (S = 0)$$

In the ground state, one of the unpaired electrons in the π_g orbital can interact with the d_{z^2} orbital of metal to form a σ bonding. The most possible mode of overlapping is illustrated in Fig. 6. The additional π bonding is accomplished by combining $\pi_g{}^x$ and d_{zx} orbitals.

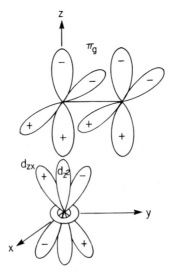

Fig. 6 Possible mode of interaction between triplet oxygen and high-spin Fe(II) heme.

3. Metal donates an electron to oxygen, with both atoms having a spin state $S = \frac{1}{2}$ on bonding, with the unpaired spins antiferromagnetically coupled:

$$\text{Fe } (d^5, S = \tfrac{1}{2}) + \text{O}_2\bar{} (S = \tfrac{1}{2}) \rightleftharpoons \text{Fe}^{3+}\!-\!\text{O}_2^- (S = 0)$$

The evidence for a superoxo formulation originally came from studies on cobalt–O_2 compexes [71]. The epr of oxygenated cobalt complexes indicated that the unpaired electron [Co(II) id d^7] had only a small spin density at the Co nucleus. It was concluded that the unpaired electron was delocalized onto the oxygen ligand with the electron coming from any of the d_{z^2} orbitals to one of the π_g orbitals of O_2 [71,72].

Similar formulation for iron–O_2 complexes was first suggested for oxyhemoglobin on the basis of the optical similarities observed between oxyhemoglobin and oxidized hemin [73]. Convincing support for this structure has recently been provided. From the heme–O_2 model studies, we [74,75] were able to show that the oxygenation of simple hemes was greatly enhanced in polar solvents, the oxygen dissociation rate being decreased 17-fold as the solvent was changed from toluene to water. This solvent effect is expected for a highly polar Fe—O_2 bond. Similar enhancement for O_2 binding had previously been observed with cobalt porphyrins [76]. Furthermore, vibrational spectroscopic data have shown that oxyhemoglobin [77,78] has a value for νO_2 of 1107 cm^{-1} and oxycobalhemoglobin [79] has a value of 1105 cm^{-1}, a frequency close to that of the well-defined inorganic superoxide (KO_2, 1145 cm^{-1} [80]). Resonance Raman spectra likewise suggest a charge-transfer for both Fe and Co complexes [81]. Moreover, hemin has been reported to form a species with O_2^- at low temperature which has close spectral similarities to that of oxyheme [82].

These observations leave little doubt that there is substantial electron delocalization from the metal center toward oxygen in the oxygen complexes. It must be emphasized, however, that the assignment of "oxidation states" for these complexes in which the Fe–O_2 bond has strong covalent character is quite artificial, and in actuality all possible resonance forms contribute to the observed properties. The degree of electron density transfered to O_2 depends on a large number of factors, including the ligand complexed *trans* to O_2, temperature, and environment.

Ligand Effect

The bonding between a metalloporphyrin and O_2, as described above, is very sensitive to factors that influence the transfer of electron density from metal to O_2. The metal acts as electron donor without being irreversibly oxidized. It is the function of soft ligands in its coordination sphere to produce a delicately balanced electronic state at the metal which controls the extent of oxygenation [83]. The influence of the axial ligand coordinated opposite the O_2 (the fifth coordination site) is termed "the *trans* effect," which perturbs directly the same orbitals (e.g., d_{z^2}) employed in O_2 binding. From model studies it is clear that strong π-base ligands such as imidazole tend to make the iron softer and thus bind O_2 stronger at the expense of decreased kinetic reversibility of the O_2 complex [74,84].

Investigation of the electronic properties of the porphyrin ring also revealed the so-called *cis* effect. In essence, the porphyrin nucleus is capable of transmitting the electronic effect of its peripheral groups to the iron and consequently influences the character of the iron–oxygen bond. Therefore, electron-withdrawing groups such as —CHO, and —COCH$_3$ would decrease oxygenation, whereas electron-donating groups such as —CH$_3$ increase the oxygen binding. This peripheral electronic effect has been documented by infrared studies [85] as well as by myoglobin and hemoglobin reconstitution studies [86,87].

Steric Effect

The axial ligand at the fifth coordination site is by far the most crucial determinant for the oxygen-binding properties of the metalloporphyrin. Maximal chelation between metal and O_2 is achieved only when certain geometric requirements of the axial ligand can be satisfied. The structural model of hemoglobin function postulated by Perutz [88] emphasizes the importance of "tension at the hemes" imposed by the proximal histidine ligand as a major cause of the low ligand affinity of the deoxy quaternary structure and the manifestation of heme–heme interaction [89]. These phenomena have been discussed in detail elsewhere [90,91].

Environment Effect

Solvent polarity plays an important role in stabilizing the highly polar metal–O_2 complex by virtue of dipole–dipole interaction or hydrogen bonding. By the same effect the distal histidine present in most vertebrate hemoglobin and myoglobin has been suggested to have a modulation effect in the kintics of heme oxygenation [74].

In acidic protic solvent, oxyheme undergoes rapid irreversible oxidation [92]. This would be consistent with reaction (5). It has been pointed out that

$$Fe^{3+}\text{—}O_2^- \longrightarrow Fe^{3+}\text{—}O_2H \longrightarrow Fe^{3+} + O_2H \qquad (5)$$

oxyheme itself is a relatively inert species [93] as a result of both σ and π bonding, and a simple superoxide dissociation may be unfavorable [Eq. (6)].

$$Fe^{3+}-O_2^- \longrightarrow Fe^{3+} + O_2^- \tag{6}$$

However, the addition of a proton to the oxy derivative would weaken both the σ and π bonding, resulting in a more facile pathway for iron oxidation.

In recent years it has been shown that under most conditions oxyheme is oxidized by forming a μ-peroxo dimer, which then undergoes irreversible oxidation to yield an oxidation product: the μ-oxo dimer [Eq. (7)] [94]. The

$$Fe^{3+}-O_2^- \xrightarrow{Fe^{2+}} Fe^{3+}-O_2-Fe^{3+} \longrightarrow [Fe^{3+}-O\cdot] \longrightarrow$$
$$Fe^{3+}-O-Fe^{3+} \tag{7}$$

addition of Fe^{2+} to oxyheme probably dictates an alternative pathway from that described above [Eq. (5)]. Since the ferric iron is still a sufficiently strong electron donor, the peroxo bridge may decompose through a homolytic cleavage leading to the $Fe^{3+}O\cdot$ species. However, this mechanism has not yet been fully elucidated.

The realization of the bimolecular oxidation process coupled with the solvent polarity effect has clarified the concept of the so-called hydrophobic crevice theory for hemoglobin [4,95], and we now know that the crevice structure and its hydrophobicity decrease the tendency for oxidation rather than increase oxygenation, a process that is favored by polar environments [74].

Mononuclear vs. Biomolecular Oxygen Complexation

Cobalt(II) chelates form both 1:1 and 2:1 complexes with O_2 [96,97]. In contrast, only 1:1 iron porphyrins have been isolated, although the natural O_2 carrier hemerythrin is considered to form a $[Fe-O_2-Fe]$ complex [98]. The dichotomous behavior between Fe and Co complexes can be reconciled as follows.

Under the theoretical framework outlined above, the 1:1 and 2:1 complexes can be visualized as superoxo [Eq. (8)] and peroxo [Eq. (9)] complexes,

$$M^{2+} + O_2 \longrightarrow M^{3+}-O_2^- \tag{8}$$
$$2M^{2+} + O_2 \longrightarrow M^{3+}-O_2-M^{3+} \tag{9}$$

respectively. To the first approximation the thermodynamic parameters for Eq. (8) are determined by those for the processes $O_2 \rightarrow O_2^-$ and $M^{2+} \rightarrow M^{3+}$, while for Eq. (9) they are determined by those for the processes $O_2 \rightarrow O_2^{2-}$ and $M^{2+} \rightarrow M^{3+}$. Since O_2 is a far better two-electron acceptor, the dimeric process is generally more favorable. Thus, if conditions allowed, the 2:1 complex would be the final product [99].

Kinetically, however, formation of the 2:1 complex has been shown to be a rather slow process. It is particularly slow in the presence of bulky ligands such as porphyrins; therefore, lower temperature, dilute solution, and bulky ligands prevent the 2:1 complex formation [97].

The principal reason that the 2:1 cobalt–O_2 complexes can be isolated whereas iron complexes cannot is that the dimer Fe—O_2—Fe would readily decompose to give the μ-oxo dimer. Interestingly, there have been no reports on the similar Co^{3+}—O—Co^{3+} dimer. There is no immediate answer for this, although it has been argued that the difference in stability between the hypothetical reaction intermediates Co^{3+}—O· and Fe^{3+}—O· may in part account [100] for the different behavior.

CYTOCHROME *P*-450

Having described our observations on the roles of metalloporphyrins in various biological processes and given a general description of reversible oxygen binding to metalloporphyrins, we now embark on our description of cytochrome *P*-450, which must for the chemist be the most fascinating of the heme proteins for not only does it combine many of the facets already described above, but this enzyme (in reality a multitude of closely related enzymes) actually mediates some complex and unusual chemistry.

Cytochrome *P*-450 belongs to a class of enzymes that incorporate one atom of molecular oxygen into substrates while concomitantly reducing the other atom to water. The stoichiometry of this process can be represented as shown in Eq. (10). These monooxygenase enzymes include heme-containing

$$RH + {}^{18}O_2 \xrightarrow{\quad NADPH^+ \quad NADP \quad} R—{}^{18}OH + H_2{}^{18}O \qquad (10)$$

cytochrome *P*-450, copper-containing enzymes, and a variety of metal-free flavoproteins [10], and the significance of monooxygenases in metabolism is evidenced by their wide variety of substrates, which include carbohydrates, lipids, amino acids, drugs, and hormones. Among the various monooxygenases, systems dependent on cytochrome *P*-450 occupy a preeminent position in that they play central roles in drug metabolism, carcinogenesis, pesticide detoxification, and steroid biosynthesis. Various aspects of cytochrome *P*-450 have been extensively covered in several reviews and symposia [101–107].

Occurrence and Properties

Cytochrome *P*-450 has a relatively short history. In 1958, the presence of a CO-binding pigment in rat liver microsomes was reported [108]. The pigment, when reduced by dithionite, was found to bind CO and give rise to a strong absorption band near 450 nm in the difference spectrum of the suspended microsomes. Since the absorption maxima of CO complexes of known hemoproteins normally occur around 420 nm, the chemical nature of this pigment was not immediately resolved. Indeed, not until 1962 was the hemoprotein nature of the CO-binding pigment established, after which the pigment was termed cytochrome *P*-450 by Omura and Sato [109]. This new cytochrome was also found in the microsome fraction of the adrenal cortex. It was from the experiments by Ryan and Engel, who showed the inhibitory effect of CO on the C-21 hydroxylation of steroids by the adrenal mitochrondrial fraction, that the catalytic role of cytochrome *P*-450 in the monooxygenation process was demonstrated [110]. The enzymatic function of cytochrome *P*-450 was then more firmly established by Estabrook and co-workers [111]. Since then, this type of enzyme has been found in microsomes from tissues of liver, kidney, lung, intestinal mucosa, testis, adrenal gland, and pancreas of various mammals. The presence of similar types of pigments in insects, plants, and microorganisms has also been established [106].

One of the most intriguing properties of this class of enzymes is that they are inducible by substrates. It was found that *in vivo* administration of certain drugs induced the synthesis of cytochrome *P*-450 and selectively enhanced certain monooxygenase activity. For example, repeated injections of sodium phenobarbital in rat increased hepatic cytochrome *P*-450 content and aminopyrine demethylation activities, while 3,4-benzpyrene hydroxylation was only slightly stimulated. The latter activity was markedly increased, however, following a single injection of 3,4-benzpyrene [112]. Apparently, a large number of cytochromes *P*-450 could be produced by induction, some being very substrate specific, others being active toward a broad range of substrates. The recent isolation and identification of distinct species of cytochrome *P*-450 with different molecular weights present evidence that different inducing agents are capable of stimulating the synthesis of distinct molecular forms of the hemoprotein [113,114].

The isolation and purification of microsomal cytochrome *P*-450 proved difficult since the enzyme was tightly bound to membrane. Early attempts at preparing cytochrome *P*-450 from membranes were thwarted by the loss of the 450 nm absorption peak (CO complex) and the appearance in its place of a 420 nm peak. This cytochrome *P*-420 pigment unfortunately has no enzymatic activity. However, recent advances in isolation techniques have resulted in preparations of liver microsomal cytochrome *P*-450 that has a purity approaching that of the soluble, crystalline cytochrome *P*-450 from *Pseudomonas putida* [103,115].

In 1968, Gunsalus' laboratory reported that a cytochrome *P*-450 type of enzyme was inducible in a strain of *P. putida* by growth on *d*-camphor as the sole carbon source [116]. The enzyme catalyzes the hydroxylation step in the camphor metabolism [Eq. (11)]. This cytochrome is soluble and is therefore

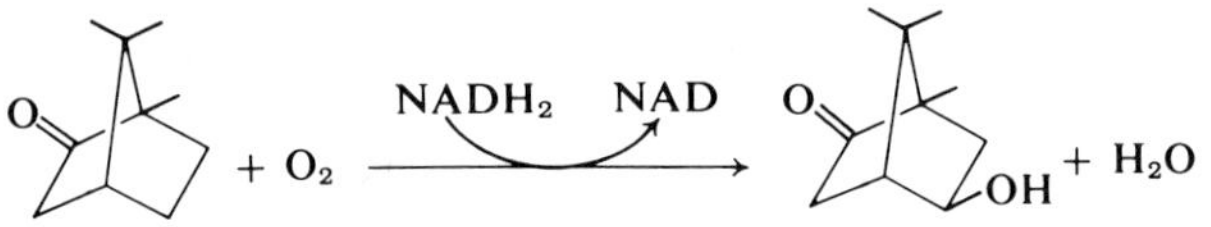

$$
\text{(camphor)} + O_2 \xrightarrow[\text{NADH}_2 \quad \text{NAD}]{} \text{(5-exo-hydroxycamphor)-OH} + H_2O \qquad (11)
$$

obtainable in a pure state. Crystalline cytochrome *P*-450 from *P. putida* has been reported [117]. The individual steps of the catalytic cycle of this resolved monooxygenase system have been characterized and often used as reference to describe the enzymatic reactions of the more complicated membrane-bound mammalian systems [105].

Components

Cytochrome *P*-450 itself functions as a terminal oxidase that converts molecular O_2 to an active form by the transfer of two electrons in discrete one-electron steps. The key components of the monooxygenase that are essential to catalytic functions consist of pyridine nucleoside–flavoprotein, proteins or cofactors that reduce cytochrome *P*-450, and cyrochrome *P*-450 itself (Scheme 2).

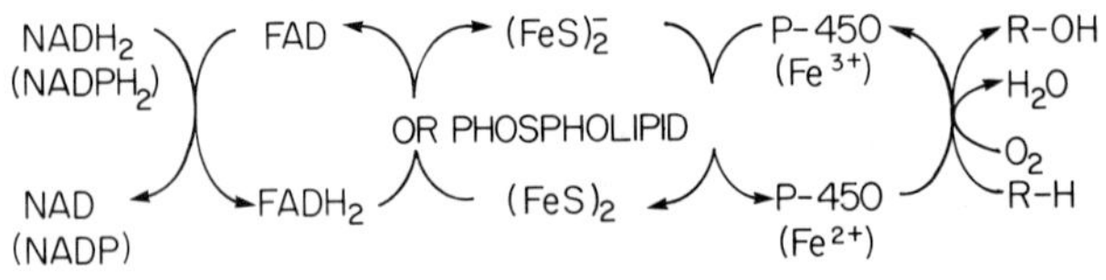

$$
\begin{array}{ccccc}
\text{NADH}_2 & \text{FAD} & (\text{FeS})_2^- & \text{P-450} & \text{R–OH} \\
(\text{NADPH}_2) & & & (\text{Fe}^{3+}) & \text{H}_2\text{O} \\
& & \text{OR PHOSPHOLIPID} & & O_2 \\
\text{NAD} & \text{FADH}_2 & (\text{FeS})_2 & \text{P-450} & \text{R–H} \\
(\text{NADP}) & & & (\text{Fe}^{2+})
\end{array}
$$

Scheme 2

Flavoprotein is common to all cytochrome *P*-450 systems. The isoalloxazine ring of the flavin allows the enzyme to undergo either one- or two-electron reductions, and as such it is well suited for its position between the two-electron carrier NADH_2 and the one-electron carrier cytochrome *P*-450 reductase.

In the *P. putida* system the reductase is putidaredoxin (MW 12,500), while in the adrenal mitochondrial system it is adrenodoxin. These are iron–sulfur proteins with a prosthetic group of the formula $\text{Fe}_2\text{S}_2\text{Cys}_4$. Although the amino acid sequence of putidaredoxin is available [105], the tertiary structure of the protein, as well as the iron–sulfur cluster, is still unknown. A model has been proposed [118] in which the complex contains an antiferromagnetically coupled pair of iron ions that are both ferric in the oxidized protein, while one is ferrous and the other ferric in the reduced protein. In hepatic microsomal systems, the corresponding iron–sulfur proteins were not found; instead, a phospholipid component, probably phosphatidylcholine [119], has been shown to be indispensable to the enzyme activity.

The hemoprotein from *P. putida* has a molecular weight of 45,000 and contains a single polypeptide chain with one molecule of protoheme as the prosthetic group. The heme is not covalently bound to the protein. Catalytic activity of cytochrome *P*-450 rapidly deteriorates, especially in the absence of the substrate camphor, with the concomitant formation of an inactive *P*-420 pigment. This decay of *P*-450 to *P*-420 can be reversed by treatment with sulfhydryl reagents such as cysteine. This observation contributes to the speculation that cytochrome *P*-450 has one axial sulfur ligand [120].

The nature of the axial ligands of the heme moiety is one of the currently more interesting questions. Amino acid analysis of the *P. putida* hemoprotein showed that there were six SH groups [121]. Titration of the SH groups with *p*-hydroxymercuribenzoate (PMB), a highly reactive SH reagent, showed a rapid reaction with the first four SH groups, in both the presence and absence of substrate, followed by a much slower rate of reaction with the two remaining SH groups. These results do not, however, provide clear-cut support for the SH coordination to iron [120]. Other, more definitive evidence that supports a sulfur ligand at the fifth coordinate site of the heme is discussed below.

The Catalytic Cycle

In recent years, a general picture of the mechanism of the cytochrome *P*-450 enzymatic action has become available [104,105]. The results obtained from the adrenal mitochondrial steroid hydroxylating systems and the *P. putida* system suggest a unified catalytic sequence independent of the source of the cytochrome (Scheme 3).

The first step, formation of substrate complexes of ferric cytochrome *P*-450, generally produces dramatic changes in the visible and epr spectra. The Soret band of bacterial cytochrome *P*-450 shifts from 418 to 392 nm on addition of camphor.* In the absence of camphor, bacterial cytochrome

---

* In liver microsomes, type I spectral change refers to a loss of the 420 nm band and the appearance of a 385 nm band on binding of substrates. Type II spectral change refers to a shift from 420 to 430 nm, which is usually elicited by addition of nitrogenous bases such as aniline and pyridine.

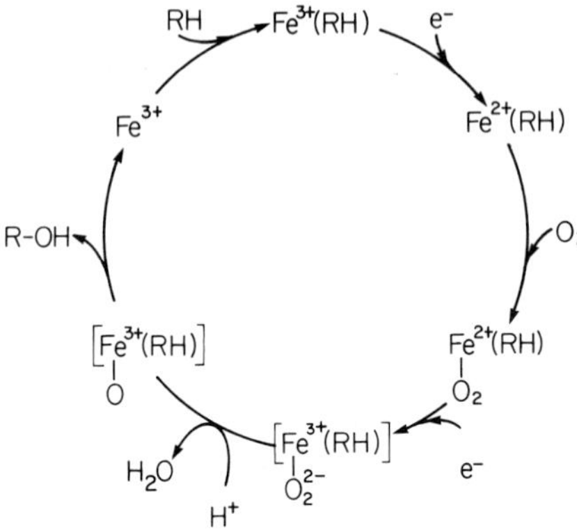

Scheme 3

P-450 exhibits an epr spectrum with g values of 2.45, 2.27, and 1.92, which are characteristic of the low-spin form of ferric porphyrins. Addition of camphor converts it to a high-spin species ($S = 5/2$) with g values of 7.8, 3.9, and 1.8 [122]. The molecular mechanism for spin change upon substrate binding is not yet clear. It is assumed that the substrate displaces a histidine residue from the sixth coordination site of the heme, with the result that formation of the high-spin complex becomes possible. Nitrogen bases such as pyridine and metyrapone cause only small perturbations in the absorption spectrum of substrate-free cytochrome P-450 when they bind to the enzyme. Presumably these strong ligands can maintain the low-spin state of the ferric porphyrin even when the original ligand is displaced [123].

The evidence from inhibition experiments with spin-labeled metyrapone suggests that both camphor and the spin-labeled analog binding sites overlap and that camphor must lie in the immediate vicinity of the heme iron [124]. This is a reasonable conclusion since one would expect to find the camphor bound close to heme, which is presumably the binding site of oxygen.

The binding of camphor to cytochrome P-450 is strong ($K = 4.7 \times 10^5\,M^{-1}$) and can be enhanced by the presence of K^+, Cs^+, Na^+, NH_4^+, Mg^{2+}, Ca^{2+}, and Mn^{2+} (but not Li^+) [123]. This effect cannot be completely attributed to increased ionic strength, which would facilitate the transfer of the lipophilic substrate from the aqueous phase to the hydrophobic environment of the active center. It is possible that the enol form of camphor, which interacts favorably with cations, may be in the proper configuration for binding to the enzyme. Alternatively, the cation may cause a change in enzyme conformation, leading to enhanced binding of camphor.

Further evidence of substrate-induced protein changes came from the kinetics of carbon monoxide binding to cytochrome P-450. Carbon monoxide

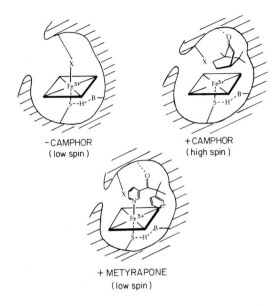

-CAMPHOR
(low spin)

+CAMPHOR
(high spin)

+ METYRAPONE
(low spin)

Fig. 7   Conceptual model of the active site of cytochrome P-450.

has been found to bind ten times stronger (K_{eq}), while the CO combination rate (k) is 140 times faster, in the absence of camphor than it is in the presence of camphor [125]. Binding of camphor decreases the accessibility of solutes to heme iron, and pulsed nuclear magnetic resonance studies of proton relaxation rates indicate that the accessibility of solvent protons to ferric iron is similarly affected by camphor binding. It is therefore conjectured that camphor serves not only as substrate, which is subsequently hydroxylated, but also as an effector for the enzyme. A plausible conceptual model of the active site is illustrated in Fig. 7.

The next step in the enzymatic reaction is the transfer of one electron to the heme protein, resulting in the reduction of the ferric to the ferrous state.* The rate of this reduction is influenced by substrates: In general, compounds that produce type I spectral changes stimulate the rate of reduction while those that give type II spectral change inhibit the reaction. In the absence of camphor, reduction of bacterial cytochrome P-450 by putidaredoxin occurs very slowly and does not proceed to completion. Chemical reductants such as dithionite reduce cytochrome P-450 at a much slower rate compared to enzymatic reduction [107].

* The reduction of purified microsomal system was reported to involve a two-electron transfer with the second electron taken up by an unknown electron acceptor [126].

The reduced ferrous *P*-450 can form a reversible oxygen adduct with molecular oxygen [127,128]. The stability of the oxygenated intermediate is dependent on a variety of conditions. In the presence of both putidaredoxin and camphor, the oxy *P*-450 instantaneously undergoes its enzymatic reaction to give the hydroxylated product. In the absence of putidaredoxin, the oxygen complex undergoes a slow autoxidation, generating the oxidized cytochrome and superoxide anion [128]. The rate of the reaction is also influenced by the substrate binding [128]. Camphor binding was found to effect an approximately 50-fold stabilization of the oxygen adduct. (At 4° C, the reported half-life of the oxygenated complex of camphor-bound cytochrome *P*-450 was 40 min [107] and 12 min [128], while that of the substrate-free oxygen complex was approximately 1 min [128].) Hydrogen ion concentration increased the autoxidation rate dramatically below pH 7. In the absence of substrate, putidaredoxin had almost no effect on this rate. The formation of similar oxygenated cytochrome intermediates in hepatic microsomes [129,130] and possibly in adrenal mitochondria [131] has been suggested.

The active O_2 species is presumably produced from the oxygenated cytochrome by transfer of a second electron. This intriguing, yet least understood step (or steps) is more than simply a one-electron reduction. Product formation requires not only an efficient reductant of oxy *P*-450 but also an effector molecule. In the bacterial cytochrome *P*-450 system, putidaredoxin serves both of these roles. Replacement of putidaredoxin by other structurally or functionally similar proteins such as adrenodoxin, spinach ferredoxin, *Clostridium pasteurianum* and *Escherichia coli* ferredoxin, *Chromatium* high-potential iron–sulfur protein, or cytochrome *c* does not catalyze the hydroxylation step [128]. Small-molecule reductants, such as dithionite, oxidized lipoic acid, and monothiols, gives less than 5% of the theoretical maximum activity. However, dihydrolipoic acid and dithiols were found to efficiently serve as effector molecules, although the rates of product formation were slow compared to the putidaredoxin-mediated step. Less detailed studies of this type have been carried out with the resolved adrenal system and indicate that adrenodoxin also plays a specific role in substrate hydroxylation. The precise molecular role of the effector as well as the mechanism for substrate hydroxylation remain to be elucidated.

ACTIVE OXYGEN SPECIES: PROPOSALS AND MODELS

The problems concerning the hydroxylation step in cytochrome *P*-450 systems are at least 3-fold: determination of the true structural identity of the oxygen derivative that attacks the substrate, the mechanism by which this active form of O_2 is generated, and the mode by which the hydroxyl group is incorporated into the substrate. Since the direct investigation of the transition state is not yet possible, many of the speculations concerning the O_2 activation are based on model hydroxylation studies promoted by simple metal ions [101,132]. Since we now know that O_2 can be activated in various ways, precautions should be taken to discriminate between "true" models and the others. For the sake of clarity we will discuss the many proposed active oxygen species according to their redox states and compare their viability against the known reaction patterns of the hydroxylations catalyzed by cytochrome P-450. An excellent review of the reactivity of reduced O_2 species has been given by Hamilton [50].

Singlet Oxygen and Biradical Process

Despite the fact that singlet oxygen cannot be easily generated under physiological conditions, it has been suggested that it might arise from the spontaneous superoxide disproportionation in solution [133]. However, in aqueous solutions evidence for this pathway is lacking, and we feel that $^1\Delta g$ O_2 cannot play an important role in the enzymatic hydroxylation because the commonly accepted reaction characteristics of monooxygenases do not identify with the reaction patterns of $^1\Delta g$ O_2. It has been brought to the authors' attention that there are disparate results concerning the enzymatic hydroxylation intermediates which seem compatible with the singlet-oxygen-like reaction patterns. Rat liver microsomal preparations were found to convert 3,5-butylhydroxytoluene (BHT) to the corresponding hydroperoxide (BHT—OOH) and phenol (BHT—3°OH), which suggests that hydroperoxide can be intermediate in this hydroxylation [134]. A similar rationale has been applied to an enzymatic conversion of tetralin to tetralol [135]. This hypothesis, if correct, suggests that an initial attack of *both* atoms of O_2 followed by reduction or other O—O bond scission might occur [Eq. (12)].

$$
\begin{array}{ccc}
\text{OH} & \text{O} & \text{O} \\
\\
\text{BHT} & \text{BHT—OOH} & \text{BHT—3°OH}
\end{array}
\tag{12}
$$

Further support for this reaction sequence comes from the isolation and identification of a *cis*-diol produced in the cytochrome P-450-dependent hydroxylation of dimethylbenzanthracene (DMBA) [136]. This bears a direct implication for attack of $^1\Delta g$ O_2 on the substrate [Eq. (13)], and these observations also suggest that the microsomal cytochromes *P*-450, in addition to their monooxygenation activities, may act as dioxygenases toward certain substrates. Recently, the formation of $^1\Delta g$ oxygen during the oxidation of

(13)

NADPH by liver microsomes has also been demonstrated during the enzyme-dependent formation of dibenzoylethylene from diphenylfuran [137]. It should be noted that since epoxides and alkyl alcohols are common products of cytochrome *P*-450-catalyzed reactions that cannot be derived by this route, these atypical reaction products impose a question as to whether there could be more than one reaction mechanism existing in the enzymatic functioning.

Superoxide and Perhydroxyl Radical

Recent evidence has shown that O_2^- is ubiquitous to all aerobic organisms and is involved in a variety of enzymatic reactions [50]. Sliger *et al.* have also demonstrated that superoxide anion is the autoxidation product of oxygenated bacterial cytochrome *P*-450 [128,138]. However, both O_2^- and HO_2^- are weak oxidants and react only with compounds that give stabilized free-radical products. Moreover, there are no known examples in which superoxide anion might react with typical substrates of cytochrome *P*-450, such as alkanes. Therefore, although O_2^- may be an intermediate in various other oxygenase reactions, it is not likely that they are involved in a direct attack on an unactivated substrate. It is not clear, however, whether the coordinated form (Fe^{3+}—O_2^-) can add as a nucleophile to electrophilic sites of the substrate [50].

Peroxide and Hydroperoxide

Hydrogen peroxide and its deprotonated species are reactive only as nucleophiles under physiological conditions. The disproportionation of H_2O_2

into HO· or HO⁺, which may be reactive toward unactivated substrates, is not a kinetically favored reaction unless high activation energy is provided [50,139]. However, peroxide can be chemically modified to give compounds (e.g., peracid) that are sufficiently reactive to be involved in various oxygenase reactions [56]. This type of mechanism, which has been termed an "oxenoid" reaction, is discussed in more detail below.

Hydroxyl Radical

In the presence of transition-metal ions of low oxidation state, peroxide readily dissociates into HO·. Fenton's reagent [Eq. (14)] is a well-documented example that catalyzes the hydroxylation of a variety of substrates [140].

$$\text{Fe(II)} + \text{H}_2\text{O}_2 \longrightarrow \text{Fe(III)} + {}^-\text{OH} + \text{HO·} \tag{14}$$

There are two principal reasons why one is reluctant to consider HO· as an intermediate in an enzymatic hydroxylation. First, the addition of the high-energy HO· to substrates is always a very exothermic reaction, and it would be uneconomical for an enzyme to waste energy when lower-energy alternatives would suffice. Second, attack by free HO· could be indiscriminate. However, one should not entirely ignore the properties of the enzyme–substrate complex, since reactions occurring inside the protein pocket are likely to be different from the chemistry of simple model systems, and it is conceivable that a system might generate a hydroxyl radical in juxtaposition to a complex substrate and transfer the HO· immediately to a suitably positioned site.

Oxenoid Reactions

The discovery in the last decade of a unique reaction characteristic of phenol formation catalyzed by monooxygenases has provided valuable insight on the nature of the active oxygen species [141,142]. The so-called NIH shift can be best illustrated in Eq. (15). Such intramolecular migrations with concomitant retentions of substituents necessitate an arene 1,2-oxide intermediate. Although most benzene oxides are too unstable to be directly characterized, the more stable naphthalene 1,2-oxide has been isolated during the enzymatic hydroxylation of napthalene by microsomal cytochrome *P*-450 [143]. The intermediacy of arene oxides is now firmly estalished and seems to be a general route for the majority of enzymatic aromatic hydroxylations.

Prior to the discovery of the NIH shift, Hamilton had noted that the reaction patterns of monooxygenases were similar to those of carbenes and

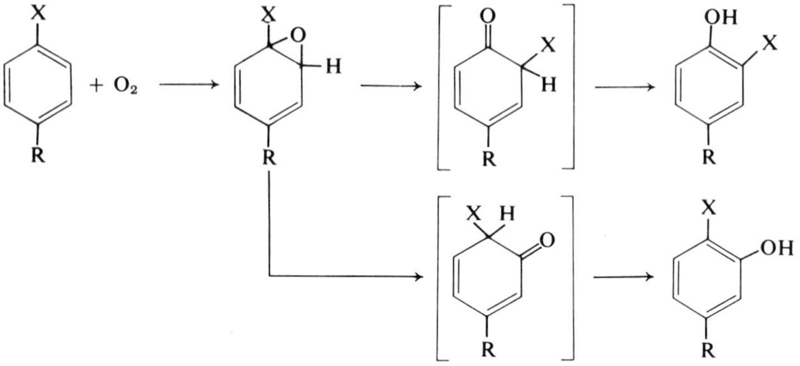

$$(15)$$

X = tritium,
   halogens,
   alkyl groups

nitrenes. By analogy, he suggested an oxene (free oxygen atom) as the reactive species [56]. Since carbenes and nitrenes are highly reactive toward completely inactivated hydrocarbon bonds, the oxygen counterpart could be expected to react similarly. Hamilton also proposed an "oxenoid" mechanism, in which the oxygen atom with six electrons is bound to enzyme and can be transferred directly to substrates. The realization of the NIH shift and the arene oxide intermediacy further substantiated the "oxene" postulate.

The generation of such an oxene species by cytochrome P-450 requires more justification. It has been pointed out that there are precedents in which peroxide derivatives are capable of hydroxylating unactivated C—H bonds. For example, the proposed mechanism for trifluoroperacetic acid oxidations is that of a concerted substitution [Eq. (16)] [144]. This reaction is facilitated

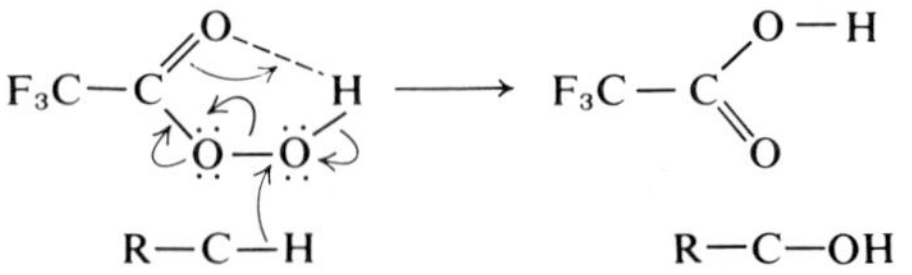

$$(16)$$

by electronegative substitutents on the α-carbon which polarize the O—O bond in addition to raising the oxidation potential. Similarly, the polarization of the O—O bond by a ferric ion might also suffice to lower the activation energy of reaction (17) [145]. Alternatively, the peroxide could disproportionate to

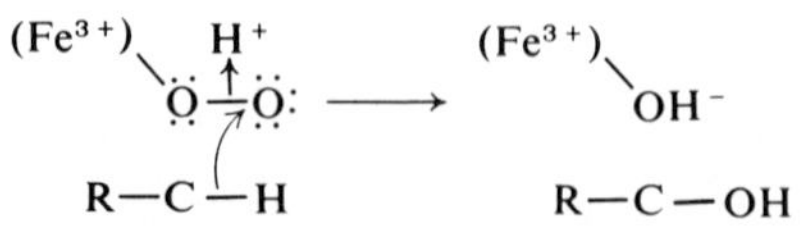

$$(17)$$

give, first, a metal-bound oxene, which then oxidizes substrate [Eq. (18)]. These various oxenoid species are identical in terms of their redox state

$$(Fe^{3+})\diagdown \quad H \qquad (Fe^{3+})\!-\!\ddot{O} \\ \ddot{O}\!-\!\ddot{O}\diagup \qquad \longrightarrow \quad (Fe^{4+})\!-\!\ddot{O}\!\cdot\! + H_2O \qquad (18) \\ \diagdown_{\rightarrow H^+} \qquad (Fe^{5+})\!=\!\ddot{O}$$

but could be discrete individual species rather than merely resonance structures of the same intermediate. As yet no metal-bound oxene species have been reported.

Clearly, we are not yet in a position to discern the true identity of oxy *P*-450 after it accepts the second electron. Formally, the intermediate could be $Fe(I)O_2$; $Fe(II)O_2^-$, $Fe(III)O_2^{2-}$, $Fe(IV)O_2^{3-}$, etc. It is also likely that the formal charge on the iron can further be distributed to the porphyrin ring, leading to π-anion radicals [9] in analogy to the π-cation radicals described above. The ferric peroxide formulation is generally used merely as a working model to describe the electronic structure of the reduced oxy *P*-450.

Theoretical considerations of the cleavage of the O—O bond in the $Fe(III)O_2^{2-}$ complex have been presented by Ochiai [93], who assumed that the coordination of peroxide ion to ferric iron could create a difference in the *p* orbital (or hydrid sp^2 orbital) energy levels between the two individual oxygen atoms. The probable mode of overlapping and the energy levels are shown in Fig. 8. The p_A^y orbital is considered to have lower energy as a result of the metal bonding, and cleavage could occur as shown in Eq. (19). The oxygen atom would thus be in the ^1D state (1.97 eV above the ground

$$Fe^{3+}\!-\!O_2^{\ 2-} \quad \longrightarrow \quad Fe^{3+}\!-\!O_A^{\ 2-} + O_B\,[(1S)^2(2S)^2(px)^2(pz)^2] \qquad (19)$$

state ^3P). It is probable that perturbations such as protonation of the peroxide ion or change of the *trans* axial ligand could alter the relative energy level of P_A^y and P_B^y. Therefore, alternative modes of electron distribution during bond cleavage to give $Fe^{3+}\!-\!O^-$ or $Fe^{3+}\!-\!O$ should also be possible.

The oxenoid mechanism seems well supported by a variety of experimental

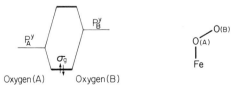

Fig. 8   Bonding of the activated oxygen complex of cytochrome *P*-450.

evidence. It has recently been shown that hepatic microsomal substrate hydroxylation mediated by cyt *P*-450 can occur in the presence of $NaIO_4$, $NaClO_2$, H_2O_2, and organic peroxides without the requirement for O_2 and NADPH [146]. Therefore, it was conjectured that the ferric cytochrome–substrate complex can be "oxidized" to an oxenoid, which then catalyzes the hydroxylation reaction [Eq. (20)].

$$
\begin{array}{c} Fe^{3+} \\ | \\ RH \end{array} \longrightarrow \left[\begin{array}{c} NaIO_4 \\ NaClO_2 \\ HOOR \end{array}\right] \longrightarrow \begin{array}{c} Fe^{3+}\!\!-\!O \\ | \\ RH \end{array} \longrightarrow Fe^{3+} + ROH \qquad (20)
$$

Model Hydroxylation Systems

The oxenoid mechanism has been considered to be operative in a number of metal–ligand model hydroxylating systems, and there have been extensive reviews of this subject [50,147,148].

1. Udenfriend System (Fe^{2+}–EDTA–ascorbic acid–O_2) [149,150]

Udenfriend and co-workers have shown that hydroxylation of organic substrates can be achieved (albeit in low yield) with the title mixture at room temperature. This system, however, is not well defined and may involve reaction of HO· radicals. Furthermore, the hydroxylated product distribution did not match that of the enzymatic system. Moreover, this system did not form epoxides nor did it demonstrate any substantial amount of the NIH shift.

2. Ullrich Systems ($SnCl_2$–quinoline–O_2 and Fe^{3+}–mercaptobenzoic acid–O_2)
   [151,152]

Both of these mixtures were able to oxidize alkanes to alcohols as well as aromatic compounds to phenols. The iron system is especially impressive in that it exhibited product distribution patterns very similar to those of rat liver microsomes. Unfortunately, these investigations did not demonstrate the NIH shift.

3. Pyridine-*N*-oxide and Related Systems

From thermodynamic considerations, one would predict that there should be a large number of organic compounds of the type $X=O$ which potentially could effect direct oxygen transfer [50]. In fact, photoactivated pyridine *N*-oxide, other related heterocyclic *N*-oxides,* and N_2O are such systems

---

* We have previously suggested that oxaziridines may be intermediates in the flavin-mediated activation of oxygen [153].

[Eq. (21)] [154]. In addition to aromatic hydroxylation, the whole range of microsomal oxidations occurs with this system, e.g., olefin epoxidation and hydrocarbon hydroxylation. This system, as anticipated, also leads to the NIH shift. Unfortunately, these systems so far have provided little information as to how "active oxygen" can be generated from molecular oxygen.

$$(21)$$

Spectroscopic Models of Cytochrome P-450

While the above systems clearly demonstrate that metals can activate oxygen for other than radical oxygenation, they are not designed to resemble the structure (or presumed structure) of the active site in the enzyme. The alternative approach toward modeling cytochrome P-450 has been focused on iron porphyrin complexes, and so far this route has yielded promising results.

FERRIC CYTOCHROME P-450 MODELS

Well-characterized ferric porphyrin–thiol compounds have been reported. Collman et al. [155] have prepared a complex of Fe(III) tetraphenylporphyrin containing 1 molar equivalent of benzenethiol. Electron paramagnetic resonance studies of this complex in toluene glass revealed g values (8.6 and 3.4) similar to those of high-spin cytochrome P-450. In the presence of Lewis bases this compound was readily reduced, yielding a metastable low-spin species with g values similar to those of low-spin cytochrome P-450. Presumably, the high-spin complex is a pentacoordinate hemin–thiolate compound, while the low-spin complex had a Lewis base at the sixth coordination site.

The X-ray structure of protohemin dimethyl ester–p-nitrobenzenethiolate complex has been reported by Koch et al. [156]. The pentacoordinate iron is about 0.5 Å above the porphyrin plane. Magnetic susceptibility and Mössbauer spectra confirmed that the complex is similar to the camphor-bound cytochrome P-450. A similar complex with p-nitrophenol as the axial ligand exhibited visible and epr spectra distinct from those of hemin–thiolate complex and cytochrome P-450. Therefore, phenolate was considered unlikely to be the axial ligand in cytochrome P-450.

FERROUS CYTOCHROME *P*-450 MODELS

Carbon Monoxide Binding. Stern and Peisach [157] first reported that the combination of reduced heme, thiol, CO, and a strong base, under stringent mixing procedures, could result in the partial appearance of a 450 nm Soret peak in the visible spectrum. The results were interpreted as evidence for a thiolate anion ligand. Their failure to produce a complete 450 nm peak as well as the inability to duplicate similar results by sequential addition of reagents were puzzling but were independently clarified later by two groups. Chang and Dolphin [158] and Collman and Sorell [159] reported the observation that 100% conversion of protoheme to a carbon monoxide complex absorbing around 450–460 nm could be realized by mixing an alkali thiolate, as its crown ether complex, with protohemin under a CO atmosphere. The position and extinction of the absorbance maximum varied somewhat depending on the solvent employed. The crown ether apparently

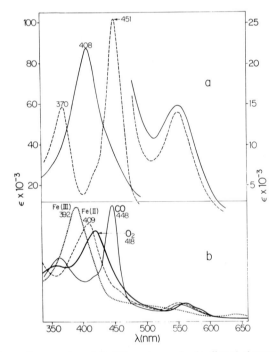

Fig. 9 (a) Absorption spectra of Fe(II) protoporphyrin dimethyl ester and mercaptide complexes in toluene at 23°C: (——), without CO; (· · ·), in the presence of 1 atm CO. Background absorption of dibenzo-18-crown-6 present in the solution has been compensated. (b) Absorption spectra of camphor-bound cytochrome *P*-450 from *P. putida* (after Peterson *et al.* [127]).

acted as a cation scavenger, thereby enhancing the coordination of the thiolate anion to heme.

We further demonstrated that, in addition to the 450 nm peak, another intense absorption band centered around 370 nm was developed upon binding of CO (Fig. 9a). Reexamination of previously published CO–cytochrome P-450 spectra from both bacterial [127] (Fig. 9b) and microsomal [126] sources revealed a similar absorption peak in the 360–370 nm region; however, its significance had not been explored in the natural systems. The presence of this absorption was firmly established from the single-crystal spectrum [160] of P. $putida$ CO–cytochrome P-450 (Fig. 10), which showed that the ultraviolet band had the same integrated intensity and polarization as the Soret band. The splitting of a single Soret band into two bands, observed in both model systems and enzymes, has been defined as a "hyper" type of spectrum and interpreted as resulting from a charge transfer of thiolate sulfur electrons in a p orbital to the porphyrin $e_g(\pi^*)$ coupled to the normal porphyrin $\pi \rightarrow \pi^*$ transition [160]. The 370 nm band in heme–thiolate–CO complex was also found in the magnetic circular dichroism spectra [161].

It has been shown [162] that, in the absence of CO, heme and thiolate form an exclusively pentacoordinate complex whose absorption spectrum is identical to that of the reduced and unliganded cytochrome P-450. The thermodynamics and kinetics of the interactions between thiolate ion, heme, and CO have been examined by both flash photolysis and static equilibria methods [162].

Oxygen Binding. We found that the pentacoordinate heme–thiolate complex was capable of reversible O_2 binding at $-45°$ C in a mixture of DMAC and 5% H_2O without extensive oxidation of the heme or the thiolate

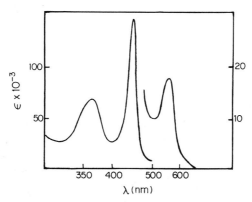

Fig. 10 Single-crystal spectra of CO complex of camphor-bound cytochrome P-450 (after Hanson *et al.* [160]).

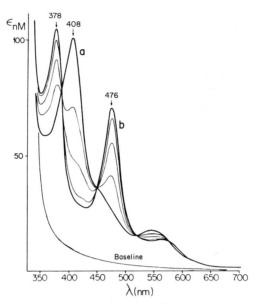

Fig. 11 Oxygenation of mercaptide–protoheme dimethyl ester complex in dimethyl-formamide (with 5% H_2O) ar $-45°C$. (a) *In vacuo*; (b) under 100 torr O_2; $[O_2]$ = 2.3, 4.7, and 7.0 torr for the intermediate curves, respectively. The baseline absorption was from potassium *n*-butylthiolate and dibenzo-18-crown-:-6.

ion [163]. The absorption spectrum of the O_2 complex, like the CO complex, is of the "hyper" type, i.e., two Soret bands (Fig. 11). Theoretical MO calculations have predicted that similar sulfur to oxyheme orbital interactions should give rise to the two peaks. It appears, then, that the "hyper" type of spectrum is a general feature of all CO– or O_2–iron(II) porphyrin–thiolate complexes (this has now been observed for meso-, proto-, and diacetyldeutero-porphyrins) as well as the O_2–cobalt(II) porphyrin–thiolate complex [164].

The apparent spectral discrepancy between the O_2–cytochrome *P*-450 (Fig. 9b) and the thiolate model systems imposes a serious question concerning the viability of thiolate as an axial ligand in any cytochrome *P*-450. Indeed, the thiolate postulate recently has been challenged by Hager's observations with chloroperoxidase. The striking optical and Mössbauer spectral similarities found between chloroperoxidase and cytochrome *P*-450 suggest that an identical axial ligand is present in both enzymes; yet there is evidence that only disulfide cystine linkage is found in both the oxidized and reduced form of chloroperoxidase, thereby opening to question the obligatory nature of a mercaptide ligand in bringing about the unique spectra [165]. In view of the striking difference in coordination properties between thiolates and thiols we have examined a number of negatively charged ligands such as alkoxide, phenoxide, carboxylate, imidazolide, and disulfide. However, none of these ligands was capable of bringing about the unique P-450 spectrum. It appears, therefore, that the thiolate–heme system is still the most viable model for the enzyme at the present stage. So far the catalytic function of these cytochrome P-450 models has not been demonstrated.

CONCLUSION

It is striking that nature employs a single prosthetic group, namely, iron protoporphyrin, to accomplish such a wide variety of important biological processes ranging from oxygen transport to drug detoxification. It is even more intriguing as we now begin to comprehend that these variations in heme protein functioning are apparently rooted in three basic factors based on ligand, environment, and steric effects.

Although advances in X-ray crystallographic techniques will enable us to "see" more and more of the functioning sites of these heme proteins, it is our contention that the construction of appropriate models that mimic the basic enzymatic functions is necessary for a complete understanding of the mechanistic aspects involved. We trust that we have shown in this review what has been achieved by this approach.

ACKNOWLEDGMENTS

This work is a contribution from the Bioinorganic Chemistry Group and was supported by operating and negotiated development grants from the National Research Council of Canada and the United States National Institutes of Health (AM 17987).

REFERENCES

1. I. Macalpine and R. Hunter, *Sci. Am.* **221**, 38 (1969).
2. R. K. Clayton, *Annu. Rev. Biphys. Bioeng.* **2**, 131 (1973).
3. M. Calvin and J. A. Bassham, "The Photosynthesis of Carbon Compounds." Benjamin, New York, 1962.
4. F. Antonini and M. Brunori, "Hemoglobin and Myoglobin and Their Reactions with Ligands." North Holland Publ., Amsterdam, 1971.
5. J. M. Rifkind, *in* "Inorganic Biochemistry" (G. L. Eichhorn, ed.), p. 832. Elsevier, New York, 1973.
6. D. F. Wilson and M. Erecinska, *in* "The Porphyrins" (D. Dolphin, ed.). Academic Press, New York (in press).
7. I. Fridovich, *Acc. Chem. Res.* **5**, 521 (1972).
8. H. J. Bright and D. J. T. Porter, *in* "The Enzymes" (P. D. Boyer, ed.), 3rd ed., Vol. 12, p. 421. Academic Press, New York, 1975.

9. D. Dolphin and R. H. Felton, *Acc. Chem. Res.* **7**, 26 (1974).
10. O. Hayaishi, *in* "Molecular Mechanisms of Oxygen Activation" (O. Hayaishi, ed.), p. 1. Academic Press, New York, 1974.
11. J. Buchler, L. Puppe, K. Rohbock, and H. H. Schnuhage, *Ann. N. Y. Acad. Sci.* **206**, 116 (1973).
12. M. Gouterman, *in* "The Porphyrins" (D. Dolphin, ed.). Academic Press, New York (in press).
13. K. Rousseau and D. Dolphin. *Tetrahedron Lett.* **48**, 4251 (1974).
14. J. B. Paine, III, W. B. Kirshner, D. W. Moskowitz, and D. Dolphin, *J. Org. Chem.* **41**, 3857 (1976).
15. J. Fajer, D. C. Borg, A. Forman, D. Dolphin, and R. H. Felton, *J. Am. Chem. Soc.* **92**, 3451 (1970).
16. J. Fajer, D. C. Borg, A. Forman, R. H. Felton, L. Vegh, and D. Dolphin, *Ann. N. Y. Acad. Sci.* **206**, 349 (1973).
17. C. Weiss, H. Kobayashi, and M. P. Gouterman, *J. Mol. Spectrosc.* **16**, 415 (1965).
18. D. Dolphin, R. H. Felton, D. C. Borg, and J. Fajer. *J. Am. Chem. Soc.* **92**, 743 (1970).
19. D. Dolphin, Z. Muljiani, K. Rousseau, D. C. Borg, F. Fajer, and R. H. Felton, *Ann. N. Y. Acad, Sci.* **206**, 177 (1973).
20. M. T. Cox, T. T. Howarth, A. H. Jackson, and G. Kenner, *J. Am. Chem. Soc.* **92**, 1232 (1969).
21. R. B. Woodward, *Ind. Chim. Belge*, p. 1293 (1962).
22. A. C. Jain and G. W. Kenner, *J. Chem. Soc.* p. 185 (1959).
23. O. T. G. Jones, *in* "The Porphyrins" (D. Dolphin, ed.), Academic Press, New York (in press).
24. R. H. Felton, D. Dolphin, D. C. Borg, and J. Fajer, *J. Am. Chem. Soc.* **91**, 196 (1969).
25. D. C. Borg, J. Fajer, R. H. Felton, and D. Dolphin, *Proc. Natl. Acad. Sci. U.S.A.* **67**, 813 (1970).
26. J. J. Katz, L. L. Shipman, T. M. Cotton, and T. R. Janson, *in* "The Porphyrins" (D. Dolphin, ed.) Academic Press, New York (in press).
27. D. H. Kohl, *in* "Biological Applications of Electron Spin Resonance" (H. M. Swartz, J. R. Bolton, and D. C. Borg, eds.), p. 213. Wiley (Interscience), New York, 1972.
28. J. D. McElroy, G. Feher, and D. Mauzerall, *Biochim. Biophys. Acta* **267**, 363 (1972).
29. W. W. Parson, *Biochim. Biophys. Acta*, **153**, 248 (1968).
30. P. A. Loach and K. Walsh, *Biochemistry* **8**, 1908 (1969).
31. J. T. Warden and J. R. Bolton, *J. Am. Chem. Soc.* **94**, 4351 (1972).
32. J. R. Bolton, R. K. Clayton, and D. W. Reid, *Photochem. Photobiol.* **9**, 209 (1969).
33. J. Fajer, D. C. Borg, A. Forman, R. H. Felton, D. Dolphin, and L. Vegh, *Proc. Natl. Acad. Sci. U.S.A.* **71**, 994 (1974).
34. D. Dolphin, T. Niem, R. H. Felton, and I. Fujita. *J. Am. Chem. Soc.* **97**, 5288 (1975).
35. D. Dolphin and E. C. Johnson, unpublished results.
36. W. R. Scheidt, *in* "The Porphyrins" (D. Dolphin, ed.) Academic Press, New York (in press).
37. B. Chance and G. R. Schonbaum, *in* "The Enzymes" (P. D. Boyer, ed.), 3rd ed., Vol. 13, p. 363. Academic Press, New York, 1975.
38. I. Yamazaki, *in* "Molecular Mechanisms of Oxygen Activation" (O. Hayaishi, ed.), p. 535. Academic Press, New York, 1974.

39. B. Chance, *Nature (London)* **161**, 914 (1948).
40. P. George, *J. Biol. Chem.* **201**, 413 (1953).
41. A. S. Brill and R. J. P. Williams, *Biochem. J.* **78**, 253 (1961).
42. M. E. Winfield, *in* "Haematin Enzymes" (J. E. Falk, R. Lemberg, and R. K. Morton, eds.), p. 245. Pergamon, Oxford, 1961.
43. R. H. Felton, G. S. Owen, D. Dolphin, A. Forman, D. Borg, and J. Fajer, *Ann. N. Y. Acad. Sci.* **206**, 504 (1973).
44. R. H. Felton, G. S. Owen, D. Dolphin, and J. Fajer, *J. Am. Chem. Soc.* **93**, 6332 (1971).
45. D. Dolphin, A. Forman, D. C. Borg, J. Fajer, and R. H. Felton, *Proc. Natl. Acad. Sci. U.S.A.* **68**, 614 (1971).
46. T. H. Moss, A. Ehrenberg, and A. J. Bearden, *Biochemistry* **8**, 4159 (1969).
47. Y. Maeda and Y. Morita, *in* "Structure and Function of Cytochromes" (K. Okunuki, M. D. Kamen, and I. Sekuzu, eds.), p. 523. Univ. of Tokyo Press, Tokyo, 1968.
48. H. Taube, *J. Gen. Physiol.* **49**, Part 2, 29 (1965).
49. P. George, *Oxidases Relat. Redox Syst., Proc. Symp., 1964* Vol. 1, p. 3 (1965).
50. G. A. Hamilton, *in* "Molecular Mechanisms of Oxygen Activation" (O. Hayaishi, ed.), p. 405. Academic Press, New York, 1974.
51. H. P. Misra and I. Fridovich, *J. Biol. Chem.* **247**, 188 (1972).
52. C. S. Foote, *Acc. Chem. Res.* **1**, 104 (1968).
53. C. S. Foote, *in* "Free Radicals in Biology" (W. A. Pryor, ed.), Vol. 2, p. 150. Academic Press, New York, 1976.
54. T. McDonagh, *in* "The Porphyrins" (D. Dolphin, ed.) Academic Press, New York (in press).
55. C. S. Foote and T.-Y. Ching, *J. Am. Chem. Soc.* **97**, 6209 (1975).
56. G. A. Hamilton, *Prog. Bioorg. Chem.* **1**, 83 (1971).
57. E. Bayer and P. Schretzmann, *Struct. Bonding (Berlin)* **2**, 181 (1967).
58. E. Bayer, P. Krauss, A. Röder, and P. Schretzmann, *in* "Oxidases and Related Model Systems" (T. E. King, M. Mason, and M. Morrison, eds.), Vol. 1, p. 227. Univ. Park Press, Baltimore, Maryland, 1974.
59. R. G. Wilkins, *Adv. Chem. Ser.* **100**, 111 (1971).
60. G. McLendon and A. E. Martell, *Coord. Chem. Rev.* **19**, 1 (1976).
61. B. R. James, *in* "The Porphyrins" (D. Dolphin, ed.) Academic Press, New York (in press).
62. T. G. Traylor, this volume, Chapter 16.
63. J. S. Griffith, *Proc. R. Soc. London, Ser. A* **235**, 23 (1956).
64. L. Pauling and C. D. Coryell, *Proc. Natl. Acad. Sci U.S.A.*, **22**, 210 (1936).
65. D. P. Mingos, *Nature (London), Phys. Sci.* **230**, 154 (1971).
66. G. A. Rodley and W. T. Robinson, *Nature (London)* **235**, 438 (1972).
67. J. P. Collman, R. R. Gagne, C. A. Reed, T. R. Halbert, G. Lang, and W. T. Robinson, *J. Am. Chem. Soc.* **97**, 1427 (1975).
68. F. Adar, M. Gouterman, and S. Aronowitz, *J. Phys. Chem.* (in press).
69. G. Lowe and S. Aronowitz, personal communication to Professor M. Gouterman.
70. A. Trautwein, H. Eicher, A. Mayer, A. Alfsen, M. Waks, J. Rosa, and Y. Beuzard, *J. Chem. Phys.* **53**, 963 (1970).
71. B. M. Hoffman, D. L. Diemente, and F. Basolo, *J. Am. Chem. Soc.* **92**, 61 (1970).
72. M. Mori and J. A. Weil, *J. Am. Chem. Soc.* **89**, 3732 (1967).
73. J. J. Weiss, *Nature (London)* **203**, 83 and 183 (1964).
74. C. K. Chang and T. G. Traylor, *Proc. Natl. Acad. Sci. U.S.A.* **72**, 1166 (1975).

75. W. S. Brinigar, C. K. Chang, J. Geibel, and T. G. Traylor, *J. Am. Chem. Soc.* **96**, 5597 (1974).
76. H. C. Stynes and J. A. Ibers, *J. Am. Chem. Soc.* **94**, 5125 (1972).
77. C. H. Barlow, J. C. Maxwell, W. J. Wallace, and W. S. Caughey, *Biochem. Biophys. Res. Commun.* **55**, 91 (1973).
78. J. C. Maxwell, J. A. Volpe, C. H. Barlow, and W. S. Caughey, *Biochem. Biophys. Res. Commun.* **58**, 166 (1974).
79. J. C. Maxwell and W. S. Caughey, *Biochem. Biophys. Res. Commun.* **60**, 1309 (1974).
80. J. S. Valentine, *Chem. Rev.* **73**, 235 (1973).
81. T. G. Spiro and T. C. Strekas, *J. Am. Chem. Soc.* **96**, 338 (1974).
82. H. A. O. Hill, D. R. Turner, and G. Pellizer, *Biochem. Biophys. Res. Commun.* **56** 739 (1974).
83. J. A. McGinnety, N. C. Payne, and J. A. Ibers, *J. Am. Chem. Soc.* **91**, 6301 (1969).
84. C. K. Chang and T. G. Traylor, *J. Am. Chem. Soc.* **95**, 8479 (1973).
85. J. O. Alben and W. S. Caughey, *Biochemistry* **7**, 175 (1968).
86. T. Asakura and M. Sono, *J. Biol. Chem.* **249**, 7087 (1974).
87. D. W. Seybert, K. Moffat, and Q. H. Gibson, *J. Biol. Chem.* **251**, 45 (1976).
88. M. F. Perutz, *Nature (London)* **228**, 726 (1970).
89. M. F. Perutz, J. E. Ladner, S. R. Simon, and C. Ho, *Biochemistry* **13**, 2163 (1974).
90. J. L. Hoard, *Science* **174**, 1295 (1971).
91. R. G. Little and J. A. Ibers, *J. Am. Chem. Soc.* **96**, 4452 (1974).
92. J. H. Wang, *Acc. Chem. Res.* **3**, 90 (1970).
93. E.-I. Ochiai, *J. Inorg. Nucl. Chem.* **36**, 2129 (1974).
94. I. A. Cohen and W. S. Caughey, *Biochemistry* **7**, 636 (1968).
95. J. H. Wang, in "Oxygenases" (O. Hayaishi, ed.), p. 469. Academic Press, New York, 1962.
96. F. A. Walker, *J. Am. Chem. Soc.* **95**, 1154 (1973).
97. D. V. Stynes, H. C. Stynes, J. A. Ibers, and B. R. James, *J. Am. Chem. Soc.* **95**, 1142 (1973).
98. I. Klotz, in "Inorganic Biochemistry" (G. L. Eichhorn, ed.), p. 320. Elsevier, New York, 1973.
99. E.-I. Ochiai, *J. Inorg. Nucl. Chem.* **35**, 3375 (1973).
100. E.-I. Ochiai, *Inorg Nucl. Chem. Lett.* **10**, 453 (1974).
101. G. S. Boyd and R. M. S. Smellie, eds., "Biological Hydroxylation Mechanisms," Biochem. Soc. Symp. No. 34. Academic Press, New York, 1972.
102. R. W. Estabrook, J. R. Gillette, and K. C. Leibman, eds. "Microsomes and Drug Oxidations." Williams & Wilkins, Baltimore, Maryland, 1972.
103. D. Y. Cooper, O. Rosenthal, R. Snyder, and C. Witmer, eds., "Cytochromes P-450 and b₅, Structure, Function, and Interactions." Plenum, New York, 1975.
104. S. Orrenius and L. Ernster, in "Molecular Mechanisms of Oxygen-Activation" (O. Hayaishi, ed.), p. 215. Academic Press, New York, 1974.
105. I. C. Gunsalus, J. R. Meeks, J. D. Lipscomb, P. Debrunner, and E. Münck, in "Molecular Mechanisms of Oxygen Activation" (O. Hayaishi, ed.), p. 561. Academic Press, New York, 1974.
106. R. H. Wickramasinghe, *Enzyme* **19**, 348 (1975).
107. B. W. Griffin, J. A. Peterson, and R. W. Estabrook, in "The Porphyrin" (D. Dolphin, ed.). Academic Press, New York (in press).
108. M. Klingenberg, *Arch. Biochem. Biophys.* **75**, 376 (1958).
109. T. Omura and R. Sato, *J. Biol. Chem.* **239**, 2370 (1964).

110. K. J. Ryan and L. L. Engel, *J. Biol. Chem.* **225**, 103 (1957).
111. R. W. Estabrook, D. Y. Cooper, and O. Rosenthal, *Biochem. Z.* **338**, 741 (1963).
112. Y. Gnosspelius, H. Thor, and S. Orrenius, *Chem.-Biol. Interact.* **1**, 125 (1969–70).
113. D. A. Haugen, T. A. van der Hoeven, and M. J. Coon, *J. Biol. Chem.* **250**, 3567 (1975).
114. A. F. Welton, F. O. O'Neal, L. C. Chaney, and S. Aust, *J. Biol. Chem.* **250**, 5631 (1975).
115. A. Y. H. Lu and W. Levin, *Biochim. Biophys. Acta* **344**, 205 (1974).
116. M. Katagiri, B. N. Ganguli, and Z. C. Gunsalus, *J. Biol. Chem.* **243**, 3543 (1968).
117. C.-A. Yu and Z. C. Gonsalus, *J. Biol. Chem.* **249**, 94 (1974).
118. J. F. Gibson, D. O. Hall, J. H. M. Thornley, and F. R. Whatley, *Proc. Natl. Acad. Sci U.S.A.* **56**, 987 (1966).
119. H. W. Strobel, A. Y. H. Lu, J. Heidema, and M. J. Coon, *J. Biol. Chem.* **244**, 3714 (1969).
120. C.-A. Yu and I. C. Gunsalus, *J. Biol. Chem.* **249**, 102 (1974).
121. K. Dus, M. Katagiri, C.-A. Yu, D. L. Erbes, and I. C. Gunsalus, *Biochem. Biophys. Res Commun.* **40**, 1423 (1970).
122. R. Tsai, C.-A. Yu, I. C. Gunsalus, J. Peisach, W. Blumberg, W. H. Orme-Johnson, and H. Beinert, *Proc. Natl. Acad. Sci. U.S.A.* **66**, 1157 (1970).
123. J. A. Peterson, *Arch. Biochem. Biophys.* **144**, 678 (1971).
124. B. W. Griffin, S. M. Smith, and J. A. Peterson, *Arch. Biochem. Biophys.* **160**, 323 (1974).
125. J. A. Peterson and B. W. Griffin, *Arch. Biochem. Biophys.* **151**, 427 (1972).
126. F. P. Guengerich, D. P. Ballou, and M. J. Coon, *J. Biol. Chem.* **250**, 7405 (1975).
127. J. A. Peterson, Y. Ishimura, and B. W. Griffin, *Arch. Biochem. Biophys.* **149**, 197 (1972).
128. J. D. Lipscomb, S. G. Sligar, M. J. Namtvedt, and I. C. Gunsalus, *J. Biol. Chem.* **251**, 1116 (1976).
129. R. W. Estabrook, A. G. Hildebrandt, J. Baron, K. J. Netter, and K. Leibman, *Biochem. Biophys. Res. Commun.* **42**, 132 (1971).
130. J. Baron, A. G. Hildebrandt, J. A. Peterson, and R. W. Estabrook, *Drug Metab. Dispos.* **1**, 129 (1973).
131. E. Bayer, P. Krauss, A. Röder, and P. Schretzmann, *in* "Oxidases and Related Model Systems" (T. E. King, M. Mason, and M. Morrison, eds.), Vol. 1, p. 81. Univ. Park Press, Baltimore, Maryland, 1974.
132. V. Ullrich, *Angew, Chem., Int. Ed. Engl.* **11**, 701 (1972).
133. A. U. Khan, *Science* **168**, 476 (1970).
134. Y.-S. Shaw and C. Chen, *Biochem. J.* **128**, 1285 (1972).
135. C.-C. Lin and C. Chen, *Biochim. Biophys. Acta* **192**, 133 (1969).
136. M.-H. Tu and C. Chen, *Fed. Proc., Fed. Am. Soc. Exp. Biol.* **34**, Abstr. No. 2303 (1975).
137. M. M. King, E. K. Lai, and P. B. McCay, *J. Biol. Chem.* **250**, 6496 (1975).
138. S. G. Sligar, J. D. Lipscomb, P. G. Debrunner, and I. C. Gunsalus, *Biochem. Biophys. Res. Commun.* **61**, 290 (1974).
139. S. Diner, *in* "Electronic Aspects of Biochemistry" (B. Pullman, ed.), p. 237. Academic Press, New York, 1964.
140. C. Walling, *Acc. Chem. Res.* **8**, 125 (1975).
141. G. Guroff, J. W. Daly, D. M. Jerina, J. Renson, B. Witkop, and S. Udenfriend, *Science* **158**, 1524 (1967).

142. D. M. Jerina and J. W. Daly, *Science* **185**, 573 (1974).
143. D. M. Jerina, J. W. Daly, B. Witkop, P. Zaltzman-Nirenberg, and S. Udenfriend, *J. Am. Chem. Soc.* **90**, 6525 (1968); *Biochemistry* **9**, 147 (1970).
144. V. Ullrich, J. Wolf, E. Amadori, and Hj. Standinger, *Hoppe-Seyler's Z. Physiol. Chem.* **349**, 85 (1968).
145. U. Frommer and V. Ullrich, *Z. Naturforsch., Teil B* **26**, 322 (1971).
146. E. G. Hrycay, J. Gustafsson, M. Ingalman-Sundberg, and L. Ernster, *Biochem. Biophys. Res. Commun.* **66**, 209 (1975).
147. A. E. Martell and M. M. Taqui Khan, *in* "Inorganic Chemistry" (G. L. Eichhorn, ed.), p. 650. Elsevier, New York, 1973.
148. G. Henrici-Olive and S. Olive, *Angew. Chem. Int. Ed. Engl.* **13**, 29 (1974).
149. B. B. Brodie, J. Axelrod, Y. R. Cooper, L. Gaudette, B. N. La Du, C. Mitoma, and S. Udenfriend, *Science* **121**, 603 (1955).
150. G. A. Hamilton, R. J. Workman, and L. Woo, *J. Am. Chem. Soc.* **86**, 3390 (1964).
151. V. Ullrich, *Z. Naturforsch., Teil B* **24**, 699 (1969).
152. U. Frommer, V. Ullrich, and Hj. Standinger, *Hoppe-Seyler's Z. Physiol. Chem.* **351**, 913 (1970).
153. H. W. Orf and D. Dolphin, *Proc. Natl. Acad. Sci. U.S.A.* **71**, 2646 (1974).
154. D. M. Jerina, D. R. Boyd, and J. W. Daly, *Tetrahedron Lett.* p. 457 (1970).
155. J. P. Collman, T. N. Sorrell, and B. M. Hoffman, *J. Am. Chem. Soc.* **97**, 913 (1975).
156. S. Koch, S. C. Tang, R. H. Holm, R. B. Frankel, and J. A. Ibers, *J. Am. Chem. Soc.*, **97**, 916 (1975); S. C. Tang, S. Koch, G. C. Papaefthymiou, S. Foner, R. B. Frankel, J. A. Ibers, and R. H. Holm, *ibid.* **98**, 2414 (1976).
157. J. O. Stern and J. Peisach, *J. Biol. Chem.* **249**, 7495 (1974).
158. C. K. Chang and D. Dolphin, *J. Am. Chem. Soc.* **97**, 5948 (1975).
159. J. P. Collman and T. N. Sorrell, *J. Am. Chem. Soc.* **97**, 4133 (1975).
160. L. K. Hanson, W. A. Eaton, S. G. Sligar, I. C. Gunsalus, M. Goutermanm and C. R. Connell, *J. Am. Chem. Soc.* **98**, 2672 (1976).
161. J. P. Collman, T. N. Sorrell, J. H. Dawson, J. R. Trudell, E. Bunnenberg, and C. Djerassi. *Proc. Natl. Acad. Sci. U.S.A.* **73**, 6 (1976).
162. C. K. Chang and D. Dolphin, *Proc. Natl. Acad. Sci. U.S.A.* **73**, 3338 (1976).
163. C. K. Chang and D. Dolphin, *J. Am. Chem. Soc.* **98**, 1607 (1976).
164. C. K. Chang, D. Dolphin, N. Farrow, and B. R. James, unpublished results.
165. R. Chiang, R. Makino, W. E. Spomer, and L. P. Hager, *Biochemistry* **14**, 4166 (1975).

CHAPTER

3

Affinity Labeling Studies on *Escherichia coli* Ribosomes

Barry S. Cooperman

INTRODUCTION*

Structural and functional studies of biological macrostructures have attracted increasing attention in recent years. The affinity labeling technique, by which one forms a covalent bond between a radioactive ligand and its receptor, permitting, on subsequent analysis, identification of the components of the ligand binding site, has emerged as a powerful tool in such studies [1]. Of the biological macrostructures currently under study, the best characterized is the *Escherichia coli* ribosome, which is also the one to which this technique has been applied most extensively. A critical evaluation of the results of affinity labeling studies on the *E. coli* ribosome should thus not only provide detailed insight into the mechanism of action of the ribosome, but also permit some conclusions to be reached as to the general utility and possible pitfalls in the application of this technique to the study of macrostructures. It is to these two goals that this review is addressed.

We begin with a brief outline of the principal features of both ribosomal structure and function [2] and the technique of affinity labeling and discuss the problems involved in the application of this technique to ribosomal

* Abbreviations: AcPhe, *N*-acetylphenylalanine; fMet, *N*-formylmethionine; SDS, sodium dodecyl sulfate; S⁴U, 4-thiouridine; Br⁵U, 5-bromouridine; PCMB *p*-chloromercuribenzoate; GPPCH₂P, guanosine 5′-(β,γ-methylene)triphosphate.

studies. This is followed by a survey of published work in this field. We conclude with a comparison of the results of affinity labeling studies with those of other techniques that have been used in structure–function mapping of the ribosome. A brief review of ribosome affinity labeling has appeared previously [1f].

GENERAL REMARKS

Ribosome Structure and Function

The *E. coli* ribosome is designated by its sedimentation coefficient as a 70 S particle (MW 2.7×10^6 daltons) and is made up of two dissociable particles. The 50 S particle (MW 1.8×10^6 daltons) is composed of two RNA chains, 23 S RNA (MW 1.1×10^6 daltons) and 5 S RNA (MW 4×10^4 daltons), and approximately 34 different proteins designated L1–L34. The 30 S particle is composed of one RNA chain, 16 S RNA (MW 0.55×10^6 daltons), and 21 different proteins designated S1–S21. Virtually all of the ribosomal proteins have molecular weights in the range 9000–30,000 daltons, the notable exception being protein S 1, having a molecular weight of 65,000 daltons.

Ribosomal protein synthesis is conveniently divided into three primary processes: initiation, elongation, and termination. A brief summary of the initiation sequence is shown in Scheme 1. Under physiological conditions,

$$70 \text{ S} \rightleftharpoons 50 \text{ S} + 30 \text{ S} \qquad (1)$$

$$30 \text{ S} + \text{fMet-tRNA}^{\text{fMet}} + \text{mRNA} + \text{GTP} \rightleftharpoons$$
$$30 \text{ S} \cdot \text{fMet-tRNA}^{\text{fMet}} \cdot \text{mRNA} \cdot \text{GTP} \qquad (2)$$

$$50 \text{ S} + 30 \text{ S} \cdot \text{fMet-tRNA}^{\text{fMet}} \cdot \text{mRNA} \cdot \text{GTP} \rightleftharpoons$$
$$70 \text{ S} \cdot \text{fMet-tRNA}^{\text{fMet}} (\text{P}) \cdot \text{mRNA} + \text{GDP} + \text{P}_\text{i} \qquad (3)$$

Initiation complex

Scheme 1   Initiation sequence.

step (1) requires two protein initiation factors designated IF-1 and IF-3; IF-1 increases the rate constants for both association and dissociation, and IF-3 shifts the equilibrium toward dissociation by binding strongly to the 30 S particle. Step (2) requires a third factor, IF-2, to direct fMet-tRNA^{fMet} binding. Step (3) is accompanied by the release of all three factors, the release of IF-2 depending on GTP hydrolysis, and results in the formation of the initiation complex. Ribosomes are currently believed to have two functional sites for tRNA binding, a P, or donor, site for peptidyl-tRNA and an A, or acceptor, site for aminoacyl-tRNA. A charged tRNA is said to occupy the P site if it is puromycin reactive, i.e., if it will transfer its amino-acyl group to the free amine of puromycin. A tRNA molecule in the A site is not puromycin reactive. At the end of initiation, fMet-tRNA^{fMet} is bound in the P site.

The elongation sequence, which requires three protein elongation factors (EF), is described in Scheme 2, where pep$_n$ denotes a peptidyl chain of n amino acid residues. In the first elongation step, n would be 1 and pep$_n$ would be fMet. Binding of aminoacyl-tRNA (aa-tRNA) to the ribosomes [step (4)] in response to the presence of the appropriate codon in the A site proceeds via prior formation of a ternary complex, aa-tRNA·EF-Tu·GTP. In step (5), the peptidyl transferase step, peptide bond formation is preceded by GTP hydrolysis and EF-Tu·GDP release. A second elongation factor, EF-G, is needed for the translocation step (6), in which the discharged tRNA

$$70\,S \cdot pep_n\text{-}tRNA(P) \cdot mRNA + aa\text{-}tRNA \cdot EF\text{-}Tu \cdot GTP \;\rightleftharpoons$$

$$70\,S \cdot pep_n\text{-}tRNA(P) \cdot aa\text{-}tRNA(A) \cdot EF\text{-}Tu \cdot GTP \cdot mRNA \quad (4)$$

$$70\,S \cdot pep_n\text{-}tRNA(P) \cdot aa\text{-}tRNA(A) \cdot EF\text{-}Tu \cdot GTP \cdot mRNA \;\rightleftharpoons$$

$$70\,S \cdot tRNA(P) \cdot pep_{n+1}\text{-}tRNA(A) \cdot mRNA + EF\text{-}Tu \cdot GDP + P_i \quad (5)$$

$$70\,S \cdot tRNA(P) \cdot pep_{n+1}\text{-}tRNA(A) \cdot mRNA + GTP \;\rightleftharpoons$$

$$70\,S \cdot pep_{n+1}\text{-}tRNA(P) \cdot mRNA + GDP + P_i + tRNA \quad (6)$$

Scheme 2. Elongation sequence.

is released, peptidyl-tRNA is transferred from the A to the P site, and the mRNA message moves by one codon. Finally, the third factor, EF-Ts, cata-lyzes the guanine nucleotide exchange reaction (7), thus recycling EF-Tu.

$$EF\text{-}Tu \cdot GDP + GTP \;\rightleftharpoons\; EF\text{-}Tu \cdot GTP + GDP \quad (7)$$

The termination process is described by Eq. (8), in which the ester linkage between protein and tRNA is cleaved in response to the presence of a non-sense codon (UAA, UAG, or UGA) in the A site. This process also requires three protein factors, RF-1, RF-2, and RF-3.

$$70\,S \cdot protein\text{-}tRNA \cdot mRNA \;\rightarrow\; protein + 70\,S \cdot mRNA \cdot tRNA \quad (8)$$

Affinity Labeling

The most common approach toward affinity labeling is illustrated by Scheme 3. Here L′X represents an electrophilic or photolabile derivative of a native ligand L, and R represents its receptor. If L′X retains high affinity for the native ligand receptor site [Eq (9)], then at low L′X con-centration site-specific covalent attachment via Eq. (10), a first-order

$$R + L'X \rightleftharpoons R \cdot L'X \qquad \text{noncovalent binding} \qquad (9)$$
$$R \cdot L'X \longrightarrow R\text{-}L' + X \qquad \text{specific covalent binding} \qquad (10)$$
$$R + L'X \longrightarrow R\text{-}L' + X \qquad \text{nonspecific covalent binding} \qquad (11)$$

Scheme 3   Affinity labeling.

process, will proceed at a much faster rate than non–site-specific attachment via Eq. (11), a second-order process. For electrophilic derivatives, covalent reaction takes place spontaneously with nucleophilic groups at or accessible from the binding site, while for photolabile derivatives covalent reaction proceeds only on illumination with an appropriate light source. A second general approach toward affinity labeling is to subject the complex of receptor with native ligand L to a nonspecific cross-linking procedure, as, for instance, by treatment with bifunctional reagents or ultraviolet (uv) irradiation [3].

The logic of an affinity labeling experiment is so evident that the simple identification of labeled receptor components by itself provides strong evidence for the presence of these components at or close to the ligand binding site. Nevertheless, it is possible to obtain misleading results, and even, as we shall see later in this review, for two research groups using the same or similar derivatives to obtain different results. Thus, we would like to consider here the inherent ambiguities in affinity labeling experiments and the control experiments that can be performed to resolve these ambiguities.

In the simplest case of a receptor with a single binding site for the ligand L, confidence that the labeling results obtained using the first approach (Scheme 3) really reflect ligand binding site components depends on the demonstrations that (1) the reagent L'X binds noncovalently to the receptor in the same manner as the native ligand, (2) noncovalent binding is a necessary prerequisite for covalent binding, (3) covalent incorporation takes place at the same site as noncovalent binding, and (4) labeling patterns obtained reflect incorporation into a native receptor site structure. The first two points are often tested by measuring whether native ligand both competes with noncovalent binding and blocks covalent incorporation of L'X, or whether an added factor important for noncovalent binding of L (for instance, mRNA for tRNA binding to ribosomes) is necessary for noncovalent and covalent binding, or whether both noncovalent and covalent binding show similar saturation behavior as a function of L'X concentration. Evidence bearing on the third point can in principle be obtained by examining whether covalent bond formation is quantitatively correlated to loss of reversible binding sites for L. This requires not only that a high fraction of receptor sites be labeled, but also that there be a reliable quantitative assay for natural ligand binding, requirements that are often difficult to fulfill. However, even a successful demonstration of such correlation leaves open the possibility that label covalently bound to a site only partially overlapping with or actually distinct from the L site blocks L binding by steric interference or allosteric effects. In appropriate cases, an alternative demonstration of this point can be obtained if it can be shown that covalently bound material behaves functionally as noncovalently bound L or inhibits a function specifically dependent on reversible binding of L, such as in enzyme turnover. Given the inherent difficulty of demonstrating this point, it is important to choose approaches maximizing the probability of accurate labeling. Electrophilic reagents, with their requirement for nucleophilic centers for reaction, are particularly liable to give misleading results, since the probability of reaction at a low-affinity, nonspecific site containing a potent nucleophilic center can be as high or higher than the probability of reaction with a high-affinity specific site lacking such a center. Photolabile reagents, because of their much lower chemical selectivity toward covalent bond formation, are clearly superior in this important respect. It is true that photoaffinity labels often afford low yields of receptor labeling, reflecting an intrinsic problem arising from reaction of the photogenerated species with solvent, but this defect can sometimes be overcome through use of iterative labeling methods [1a]. One can distinguish between two groups of photolabile reagents, one that is irreversibly photoactivated and either reacts covalently with receptor or is discharged via reaction with solvent, and another that is reversibly converted to an excited electronic state, from which it can either react with receptor or solvent or simply return to the ground state [4]. In the first group are diazo and azide reagents, which on photolysis give, respectively, the electron-deficient species carbenes and nitrenes, which are quite reactive toward covalent bond insertion. In the second group are reagents such as aryl ketones, which can be converted to excited-state triplets, capable of hydrogen atom abstraction, especially from tertiary carbons, and eventual carbon–carbon bond formation. For all photolabile reagents, the dependence of incorporation on light flux is easily demonstrated via a comparison of labeling obtained with and without irradiation. For the irreversible reagents, proof that labeling is due to carbene or nitrene formation can be obtained by a comparison of labeling results obtained with or without separate prephotolysis of the reagent with a light fluence just high enough for complete nitrogen loss.

Concern over possible changes in receptor site structure during the course of a photoaffinity labeling experiment arises because of the possibility that the ligand acts as a photosensitizer [1a]. In a similar way, conformational changes can occur in a receptor when electrophilic affinity labels are used, if long incubation times are required in order to get a significant yield of covalent reaction. In both cases, this point can be examined directly by looking for changes in the labeling pattern as a function of either total radiation fluence [6] or time of incubation.

The approach of using native ligand and a nonspecific cross-linking procedure presents the distinct advantages of obviating the need both for the synthesis of a derivative and for demonstrating correct noncovalent binding. On the other hand, since in general the native ligand has no inherent special reactivity, the chemical specificity toward covalent bond formation inherent in the first approach is lost. It is therefore likely that some destruction of the receptor is occurring during the labeling experiment, and it is thus particularly important that the labeling pattern be extrapolated back to either zero irradiation fluence or zero incubation time. Thus, practically speaking, the approach is confined to ligands available in highly radioactive form, for which labeling patterns can be obtained at low stoichiometric yields of incorporation.

Affinity Labeling of Ribosomes

Affinity labeling studies on a receptor as complex as the ribosome pose problems of interpretation in addition to those discussed above. First, it is not uncommon for a ribosomal ligand to have one specific binding site of high affinity and one or more secondary sites of lower affinity [5], creating an ambiguity as to whether the labeling pattern seen at any one concentration of reagent arises from reagent bound to a tight or a weak site. Measuring the labeling pattern as a function of reagent concentration is a generally valid way of resolving this ambiguity [6]. Another possibility for reagents forming tight 1:1 noncovalent complexes with the ribosome is to isolate the complex before allowing covalent bond formation to proceed. This procedure can be carried out with some electrophilic reagents by isolating the complex at low temperature and obviously poses no problem for photolabile reagents. A third method is simply to use an excess of ribosomes over ligand, since this favors tight site binding [7a]. The latter two approaches, although simpler experimentally, must be used with caution. For example, in cases where low yields of incorporation are found and where the difference in affinities between tight and weak sites is within one or two orders of magnitude, the low overall yield of reagent incorporation could mask a high yield at a site of low occupancy. Moreover, with electrophilic and some reversible photolabile reagents having high chemical specificity for covalent bond formation, high incorporation yields could be obtained at a low-affinity site on prolonged incubation or photolysis.

A second problem is the lack of a standard method for preparing homogeneous, active ribosomes. Thus, ribosomes prepared from different stages of the growth phase, or using a different method of cell breaking, or including or excluding high-salt washing, show some variation in both their protein compositions and functional properties [8,9]. Thus, it is hardly surprising that research groups have reported significant variability in their labeling data, particularly in comparing results obtained with different ribosomal preparations (although using the same procedure) [7] or that different laboratories using similar or even identical reagents obtain different labeling results [7,10–13]. Given the considerable effort it would take to control, and demonstrate that one has controlled, all of the possible variables in ribosome preparation, the simplest way to resolve this problem is to examine the variation in labeling pattern obtained as one varies the method of ribosome preparation.

Finally, the ribosome is known to be a conformationally flexible particle, particularly as a function of Mg^{2+} and monocation concentration [14]. Until now, the tendency has been for each laboratory to use a single reaction medium for all of its labeling experiments. However, these reaction media are often significantly different from one laboratory to another, providing another possible explanation for differences in the results obtained. Again, as above, it would seem advisable to systematically examine variation in the labeling pattern as a function of reaction medium.

It should be noted that in few of the ribosomal affinity labeling studies we will be examining have all or even most of the controls discussed above been reported. This is in large measure a result of the newness of the field, the first study having been published only five years ago [23c], and the technically demanding nature of some of the control experiments. In addition, it probably also reflects a certain naiveté toward the possible difficulties in affinity labeling experiments, which have become manifest only with the publication by different laboratories of apparently contradictory results [7,10,11–13]. Hopefully, more detailed analyses of affinity labeling results will become available in the coming years.

Mapping Studies

Very powerful and straightforward methods are currently in use for identifying labeled ribosomal proteins. Using a method introduced by Kaltschmidt and Wittman [15], or a more recent variant thereof [16], essentially all of the ribosomal proteins can be resolved by two-dimensional polyacrylamide gel electrophoresis. If the attached ligand is small and not highly charged, finding the radioactivity coincident or close to coincident [17] with a ribosomal protein position suffices. When this is not the case, attachment of the ligand can cause substantial shifts with respect to the normal protein position. Here labeled proteins have been identified through specific antigen–antibody reaction with antibodies directed against individual ribosomal proteins. Alternatively, one can first separate proteins from a single subunit into several different groups on the basis of how tightly they are bound to RNA and then analyze the groups of proteins on SDS or Sarkosyl containing polyacrylamide gels, in which proteins are resolved on the basis of their molecular weights [18].

Under certain conditions of enzymatic hydrolysis, the 16 S RNA can be broken into large pieces: a 12 S fragment containing the 5′ end and comprising some 1000 nucleotides, and an 8 S fragment containing the 3′ end and comprising some 500 nucleotides. Similarly, 23 S RNA can be hydrolyzed into a 13 S fragment (5′ end, 1300 nucleotides) and an 18 S fragment (3′ end, 1700 nucleotides). Localization of radioactive label into one of these fragments is a reasonable first step in RNA mapping studies [13c,19]. More precise localization requires the isolation and identification of labeled oligonucleotides resulting from carefully controlled RNase (T1 and A) hydrolysis of the parent RNA chains, a procedure that is much more laborious than that used for identifying labeled proteins. As yet, not a single precise localization of an affinity-labeled RNA site has been reported, although one can expect such results in the near future [10].

Ribosomal Affinity Labeling Studies

AFFINITY LABELS FOR tRNA

Fourteen different tRNA affinity labels have been used in *E. coli* ribosomal studies. Of these, thirteen are acylaminoacyl-tRNA's including eleven derivatives of Phe-tRNA^{Phe} and two derivatives of fMet-tRNA^{fMet}. The fourteenth is an aminoacyl-tRNA, Val-tRNA^{Val}, derivatized at the 8-thiouridine position. Structures of the reactive portion of the acylaminoacyl-tRNA labels are given in Table 1.

N-AcetylPhe-tRNA^{Phe} has been shown to be functionally similar to peptidyl-tRNA in its binding to the ribosome; that is, in the presence of poly(U) it binds primarily to the P site [20]. Given the structural similarity between the N-acylPhe-tRNA^{Phe} affinity labels and N-acetylPhe-tRNA^{Phe} most of the experimental results have been interpreted as reflecting labeling of the 3′-aminoacyl binding locus of the P site. The tests that have been used for proper noncovalent binding are (1) that it is stimulated by poly(U), (2) that it is puromycin reactive, and (3) that P-site ligands, such as deacylated tRNA [7a] and certain antibiotics [7d], inhibit binding. The necessity of noncovalent binding for covalent binding has been demonstrated by showing a parallelism between the two; that is, covalent labeling has been shown to be stimulated by poly(U) and inhibited by puromycin (added prior to the labeling reaction) [17] and P-site ligands. It is important to note that noncovalent binding is a necessary but not sufficient condition for covalent binding. Thus, some antibiotics have been shown to strongly inhibit covalent reaction without

TABLE 1

tRNA Affinity Label Reagents

A. Reactive group attached to the α-amino group of Phe-tRNA^{Phe}

General structure:
$$3'\text{-O}\underset{\overset{\|}{O}}{\text{C}}\underset{\overset{|}{CH_2C_6H_5}}{\text{CH}}\text{NHR}$$

Reagent	R	References
1	—CCH$_2$Br (with C=O)	7,10,13e,20
2	—CCH$_2$I (with C=O)	21
3	—C(=O)—O—⟨benzene⟩—NO$_2$	17
4	—C(=O)—(CH$_2$)$_3$—⟨benzene⟩—N(CH$_2$CH$_2$Cl)$_2$	23
5	—(glycine)$_n$CCH$_2$Br($n = 1$–17) (with C=O)	22
6	—C(=O)—CH$_2$—⟨benzene⟩—O—⟨benzene (NO$_2$)⟩—N$_3$	13b
7	—C(=O)—CH$_2$—NH—⟨benzene (NO$_2$)⟩—N$_3$	13a
8	—C(=O)—⟨benzene (NO$_2$)⟩—N$_3$	13d

(*continued*)

TABLE 1—*continued*

A. Reactive group attached to the α-amino group of Phe-tRNA^{Phe}

Reagent	R	References
9		13c
10	$-\overset{\overset{\displaystyle O}{\|\|}}{C}CN_2CO_2Et$	24
11		19

B. Reactive group attached to the α-amino group of Met-tRNA^{fMet}

General Structure: $3'-O-\overset{\overset{\displaystyle O}{\|\|}}{C}-\overset{\overset{\displaystyle CH_2CH_2SCH_3}{\|}}{CH}-NHR$

Reagent	R	References
12	$-\overset{\overset{\displaystyle O}{\|\|}}{C}CH_2Br$	32
13		31

having much of an effect on noncovalent binding. This result can be interpreted as arising from a local competition of the antibiotic for the 3′-aminoacyl end binding locus, while the remainder of the tRNA molecule remains bound to the ribosome through the codon and other interactions [7d,19].

Another very important control, which has been reported only for reagents 1, 4, 6–8, and 11 (Table 1) has been to show that the covalently bound affinity label retains the ability to act as a peptidyl donor [7a,13a,b,e,19,23b]. This control is discussed more fully below.

P-Site Labeling. The results of experiments using N-acylPhe-tRNA^{Phe} derivatives directed toward the P site are summarized in Table 2. Although all of the reagents used react either predominantly or exclusively with the 50 S particle, once past this very low level of resolution there is an apparent scatter in the results, the principal products being the proteins L2, L11, L18, and L27, as well as what are at present rather ill-defined regions of 23 S RNA. These results raise the following questions. First, do the incorporation products found fall in a single, well-defined region on the 50 S particle? Second, what are the reasons for the differences obtained? Third, is there any reason to believe that some results are more significant than others? It should be said at the outset that it is at present not possible to provide definitive answers to any of these questions. This concession to prudence made, it appears that, using a rather loose definition of "region," the answer to the first question is yes. Thus, L11 and L18 have been found to be close neighbors in immune electron microscopy studies [25a] (Fig. 1a), L2 forms a complex with 23 S RNA that facilitates L11 binding [26], and L2 and L27 have been shown to form part of a single cross-linked complex [27a]. In addition, of the seven studies reporting 23 S RNA as the major labeled site, two have been investigated further and both have shown labeling to be located in the 18 S fragment, to which all four of the above-mentioned proteins are bound [28]. Studies are currently underway to more precisely define the sites of RNA incorporation, the results of which should allow a less qualified response to the first question.

As noted above, the ribosome is a flexible structure whose conformation varies with pH, Mg^{2+} concentration, monocation concentration, temperature, and source and method of preparation. Thus, a possible explanation for the scatter in the results obtained is that, in addition to the differences in chemical reactivity of the various reagents used, different workers have been labeling ribosomes in different conformational states. A more emphatic conclusion awaits the results of an experiment systematically testing the effect of variation of these parameters on the labeling pattern. However, a comparison of results already published (Table 3) leads to the conclusion, which must be considered

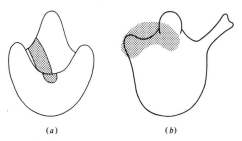

(a)   (b)

Fig. 1   Structural models for the 50 S particle. (a) Tischendorf *et al.* [25a]. The shaded area contains all or part of proteins L11, L16, L18, and L23. (b) Lake [25b]. The shaded area is proposed as the binding site for the 3′ end of tRNA.

TABLE 2

AcylPhe-tRNA^{Phe} Affinity Labeling Studies

Reagent	Target particle	Noncovalent binding				Covalent binding								Label distribution and comments	Major labeling sites	Minor labeling sites	References
		Poly(U) stimulated	Puromycin reactive	tRNA inhibited	Antibiotic inhibited	Poly(U) stimulated	Puromycin inhibited	tRNA inhibited	Antibiotic inhibited	Donor capability	Yield (%)	No photolysis control	Prephotolysis control				
1	70 S	+	+	+	−	+		+	+	+	3			RNA <10% of total	L2, L26/27, S18^a	L6 and/or L11, L14–17, L20	7
1	70 S	+				+				+	12			Protein <10% of total	23S RNA		10
1	30 S	+	+			+									S13		13f
2	70 S	+				+		−		+	20			Protein labeling high but poly(U) independent	23S RNA		21
3	70 S	+	+			+	+^{b}		+	+	10			No RNA labeling	L27	L2, L14–L16	17
4	70 S	+	+			+	+		+	+	30			Protein labeling 10–20%	23S RNA		23a,b

								Ref.
5	70 S		~5			S18, L27 labeling independent of chain length		22
						L2, L16 best labeling with short chain length		22
						L24, L32/33 favored by long chain lengths	L2, L5, L27, L33	22
6	70 S	+ +[b]	+	+ ~1 + +	50 S protein 80% of total	L1,[c] L11, L18, L22[c]		13b
7	70 S	+ +	+	+	50 S protein >90% of total	L11, L18	L1, L5, L27, L33	13a
8	70 S	+	+	+ 15 +	50 S protein 1/3 of total	23 S RNA		13d,e
9	30 S			25	RNA <10% of total	S3, S7, S14	S1, S4, S9, S11, S13	
	70 S	+ +	+ +	10 +	23 S RNA 80–90% of total	18 S fragment of 23 S RNA		13c
10	70 S	+	+	4 +	Protein <10% of total	23 S RNA		24
11	70 S		+	+ 15 +	No protein labeling	18 S fragment of 23 S RNA		19

[a] Not poly(U) dependent.
[b] Only partial.
[c] Labeled L1 and L22 do not have donor capacity.

TABLE 3

Reaction Medium Effect on Labeling

Reagent	Total monocation concentration (mM)[a]	Total Mg^{2+} concentration (mM)	Principal labeled moiety	Reference
1	40	10	Protein	7
1	150	10	RNA	10
2	250	25	RNA	21
3	80	20	Protein	17
4	150	40	RNA	23
6	40	10	Protein	13b
7	40	10	Protein	13a
8	110	20	RNA	13d
9	200	30	RNA	13c
10	70	10	RNA	24
11	100	10.5	RNA	19

[a] K^+ plus NH_4^+ plus buffer cation (usually Tris).

preliminary, that a major factor controlling the partitioning of label between protein and RNA is the monocation concentration in the labeling medium. This is indicated most clearly by comparing results obtained using either chemically identical or very similar reagents, such as reagents 1 and 2 and reagents 6–9. In both cases, as the monocation concentration rises past 70–80 mM, ribosomal RNA becomes more available for reaction. Of course, this effect might be overridden if a reagent with a strong preference for protein labeling were used, and this is a possible explanation for the fact that reagent 3, which is quite lysine specific, labels only protein when used at a total monocation concentration of 80 mM.

From Table 2 it is clear that in only a very few cases have all or even most of the proper controls been examined. Given the structural similarity of the derivatives, it is likely that they will all show similar behavior in control experiments for noncovalent binding and for covalent binding dependence on noncovalent binding, so that, in the absence of a contraindication, the demonstration of successful controls of these types for even only one derivative makes it probable that such controls would be successful for all of the derivatives. However, this line of reasoning cannot be applied to the control demonstrating the functionality of the covalently incorporated tRNA. Given the differences in chemical selectivity of the reagents used, it is to be expected that many different sites will be labeled, of which some may be functional and some not. It is for these reasons that, in comparing the labeling results, those for which this important control has been demonstrated must be given special weight.

The control consists of adding a radioactive peptidyl acceptor (puromycin or aminoacyl-tRNA) to a ribosome that has been labeled with a nonradioactive affinity label and comparing the labeling pattern obtained in this manner with the pattern obtained when a radioactive affinity label is used. For the aryl azide reagents 6 and 7, the timing of covalent bond formation to the ribosome can be easily controlled by the absence or presence of a light fluence, and the results are clear and positive; i.e., L11 and L18 are labeled both by radioactive affinity label reagent and by radioactive acceptor. For the bromoacetyl reagent 1 incorporation of acceptor into ribosome was obtained, the principal products being L2, L16, and L27, but the experiment was done in such a manner as to make it ambiguous whether covalent bond formation preceded or followed peptide bond formation. These results can be interpreted in two different ways, depending on the validity of the latter experiment. If, with the bromoacetyl reagent, covalent bond formation to the ribosome preceded peptide bond formation, then acylPhe-tRNA^{Phe} covalently bound to four or perhaps five different proteins (L2, L11, L18, L27, and perhaps L16) is functionally active as a donor substrate, which implies that the peptidyl transferase center is rather large and diffuse. If, on the other hand, covalent bond formation to the ribosome follows peptide bond formation, then there is no evidence for direct participation of L2 and L27. Given the inherent advantages of photoaffinity labels as compared to electrophilic affinity labels (as discussed elsewhere in this review), emphasized in this case by the known high intrinsic nucleophilicity of L2 and L27 [11a,12a,29], the most likely interpretation would be that L11 and L18 are directly involved in the P site, with L2 and L27 being found in the general vicinity. It should be pointed out that the ambiguity in the experiments testing the donor activity with the electrophilic reagent is not inherent and could be resolved if all noncovalently bound tRNA affinity label were removed from the ribosome before addition of acceptor substrate. Reagents 4, 8, and 11, all of which label 23 S RNA principally, have also been reported to function as peptidyl donors following covalent attachment [13e,19,23b].

A-Site Labeling.   Under conditions in which the binding of acylPhe-tRNA^{Phe} is predominantly to the A site (30 mM Mg^{2+}, excess deacylated tRNA), reagent 1 labels L16 in addition to L2 and L27, and L16 labeling is inhibited by the A-site-directed antibiotic inhibitor tetracycline to a much greater extent than either L2 or L27 [30]. Furthermore, in labeling experiments coupled to peptide bond formation, as described above, with acceptor added before (to favor A-site labeling) or after a preincubation of the donor affinity label with ribosome (to favor P-site labeling), the major products are in both cases L2, L16, and L27 [7a]. These results have been interpreted as showing L16 to be present in the A-site region. Here it is worth noting that, in contrast to L2 and L27, L16 apparently does not have a site of high intrinsic nucleophilicity.

fMet-tRNA. Two electrophilic acylMet-tRNA[fMet] affinity labeling reagents (reagents 12 and 13) have been found to label the ribosome [31,32]. In both cases, viral mRNA and initiation factors were found to be required for noncovalent binding and/or covalent binding. Bound reagent 12 is puromycin reactive, but only to the extent of 50% (compared to 80% with fMet-tRNA[fMet]), thus indicating appreciable heterogeneity in noncovalent binding, whereas addition of puromycin prior to incubation of derivative 13 with the ribosome reduces covalent binding by more than 80%. The results, using a more natural protein-synthesizing system [natural messenger instead of poly(U), factor-dependent tRNA binding], are quite similar to the results found with acylPhe-tRNA[Phe]; reagent 12 labels chiefly L2 with L27 as one of several minor products, and reagent 13 labels chiefly L27 with L15 as a minor product. For reagent 12, as with reagent 1, the attempt was made to (1) examine the functionality of the P-site labeling by looking at the labeling with radioactive Ala-tRNA[Ala] after preincubation of reagent 12 with the ribosome (but, unfortunately, without prior removal of noncovalently bound tRNA) and (2) look at A-site labeling by adding radioactive Ala-tRNA[Ala] before such a preincubation. In both cases, L2 was found as the principal labeled protein, with varying amounts of labeled L27, the main difference with the AcPhe-tRNA[Phe] experiments being the absence of significant L16 labeling. Thus, these experiments confirm the presence of L2 and L27 in the general vicinity of the peptidyl transferase center but provide no further evidence, in addition to what was shown in the acylPhe-tRNA[Phe] experiments, as to whether these proteins are directly involved in the center.

Val-tRNA[Val]. tRNA[Val] has a 4-thiouridine residue at position 8. This highly reactive, nucleophilic position has been alkylated by 4-azidophenacyl bromide, and the resulting photolabile derivative (reagent 14) has been used in the charged form (Val-tRNA[Val] or N-acetylVal-tRNA[Val]) as a ribosomal affinity label [33]. The derivative has been shown to mimic native N-acetylVal-tRNA[Val] in its nonenzymatic P-site binding and native Val-tRNA[Val] in its enzymatic (Tu-dependent) A-site binding. Covalent binding to both the A and P sites has been shown to be highly stimulated by the messenger poly(U_2G). In addition, covalent binding to the A site follows noncovalent binding in being Tu dependent. The covalent reaction is complex. As compared with the incorporation found on irradiation with the azide derivative, there is some covalent incorporation found in the absence of irradiation (10%) and on irradiation when the azidophenyl group is replaced by a simple phenyl group (22%). The experiment using prephotolyzed derivative was not reported. Despite these problems and the fact that detailed mapping results have not as yet been discussed, the very interesting results have already been obtained that quite different incorporation products are obtained when the reagent occupies the A or the P site. On irradiation of the reagent in the P site all incorporation is found in 16 S RNA, while on irradiation in the A site 35% of incorporation is in 16 S RNA with 15% in 30 S protein and 50% in 50 S protein. This contrasts with the previously discussed results with both electrophilic and photolabile derivatives at the 3′-aminoacyl end

TABLE 4

Oligonucleotide Affinity Label Reagents

Reagent	Structure	References
15	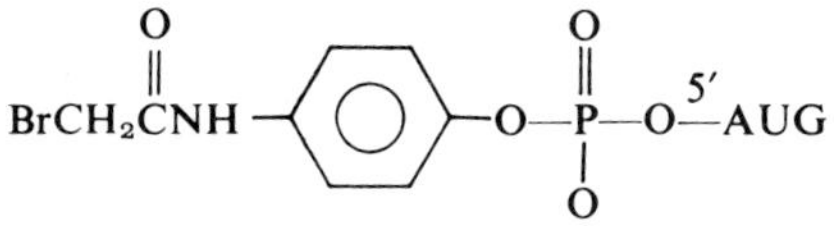	18a–c
15a	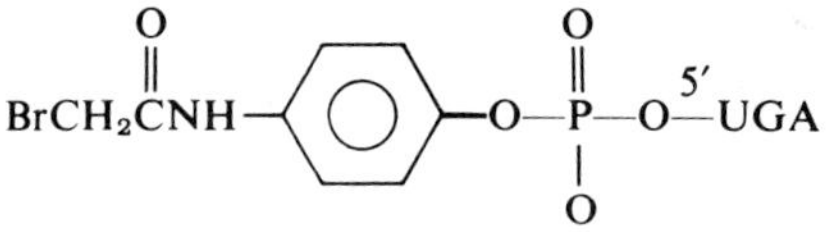	37a,b
16	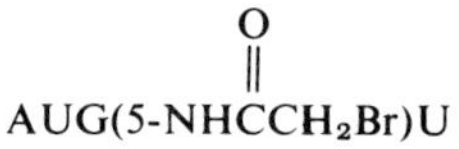 $AUG(5\text{-}NHCCH_2Br)U$	18d
17	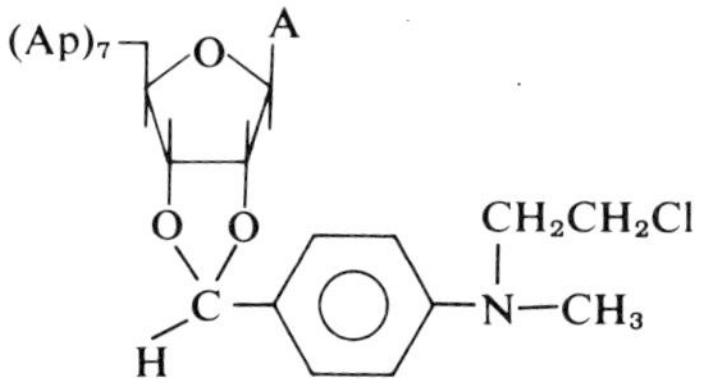	38
18	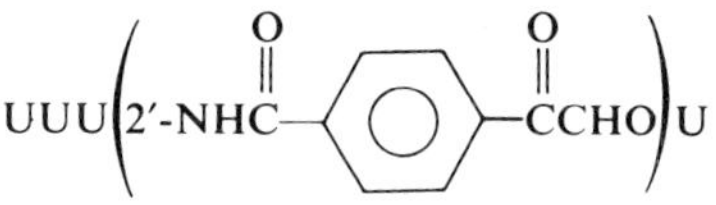	39a
19		39b

of tRNA, where labeling from either the A or P site produced similar incorporation products.

AFFINITY LABELS FOR mRNA

Work with mRNA affinity labeling reagents is rapidly evolving. The first such study appeared in 1973 [38], and within the last two years additional studies involving eight different reagents have appeared. Three different polynucleotide reagents, poly(U) [34], poly(S⁴U) [35], and poly(Br⁵U) [36], all of which have been cross-linked to ribosomes by photoinduced reactions, and six electrophilic derivatives of the oligonucleotides AUG [18a–d], UGA [37], octaadenylic acid [38], UUUU [39a], and GU_3 [39b] (Table 4), have been used. In accord with the general model of protein synthesis presented above, the mRNA affinity labeling reagents appear to label mainly or exclusively the 30 S particle, with poly(Br⁵U) and reagents 15, 16, and 19 reacting primarily with 30 S protein, reagents 17 and 18 reacting primarily with 16 S RNA, particularly region C (reagent 18), poly(U) reacting with both, and poly(S⁴U) reacting certainly with protein and possibly with 16 S RNA. The protein labeling results are summarized in Table 5 and are striking in the degree of overlap obtained with different reagents. Thus, S18 is found as a major or minor labeled protein with six different reagents, S1 with four, S4 and S21 with three, and S12 with two. Furthermore, there is other evidence to suggest that these proteins are in mutual proximity on the 30 S particle. The following pairs of proteins have been found cross-linked: S4–S12, S12–S21, S18–S21 [47a,b]. In addition, both proteins S1 and S21 can be cross-linked to 16 S RNA on oxidation of a 30 S particle with periodate and subsequent reduction with $NaBH_4$, suggesting that they are neighbors at or near the 3′ end of 16 S RNA [47c]. Also, treatment of 30 S·IF-3 complexes

TABLE 5

Proteins Labeled by Oligo- and Polynucleotide Affinity Labels

Reagent	Labeled 30 S proteins[a]					References
	S1	S4	S12	S18	S21	
15		+ +	+ +	+ +	+	18a–c
16	+ +			+		18d
17		+ +		+ +		37a,b
19	+ +		+	+ +		39b
poly(U)	+ +					34
poly(S⁴U)	+ +			+ +[b]	+ +[b]	35
poly(Br⁵U)		+ +		+ +	+ +	36

[a] + +, major labeled protein; +, minor labeled protein.

[b] Isolated as part of RNA–protein complex; no evidence for direct covalent binding

with cross-linking reagents results in the cross-linking of IF-3 to S1, S12, and S21. Similarly, IF-1 and IF-2 are cross-linked to both S1 and S12 [47d]. Corroborating evidence from immune electron microscopy studies is discussed elsewhere in this review.

A promising start has been made on demonstrating the functional properties of the covalently incorporated material. Labeling by reagent 15 is blocked by $AUG(U)_{40}$, labeling by reagent 17 is blocked by poly(A), and labeling by reagent 19 is blocked by a combination of poly(U) and Phe-tRNA^{Phe}, thus providing evidence of the dependence of covalent binding on noncovalent complex formation. Furthermore, ribosomes covalently labeled with reagents 15–19 all show the ability to bind tRNA's having the appropriate anticodon in the absence of added messenger. Ribosomes labeled by the closely similar reagents 15 and 15a principally in proteins S4 and S18 bind charged tRNA in a puromycin-unreactive site, as do ribosomes labeled by reagents 16 and 19, where the principal labeled products are S1 and S18. Also, charged tRNA binding to ribosomes labeled with reagents 15a and 16 is stimulated by protein elongation factors. Reaction of reagent 15 (but not 15a) with 30 S particles that have been first subjected to a freeze-thaw treatment results in a change of the labeling pattern, S4 and S12 becoming the principal labeled proteins. This change has apparent functional significance, since fMet-tRNA^{fMet} bound to such labeled 30 S particles becomes, on addition of 50 S particles, at least partially puromycin reactive. These results can be explained by postulating two binding sites for the triplet codon, an A site in which are found S1, S4, and S18, with S1 being closer to the 3' end of the codon and S4 and S18 closer to the 5' end, and a P site in which is found S12. Alternatively, one could posit a single codon binding site, which via conformational changes, possibly induced by the freeze–thaw treatment and of which the labeling patterns are a reflection, can direct tRNA into either the A or the P site.

Although these experiments already provide strong indications as to the sites of mRNA interaction with the ribosome, the drawing of firm conclusions will necessitate more detailed studies. For example, at present the strongest evidence for A-site binding is of a negative kind (the lack of puromycin reactivity), and most of what has been considered to be A-site binding could just as well represent tRNA bound in a nonfunctional manner. A more valid demonstration of A-site binding would be that the tRNA became puromycin reactive under translocation conditions. This experiment has been tried with reagent 19 with negative results, although it should be pointed out that the lack of puromycin reactivity could be accounted for by reasons other than nonfunctional binding. It would also be of obvious interest to compare labeling patterns obtained in the presence and absence of tRNA under conditions favoring both A- and P-site binding. A start along these lines has already been made, Luhrmann *et al.* [39b] finding, with reagent 19, more labeling of S12 in the absence than in the presence of tRNA. Finally, the reagents used so far clearly favor reaction with nucleophiles (reagents 15–19 are electrophilic, S^4U becomes electrophilic on irradiation [40], and uridine itself shows preferential reaction with thiols and amines on irradiation [41]), and the labeled proteins S4, S12, S18, and S21 are known to contain highly nucleophilic centers. Thus, as with the tRNA

TABLE 6
Antibiotic and Guanine Nucleotide Affinity Labels

A. Chloramphenicol

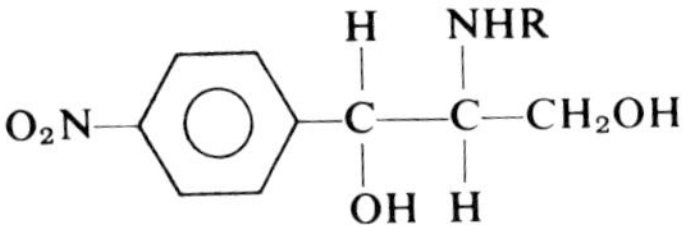

Reagent	R	References
Natural antibiotic	$\overset{O}{\overset{\|}{C}}CHCl_2$	12b
21	$\overset{O}{\overset{\|}{C}}CH_2I$	11a,b
22	$\overset{O}{\overset{\|}{C}}CH_2Br$	12a

B. Puromycin

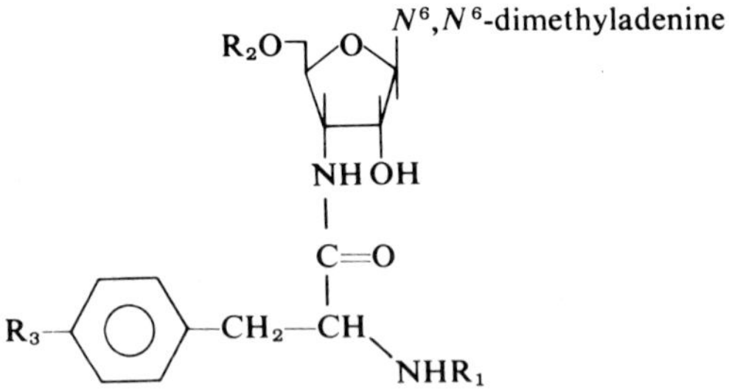

Reagent	R_1	R_2	R_3	References
Natural antibiotic	H	H	CH_3O-	6,42a
23	$-\overset{O}{\overset{\|}{C}}CN_2CO_2Et$	H	CH_3O-	42a
24	$-\overset{O}{\overset{\|}{C}}CH_2I$	H	CH_3O-	42b
25	H	$-PO_3-$⟨O⟩$-NH\overset{O}{\overset{\|}{C}}CH_2Br$	H	42c,d

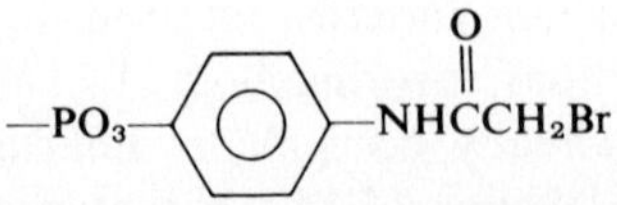

TABLE 6—*continued*

C. Streptomycin

Reagent	R	References
Natural antibiotic	$-\overset{\overset{\text{O}}{\|}}{\text{C}}\text{H}$	—
26	$-CH_2NHNH\overset{\overset{\text{O}}{\|}}{\text{C}}$ —[2-NO₂, 4-N₃ phenyl]	18c
27	$-CH=NNH$—[phenyl]—$NH\overset{\overset{\text{O}}{\|}}{\text{C}}CH_2I$	43

D. Guanine nucleotides

Reagent		Comments	Ref.
28	Modified ribose of GTP	guanine, OH, $O_3PO_3PO_3PO$, OH; $N-NH\overset{\overset{\text{O}}{\|}}{\text{C}}$—[3-NO₂, 4-N₃ phenyl]	44
29	Modified GTP	guanosine$-OPO_3\,PO_3PO_2NHCH_2$—[phenyl]—N_3	44b
30	Modified ribose of GTP	Same as 28, but with $-CH_2-$ in place of $-O-$ between the β and γ P nuclei.	44b
31	Modified GDP	guanosine$-OPO_3PO_3$—[phenyl]—N_3	45a,b.

studies, it would be interesting to compare the results already obtained with results using reagents of lower and/or different chemical selectivity, such as aryl azides, α-diazocarbonyls, and aromatic ketones.

AFFINITY LABELS FOR ANTIBIOTICS AND GUANINE NUCLEOTIDES

Antibiotics, by the diversity of their structures and of their effects on ribosomal function, are excellent ligands for affinity labeling work. Several studies, using derivatives of chloramphenicol [12], puromycin [42], and streptomycin [18e], have already been published (Table 6), and work with other antibiotics is currently underway.

Two electrophilic analogs of chloramphenicol (reagents 21 and 22) have been used in affinity labeling studies on ribosomes, and the direct incorporation of chloramphenicol itself on irradiation at wavelengths greater than 300 nm has also been reported [12b]. The latter study shows no evidence for specific labeling and will not be discussed further. The two studies with electrophilic derivatives give similar results up to a point. The noncovalent binding of both is competitive with chloramphenicol, with somewhat lower affinities than chloramphenicol itself. In both cases, only protein is significantly labeled, and covalent incorporation reduces noncovalent chloramphenicol binding. In the study of Pongs et al., with 70 S particles as the target, covalent incorporation of reagent 21 is accompanied by a partial loss of poly(U)-directed polyphenylalanine synthetic activity whereas, with 50 S particles as the target, Sonenberg et al. found incorporation of reagent 22 to be accompanied by partial loss of peptidyl transferase activity. However, with respect to the identity of the proteins labeled the two studies differ markedly. Pongs et al. found L16 to be the major labeled protein and L24 to be a secondary labeling site, with either 70 S or 50 S particles as the target, whereas Sonenberg et al. found L2 and L27 with 50 S particles as the target. It is noteworthy that Pongs et al. found very little labeling of either L2 or L27, and Sonenberg et al. found very little labeling of L16. Following our discussion of the tRNA studies, these divergent results may be a consequence of the quite different reaction media used for the labeling reaction in the two studies.

Sonenberg et al. found that chloramphenicol in 10-fold molar excess had no effect on the labeling reaction. There is some evidence that chloramphenicol has both high- and low-affinity sites on the ribosome, binding to either of which results in inhibition of peptidyl transferase. The authors offer as a possible explanation for their results that the covalent labeling is taking place at the low-affinity site, which would not have been saturated at the concentration of chloramphenicol used. The difficulty with this explanation is that it would require, given the reduction of noncovalent binding mentioned above, that chloramphenicol binding to the tight site be inhibited by covalent labeling without the reciprocal effect being true. Given failure of this very important control experiment and the high intrinsic nucleophilicity of the two labeled proteins L2 and L27, the significance of this study in defining the chloramphenicol binding site must be considered questionable. The significance of the Pongs *et al.* study is also somewhat unclear. First, the effect of added chloramphenicol on covalent incorporation of reagent 21 is not reported, so it is not known whether this study has the same problem as that discussed above. Second, in more recent studies in which reagent 21 was added to a growing *E. coli* culture, the major labeled protein was found to be S6, with L16 and L24 as secondary incorporation sites. This apparent discrepancy in the *in vivo* vs. *in vitro* results may arise from a procedural artifact. The *in vitro* studies were done using protein samples extracted from 30 S and 50 S particles separated on sucrose gradients, a procedure in which S6 can dissociate from the ribosome, whereas the *in vivo* studies were done on proteins prepared omitting the sucrose gradient step. As yet, only the plausibility of this explanation has been demonstrated. However, it should be noted that the placing of L16 at the chloramphenicol binding site is consistent with reconstitution studies showing a dependence of chloramphenicol binding on the presence of L16 [51] and with the apparent correlation (see above) of affinity labeling of L16 by bromoacetylPhe-tRNA^{Phe} with A-site occupancy, chloramphenicol having been shown to be an A-site inhibitor. Moreover, as noted above, L16 is not a protein of high intrinsic nucleophilicity.

Affinity labeling studies have been reported for three derivatives of puromycin, two electrophilic and one photolabile (reagent 23–25), and puromycin itself has been shown to incorporate into ribosomes on irradiation with uv light. It is the latter study that has so far yielded the most significant results. In work already published [42a], incorporation into both the RNA and protein fractions of 70 S ribosomes was found to proceed via both site-specific and nonspecific processes, with the dissociation constant calculated for the specific process being quite similar to K_m values found for puromycin in peptidylpuromycin assays. By far the major labeled protein is L23, while S14 is a secondary labeled protein. More recently [6], direct evidence has been obtained that L23 is labeled by an affinity labeling process. Thus, incorporation of puromycin into L23 followed a simple saturation curve as a function of puromycin concentration, with a dissociation constant equal to that found for the site-specific labeling of ribosomes, and structural analogs of puromycin blocked L23 labeling. Protein L23 remained the major labeled protein on variation of both ribosomal preparation and wavelength of irradiation, and by studying the labeling pattern as a function of light dose it could be shown that incorporation into L23 did not proceed via a uv-irradiation-induced structural change in the ribosome. In addition, some evidence was obtained for affinity labeling of 23 S RNA, but no mapping studies are yet available.

Reagent 25 is at least a poor substrate for peptidyl transferase. Incubation of reagent 25 with 70 S ribosomes leads to a partial loss of peptidyl transferase activity, but the overall covalent labeling reaction is quite non-specific, 21 molecules of label being incorporated in a ribosome (in both 30 S and 50 S subunits) that retained 47% of its activity. However, under certain conditions, chloramphenicol, which has very little effect on overall labeling, specifically blocks RNA labeling, especially that of 23 S RNA, while at the same time protecting the ribosome against loss of peptidyl transferase activity. Puromycin has a similar protective effect on the peptidyl transferase activity, although its effect on labeling was not reported. Furthermore, covalently labeled ribosomes washed free of excess, noncovalently bound reagent show acceptor activity in a peptidyl transferase assay, although to only a very low level, approximately one transfer being effectuated for every 400 molecules covalently bound to 23 S RNA. From these results the authors concluded that they are labeling a portion of the 23 S RNA that makes up part of the A site. This conclusion is based mostly on the chloramphenicol effects and is not necessarily contradicted by the low yield of acceptor activity obtained, since this might merely reflect a microheterogeneity in the reagent attachment sites. Detailed studies on both the reagent and acceptor activity labeling patterns should provide a clear test for A-site labeling.

The photolabile puromycin derivative reagent 23 was found to incorporate into ribosomes on uv irradiation by both carbene-dependent and carbene-independent processes. Limited solubility and the known weak affinity of N-acylated puromycin derivatives for ribosomes prevented any direct demonstration that incorporation proceeded via affinity labeling, although reagent 23 was shown to be a ribosomal ligand since it partially inhibited both poly(U)-directed polyphenylalanine synthesis and peptidylpuromycin synthesis. The labeling pattern obtained with reagent 23 is quite different from that obtained with puromycin itself, 30 S incorporation being favored over 50 S incorporation, and S18 and S14 (with very little L23) being the principal proteins labeled. Reagent 23 is almost certainly not a good puromycin analog, as shown by the observed differences in labeling pattern, the failure of 1 mM puromycin to block its incorporation, and the known importance of a free α-amino group for proper A-site binding. It remains possible that incorporation into S14 and S18 might reflect labeling of the tRNA 3' terminus portion of the P site. Results with the structurally similar electrophilic reagent 24 were entirely analogous. Although partial inhibition of labeling by this reagent in the presence of 10mM puromycin was reported, such a high concentration of puromycin has the effect of precipitating ribosomes, so that it is not clear that the protective effect arises from competitive binding.

The aryl azide derivative of streptomycin, reagent 26, has been shown to mimic native streptomycin in inhibiting nonenzymatic, PCMB-stimulated, poly(U)-dependent polyphenylalanine synthesis and in binding (noncovalently) preferentially to 30 S as opposed to 50 S particles. Furthermore, the noncovalent binding is blocked by added streptomycin, and this effect is not merely due to a nonspecific polycation interaction since spermidine is without effect on the binding. With 30 S particles as the target almost all of the labeling is in the protein fraction, the major labeled protein being S7, with S14 and S16/S17 as secondary incorporation sites. With 70 S particles as the target, labeling is again confined to the protein fraction, and both 30 S and 50 S particles are labeled to similar extents, although the detailed labeling pattern has not as yet been established. This labeling of 50 S particles is interpreted as providing evidence that the noncovalent binding of reagent 26 occurs at the interface between 30 S and 50 S particles. Streptomycin has been shown to have one tight binding site on both 30 S and 50 S ribosomes, and several weak binding sites [5c,d], probably arising from electrostatic interaction between the streptomycin trication and RNA. As the labeling experiment has thus far only been done at a single concentration, it is unclear whether the results obtained arise from tight or weak site binding.

We also note a preliminary report [43] that a streptomycin derivative whose structure was claimed to be that of reagent 27 reacted with 70 S particles to label mostly 30 S protein. The uncertainty in the structural assignment comes from the method of synthesis, which involved bromacetylation of the parent aniline under conditions in which the methylamine moiety of streptomycin would also be expected to react.

Using three different aryl azide derivatives of GTP (reagents 28–30), Girshovich and his co-workers [44] have obtained strong evidence that EF-G provides the guanosine nucleotide binding site in the ternary complex 70 S·EF-G·guanine nucleotide. All three reagents form such ternary complexes. Only complex formation with reagent 28 is fusidic acid dependent, the explanation being that this reagent is hydrolyzed on formation of the ternary complex and, as a GDP analog, requires fusidic acid to bind. Reagents 29 and 30 are not substrates for GTPase. Reagent 28 labels EF-G exclusively, and the labeling is blocked in the presence of GDP (but not of GTP). Reagent 29 labels EF-G mostly, with about 15% of the label incorporating into ribosomes. Only the EF-G labeling is inhibited by GTP (or GPPCH$_2$P), however. Fusidic acid reduces EF-G labeling without affecting ribosome labeling. Reagent 30 labels EF-G and ribosomes about equally, but again only EF-G labeling is blocked by GPPCH$_2$P. Evidence that EF-G is also the functional GTPase center in the ternary complex comes from a recent [44c] experiment in which reagent 28 is incorporated (as a triphosphate) into EF-G on irradiation of a solution containing just these two components. On addition of 70 S ribosomes to form the ternary complex, inorganic phosphate is released into the medium.

These results are in accord with studies showing that EF-G forms a binary complex with GTP [46] but are in apparent disagreement with the earlier work of Maassen and Möller [45], who studied the incorporation of the GDP derivative reagent 31 into both 50 S and 70 S particles in the presence of EF-G. These authors found labeling to occur principally in the ribosome, with little in EF-G. Ribosomal protein labeling was found to be rather nonspecific, but addition of fusidic acid led to a selective enhancement in the labeling of proteins L5, L11, L18, and L30, as well as in the disappearance of protein L16. The latter effect may result from a peptide bond cleavage induced by a nitrene insertion reaction. When the same experiment was carried out with 70 S particles derived from *Bacillus stearothermophilus*, fusidic acid was found to specifically stimulate labeling of proteins L10 and L22 [45b], which correspond to *E. coli* proteins L11 and L18. A major difference in the experimental protocol used by the two groups is that during photolysis Girshovich *et al.* [44b] use a 7.5-fold excess of ribosomes over reagent, while Maassen and Möller [45a] use a 60-fold excess of reagent over ribosomes. Since increasing the concentration of the labeling reagent favors labeling of sites of little or no specificity over labeling of specific sites, this difference in protocol could account for the observed differences in labeling pattern. The most plausible interpretation of the current results is that EF-G is labeled via tight site binding, while the ribosomal proteins are labeled via weak site binding. A systematic study of labeling pattern vs. affinity label concentration would provide a stringent test for this interpretation.

Correlation of Affinity Labeling and Other Results

THE 50 S PARTICLE

As can be seen in Table 7, there is considerable overlap in the proteins labeled by affinity labeling reagents that incorporate into the 50 S particle. Thus, L16 reacts with derivatives of aminoacyl-tRNA (in the A site), chloramphenicol, and GDP; L2, L24, and L26/27 react with derivatives of aminoacyl-tRNA and chloramphenicol; and L11 and L18 react with derivatives of aminoacyl-tRNA and GDP. We have already discussed evidence that L2, L11, L18, and L27 are found in a single region of the 50 S particle. In addition, other studies provide evidence that L16, L23, L32/33, L5, and L30 can also be added to this group. Thus, immune electron microscopy studies have shown L23 to be close to parts of L11 and L18 and L16 to be close to a part of L11 (Fig. 1), and L32/33 has been shown to be part of the same cross-linked complex as L2 and L27 [27a]. Moreover, L5, L18, and L30 bind directly to 5 S RNA, and L2 is required for the binding of the 5 S RNA · protein complex to the rest of the 50 S particle [48]. Furthermore, evidence

TABLE 7

Correlation of Affinity-Labeled 50 S Proteins

L protein	Affinity label reagent type[a]		
	tRNA	Chloramphenicol or puromycin	GDP
2	+ +	+ +	
5	+		+ +
11	+ +		+ +
16	+ +[b]	+ +	+ +[c]
18	+ +		+ +
23		+ +	
24	+ +	+ +	
26/27	+ +	+ +	
30			+ +
32/33	+ +		

[a] + +, major labeled protein; +, minor labeled protein.
[b] A site.
[c] Missing protein.

has been obtained that L11 is at least important and perhaps vital for peptidyl transferase [49] and is the site of thiostrepton (an A-site-directed antibiotic) binding [50], that L16 is the site of chloramphenicol binding [11,51], and that L11 and L16 have mutual synergistic effects on both peptidyl transferase and chloramphenicol binding [49,51]. Taken together, these results provide strong evidence that chloramphenicol and puromycin bind either at or close to the same region in the 50 S particle as the 3' end of tRNA bound in either the A or P site, which is in accord with the inhibition by chloramphenicol of 3'-aminoacyl oligonucleotide binding [5b] and the peptide acceptor activity of puromycin.

THE 30 S PARTICLE

Affinity labeling results with the 30 S particle also indicate an overlap between the binding sites on this particle for the 3' terminus of tRNA and chloramphenicol and puromycin. Reagents 1 [13f] and 8 react with isolated 30 S particles in poly(U)-dependent reactions to label S13 and S3 and S7 and S14, respectively, while the 70 S particle reacts with bromoacetyl derivatives of a series of oligoglycylPhe-tRNAs[Phe] (reagent 5) to give S18 as the principal labeled 30 S protein, although this reaction is not poly(U) dependent. Of these five proteins, one, S3, is labeled during reaction of reagent 21 with 70 S particles, and two others, S14 and S18, are the major proteins labeled on reaction of 70 S particles with reagent 23. Protein S6, labeled by reaction of reagent 21 with ribosomes of growing cells, completes the list of six 30 S proteins labeled by reagents linked to the 3′ terminus of tRNA. These proteins are also linked by structural and functional studies. Thus, S6 can be cross-linked to S14 and S18, and S7 can be cross-linked to S13 [27b,47b]. In addition, S3, S6, S14, and S18 have all been implicated in tRNA binding, both enzymatic and nonenzymatic [52], while the 30 S assembly map indicates that there is a strong S6–S18 interaction and a weaker interaction between S3, S7, and S14 [53].

Table 8 lists the ten proteins labeled by tRNA 3′ terminus, mRNA, and streptomycin affinity labels. Attempts to correlate affinity labeling results with those obtained by immune electron microscopy (Fig. 2) are at present somewhat uncertain for two major reasons. In the first place, the two groups currently using this technique [25] have significant differences in their respective models for ribosome structure. We did not explicitly consider these differences in discussing the 50 S particle because Lake's group has not yet published information on the localization of 50 S proteins. However, both groups have localized many of the 30 S proteins, and their results will have to be considered in turn. Second, several of the 30 S proteins of interest appear to have elongated conformations, raising the possibility that two or several proteins may be near neighbors in more than one region of the 30 S particle. Of the five proteins implicated at the mRNA binding site (Table 5) Tischendorf et al. [25a] have localized four: S4, S12, S18, and S21. According to these authors, three of these proteins, S4, S12, and S18, have elongated conformations and are close neighbors in two different regions of the 30 S particle, designated I and II (Fig. 2a). The fourth, S21, appears to be localized only in region I. Lake and his co-workers have also localized four proteins: S1, S4, S12, and S18. Of these, a part of S4 and S12 fall in region I′ (Fig. 2b), corresponding roughly to region I, while S1, another part of S4, and a part of S18 fall in region II′ (Fig. 2b), which may correspond to region II. At present it

(a)   (b)

Fig. 2 Structural models proposed for the 30 S particle. (a) Tischendorf et al. [25a]. Region I contains all or parts of proteins S4, S7, S12, S13, S18, and S21; region II contains parts of proteins S4, S12, and S18; region III contains S6 and part of S18. (b) Lake [25b,c]. Region I′ contains a part of S4, S7, S12, and S13; region II′ contains S1 and parts of S4 and S18; region III′ contains S6 and part of S18.

TABLE 8

Correlation of 30 S Affinity-Labeled Proteins

	Affinity label reagent type[a]			
S protein	tRNA	Chloramphenicol or puromycin	mRNA	Streptomycin
1	+		+ +	
3	+ +			
4	+		+ +	
6		+ +		
7	+ +			+ +
12			+ +	
13	+ +		+	
14	+ +	+ +		+
18	+ +	+ +	+ +	
21			+ +	

[a] + +, major labeled protein; +, minor labeled protein.

seems certain only that the mRNA binding site will fall within either site I/I' or II/II'. For the 3' terminus of tRNA, the model of Tischendorf *et al.* [25a] places all or parts of the four (S7, S13, S14, S18) of the six implicated proteins in region I, and part of protein S3 is near this region. By contrast S6 is found in region III, as a neighbor of part of S18. In Lake's model [25b,c,d] only S7 and S13 fall within region I', with S14 and S3 being substantially removed, although still part of the "head" segment. Protein S6 and part of S18 are near neighbors in region III', which corresponds closely to region III of Tischendorf *et al.* At present the weight of the evidence favors placing the binding site for the 3' terminus of tRNA in the head segment of the 30 S particle but not necessarily coincident with region I/I'. The same may be said for streptomycin, the two major proteins labeled by reagent 26 being S7 and S14.

Future experimentation should permit many of the present ambiguities to be resolved. Thus, identification of covalently labeled peptides within an affinity labeled protein combined with an extension of the immune electron microscopy technique through use of antibodies to ribosomal protein fragments should provide a clear indication as to whether affinity labeling of a given protein takes place in more than one region. Furthermore, investigation of the labeling pattern as a function of reagent concentration should reveal whether different labeled proteins are part of a single site of given affinity or of different sites with different affinities. Finally, investigation of the labeling pattern as a function of reaction medium should reveal whether the labeling pattern is sensitive to conformational change, i.e., whether different binding sites are favored under different experimental conditions.

Region I/I′ is just barely large enough to accommodate both the anticodon and the 3′ terminus of tRNA (separated in the crystalline form by 77 Å [54]), so that a tRNA molecule confined to region I/I′ might be in a strained conformation. Such strain would plausibly be relieved in the 70 S particle, where the anticodon region remained bound to the 30 S particle, while the 3′ terminus moved over to the 50 S particle.

30 S–50 S INTERFACE

As we have noted, the 3′ terminus of tRNA has binding sites on both the 30 S and 50 S particles, although binding to the 50 S particle appears to be stronger, as evidenced by the general result that acylaminoacyl-tRNA affinity label derivatives react principally with the 50 S particle when the 70 S particle is the target. The results we have discussed above support the notion that these two sites are in close proximity in the 70 S particle. Thus, several experimental approaches [55], including partial reconstitution, direct binding of 30 S proteins to 50 S particles or 23 S RNA, cross-linking of 30 S protein to 50 S particles, and chemical modification and antibody inhibition of 30 S and 50 S association, show four of the affinity-labeled 50 S proteins, L2, L18, L23, and L26/27, and six of the affinity-labeled 30 S proteins, S3, S12, S14, S16, S18, and S21, to be at the 30 S–50 S interface. Also, as noted above, the reactions of several affinity label reagents (e.g., 14 and 26) with 70 S particles result in the labeling of both 30 S and 50 S subunits. A particularly interesting example is provided by puromycin. On photolysis, puromycin incorporates mainly into 50 S particles, but, on acylation of the α-amino group, incorporation occurs mainly in the 30 S particle. Moreover, it has been shown recently [56] that in the presence of chloramphenicol, puromycin itself incorporates principally in the 30 S particle (S14 is the major product labeled), as though it had been pushed over from the 50 S particle. Finally, the

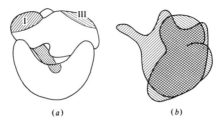

(a)   (b)

Fig. 3 Structural models proposed tor the 70 S particle. (a) Tischendorf *et al.* [25a]. (b) Lake [25b]. Note the proximity of region I/I′ (30 S) to the putative peptidyl transferase center (50 S).

models of the 70 S particle presented by both Tischendorf *et al.* and Lake [25] place region I/I' of the 30 S particle in close proximity to the peptidyl transferase center of the 50 S particle (Fig. 3).

FINAL REMARKS

Has affinity labeling been successfully applied to the study of ribosomes? If we consider it to be a low-resolution technique and are content with using it to define quite large "regions" surrounding a ligand binding site, then, given the strong overlap among the affinity labeling studies herein reviewed and between these results and those of related approaches, the answer would appear to be an unqualified yes. The response is much less clear for use of affinity labeling as a high-resolution technique capable of identifying those labeled proteins (and eventually RNA oligonucleotides) that are actually in contact with the ligand at the binding site. Such identifications should be possible through systematic study of the effects of reagent concentration, reaction medium, ribosomal preparation, time of incubation or irradiation, and reagent type on the labeling pattern, as discussed in this review, but as of now there is not a single ribosomal ligand for which the ensemble of this information exists. However, there are no technical barriers to obtaining such information, so that future work should allow a judgment to be made as to the utility of affinity labeling as a high-resolution technique for mapping biological macrostructures.

ACKNOWLEDGMENT

The author wishes to thank the American Cancer Society (NP-176) and the Alfred P. Sloan Foundation for financial support and Dr. Marianne Grunberg-Manago (Institut de Biologie Physico-Chimique, Paris, France) for her warm hospitality during the writing of this review.

REFERENCES

1a.* B. S. Cooperman, *in* "Aging, Carcinogenesis and Radiation Biology. The Role of Nucleic Acid Addition Reactions" (K. C. Smith, ed.), p. 315. Plenum, New York, 1976.
1b. D. Creed, *Photochem. Photobiol.* **19**, 459 (1974).
1c. J. R. Knowles, *Acc. Chem. Res.* **5**, 155 (1972).

* For recent reviews of affinity labeling see 1a–1f.

1d. S. J. Singer, *Adv. Protein Chem.* **22**, 1 (1967).

1e. E. Shaw, *in* "The Enzymes" (P. D. Boyer, ed.), 3rd ed., Vol. 1, p. 91. Academic Press, New York, 1970.

1f. C. R. Cantor, M. Pellegrini, and H. Oen, *in* "Ribosomes" (M. Nomura and A. Tissières, eds.), p. 573. Cold Spring Harbor Lab., Cold Spring Harbor, New York, 1974.

2a.† Several articles *in* "Ribosomes" (M. Nomura and A. Tissières, eds.), Monogr. Ser., Cold Spring Harbor Lab., Cold Spring Harbor, New York, 1974.

2b. H. G. Wittman, *Eur. J. Biochem.* **61**, 1 (1976).

2c. R. A. Garrett and H. G. Wittmann, *Adv. Protein Chem.* **27**, 277 (1973).

2d. M. Nomura, *Science* **179**, 864 (1973).

2e. R. Haselkorn and L. B. Rothman-Denes, *Annu. Rev. Biochem.* **42**, 379 (1973).

2f. C. G. Kurland, *Annu. Rev. Biochem.* **41**, 377 (1972).

2g. O. Pongs, K. H. Nierhaus, V. A. Erdmann, and H. G. Wittmann, *FEBS Lett.* **40**, 528 (1974).

3a. R. R. Traut, R. L. Heimark, T.-T. Sun, J. W. B. Hershey, and A. Bollen, *in* "Ribosomes" (M. Nomura and A. Tissières, eds.), p. 271. Cold Spring Harbor Lab., Cold Spring Harbor, New York, 1974.

3b. B. S. Cooperman, M. Grunberg-Manago, J. Dondon, J. Finelli, and A. M. Michelson, *FEBS Lett.* **76**, 59 (1977).

3c. C. Hélène, *in* "Aging, Carcinogenesis and Radiation Biology. The Role of Nucleic Acid Addition Reactions" (K. C. Smith, ed.), p. 149. Plenum, New York, 1976.

3d. P. R. Schimmel, G. P. Budzik, S. S. M. Lam, and H. J. P. Schoemaker, *in* "Aging, Carcinogenesis and Radiation Biology. The Role of Nucleic Acid Addition Reactions" (K. C. Smith, ed.), p. 123. Plenum, New York, 1976.

4. J. A. Katzenellenbogen, H. J. Johnson, Jr., K. E. Carlson, and H. N. Meyers, *Biochemistry* **13**, 2986 (1974).

5a. R. Harris and S. Pestka, *J. Biol. Chem.* **248**, 1168 (1973).

5b. J. L. Lessard and S. Pestka, *J. Biol. Chem.* **247**, 6909 (1972).

5c. J. G. Flaks and F. N. Chang, *Antimicrob. Agents & Chemother.* **2**, 294 (1972).

5d. G. Schreiner and K. H. Nierhaus, *J. Mol. Biol.* **81**, 71 (1973).

6. E. N. Jaynes, Jr., P. G. Grant, R. Wieder, G. Giangrande, and B. S. Cooperman, submitted for publication.

7a. M. Pellegrini, H. Oen, D. Eilat, and C. R. Cantor, *J. Mol. Biol.* **88**, 809 (1974).

7b. H. Oen, M. Pellegrini, D. Eilat, and C. R. Cantor, *Proc. Natl. Acad. Sci. U.S.A.* **70**, 2799 (1973).

7c. M. Pellegrini, H. Oen, and C. R. Cantor, *Proc. Natl. Acad. Sci. U.S.A.* **69**, 837 (1972).

7d. H. Oen, M. Pellegrini, and C. R. Cantor, *FEBS Lett.* **45**, 218 (1974).

8a. C. G. Kurland, *in* "Methods in Enzymology" (K. Moldave and L. Grossman, eds.), Vol. 20, p. 379. Academic Press, New York, 1971.

8b. P. Traub, S. Mizushima, C. V. Lowry, and M. Nomura, *in* "Methods in Enzymology" (K. Moldave and L. Grossman, eds.), Vol. 20, p. 391. Academic Press, New York, 1971.

8c. T. Staehelin, and D. R. Maglott, *in* "Methods in Enzymology" (K. Moldave and L. Grossman, eds.), Vol. 20, p. 449. Academic Press, New York, 1971.

9a. H. Noll, M. Noll, B. Hapke, and G. Van Dieijen, *Colloq. Ges. Biol. Chem.* **24**, 257 (1973).

† For recent reviews of ribosomal structure see 2a–2g.

9b. P. Debey, G. Hui Bon Hoa, P. Douzou, T. Godefroy-Colburn, M. Graffe, and M. Grunberg-Manago, *Biochemistry* **14**, 1553 (1975).

9c. O. P. Van Diggelen and L. Bosch, *Eur. J. Biochem.* **39**, 499 (1973).

9d. O. P. Van Diggelen, H. Oostrom, and L. Bosch, *Eur. J. Biochem.* **39**, 511 (1973).

9e. S. Ramagopal and A. R. Subramanian, *J. Mol. Biol.* **94**, 633 (1975).

9f. E. Deusser, *Mol. Gen. Genet.* **119**, 249 (1972).

9g. H. J. Weber, *Mol. Gen. Genet.* **119**, 233 (1972).

10. J. B. Breitmeyer and H. F. Noller, *J. Mol. Biol.* **101**, 297 (1976).

11a. O. Pongs and W. Messer, *J. Mol. Biol.* **101**, 171 (1976).

11b. O. Pongs, R. Bald, and V. A. Erdmann, *Proc. Natl. Acad. Sci. U.S.A.* **70**, 2229 (1973).

12a. N. Sonenberg, M. Wilchek, and A. Zamir, *Proc. Natl. Acad. Sci. U.S.A.* **70**, 1423 (1973).

12b. N. Sonenberg, M. Wilchek, and A. Zamir, *Biochem. Biophys. Res. Commun.* **59**, 663 (1974).

13a. N. Hsiung and C. R. Cantor, *Nucleic Acids Res.* **1**, 1753 (1974).

13b. N. Hsiung, S. A. Reines, and C. R. Cantor, *J. Mol. Biol.* **88**, 841 (1974).

13c. M. Sonenberg, M. Wilchek, and A. Zamir, *Proc. Natl. Acad. Sci. U.S.A.* **72**, 4332 (1975).

13d. A. S. Girshovich, E. S. Bochkareva, V. A. Kramarov, and Yu, A. Ovchinnikov, *FEBS Lett.* **42**, 213 (1974).

13e. A. S. Girshovich, E. S. Bochkareva, and V. A. Pozdnyakov, *Acta Biol. Med. Ger.* **33**, 639 (1974).

13f. A. S. Girshovich and E. S. Bochkareva, *Dokl. Akad. Nauk SSSR* **217**, 1201 (1974).

14a. I. Guinzburg and A. Zamir, *J. Mol. Biol.* **93**, 465 (1975).

14b. I. Guinzburg, R. Miskin, and A. Zamir, *J. Mol. Biol.* **79**, 481 (1973).

14c. R. Miskin and A. Zamir, *J. Mol. Biol.* **87**, 121 (1974).

14d. R. Miskin and A. Zamir, *J. Mol. Biol.* **87**, 135 (1974).

14e. M. Barbacid and D. Vazquez, *J. Mol. Biol.* **93**, 449 (1975).

14f. H. U. Petersen, A. Danchin, and M. Grunberg-Manago, *Biochemistry* **15**, 1362 (1976).

14g. D. J. Litman, C. C. Lee, and C. R. Cantor, *FEBS Lett.* **47**, 268 (1974).

15. E. Kaltschmidt and H. G. Wittmann, *Anal. Biochem.* **36**, 401 (1970).

16a. G. A. Howard and R. R. Traut, *FEBS Lett.* **29**, 177 (1973).

16b. D. Barritault, A. Expert-Bezançon, M. Milet, and D. H. Hayes, *Anal. Biochem.* **70**, 600 (1976).

16c. A. Lin, E. Collatz, and I. G. Wool, *Mol. Gen. Genet.* **144**, 1 (1976).

16d. U. C. Knopf, A. Sommer, J. Kenny, and R. R. Traut, *Mol. Biol. Rep.* **2**, 35 (1975).

17a. A. P. Czernilofsky, E. E. Collatz, G. Stöffler, and E. Kuechler, *Proc. Natl. Acad. Sci, U.S.A.* **71**, 230 (1974).

17b. K. Bauer, A. P. Czernilofsky, and E. Kuechler, *Biochim. Biophys. Acta* **395**, 146 (1974).

18a. O. Pongs, G. Stöffler, E. Lanka, *J. Mol. Biol.* **99**, 301 (1975).

18b. O. Pongs and E. Lanka, *Proc. Natl. Acad. Sci. U.S.A.* **72**, 1505 (1975).

18c. O. Pongs and E. Lanka, *Hoppe Syeler's Z. Physiol. Chem.* **356**, 449 (1975).

18d. O. Pongs, G. Stöffler, and R. W. Bald, *Nucleic Acids Res.* **3**, 1635 (1976).

18e. A. S. Girshovich, E. S. Bochkareva, and Y. A. Ovchinnikov, *Mol. Gen. Genet* **144**, 205 (1976).

19. A. Barta, E. Kuechler, C. Branlant, J. Sriwadada, A. Krol, and J. P. Ebel, *FEBS Lett.* **56**, 170 (1975).

20a. T. Nakomoto and P. Kolakofsky, *Proc. Natl. Acad. Sci. U.S.A.* **55**, 606 (1966).

20b. A.-L. Haenni and F. Chapeville, *Biochim. Biophys. Acta* **114**, 135 (1966).

21. M. Yukioka, T. Hatayama, and S. Morisawa, *Biochim. Biophys. Acta* **390**, 192 (1975).

22. D. Eilat, M. Pellegrini, H. Oen, Y. Lapidot, and C. R. Cantor, *J. Mol. Biol.* **88**, 831 (1974).

23a. E. S. Bochkareva, V. G. Budker, A. S. Girshovich, D. G. Knorre, and N. M. Teplova, *Mol. Biol. (Moscow)* **7**, 278 (1973).

23b. V. G. Budker, A. S. Girshovich, and L. M. Skobettsyna, *Dokl. Akad. Nauk SSSR* **207**, 215 (1972).

23c. E. S. Bochkareva, V. G. Budker, A. S. Girshovich, D. G. Knorre, and N. M. Teplova, *FEBS Lett.* **19**, 121 (1971).

24. L. Bispink and H. Matthaei, *FEBS Lett.* **37**, 291 (1973).

25a. G. W. Tischendorf, H. Zeichhardt, and G. Stöffler, *Proc. Natl. Acad. Sci. U.S.A.* **72**, 4820 (1975).

25b. J. A. Lake, *J. Mol. Biol.* **105**, 131 (1976).

25c. J. A. Lake, private communication (1976).

25d. J. A. Lake and L. Kahan, *J. Mol. Biol.* **99**, 631 (1975).

26. P. N. Gray and R. Monier, *Biochimie* **54**, 41 (1972).

27a. D. Barritault, A. Expert-Bezançon, and M. Milet, *C. R. Hebd. Seances Acad. Sci.* **281**, 1043 (1975).

27b. D. Barritault, A. Expert-Bezançon, M. Milet, and D. Hayes, *FEBS Lett.* **50**, 114 (1975).

28. P. Spierer and R. A. Zimmermann, *J. Mol. Biol.* **103**, 647 (1976).

29. L. Kahan and C. Kaltschmidt, *Biochemistry* **11**, 2691 (1972).

30. D. Eilat, M. Pellegrini, H. Oen, N. DeGroot, Y. Lapidot, and C. R. Cantor, *Nature* **250**, 514 (1974).

31a. R. Hauptmann, A. P. Czernilofsky, H. O. Voorma, G. Stöffler, and E. Kuechler, *Biochem. Biophys. Res. Commun.* **56**, 331 (1974).

31b. E. Collatz, E. Kuechler, G. Stöffler, and A. P. Czernilofsky, *FEBS Lett.* **63**, 283 (1976).

32c. A. P. Czernilofsky, G. Stöffler, and E. Kuechler, *Hoppe-Seyler's Z. Physiol. Chem.* **355**, 89 (1974).

32. M. Sopori, M. Pellegrini, P. Lengyel, and C. R. Cantor, *Biochemistry* **13**, 5432 (1974).

33a. I. Schwartz, E. Gordon, and J. Ofengand, *Biochemistry* **14**, 2907 (1975).

33b. I. Schwartz and J. Ofengand, *Proc. Natl. Acad. Sci. U.S.A.* **71**, 3951 (1974).

34a. M. L. Schenkman, D. C. Ward, and P. B. Moore, *Biochim. Biophys. Acta* **353**, 503 (1974).

34b. I. Fiser, P. Margaritella, and E. Kuechler, *FEBS Lett.* **52**, 281 (1975).

35a. I. Fiser, K. H. Scheit, G. Stöffler, and E. Kuechler, *Biochem. Bipophys. Res. Commun.* **60**, 1112 (1974).

35b. I. Fiser, K. H. Scheit, G. Stöffler, and E. Kuechler, *FEBS Lett.* **56**, 226 (1975).

36. O. Pongs, E. Lanka, and R. Bald, *Fed. Eur. Biochem. Soc. Meet., 10th, 1975* Abstract No. 447 (1975).

37a. O. Pongs and E. Rossner, *Hoppe Seyler's Z. Physiol. Chem.* **356**, 1297 (1975).

37b. O. Pongs and E. Rossner, *Nucleic Acids Res.* **3**, 1625 (1976).

38. V. G. Budker, A. S. Girshovich, N. I. Grineva, G. G. Karpova, D. G. Knorre, and N. D. Kobets, *Dokl. Akad. Nauk SSSR* **211**, 725 (1973).

39a. R. Wagner and H. G. Gassen, *Biochem. Biophys. Res. Commun.* **65**, 519 (1975).

39b. R. Luhrmann, H. G. Gassen, and G. Stöffler, *Eur. J. Biochem.* **66**, 1 (1976).

40. A. M. Frischauf and K. H. Scheit, *Biochem. Biophys. Res. Commun.* **53**, 1227 (1973).
41. A. J. Varghese, *in* "Aging, Carcinogenesis and Radiation Biology. The Role of Nucleic Acid Addition Reactions" (K. C. Smith, ed.), p. 207. Plenum, New York, 1976.
42a. B. S. Cooperman, E. N. Jaynes, D. J. Brunswick, and M. A. Luddy, *Proc. Natl. Acad. Sci. U.S.A.* **72**, 2974 (1975).
42b. O. Pongs, R. Bald, T. Wagner, and V. A. Erdmann, *FEBS Lett.* **35**, 137 (1973).
42c. P. Greenwell, R. J. Harris, and R. H. Symons, *Eur. J. Biochem.* **49**, 539 (1974).
42d. R. J. Harris, P. Greenwell, and R. H. Symons, *Biochem. Biophys. Res. Commun.* **55**, 117 (1973).
43. O. Pongs and V. A. Erdmann, *FEBS Lett.* **37**, 47 (1973).
44a. A. S. Girshovich, V. A. Pozdnyakov, and Yu. A. Ovchinnikov, *Dokl. Akad. Nauk SSSR* **219**, 481 (1974).
44b. A. S. Girshovich, V. A. Pozdnyakov, and Yu. A. Ovchinnikov, *Eur. J. Biochem.* **69**, 321 (1976).
44c. A. S. Girshovich, T. V. Kurtskhalia, V. A. Pozdnyakov, and Yu. A. Ovchinnikov, *FEBS Lett.* **80**, 161 (1977).
45a. J. A. Maassen and W. Möller, *Proc. Natl. Acad. Sci. U.S.A.* **71**, 1277 (1974).
45b. J. A. Maassen and W. Möller, *Biochem. Biophys. Res. Commun.* **64**, 1175 (1975).
46a. R. C. Marsh, G. Chinali, and A. Parmegianni, *J. Biol. Chem.* **250**, 8344 (1975).
46b. N. Arai, K. I. Arai, and Y. Kaziro, *J. Biochem. (Tokyo)* **78**, 243 (1975).
47a. L. C. Lutter, H. Zeichhardt, C. G. Kurland, and G. Stöffler, *Mol. Gen. Genet.* **119**, 357 (1972).
47b. A. Sommer and R. Traut, *J. Mol. Biol.* **106**, 995 (1976).
47c. A. P. Czernilofsky, C. G. Kurland, and G. Stöffler, *FEBS Lett.* **58**, 281 (1975).
47d. R. Traut, private communication (1976).
48. J. R. Horne and V. A. Erdmann, *Mol. Gen. Genet.* **119**, 337 (1972).
49a. K. H. Nierhaus, *Prog. Nucleic Acid Res. Mol. Biol.* (in press).
49b. K. H. Nierhaus and V. Montejo, *Proc. Natl. Acad. Sci. U.S.A.* **70**, 1931 (1973).
50. J. H. Highland, E. Ochsner, J. Gordon, J. Bodley, R. Hasenbank, and G. Stöffler, *J. Biol. Chem.* **250**, 1141 (1975).
51. D. Nierhaus and K. H. Nierhaus, *Proc. Natl. Acad. Sci. U.S.A.* **70**, 2224 (1973).
52a. J. Van Duin, P. H. Van Knippenberg, M. Dieben, and C. G. Kurland, *Mol. Gen. Genet.* **116**, 181 (1972).
52b. L. L. Randall-Hazelbauer and C. G. Kurland, *Mol. Gen. Genet.* **115**, 234 (1972).
52c. C. Lelong, D. Gros, F. Gros, A. Bollen, R. Maschler, and G. Stöffler, *Proc. Natl. Acad. Sci. U.S.A.* **71**, 248 (1974).
52d. M. Nomura, S. Mizushima, M. Ozaki, P. Traub, and C. V. Lowry, *Cold Spring Harbor Symp. Quant. Biol.* **34**, 49 (1969).
52e. D. P. Rummel and H. F. Noller, *Nature (London), New Biol.* **245**, 72 (1973).
53. W. Held, B. Ballar, S. Mizushima, and M. Nomura *J. Biol. Chem.* **249**, 3103 (1974).
54. S. H. Kim, G. J. Quigley, F. L. Suddath, A. McPherson, D. Sneden, J. J. Kim, J. Weinzierl, and A. Rich, *Science* **179**, 285 (1973).
55. G. Stöffler, *in* "Ribosomes" (M. Nomura and A. Tissières, eds.), p. 615, and references therein. Cold Spring Harbor Lab., Cold Spring Harbor, New York, 1974.
56. P. G. Grant, E. N. Jaynes, Jr., and B. S. Cooperman, in preparation.

4

Mechanisms of Electron Transfer by High-Potential *c*-Type Cytochromes

Michael A. Cusanovich

INTRODUCTION

Cytochrome *c* has been the subject of a number of reviews [1–4] in recent years as considerable structural and chemical information has become available. Myriad approaches have been employed in attempts to solve the question of the mechanisms of action of cytochrome *c*, and a variety of mechanisms have been proposed [5–10]. This chapter focuses on kinetic studies of the mechanism of electron transfer by cytochrome *c* in order to illustrate the characteristics of this reaction that must be taken into account in any chemical mechanism. As will become clear, the mechanism cannot be written at this time, but considerable information is now available which bears on the problem, the solution of which will ultimately provide conceptual insight into a number of important related problems, particularly in the realm of protein–protein interactions.

Before proceeding to a discussion of the kinetic data, it is useful to summarize briefly the major characteristics of cytochrome *c* and define the philosophy of this chapter. Based on chemical, physical, and functional properties, nine classes of *c*-type cytochromes can be defined [11]. However, examples of only two of these classes have been characterized extensively

TABLE 1

**Properties of Horse Heart Cytochrome c and *Rhodospirillum rubrum*
Cytrochrome c_2**

Property	Cytochrome c	Cytochrome c_2
$E_{m,7}$ (mV)	260	320
Isoelectric point	10.1	6.3
Function	Mitrochondrial electron transport	Photosynthetic electron transfer
Electron donor	Complex III (cytochrome c_1)	Cytochrome b
Electron acceptor	Cytochrome oxidase (cytochrome a)	P-870 (active center bacteriochlorophyll)
No. of amino acid residues	104	112

in terms of chemical and structural information. These are respiratory cytochrome c and bacterial cytochrome c_2. The amino acid sequences of over 60 examples of respiratory cytochrome c have been determined, providing a group of functionally identical proteins with different primary structures [3]. Among these cytochromes c, 28 amino acid residues are absolutely invariant [3]. Table 1 summarizes the properties of a typical respiratory cytochrome c obtained from horse heart. In the discussion to follow, the term "cytochrome c" refers to horse cytochrome c unless otherwise noted. Examples of cytochrome c_2 have been isolated from most purple photosynthetic bacteria. The amino acid sequences of cytochrome c_2 from eight different sources have been determined [3,12] with substantial homology detectable among the various examples [3]. As shown in Table 1, a typical example has some properties distinct from cytochrome c. Hereafter, the term "cytochrome c_2" refers to the protein isolated from *Rhodospirillum rubrum* unless otherwise noted. In addition to the reported sequence similarities between cytochrome c and cytochrome c_2 [13], a comparison of the three-dimensional structure of cytochrome c_2 [14] and cytochrome c [15] dramatically demonstrates the extensive structural homology between the two classes of cytochromes. The structural similarities and differences between the two proteins have been discussed [6] and are not repeated in detail here. However, in terms of general features, both cytochromes have a heme chromophore (protoheme IX) buried in the interior of the molecule, with methionine and histidine serving as out-of-plane ligands. It appears that only pyrrole ring 2 and the ring 2 to ring 3 edge are solvent accessible [3]. In both cytochromes, the bulk of the aromatic amino acid side chains are found packed around the heme, and a large fraction of the lysine side chains are near the solvent-accessible heme edge [6]. For discussion purposes, the "front" of the mole-

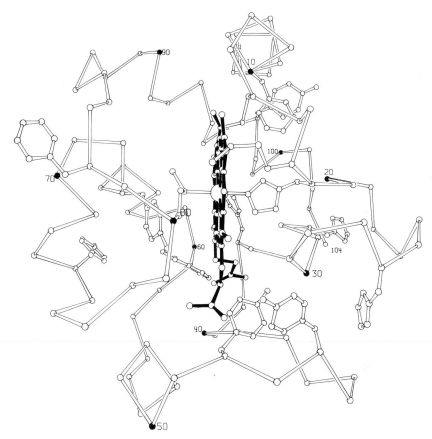

Fig. 1 Structure of horse heart cytochrome c as viewed facing the front heme edge. Residue 82 (phenylalanine) should be rotated into the unoccupied space at the upper left of the heme.

cule is defined as the side at which pyrrole rings 2 and 3 are found, with the heme propionic acid side chains at the "bottom." Figure 1 depicts the structure of ferricytochrome c as approached from the front heme edge; Fig. 2 presents the same view for ferricytochrome c_2. These figures are presented to orient the reader and are referred to throughout this discussion. The structure in Fig. 1 is not entirely correct, as more recent studies [3] indicate that Phe-82 is actually rotated into the heme crevice in a postion similar to that for Phe-93 in the cytochrome c_2 structure (Fig. 2). This point should be kept in mind, as mechanisms for cytochrome c involving movement of Phe-82 on reduction are no longer viable.

Table 2 summarizes properties of several examples of cytochrome c_2 and

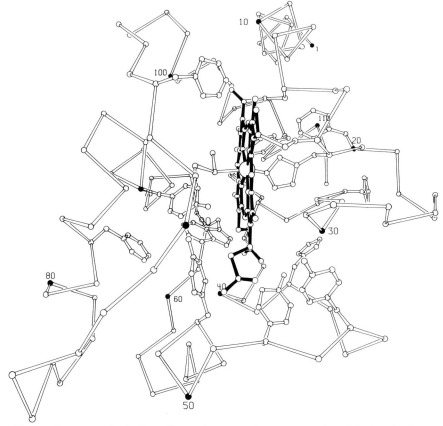

Fig. 2 Structure of *Rhodospirillum rubrum* cytochrome c_2 as viewed facing the front heme edge.

reveals the substantial variation of these properties within this class of cytochromes. This is in sharp contrast to the respiratory cytochromes c, which show little variation in isoelectric point and oxidation–reduction potential among the numerous examples studied. Considering the sequence and structural homology between cytochrome c and cytochrome c_2, it seems likely that differences and similarities in reactivity will be explainable in terms of specific amino acid substitutions.

In terms of information on the mechanism of electron transfer derived from the kinetics of oxidation and reduction, several general comments can be made. Relative to simple inorganic oxidation–reduction reactions, three mechanistic possibilities are likely: inner-sphere electron transfer, outer-sphere electron transfer, and electron tunneling. Inner-sphere electron

TABLE 2

Properties of Cytochrome c_2 from Various Sources

Source	$P_I{}^a$	$E_{m,7}$	No. of amino acid residues [b]
Rhodospirillum rubrum	6.3	320^a	112
Rhodopseudomonas capsulata	Acidic	368^c	117
Rhodopseudomonas spheroides	5.5	351^c	123
Rhodopseudomonas palustris (strain 2.1.37)	9.7	377^d	118
Rhodomicrobium vanniellii	7.9	356^c	105

[a] Meyer [11].
[b] Ambler and Meyer [12] and Dus *et al.* [13].
[c] Pettigrew *et al.* [16].
[d] M. A. Cusanovich and C. B. Post, unpublished observations.

transfer is operative when an oxidation state change occurs by a path involving the mutual sharing of a "bridging" ligand in the coordination sphere of both reactants. This is in contrast to outer-sphere electron transfer, in which the bridging ligand is not shared. Electron tunneling can take on a variety of forms [9,10], but for the purposes of this discussion it is defined as the reaction of cytochrome *c* or c_2 with an oxidant or reductant that does not involve direct interaction of the chromophores. In more descriptive terms, electron transfer can take place by adjacent attack, that is, direct contact with the heme iron, or by remote attack, at some site remote from the heme iron [17]. Remote electron transfer as defined here can be at the exposed heme edge or any other site on the surface of the molecule. Adjacent electron transfer is required for inner-sphere electron transfer, while remote attack can be either outer-sphere electron transfer or tunneling. With each possible mechanism, it can be expected that the protein moiety will significantly influence the electron transfer with regard to electrostatics, steric restrictions, and nonpolar interactions. In addition to providing the chemical environment of the heme necessary for optimum activity, the protein moiety also establishes the chemistry encountered by an approaching reactant, hence inducing specificity on the reaction. This last point is important, as the large number of possible reactants available *in vivo* imputes a requirement for specificity so that optimum rates of electron transfer will be obtained via the appropriate physiological carriers.

In the discussion that follows, studies on the interaction of a variety of oxidants and reductants with cytochrome *c* and cytochrome c_2 are reviewed and the implications of these studies examined. Three general types of reactants have been used: nonphysiological oxidants and reductants, such as the iron hexacyanides; biological oxidants and reductants (molecules of biological origin that do not naturally interact with cytochrome c); and physiological oxidants and reductants, the natural electron donors and acceptors. All three types of reactants are discussed, although the studies employing the latter two groups are less defined in quantitative terms. The reaction of cytochrome c with reactants produced by pulsed radiolysis has been extensively studied [18–22] but is not discussed here due to the exceptional reactivities of these reactants and lack of reversibility of the reaction. In addition, studies at pH values outside of the physiological range (less than 5, greater than 8) have been omitted since the molecular species that exist at these pH values are not well defined structurally.

NONPHYSIOLOGICAL OXIDANTS AND REDUCTANTS

The iron hexacyanides have been the most extensively studied of the non-physiological oxidants and reductants. Sutin and Christman [23] and subsequently Havsteen [24] established that the iron hexacyanides readily interact with cytochrome c. These studies were interpreted in terms of a second-order reaction as given in Eq. (1).

$$\text{cyt(III)} + \text{Fe(CN)}_6{}^{4-} \underset{k_{21}}{\overset{k_{12}}{\rightleftharpoons}} \text{cyt(II)} + \text{Fe(CN)}_6{}^{3-} \tag{1}$$

In 1973, Stellwagen and Shulman followed the iron hexacyanide reaction with cytochrome c using nuclear magnetic resonance line widths [25]. These data could be interpreted only in terms of the complex mechanism given in Eq. (2), where C_1 is a ferricytochrome c–ferrocyanide complex and

$$\text{cyt(III)} + \text{Fe(CN)}_6{}^{4-} \underset{k_{21}}{\overset{k_{12}}{\rightleftharpoons}} C_1 \underset{k_{32}}{\overset{k_{23}}{\rightleftharpoons}} C_2 \underset{k_{43}}{\overset{k_{34}}{\rightleftharpoons}} \text{cyt(II)} + \text{Fe(CN)}_6{}^{3-} \tag{2}$$

C_2 is a ferrocytochrome c–ferricyanide complex. The nmr data gave values of 208 sec^{-1} for k_{23}, 400 M^{-1} for k_{12}/k_{21}, and 2.1×10^4 sec^{-1} for k_{32}. Recently, a stopped-flow study of ferrocyanide reduction of cytochrome c [26] yielded very similar results ($k_{23} = 132$ sec^{-1}; $k_{12}/k_{21} = 2400$ M^{-1}). Considering the different approaches and experimental conditions, the agreement is excellent and confirms the suggestion that the iron hexacyanides and cytochrome c form complexes during their reaction. The effect of ionic strength on the iron hexacyanide–cytochrome c reaction was analyzed in terms of Debye–Hückel theory for ionic reactants [27] and yielded an apparent charge at the site of electron transfer of $+1.3$ for both ferricyanide oxidation [28] and ferrocyanide reduction [26]. As might be expected, neither k_{23} nor k_{21} are significantly affected by changes in ionic strength [26]. The discrepancy between the net protein charge of $+7$ to $+8$ on cytochrome c [1] and the apparent charge $(+1.3)$ as discussed above suggests that the site of electron transfer does not involve the net protein charge but that a specific site of electron transfer exists.

All six rate constants given by Eq. (2) can be deduced from the measured rate constants and the overall equilibrium constant and are summarized in Table 3. Stellwagen and Cass [29] investigated the nature of the interaction of the iron hexacyanides and cytochrome c from the equilibrium standpoint. These studies indicated two iron hexacyanide binding sites on ferricytochrome c ($K_{app} \sim 300$ and ~ 1000 M^{-1}) but multiple sites on ferrocytochrome c [29]. Based on correlation of the loss of affinity of cytochrome c for iron hexacyanides with the loss of the 695 nm absorption band at elevated pH values, it was suggested that Lys-79 was involved in the binding of at least one of the iron hexacyanides [29]. However, it was recently pointed out that, if this were the case, the pK value of the lysine residue bound to ferricyanide would be altered by at least 1 pK unit, and this was not observed [30]. Thus, it is not possible to pinpoint the iron hexacyanide binding sites at this time.

TABLE 3

Rate Constants for the Interaction of the
Iron Hexacyanides with Cytochrome c and
Cytochrome c_2 at pH 7.0

Rate constant[a]	Cytochrome c	Cytochrome c_2[b]
k_{12} (M^{-1} sec^{-1})	4×10^{4}[c]	6.6×10^4
k_{21} (sec^{-1})	7[c]	Not measured
k_{23} (sec^{-1})	132[c]	250
k_{32} (sec^{-1})	2.1×10^{4}[d]	350
k_{34} (sec^{-1})	557[e]	Not measured
k_{43} (M^{-1} sec^{-1})	8.7×10^{6}[f]	2.2×10^6

[a] Rate constant nomenclature from Eq. (2).
[b] $I = 0.135$, 20°C [31].
[c] $I = 0.18$, 20°C [26].
[d] $I = 1.0$, 25°C [25].
[e] Calculated from $K = k_{12}/k_{21} \times k_{23}/k_{32} \times k_{34}/k_{43}$ using the rate constants in the table and a value of 2.3×10^{-3} for K, which was obtained from $E_{m,7} = 260$ mV for cytochrome c and $E_{m,7} = 418$ mV for the iron hexacyanides.
[f] $I = 0.18$, 22°C [32].

The reaction of iron hexacyanides with cytochrome c_2 has also been studied in detail [31]. As with cytochrome c, the reaction of iron hexacyanides with cytochrome c_2 was found to proceed via complex formation and showed ionic strength dependence consistent with a positive charge at the site of electron transfer. For ferricytochrome c_2 the apparent charge was +1.3, and for ferrocytochrome c_2 the apparent charge was +0.7 [31]. The individual rate constants are summarized in Table 3 and are of the same order of magnitude as those found for cytochrome c. It is significant that the charge at the site of electron transport was positive, as cytochrome c_2 is negatively charged at pH 7.0. These data unequivocally point to a specific site of electron transfer not involving the net protein charge.

Over the pH range 7–9, no change in the kinetic parameters k_{12} and k_{43} have been noted for cytochrome c [32]. In contrast, the oxidation–reduction potential of cytochrome c_2 is pH dependent, and k_{12} and k_{43} show pH-induced changes consistent with the behavior of the midpoint potential [31]. These data have been interpreted in terms of contributions of at least three pK values in each oxidation state, implicating the involvement of at least three different amino acid side chains [31], which implies participation of a significant portion of the protein moiety.

Ferri–ferrocyanide is substitution inert on oxidation and reduction and hence cannot undergo inner-sphere electron transfer. Studies on ligand binding by ferricytochrome c established that ionic ligands (imidazole, CN^-, N_3^-) can displace the out-of-plane heme ligand methionine [33–35]. This ligand displacement reaction has been interpreted in terms of penetration into the pocket to the left of the heme crevice (Fig. 1). Kinetic studies on the ligand binding process have established a rate constant of 30–60 sec^{-1} for crevice opening to permit ligand binding [35]. As the limiting rate for ferrocyanide reduction of cytochrome c is greater than 100 sec^{-1} (Table 3), adjacent attack is not feasible. Finally, the formation of iron hexacyanide complexes as described by Eq. (2) requires that oxidation and reduction take place at the same site. With the above discussion in mind, it can be concluded that electron transfer takes place at a site remote from the heme iron. Moreover, the relatively rapid electron transfer rates, the presence of a positive charge at the site of electron transfer, and the requirement for contact for outer-sphere electron transfer all point to the exposed heme edge as the likely site of electron transport. Although this is not irrefutably established, it has been proposed by a number of authors [6,26,31,36,37] and will serve as a focus for further discussion.

For outer-sphere electron transfer, application of relative Marcus theory [38] should yield the observed rate constant (k_{12} or k_{43}) as given by Eq. (3),

$$k_{\mathrm{Obs}} = (k_{11}k_{22}K_{12})^{1/2} \qquad (3)$$

where k_{11} is the cytochrome self-exchange rate constant $(0.2\text{–}1 \times 10^3 \text{ M}^{-1}$ sec^{-1}, pH 7.0, $I = 0.1$, 25°C [39,40]), k_{22} is the iron hexacyanide self-exchange rate constant $(9.6 \times 10^3 \text{ M}^{-1} \text{sec}^{-1}$ [41]), and K_{12} is the equilibrium constant (435, see footnote to Table 3) for the reaction. This calculation yields $2.9\text{–}6.6 \times 10^4 \text{ M}^{-1} \text{sec}^{-1}$ for ferricyanide oxidation and $0.6\text{–}1.4 \times 10^2$ $\text{M}^{-1} \text{sec}^{-1}$ for ferrocyanide reduction [42]. The calculated rate constants tend not to support outer-sphere electron transfer, as agreement with the measured constants (Table 3) is poor. However, electrostatic interactions or steric rectrictions can be invoked to explain the lack of agreement between the calculated and measured rates, leaving the question of outer-sphere electron transfer unanswered for the iron hexacyanides.

It is well established that ferricytochrome c is capable of tightly binding a number of anions [43–45], and it has been shown that anion binding can influence the interaction of cytochrome c with its natural electron donors and acceptors [45]. It is difficult to separate the effects of specific anion binding from those of a general ionic strength effect when observing rates of oxidation and reduction. However, recent work [46] using the nonbinding buffer Tris–cacodylate [43] demonstrates dramatic effects of anion binding on the kinetics of ferrocyanide reduction. In these experiments, constant ionic strength was obtained by adjusting the Tris–cacodylate concentration, and the influence of phosphate, ATP, and CTP on k_{12} and k_{23} was investigated. From the data presented in Table 4, it can be seen that only ATP and CTP have a substantial effect at the concentrations used, and this effect was manifest on both k_{12} and k_{23}. At higher concentrations of phosphate, a substantial effect is noted; for example, in 0.06 M Tris–cacodylate with $I = 0.048$, $k_{12} = 1 \times 10^6 \text{ M}^{-1} \text{sec}^{-1}$ and $k_{23} = 167 \text{ sec}^{-1}$ [46], whereas in 0.02 M potassium phosphate $(I = 0.035)$, $k_{12} \sim 2 \times 10^5 \text{ M}^{-1} \text{sec}^{-1}$ and $k_{23} = 133$ sec^{-1} [26]. Thus, phosphate has a specific effect but is manifest only at much higher concentrations than found for the nucleotide triphosphates. This

TABLE 4

Effect of Specific Anions on the Reduction of Cytochrome c by Ferrocyanide[a]

Anion	Concentration (mM)	I	k_{12} (M^{-1} sec^{-1})	k_{23} (sec^{-1})
None	—	0.0492	6.9×10^5	161
PO_4^{2-}	1	0.0492	5.3×10^5	137
ATP	1	0.0492	2.2×10^5	125
CTP	1	0.0492	1.8×10^5	125

[a] Tris–cacodylate buffer, pH 7.0, 20°C.

probably results from different affinities for the different anions, but no experimental evidence is available to confirm this point. The observation that anion binding affects both k_{12} and k_{23} argues that more is involved than simple competition for the binding site by the ferrocyanide and the anion, as this should affect only k_{12}. This latter point suggests that specific anion binding can affect the oxidation–reduction potential and/or the conformation of the protein at the site of electron transfer. Indeed, it has been demonstrated that varying the concentration of a nonbinding buffer has a much different effect on the oxidation–reduction potential of cytochrome c than does varying the concentration of binding buffers [44]. The binding of ATP by cytochrome c has been demonstrated by nmr studies [47], and it has been suggested that His-26 is involved in the binding process [48]. If this is the case, ATP would be bound to the right of the heme crevice and in such a position should not interact with ferrocyanide approaching the exposed heme edge.

As various examples of cytochrome c have the same oxidation–reduction potential [49], a study of the relative rates of reduction of cytochrome c by ferrocyanide should show differences due primarily to amino acid substitutions at the site of electron transfer (assuming the same three-dimensional structure). Table 5 presents a comparison of the reduction of cytochrome c, at identical ionic strengths, from horse, cow, chicken, tuna, and the yeast *Candida krusei* [50]. Tuna, chicken, and bovine cytochrome c are identical but differ from horse and *C. krusei* cytochrome c, which are different from each other. Table 6 summarizes the amino acid sequences of the various cytochromes, indicating the differences from horse cytochrome c (blanks indicate the same sequence as horse cytochrome c). The situation is more complicated with *C. krusei*, in which an additional 25 positions (beyond those in Table 6) contain side chains different from those in horse cytochrome c

TABLE 5

Kinetics of Reduction of Cytochrome c from Different Sources by Ferrocyanide[a]

Source	k_{12} (M^{-1} sec^{-1})	k_{23} (sec^{-1})
Horse heart	3.2×10^5	139
Bovine	2.4×10^5	61
Chicken	2.4×10^5	83
Tuna	2.4×10^5	62
Candida krusei	6×10^5	304

[a] Buffer 0.02 M Tris–cacodylate/20.1 N NaCl, pH 7.0, 20°C.

TABLE 6

Sequence Differences among Various Examples of Cytochrome c

Source	Sequence position																
	4	9	28	33	44	46	47	54	58	60	62	89	92	95	100	103	104
	Amino acid residue																
Horse	E	I	T	H	P	F	T	N	T	K	E	T	E	I	K	N	E
Bovine							S			G		G					
Chicken			E				S			G	D	S	V		D	S	K
Tuna	A	T	V	N	E	O	S	S	V	N	D	G	Q	V	S	S	E
Candida krusei	K	L	V		E	O	S	R	E	A	D	K	N	N	E	K	

[3]. Only three of the positions that differ from equivalent positions in horse cytochrome c are identical for tuna, bovine, and chicken cytochrome c. Thus, it can be proposed that at least one of these residues is responsible for the observed kinetic differences. Position 47, in which a threonine is replaced by a serine residue, is at the lower front heme edge, and from a structural standpoint could be in an important position if electron transport takes place at the heme edge. However, the substitution of a serine for a threonine would not be expected to substantially distort hydrogen bonding or affect the net charge. Position 60 has a lysine residue in horse cytochrome c replaced by uncharged residues in bovine, tuna, and chicken cytochrome c. Clearly the loss of a positive charge could explain the reduced value of k_{12}; however, position 60 is on the back side of the molecule, well removed from the heme edge (near Trp-59) and would not be consistent with electron transfer at the heme edge. Interestingly, from the reported structure [15], Lys-60 is near Lys-39 and Arg-38, possibly providing a cluster of positive charge. Position 89 also shows substitution, although this position is at the top back left of the molecule, and the replacement of threonine by serine or glycine would not be expected to be significant. To summarize, two possibilities appear likely to explain the differences between horse, bovine, chicken, and tuna cytochrome c. (1) Electron transport does not take place at the heme edge but at some other position, for example, in the vicinity of Lys-60. (2) The amino acid substitutions (for example, Gly for Lys) can induce conformational differences or perturbations of hydrogen-bonding networks important for the function of the molecule. In the latter case, the perturbation would not necessarily have to be at the site of electron transfer.

The differences between the rate of reduction of horse and Candida cytochrome c present some interesting problems. In addition to the data presented here (Table 5), it has been found [42] that ferricyanide oxidizes C. krusei cytochrome c with a rate constant twice that found for horse cytochrome c. Further, the rate constant for oxidation of C. *krusei* cytochrome c by tris(1,10-phenanthroline) cobalt [51] has been determined and is 1.8 times that found for horse cytochrome c. As this reagent is positively charged $(+3)$ and the iron hexacyanides are negative, from electrostatic considerations the oxidation of C. *krusei* cytochrome c should be considerably slower with one and faster with the other of these reagents than horse cytochrome c. Finally, the self-exchange rate for C. *krusei* cytochrome c has been reported to be only 10% of that found for horse cytochrome c [39], thus predicting that the reaction of iron hexacyanide and C. *krusei* cytochrome should be slower than the analogous reaction of horse cytochrome c by a factor of 3.3 [Eq. (3)]. Although the discrepancies noted above cannot be explained at this time, they do suggest the participation of factors not accounted for by Marcus theory and direct heme edge attack but will most likely require structural information for solution.

The cytochromes c_2 from different sources show considerably more amino acid substitutions in their sequences as compared to the mammalian cytochromes [3]. Moreover, the wide range of oxidation–reduction potentials and isoelectric points (Table 3) suggest that the sequence variations may affect the kinetics of electron transport more extensively than observed among various cytochromes c. Table 7 presents k_{12} at infinite dilution for five examples of cytochrome c_2 [52]. Also presented are the values of k_{23}, which, with the

TABLE 7

Kinetics of Reduction of Cytochrome c_2 from Different Sources by Ferrocyanide[a]

Source	k_{12} $(M^{-1} sec^{-1})$	k_{23} (sec^{-1})	Apparent charge
Rhodospirillum rubrum	5×10^6	250	$+1.3$
Rhodopseudomonas capsulata	1.3×10^6	500	$+0.7$
Rhodopseudomonas spheroides	4.2×10^7	270	$+1.6$
Rhodopseudomonas palustris (strain 2.1.37)	1.3×10^6	$90 -200$[b]	$+0.7$
Rhodomicrobium vanniellii	4.0×10^5	32.4	$+1.0$

[a] Experiments were conducted in 0.02 M potassium phosphate supplemented with various amounts of NaCl, pH 7.0, 20°C. The reported values of k_{12} were obtained from Debye–Hückel plots [27] and represent k_{12} at infinite dilution. The apparent charge was calculated from the slope of the Debye–Hückel plots.

[b] Rate constant k_{23} was dependent on ionic strength, increasing with decreasing ionic strength; in the remainder of the cases, k_{23} was independent of ionic strength.

exception of cytochrome c_2 from *Rhodopseudomonas palustris*, are independent of ionic strength. It was found that k_{12} and k_{23} varied over a wide range with no apparent correlation between the rate constants and either the isoelectric points or the oxidation–reduction potentials of the various cytochromes c_2. In all cases a positive site of electron transfer was indicated and, with the exception of cytochrome c_2 from *Rhodomicrobium vanniellii*, the apparent charge correlated with the value of k_{12}. It should be pointed out that the absolute values of the apparent charge are of dubious value due to the uncertainties in applying Debye–Hückel theory to polyvalent ions [27]; however, in a relative sense they probably represent a reasonable parameter for comparative purposes. Electron transfer at a site distant from the heme iron is the only viable choice of mechanism for the cytochrome c_2–iron hexacyanide reaction for the same reasons put forth earlier for cytochrome c. Further, attack at the heme edge has been proposed for cytochrome c_2 [6,31] and certainly is consistent with accessibility and the extensive clustering of positively charged residues about the heme crevice [6]. With this in mind, it is useful to attempt to analyze the sequence substitutions in the vicinity of the heme edge among the examples of cytochrome c_2 discussed here. From the reported structure of *Rhodospirillum ruburum* cytochrome c_2 [14], the amino acid side chains are compared for the various cytochromes c_2 in Table 8. It is interesting to note the large number of charged residues and residues potentially capable of hydrogen bonding in this structural region (Table 9). Although not enough quantitative data are available to predict the charge distribution and hydrogen bonding that would be encountered by an approaching reactant, a qualitative assessment can be made. By way of

TABLE 8

Amino Acids in the Vicinity of the Heme Edge of Various Cytochromes c_2

Source	Sequence position														
	16	27	28	29	46	47	48	49	50	51	52	88	89	90	92
	Amino acid residue														
Rhodospirillum rubrum	A	K	V	G	Y	A	Y	S	D	S	Y	K	S	K	T
Rhodopseudomonas capsulata	T		T		F	K		K			I		T	G	A
Rhodopseudomonas spheroides	T		T		F	K		G	E	G	M	V	G		
Rhodopseudomonas palustris (strain 2.1.37)	T	[a]			F	T		P	L	N			T		
Rhodomicrobium vanniellii	I	G			F	N			A	M	[a]		T		V

[a] Deletion.

TABLE 9

Chemical Character of Amino Acid Residues in the Vicinity of
the Heme Crevice of Cytochrome c_2

Source	Total residues	Basic	Acidic[a]	Potential hydrogen-bonding residue
Rhodospirillum rubrum	15	3	2	7
Rhodopseudomonas capsulata	15	4	2	5
Rhodopseudomonas spheroides	15	3	2	4
Rhodopseudomonas palustris (strain 2.1.37)	14	2	1	7
Rhodomicrobium vanniellii	14	1	2	4

[a] Included in this column is the propionic side chain at the lower front edge of the heme crevice and present in all of the cytochromes c_2.

example, a comparison of *Rhodospirillum rubrum* and *Rhodopseudomonas capsulata* cytochrome c_2 is shown in Fig. 3. *Rhodopseudomonas capsulata* is particularly interesting, as position 90 (analogous to position 79 in cytochrome c) is occupied by glycine. Since this position is always occupied by a lysine residue in other cytochromes c and c_2 [3] and is the lysine nearest the heme edge [6] it can be predicted to play a key role if electron transport takes place at the heme edge. Figure 3 depicts the relative positions of a number of amino acid side chains at the heme edge as derived from available co-ordinates [53] for *Rhodospirillum rubrum* cytochrome c_2. The *Rhodopseudomonas capsulata* positions were assigned from the analogous positions in the *Rhodospirillum rubrum* cytochrome c_2. The solid line connecting residues indicates a probable hydrogen bond [6,14]; crosshatching to the right indicates a positive charge and crosshatching to the left a negative charge. Qualitatively, it appears that replacement of Lys-90 by glycine in *Rhodopseudomonas capsulata* is compensated for by the substitution of lysine at positions 47 and 49. Further, distortion of the hydrogen bond network about the heme crevice is required, although the significance of this cannot be assessed. An exercise similar to that described above can be applied to other cytochromes c_2 discussed here to relate the structure in the vicinity of the heme edge to the observed values of k_{12}. However, such an approach is speculative and will be deferred until more structural and kinetic information is available.

The foregoing remains to be demonstrated experimentally. However, it is useful in exploring possible explanations for the observed kinetics and can serve to stimulate further experiments. Clearly the data reviewed concerning the reaction of cytochrome c and c_2 with the iron hexacyanides point to a

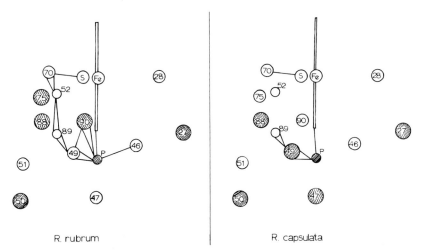

R. rubrum   R. capsulata

Fig. 3 Structure of *Rhodospirillum rubrum* and *Rhodopseudomonas capsulata* in the vicinity of the heme edge. Solid lines represent probable hydrogen bonds [6]. Circles crosshatched to the right represent ε-amino groups of lysine residues; crosshatching to the left indicates a carboxyl oxygen. The methionine sulfur is denoted by S; the heme iron by Fe.

significant participation of the protein moiety and suggest a relationship between the primary structure of the protein and the measured kinetics of oxidation and reduction.

The interactions of a number of nonphysiological oxidants and reductants with cytochrome *c*, in addition to the iron hexacyanides, have been investigated, and the results of these studies are summarized in Table 10 [17,26,36–38,42,51,54,55]. The reactants in Table 10 generally yield results similar to those discussed with the iron hexacyanides in that the reactions are strongly dependent on ionic strength, are consistent with a positive site of electron transfer on cytochrome *c*, and demonstrate an apparent charge less than that of the net protein charge. Application of relative Marcus theory in experiments employing $Ru(NH_3)_6^{2+}$ [36] and tris(1,10-phenanthroline) cobalt(III)$^{3+}$ [51] yields calculated values of the rate constants in excellent agreement with the measured values and supports the proposal of outer-sphere electron transfer [26,42,51]. No evidence of any intermediate complex formation was obtained in any of the reactions given in Table 10. This point does not preclude complex formation; however, it does mean that, if complexes are formed, they must have lifetimes of less than 5–10 msec, the mixing time required by the methods employed. It is interesting that driving force (difference in oxidation–reduction potential) has no notable influence on the rates.

TABLE 10

Kinetic Parameters Describing the Reaction of a Variety of Nonphysiological Oxidants and Reductants with Cytochrome c

Reactant	$E_{m,7}$ (mV)	k_{12} (M⁻¹ sec⁻¹)	I	pH	Temp (°C)	Calculated[a] rate constant (M⁻¹ sec⁻¹)	Apparent charge	References
Reductants								
Ru(NH₃)₆²⁺	78	3.8 × 10⁴	0.10	7.0	25	3.6 × 10⁴	+	36
SO₂⁻	−740	5.8 × 10⁷	0.10	7.0	20	—	+1.3	17,26
S₂O₄²⁻	180	9.0 × 10⁴	0.088	7.0	20	—	+2.4	17,26
Ru(NH₃)₅benzimidazole²⁺	150	5.8 × 10⁴	0.10	6.9	25	?	+	42
Sodium ascorbate	58	34	0.035	7.0	20	—	+	54,55
Fe(EDTA)²⁻	117	2.5 × 10⁴	0.10	7.0	25	?	+1.7	37,54
Oxidants								
Tris(1,10-phenanthroline) cobalt(III)³⁺	420	1.5 × 10³	0.10	7.2	25	1.2–2.7 × 10³	+0.4	51

[a] Calculated from relative Marcus theory [38].

Some disagreement exists concerning the reaction of dithionite with cytochrome c because of the complexity of the reaction. Lambeth and Palmer [56] proposed the mechanism given by Eqs. (4)–(6), and this was the form used to obtain the data presented in Table 10. Here, k_1 and k_2 are the rate

$$S_2O_4^{2-} \underset{}{\overset{K_{eq}}{\rightleftharpoons}} 2SO_2^- \qquad (4)$$

$$S_2O_4^{2-} + cyt(III) \overset{k_1}{\rightleftharpoons} products \qquad (5)$$

$$SO_2^- + cyt(III) \overset{k_2}{\rightleftharpoons} products \qquad (6)$$

constants for $S_2O_4^{2-}$ and SO_2^- reduction, respectively. This mechanism derives from the fact that $S_2O_4^{2-}$ and SO_2^- are in rapid equilibrium; hence, formation of SO_2^- is not rate limiting. Creutz and Sutin [17], on the other hand, have used the mechanism given by Eqs. (7)–(9), which yields a rate law of similar form to that derived from Eqs. (4)–(6). Here, cyt*(III) represents

$$cyt(III) \underset{k_{-1}}{\overset{k_1}{\rightleftharpoons}} cyt^*(III) \qquad (7)$$

$$cyt^*(III) + S_2O_4^{2-} \overset{k_2}{\rightleftharpoons} products \qquad (8)$$

$$cyt^*(III) + S_2O_4^{2-} \overset{k_3}{\rightleftharpoons} products \qquad (9)$$

ferricytochrome c with an open heme crevice, and cyt(III) represents cytochrome c with a closed crevice (Fig. 1). Hence, k_2 describes adjacent attack and k_3 remote attack (parallel pathways) [17]. Experimentally, it is difficult to decide between the two possibilities; however, SO_2^- is a vigorous reducing agent in a variety of systems [56] including cyanoferricytochrome c [42], where $S_2O_4^{2-}$ is inactive. Thus, in our view it is likely that SO_2^- reduces cytochrome c, and the mechanism given by Eqs. (4)–(6) is operative.

To test the importance of the integrity of the protein moiety of cytochrome c, the reactions of sodium dithionite with a variety of denatured forms of cytochrome c have been investigated [26]. Sodium dithionite was used for these studies rather than the iron hexacyanides because denatured cytochrome c has a lower oxidation–reduction potential relative to that of the native cytochrome, and thus differences in rates due to changes in the driving force could be expected with the iron hexacyanides. It was found that a variety of treatments including NaOH, heat, ethanol, and 6 M guanidine (all of which yield a form of cytochrome c capable of binding CO) yield molecular species that react readily with both $S_2O_4^{2-}$ and SO_2^- and yield rate constants that do not differ from those of native cytochrome c by more than a factor of 5 (a factor of 2 was typical) [26]. These results argue against the participation of a highly organized arrangement of amino acid side chains and point to the heme edge as the site of electron transfer.

All of the reactants discussed to this point interact with cytochrome c in a manner consistent with remote attack. One exception exists, and that is chromous ion. Chromous ion is substitution labile in its coordination sphere, and chromic ion is substitution inert. Hence, any ligand bound to chromous ion at the time of electron transfer remains after the transfer in the coordination shell of chromic ion [57]. Kowalsky [58] demonstrated the binding of chromic ion to cytochrome c following reduction, and subsequently the kinetics of chromous reduction were investigated [59,60]. These studies established that chromous reduction involves adjacent attack on the iron(III) center, although in the presence of anions (I^-, N_3^-, SCN^-) a remote electron transfer pathway is apparently operative [60].

BIOLOGICAL OXIDANTS AND REDUCTANTS

The studies discussed above with nonphysiological reactants suffer occasionally from the criticism that any mechanism proposed has no necessary relation to the physiological situation. However, studies on the interaction of either cytochrome c or cytochrome c_2 with their natural donors and acceptors are difficult, as these components are not freely water soluble and hence must be solubilized with detergents or studied in the membrane environment. In order to circumvent these problems, some studies have been conducted on the interaction of cytochrome c and cytochrome c_2 with soluble oxidation–reduction proteins.

Pseudomonas aeruginosa cytochrome c_{551} is a high-potential (250 mV), small molecular weight ($\sim$ 10,000 daltons), acidic protein whose reaction with cytochrome c has been investigated [28]. The rate of oxidation of cytochrome c by cytochrome c_{551} was shown to be equal to the rate of reduction (4.9×10^4 M^{-1} sec^{-1}), and no evidence for complex formation was obtained [28]. As cytochrome c is basic and cytochrome c_{551} is acidic, increasing ionic strength would be expected to decrease the rate of electron transfer if net portein charge were controlling the reaction. Interestingly, no effect of ionic strength was found [28], indicating that at least one of the two proteins is uncharged at the site of electron transfer. Thus, specific sites of electron transfer are implied, although their charge cannot be deduced from available data.

The reaction of solubilized cytochrome c_1 with cytochrome c has been studied [61]. These data are presented in Table 11 along with the *Pseudomonas* cytochrome c_{551} results. The reaction of cytochrome c with cytochrome c_1 is fast and the ratio of the rate constants is consistent with the measured equilibrium constant [61], suggesting that no complex formation occurs. Since the midpoint potential of cytochrome c_1 is 228 mV [61] as

TABLE 11

Reaction of Cytochrome c with Biological Oxidants
and Reductants

Reductant	k (M^{-1} sec^{-1})
P. aeruginosa Cytochrome c_{551}	4.9×10^4 [a]
Cytochrome c_1	3.3×10^6 [b]

Oxidant	k (M^{-1} sec^{-1})
P. aeruginosa Cytochrome c_{551}	4.9×10^4 [a]
Cytochrome c_1	1×10^6 [b]
Laccase	1.7×10^4 [c]

[a] $I = 0.2$, 20°C, pH 7.0 [28].
[b] 0.05 M potassium phosphate, pH 7.4, 10° [61].
[c] $I = 0.1$, pH 5.3, 25°C [62].

compared to 260 mV for cytochrome c, it would appear that driving force is not the explanation for the very rapid rates compared to that of *Pseudomonas* cytochrome c_{551} with cytochrome c. Increasing ionic strength decreases the rate of reduction of cytochrome c by cytochrome c_1, suggesting a plus–minus interaction at the site of electron transfer [61]. However, the rate of the reaction of ferrocytochrome c_1 with ferricyanide also decreases with increasing ionic strength, suggesting a positive site of electron transfer on cytochrome c_1 [61]. This observation raises an interesting and significant question, namely: Is the site of electron transfer on cytochrome c_1 the same for ferricyanide and cytochrome c? Clearly the reaction with the iron hexacyanides could be at a nonphysiological site. However, the discrepancy could be equally well explained by proposing that the reaction of cytochrome c and cytochrome c_1 establishes more extensive interactions over a larger surface area of each protein, thus producing the net effect of a change in the apparent charge at the site of electron transfer relative to that found for the iron hexacyanides. This possibility does not appear unlikely considering the size of the two reactants (cytochrome c and cytochrome c_1). Available information and the lack of structural data on cytochrome c_1 do not allow the nature of the interaction of the two cytochromes to be specified. Nevertheless, in view of the small driving force and the steric restrictions probably imposed by the size of the two macromolecules, the rapid rates observed are impressive and point to considerable specificity.

Also presented in Table 11 is the rate of oxidation of cytochrome c by the copper-containing protein laccase [62]. This reaction was not studied in detail since anions tend to inhibit laccase [62], but it does demonstrate a reasonably rapid electron transfer rate.

The reaction of cytochrome c_2 with *Chromatium vinosum* high-potential iron–sulfur protein (HIPIP) has been reported [63]. Studies on this reaction were initiated because the structure of HIPIP is known [64,65], and it was felt that a study of the interaction of two proteins with known structures might be more amenable to interpretation. The rate of reduction of cytochrome c_2 by HIPIP was found to decrease with increasing ionic strength and was reasonably rapid ($k = 5 \times 10^4$ M^{-1} sec^{-1} at $I = 0.085$, 20°C, pH 7.0). However, the rate of oxidation of ferrocytochrome c_2 by oxidized HIPIP was independent of ionic strength ($k = 2.2 \times 10^5$ M^{-1} sec^{-1}, 20°C, pH 7.0). Taking the site of electron transfer on both ferri- and ferrocytochrome c_2 as positive (as was shown with the iron hexacyanide reactions [31]), the inter-action of the two proteins suggests a negative site of electron transfer on reduced HIPIP and a site with zero net charge on oxidized HIPIP [63]. Similarly, a study of the interaction of HIPIP with the iron hexacyanides [63] indicates a negative site on reduced HIPIP and a neutral site on oxidized HIPIP. Thus, the cytochrome c_2–HIPIP results are consistent with the predictions made by the interactions of each protein with the iron hexa-cyanides.

The HIPIP–cytochrome c_2 reaction presents some interesting problems in interpretation due to the constraints applied by the structure of HIPIP. The iron–sulfur cluster, the electron-transferring chromophore of HIPIP, is buried in the interior of the molecule with the closest point of approach to the solvent 4.5 Å and is insulated from the solvent by the side chains of Leu-17, Phe-48, Leu-65, Phe-66, and Ser-79 [65]. In terms of available data, three possibilities are likely explanations of the rapid electron transfer between HIPIP and cytochrome c_2, and these are presented diagramatically in Fig. 4. Option A, proposed by Carter *et al.* [65], depicts cytochrome c_2, [R], approaching the iron–sulfur cluster in the vicinity of a tyrosine residue (Tyr-19) and electron transfer being facilitated via a phenoxy radical transition state. This possibility is unlikely on energetic grounds, as the reduction poten-tials of phenols are low (~ -2 V [36]) and sufficient energy is not available to drive the reaction. Option B in Fig. 4 suggests that the cytochrome must approach only to within some reasonable distance of the iron-sulfur cluster, and the electron "hops" through a nonaqueous medium provided by the amino acid side chains to or from the chromophore. This type of mechanism can be termed oriented electron tunneling and suggests that the two proteins, when in the appropriate orientations, can transfer electrons through some minimum-energy pathway. Option C places the iron–sulfur cluster at the surface of the molecule and would permit outer-sphere electron transfer by direct contact between the heme edge and the cluster. This option requires

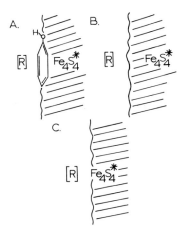

Fig. 4 Possible models for the reaction of cytochrome c_2 and HIPIP ($Fe_4S_4^*$ represents the iron–sulfur cluster; [R], ferri- or ferrocytochrome c_2; and the crosshatched areas the protein interior).

that the solution structure and crystal structure be different, a requirement difficult to approach experimentally. Unfortunately, available data do not allow one to decide between the possibilities discussed.

To summarize, a variety of proteins are able to rapidly interact with cytochrome c or cytochrome c_2, with available evidence pointing to an important role for the protein moiety in these reactions and particularly for the charged surface residues. The available data (nonphysiological and biological reactants) indicate that driving force and steric restrictions are not major controlling factors in the observed rates. This analysis suggests by deduction that electrostatics, orientation factors, and possibly nonpolar interactions must be the controlling factors.

PHYSIOLOGICAL OXIDANTS AND REDUCTANTS

The reaction of cytochrome c with its physiological electron donors and acceptors has been studied to some extent. Rates of oxidation and reduction have been measured in mitochondria [66,67]; however, these rates have not been studied thoroughly, as the presence of the intact membrane limits the application and interpretation of external perturbations. Considerably more work has been done using the solubilized cytochrome c reductase (complex I-III, complex III) [68,69] and solubilized cytochrome oxidase [70–72].

Solubilized donors and acceptors have been studied primarily via the application of steady-state kinetics and, because the studies were not designed to measure the individual rate constants, they provide information on "activity" but not (directly) on the mechanism.

Some transient-state kinetic studies with the solubilized donors and acceptors have been reported. The rate of reduction of cytochrome c by purified succinate-cytochrome c reductase was found to be 2.8×10^6 M^{-1} sec^{-1} [61], a value in excellent agreement with that found for solubilized cytochrome c_1 (Table 11). The reaction of ferrocytochrome c with purified cytochrome oxidase has been reported to have a rate constant of 4×10^7 $(M^{-1} sec^{-1})$ [72]. Finally, substantial evidence exists that cytochrome c and cytochrome oxidase form complexes (see Nicholls [2] for a review). Thus, it can be established that physiological oxidants and reductants have rate constants equal to or much greater than those observed for nonphysiological or biological oxidants and reductants.

A number of observations have been made which bear on the reaction of cytochrome c with its natural electron donors and acceptors. (1) Cytochrome c trinitrophenylated at Lys-13 was 50% as active as the native protein with cytochrome oxidase [73]. (2) Polylysine and other polycations inhibit the reaction of cytochrome c with cytochrome oxidase and NADH-cytochrome c reductase [68,70,74]. (3) Cytochrome c_2 reacts with bovine NADH-cytochrome reductase (with 60% of the activity of cytochrome c) and bovine cytochrome oxidase (with 5% of the activity of cytochrome c) [69]. Both the oxidase and reductase activities are sensitive to polylysine, and the reductase activity is also antimycin A sensitive [69]. Thus, the participation of positively charged groups on both cytochrome c and cytochrome c_2 is suggested, and the results are consistent with both cytochromes acting at the same site on the oxidase and reductase [6].

Evidence for separate pathways for the oxidation and reduction of cytochrome c by the reductase and oxidase has been presented. (1) Antibodies against cytochrome c were prepared and different antibodies inhibited either the oxidase or reductase reaction, suggesting two different pathways [75]. (2) A number of chemically modified forms of cytochrome c have been prepared which show differential activities with the oxidase and reductase [7]. However, these studies are difficult to interpret, as the relationship of the modified forms to native cytochrome c has not been established. (3) Pyridoxal phosphate binds tightly to cytochrome c and affects the enzymatic reducibility of cytochrome c without affecting the reaction with the oxidase [45]. The difficulty with the experiments discussed above is that they do not directly address the question of the site of electron transport. That is, the physiological oxidant and reductant may bind to cytochrome c in different fashions, interacting with different amino acid side chains away from the site of electron transfer itself.

Thus, perturbations away from the site of electron transfer would be expected to affect complex formation and the apparent rate but not necessarily the rate of electron transfer once the complex is formed.

The interaction of cytochrome c_2 with reaction-center particles (RCP) derived from *Rhodopseudomonas spheroides* has been investigated [76]. Reaction-center particles devoid of cytochromes are detergent extracted and are fully capable of undergoing photooxidation, retaining the ability to photooxidize added cytochrome *c* [77–79]. It was reported by Prince *et al.* [76] that the oxidation of ferrocytochrome c_2 takes place rapidly upon illumination ($k = 8 \times 10^8$ M^{-1} sec^{-1}) and is second-order with no evidence of complex formation. They found that the rate of oxidation of cytochrome c_2 decreased with increasing ionic strength and that the reaction was inhibited by polylysine, data consistent with a positive charge for the cytochrome site of electron transfer and a negative charge on the RCP. These observations are compatible with the results obtained from the interaction of iron hexacyanides with cytochrome c_2 (Table 7). The question of complex formation by cytochrome and RCP is still open, however, as Ke *et al.* [79] found evidence for complex formation between RCP and cytochrome *c* in a different detergent. Further, the data of Prince *et al.* are ambiguous in that their plots of $t_{1/2}$ vs. [cytochrome c_2] do not extrapolate through zero, suggesting a change in rate-limiting step at high cytochrome concentrations and hence complex formation.

To summarize, the reactions of cytochromes *c* and c_2 with physiological oxidants and reductants are in agreement with those obtained with nonphysiological reactants in terms of electrostatics and rapid rates of oxidation and reduction. Considering the data available and the probable steric restrictions, it seems reasonable to conclude that remote electron transfer is involved in the physiological cases and that the rapid rates observed result from a great deal of specificity conferred on the reaction by specific interactions between the reactants. Unfortunately, quantitative analysis of the available data with physiological reactants is greatly limited by the presence of detergents and the lack of structural and chemical information concerning the cytochrome donor and acceptor.

CONCLUSIONS

All of the reactions reviewed here, with the exception of those of chromous ion, can best be accommodated by a mechanism of electron transfer by cytochromes *c* and c_2 involving remote attack of the oxidant or reductant. The available data point to a specific site of electron transfer as opposed to multiple or random sites. The pseudo first-order kinetics of all reactions studied at neutral pH are homogeneous; that is, multiple kinetic processes are not observed. Further, the net protein charge is not a controlling factor, as the apparent charge at the site of electron transfer is always less than or different in sign from the overall protein charge. Although unequivocal proof is lacking, the proposal that the exposed heme edge is the site of electron transfer is most consistent with available data. The clustering of positive charges about the heme edge in both cytochrome c and cytochrome c_2 [6], the effect of polylysine on the interaction of the cytochromes with their physiological oxidants and reductants [69,70], the rapid rates of electron transfer, the minimal steric restrictions in the proximity of the heme edge, and the lack of any convincing evidence for electron transport through an organized array of amino acid side chains [26] all argue for the heme edge. Some data, for example, the comparative study of tuna, chicken, bovine, and horse cytochrome c (Table 5), tend to argue against the heme edge. However, this interpretation is not supported by any other data and is presently outweighed by arguments for the heme edge.

The studies employing antibodies and pyridoxal phosphate binding discussed with regard to the interaction of cytochrome c with its physiological reactants have been interpreted in terms of different sites on cytochrome c for oxidation and reduction [45,75]. However, this approach does not address the problem directly in that complex formation and electron transfer are treated as a single problem. Complexes have been identified for the reaction of cytochrome c and iron hexacyanides [25,26,31] and cytochrome c and cytochrome oxidase [2]. Further, no evidence is available which indicates that complexes are not formed with the anionic reactants discussed here. In terms of driving force, ferrocyanide is the weakest of all the reductants studied, yet the rate of electron transfer [k_{23} in Eq. (2)] is rapid (Table 3). Probably a more typical case would be ferricyanide oxidation in which the rate of electron transfer (k_{32}) is 2.1×10^4 sec^{-1} [25], a value well beyond the limits of the mixing methods used in these studies. Intuitively, it would appear that the chemical and structural properties of the reactant and the cytochrome would control the rate of complex formation, and the difference in midpoint potentials would be an important factor in determining the rate of electron transfer once the complex is formed [38]. Applying this logic, the following relative rates of reduction in terms of k_{23} can be predicted: $SO_2^- \gg$ ascorbate $>$ Fe(EDTA)$^{2-}$ $>$ $S_2O_4^{2-}$ $\gg$ cytochrome c_1 $\gg$ ferrocyanide. For ferrocyanide reduction k_{23} is approximately 132 sec^{-1} [26], which is within a factor of 3 or 4 of the limit of mixing methods, complexes with reactants other than ferrocyanide would not be expected to be kinetically detectable. This analysis of electron transfer has omitted consideration of cationic reactants, as no long-lived complexes would be expected; indeed, the two investigated to date, $Ru(NH_3)_6^{2+}$ and tris(1,10-phenanthroline) cobalt(III)$^{3+}$, yield rate constants as predicted by Marcus theory (Table 10). A similar scheme can be presented for oxidation of ferrocytochrome c, although only a small number of oxidants have been studied. Interestingly, k_{23} for cytochrome c_2 is greater than k_{23} for cytochrome c and k_{32} is less, as predicted from the decreased driving force between cytochrome c_2 and the iron hexacyanides. Taking the arguments presented here a step further, any discussion of the difference in reactivity of a particular form of cytochrome c with its physiological reactants has to be analyzed in terms of the effect of the modification on complex formation (k_{12}/k_{22}, k_{34}/k_{43}) and electron transfer (k_{23}, k_{32}). In light of the foregoing discussion it can be seen that no convincing evidence exists for separate sites for oxidation and reduction of cytochrome c.

A rigorous demonstration that oxidation and reduction of cytochrome c take place at the same site with physiological and nonphysiological reactants is not possible at this time. However, all available information is consistent with a positive charge at the site of electron transfer on cytochromes c and c_2 irrespective of the origin of the reactant, implying a similar site of action for all reactants. The reaction of cytochrome c with cytochrome c_1 represents an anomaly in that the reaction of the respective cytochromes with ferricyanide indicates a positive charge at the site of electron transfer for both proteins, yet a plus–minus reaction is indicated when the two proteins react [61]. This reaction may represent an example in which interactions at the site of electron transfer are outweighed by interactions in other regions of the two molecules. A more detailed analysis requires some structural information concerning cytochrome c_1, but the cytochrome c–cytochrome c_1 interaction may provide a most profitable area for future studies.

Recently, the distance of closest approach of the heme groups of cytochrome c and cytochrome b_5 has been calculated from available coordinates [80]. Cytochrome b_4 has an exposed heme edge and a necklace of negative charges about the heme crevice [81], allowing orientation of complementary charges on the two cytochromes [80]. It was found that the closest approach of the heme edges of each cytochrome was approximately 8 Å due to steric restrictions. Although it is not known whether cytochrome b_5 has any structural resemblance to the physiological reactants of cytochromes c and c_2, the predicted steric restrictions [80] illustrate the potential problem of electron transfer between macromolecules and suggest that outer-sphere electron transfer, which requires close contact between redox centers, may be difficult or impossible to achieve. Hopfield has recently proposed a mechanism for electron transfer between two sites in fixed geometry [10]. He proposed that electron transfer takes place by electron tunneling, that this tunneling can be temperature dependent, and that the separation linking the sites of electron transfer can be as small as 8–10 Å [10]. Although Hopfield evolved this mechanism in terms of low-potential cytochrome from *Chromatium vinosum* (not cytochrome c or c_2), the concept of oriented electron tunneling is most attractive. In the context of the Hopfield mechanism, it can be proposed that electron transfer by cytochrome c or c_2 can occur by oriented electron tunneling or outer-sphere electron transfer, depending on the distance of closest approach that is controlled by steric restriction of the two reactants. Orientation would be provided by electrostatic and possibly nonpolar interactions both at and away from the site of electron transfer. This proposal is particularly attractive in explaining the rapid electron transfer between cytochrome c_2 and HIPIP [63] in which the HIPIP chromophore is buried in the interior of the molecule [63].

Salemme *et al.* [6] presented a rather detailed mechanism for the physiological oxidation–reduction of cytochrome c_2 involving a front side (heme edge) attack of the cytochrome molecule by its oxidase and reductase coupled to destabilization of the existing heme oxidation state. Based on the effect of pH on the oxidation–reduction potential of cytochrome c_2, a hydrogen-bonding network involving Ser-89–Tyr-52–Tyr-70 was invoked to stabilize the positive charge on the ferriheme iron of cytochrome c_2. Although the generality of this proposal may no longer be valid due to the sequence variations since found among the cytochromes c_2 (Table 8), the idea has considerable utility. The large number of potential hydrogen-bonding side chains in the vicinity of the heme edge of cytochrome c_2 (Table 9) and the correlation between the differences in kinetics of cytochrome c_2 from various sources with the sequence differences (Table 7 and 8) suggest that hydrogen bonding could play an important role in determining the chemistry in the region of heme edge. More detailed structural, chemical, and kinetic information will have to be available before specific proposals can be made, but studies in this area may prove fruitful in the future.

The discussion to this point has omitted any direct implication of nonpolar interactions, as no data bearing on this point exist. However, it does not seem inconceivable that nonpolar interactions could facilitate both orientation and complex formation between reacting macromolecules, thus providing another area for future investigations.

Overall, considerable progress has been made in the last few years with regard to our understanding of the mechanism of action of cytochromes c and c_2. The inquiry now focuses on the specific chemical and structural parameters mediating the interactions of these molecules with their biological donors and acceptors. Future investigations should extend this understanding and provide some basic ground rules for protein–protein interactions in biological electron transport, which will most likely prove to be applicable to studies of other biological processes.

ACKNOWLEDGMENTS

The author gratefully acknowledges the support of U.S. Public Health Service Research Grant GM-21277 and Career Development Award EY-00013. The author is indebted to Dr. R. G. Bartsch, Dr. M. J. Halonen, Dr. M. D. Kamen, Dr. T. E. Meyer, Dr. G. W. Pettigrew, and Dr. F. R. Salemme for many valuable discussions.

REFERENCES

1. E. Margoliash and A. Schejter, *Adv. Protein Chem.* **21**, 113 (1966)
2. P. Nicholls, *Biochim. Biophys. Acta* **340**, 261 (1974).
3. R. E. Dickerson and R. Timkovich, *in* "The Enzymes" (P. D. Boyer, ed.), Vol. 11, p. 397. Academic Press, New York, 1975.
4. R. Lemberg and J. Barrett, "The Cytochromes." Academic Press, New York, 1972.
5. T. Takano, R. Swanson, A. B. Kallai, and R. E. Dickerson, *Cold Spring Harbour Symp. Quant. Biol.* **36**, 397 (1971).
6. F. R. Salemme, J. Kraut, and M. D. Kamen, *J. Biol. Chem.* **248**, 7701 (1973).
7. E. Margoliash, S. Ferguson-Miller, J. Tulloss, C. H. Kee, B. A. Feinberg, D. L. Brautigan, and M. Morrison, *Proc. Natl. Acad, Sci. U.S.A.* **70**, 3245 (1973).
8. D. W. Urry and H. Erying, *Proc. Natl. Acad. Sci. U.S.A.* **49**, 253 (1963).
9. D. Devault and B. Chance, *Biophys. J.* **6**, 825 (1966).
10. J. Hopfield, *Proc. Natl. Acad. Sci. U.S.A.* **71**, 3640 (1974).
11. T. E. Meyer, Ph.D. Thesis, University of California, San Diego (1970).
12. R. P. Ambler and T. E. Meyer, private communication.
13. K. Dus, K. Sletten, and M. D. Kamen, *J. Biol. Chem.* **243**, 5507 (1968).
14. F. R. Salemme, S. T. Freer. Ng. H. Yuong, R. A. Alden, and J. Kraut, *J. Biol. Chem.* **248**, 3910 (1973).
15. R. E. Dickerson, T. Takano, D. Eisenberg, O. B. Kallai, L. Samson, A. Cooper, and E. Margoliash, *J. Biol. Chem.* **246**, 1511 (1971).
16. G. W. Pettigrew, T. E. Meyer, R. G. Bartsch, and M. D. Kamen, *Biochim. Biophys. Acta* **430**, 197 (1975).
17. C. Creutz and N. Sutin, *Proc. Natl. Acad. Sci. U.S.A.* **70**, 1701 (1973).
18. E. J. Land and A. J. Swallow, *Arch. Biochem. Biophys.* **145**, 365 (1971).
19. J. Wilting, R. Braans, H. Nauta, and K. J. H. Van Buuren, *Biochim. Biophys. Acta* **283**, 543 (1972).
20. N. N. Lichtin, A. Shafferman, and G. Stein, *Biochim. Biophys. Acta* **314**, 117 (1973).
21. E. J. Land and A. J. Swallow, *Biochim. Biophys. Acta* **368**, 86 (1974).
22. J. Wilting, K. J. H. Van Buuren, R. Braams, and B. F. VanGelder, *Biochim. Biophys. Acta* **376**, 285 (1975).
23. N. Sutin and D. R. Christman, *J. Am. Chem. Soc.* **83**, 1773 (1961).
24. B. H. Havsteen, *Acta Chem. Scand* **19**, 1227 (1965).
25. E. Stellwagen and R. G. Shulman, *J. Mol. Biol.* **80**, 559 (1973).
26. W. G. Miller and M. A. Cusanovich, *Biophys. Struct. Mech.* **1**, 97 (1975).
27. A. A. Frost and R. G. Pearson, "Kinetics and Mechanism." Wiley, New York, 1961.
28. R. A. Morton, J. Overnell, and H. A. Harbury, *J. Biol. Chem.* **245**, 4653 (1970).
29. E. Stellwagen and R. D. Cass, *J. Biol. Chem.* **250**, 2095 (1975).
30. S. D. Power, A. Choucair, and G. Palmer, *Biochem. Biophys. Res. Commun.* **66**, 103 (1975).

31. F. E. Wood and M. A. Cusanovich, *Bioinorg. Chem.* **4**, 337 (1975).
32. K. B. Brandt, P. C. Parks, G. H. Czerlinski, and G. P. Hess, *J. Biol. Chem.* **241**, 4180 (1966).
33. P. George and C. L. Tsou, *Biochem. J.* **50**, 440 (1952).
34. C. Greenwood and G. Palmer, *J. Biol. Chem.* **240**, 3660 (1965).
35. N. Sutin and J. K. Yandell, *J. Biol. Chem.* **247**, 6932 (1972).
36. R. X. Ewall and L. E. Bennett, *J. Am. Chem. Soc.* **96**, 940 (1974).
37. H. L. Hodges, R. A. Holwerda, and H. B. Gray, *J. Am. Chem. Soc.* **96**, 3132 (1974).
38. R. A. Marcus, *J. Phys. Chem.* **67**, 853 (1963).
39. R. K. Gupta, *Biochim. Biophys. Acta* **292**, 291 (1973).
40. A. G. Redfield and R. K. Gupta, *Cold Spring Harbour Symp. Quant. Biol.* **36**, 405 (1971).
41. R. J. Campion, C. F. Deck, P. King, Jr., and A. C. Wahl, *Inorg. Chem.* **6**, 672 (1967).
42. C. Creutz and N. Sutin, *J. Biol. Chem.* **249**, 6788 (1974).
43. E. Margoliash, G. H. Barlow, and V. Byers, *Nature (London)* **228**, 723 (1970).
44. R. Margalit and A. Schejter, *Eur. J. Biochem.* **32**, 500 (1973).
45. I. Aviram and A. Schejter, *FEBS Lett.* **36**, 174 (1973).
46. M. A. Cusanovich and D. E. Stringer, unpublished observations.
47. L. P. Kayushin and Y. I. Ajipa, *Ann. N.Y. Acad. Sci.* **222**, 255 (1973).
48. R. G. Shulman, *Ann. N.Y. Acad. Sci.* **222**, 265 (1973).
49. R. Margalit and A. Schejter, *FEBS Lett.* **6**, 278 (1970).
50. M. A. Cusanovich, unpublished observations.
51. J. V. McArdle, H. B. Gray, C. Creutz, and N. Sutin, *J. Am. Chem. Soc.* **96**, 5739 (1974).
52. M. A. Cusanovich and C. B. Post, unpublished observations.
53. F. R. Salemme, S. T. Freer, R. A. Alden, and J. Kraut, *Biochem. Biophys. Res. Commun.* **54**, 47 (1973).
54. W. M. Clark, "Oxidation-Reduction Potentials of Organic Systems." Williams & Wilkins, Baltimore, Maryland, 1960.
55. W. G. Miller, Ph.D. Thesis, University of Arizona, Tucson (1974).
56. D. O. Lambeth and G. Palmer, *J. Biol. Chem.* **248**, 6095 (1973).
57. H. Taube, H. Meyers, and R. Rich, *J. Am. Chem. Soc.* **75**, 4118 (1953).
58. A. Kowalsky, *J. Biol. Chem.* **244**, 6619 (1969).
59. J. W. Dawson, H. B. Gray, R. A. Holwerda, and E. W. Westhead, *Proc. Natl. Acad. Sci. U.S.A.* **69**, 30 (1972).
60. J. K. Yandell, D. P. Fay, and N. Sutin, *J. Am. Chem. Soc.* **95**, 1131 (1973).
61. C. A. Yu, L. Yu, and T. E. King, *J. Biol. Chem.* **248**, 528 (1973).
62. B. G. Malmström, A. F. Agrò, C. Greenwood, E. Antonini, M. Brunori, and B. Mondoui, *Arch. Biochem. Biophys.* **145**, 349 (1971).
63. I. A. Mizrahi, F. E. Wood, and M. A. Cusanovich, *Biochemistry* **15**, 343 (1976).
64. C. W. Carter, Jr., J. Kraut, S. T. Freer, Ng. H. Yuong, R. A. Alden, and R. G. Bartsch, *J. Biol. Chem.* **249**, 4212 (1974).
65. C. W. Carter, Jr., J. Kraut, S. T. Freer, and R. A. Alden, *J. Biol. Chem.* **249**, 6339 (1974).
66. B. Chance and G. R. Williams, *J. Biol. Chem.* **217**, 429 (1955).
67. L. Smith, D. C. White, P. Sinclair, and B. Chance, *J. Biol. Chem.* **245**, 5096 (1970).
68. L. Smith and K. Minnaert, *Biochim. Biophys. Acta* **105**, 1 (1965).
69. K. A. Davis, Y. Hatefi, F. R. Salemme, and M. D. Kamen, *Biochem. Biophys. Res. Commun.* **49**, 1329 (1972).

70. H. C. Davies, L. C. Smith, and A. R. Wasserman, *Biochim. Biophys. Acta* **85**, 238 (1964).
71. B. Errede, G. P. Haight, Jr., and M. D. Kamen, *Proc. Natl. Acad. Sci. U.S.A.* **73**, 113 (1976).
72. Q. H. Gibson, C. Greenwood, D. C. Wharton, and G. Palmer, *J. Biol. Chem.* **240**, 888 (1965).
73. K. Wada and K. Okunuki, *J. Biochem. (Tokyo)* **66**, 249 (1969).
74. L. Smith and H. Conrad, *Arch. Biochem. Biophys.* **63**, 403 (1956).
75. L. Smith, H. C. Davies, M. Reichlin, and E. Margoliash, *J. Biol. Chem.* **248**, 237 (1972).
76. R. C. Prince, R. J. Cogdell, and A. R. Crofts, *Biochim. Biophys. Acta* **347**, 1 (1974).
77. G. Feher, *Photochem. Photobiol.* **14**, 373 (1971).
78. R. K. Clayton, H. Fleming, and E. Z. Szuts, *Biophys. J.* **12**, 46 (1972).
79. B. Ke, T. H. Chaney, and D. W. Reed, *Biochim. Biophys. Acta* **216**, 373 (1970).
80. F. R. Salemme, *J. Mol. Biol.* **102**, 563 (1976).
81. P. Argos and F. S. Mathews, *J. Biol. Chem.* **250**, 747 (1975).

CHAPTER

5

Structural and Mechanistic Aspects
of Catalysis by Thiamin

Anthony A. Gallo, John J. Mieyal,
and Henry Z. Sable

INTRODUCTION

General

When thiamin (vitamin B_1) was isolated in the early part of this century [1], the chief concern was its role in nutrition. Since then its structure* has been elucidated [2], and its pyrophosphate ester **1b** was identified as the cofactor for the enzyme pyruvate decarboxylase [3]. Subsequently, TPP†

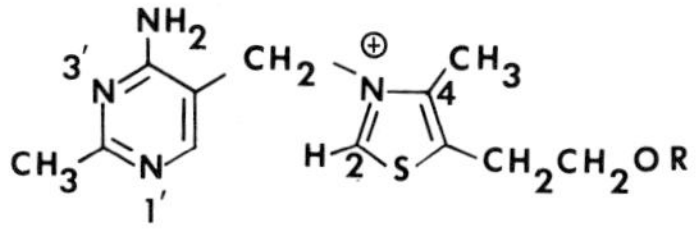

1a, THIAMIN R = H
1b, THIAMIN PYROPHOSPHATE R = $P_2O_6^{3-}$

* The pyrimidine ring is numbered as shown in **1**. The primes are used only when senior groups or rings are present, as in thiamin.

† Abbreviations: nmr, nuclear magnetic resonance; pmr, proton magnetic resonance; cmr, carbon magnetic resonance; TPP, thiamin-PP; HET, α-hydroxyethylthiamin; HBT, α-hydroxybenzylthiamin; TFT, 2′-CF_3-thiamin; HBTFT, α-hydroxybenzyl-TFT.

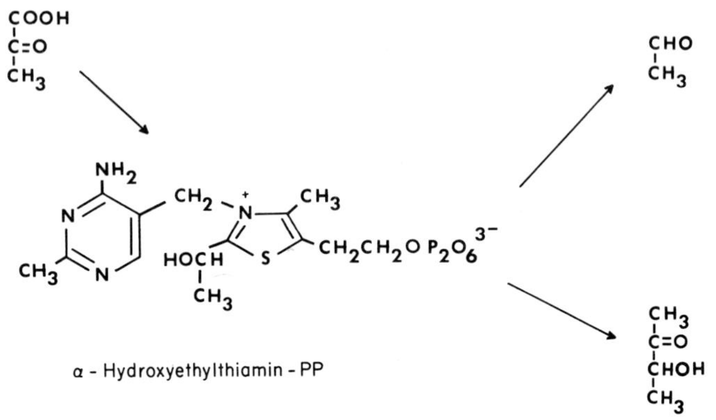

COOH
|
C=O
|
CH$_3$

CHO
|
CH$_3$

NH$_2$

CH$_2$—N$^+$—CH$_3$

CH$_3$

HOCH   S—CH$_2$CH$_2$O P$_2$O$_6$$^{3-}$
|
CH$_3$

CH$_3$
|
C=O
|
CHOH
|
CH$_3$

α - Hydroxyethylthiamin - PP

Fig. 1   Two reactions of pyruvate.

was found to be a cofactor for many single enzymes and multienzyme systems that catalyze a variety of reactions, all of which involve the cleavage of a carbon–carbon covalent bond adjacent to a carbonyl group. Figure 1 shows two reactions catalyzed by TPP enzymes acting on pyruvate, i.e., cleavage to acetaldehyde and formation of acetoin. Each of these reactions involves prior decarboxylation of pyruvate and formation of the common intermediate hydroxyethylthiamin pyrophosphate (HET-PP).

Most of the reactions carried out by TPP-requiring enzymes can be catalyzed by thiamin, although less efficiently, in the absence of enzymes; consequently, much of what is known about the mechanism of thiamin-catalyzed reactions has been learned from model systems from which enzymes have been omitted. This review is concerned mainly with the inherent catalytic activity of thiamin. Related subjects such as the biosynthesis and metabolism [4] of thiamin, the nutritional requirements for thiamin of various species and the possible role of thiamin in the function of the nervous system [5], the binding of thiamin to apoenzymes [6–8], the characteristics of reactions catalyzed by TPP-requiring enzymes [7, 9], and the structure–activity correlations of analogs of TPP with regard to their coenzyme function [8] are not reviewed. The discussion is restricted to considerations of the thiamin molecule and how it is activated as the catalyst in reactions involving carbonyl compounds.

Mechanism of Catalysis by Thiamin

The nature of the thiamin molecule itself, with its unique combination of an aromatic aminopyrimidine ring and an aromatic thiazolium ring having a hydroxyethyl side chain, does not immediately suggest a unique mode of reactions with substrates. Nevertheless, various predictions were made and tested until the currently accepted mechanism for catalysis by thiamin evolved. Langenbeck [10,11] has studied the ability of a variety of amines to catalyze both decarboxylation and acyloin condensation of α-keto acids, and he considered these reactions as models for enzymatic decarboxylations. Many years later, Wiesner and Velenta [12] proposed that the 4'-amino group of thiamin acted in an analogous manner in thiamin-catalyzed reactions; they postulated Schiff base (imine) formation followed by formation of a carbanion on the bridge carbon with subsequent rearrangement to form a readily decarboxylated β-unsaturated acid (Scheme 1). Both of the key

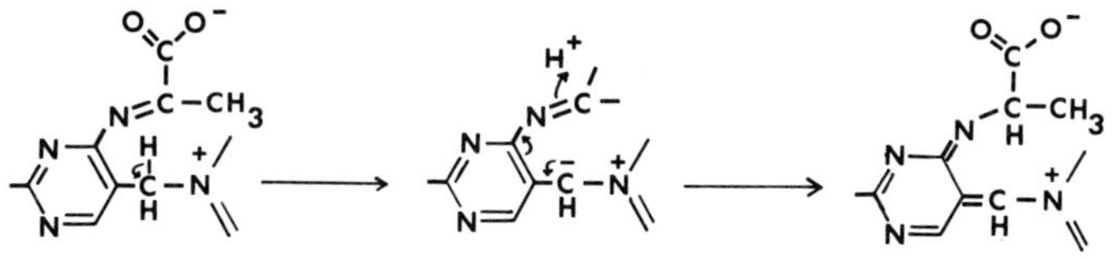

Scheme 1

points (imine formation and an intermediate bridge methylene carbanion) in the suggested mechanism have been shown to be highly unlikely. Stern and Melnick [13] found that thiamin had no catalytic properties under the Langenbeck conditions. In addition, the 4'-amino group is a weak base, as shown by the difficulty with which it is diazotized and by its relative inertness in the Schotten–Baumann acylation reaction. The second point, the bridge methylene carbanion, is discussed below in another connection.

Catalysis of acyloin condensations by cyanide, such as the formation of benzoin [14,15], was long known, as was the condensation of quaternary pyridinium compounds (e.g., **2**) with aldehydes to yield adducts at the position

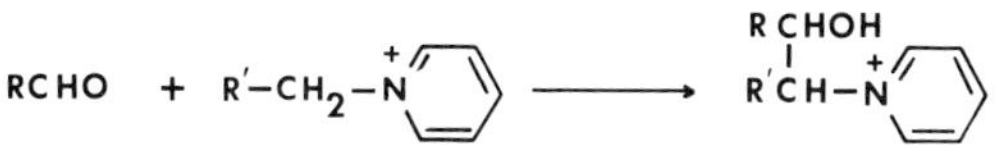

2

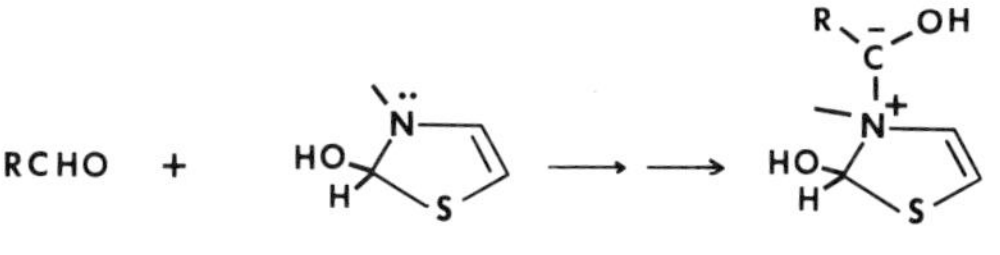

PSEUDO BASE

α to the quaternary nitrogen atom [16]. The similarity between the thiazolium ring and the pyridinium ring was recognized by Ugai *et al.* [16,17]. These authors found, however, that instead of forming adducts, the thiazolium derivatives catalyzed acyloin condensations. These findings led Mizuhara

[18,19] and Breslow [20] to suggest different mechanisms of catalysis by thiamin. Mizuhara *et al.* [18] repeated Ugai's experiments and found an optimum pH of 8.1 for the production of acetoin and CO_2 from pyruvate and acetaldehyde. They proposed that the carbonyl carbon atom of the substrate is attacked by the tertiary nitrogen atom of the pseudo base, forming a zwitterionic adduct that can cleave to give free aldehyde or condense with another molecule of aldehyde to form an acyloin. However, the acid–base chemistry of thiamin has been described in detail by Metzler [21] and by Duclos and Haake [22,23], and they found no appreciable amount of pseudo base present in solution at any pH. Hence, the Mizuhara mechanism appears not to explain thiamin catalysis.

Breslow [20] studied various thiazolium compounds under the experimental conditions of Mizuhara and Handler [18] as a model system for catalysis by thiamin, measuring the formation of acetoin from acetaldehyde and pyruvate. He found, for example, that thiamin and *N*-benzyl-4-methylthiazolium ion were much more active than *N*-ethyl-4-methylthiazolium ion. He concluded that the bridge methylene group must be activated both by the the adjacent quaternary nitrogen atom and by an adjacent aromatic ring, and he proposed a mechanism whereby the electron-deficient carbonyl carbon atom of the substrate reacts with a carbanion formed by dissociation of one of the protons of the bridge methylene group (Scheme 2). The adduct **A** (Scheme 2)

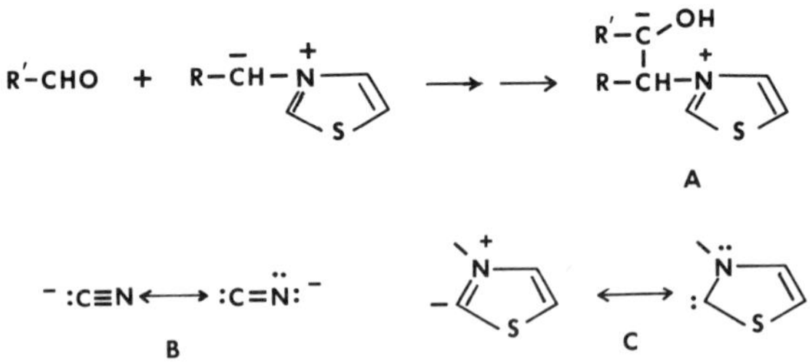

Scheme 2

would be cleaved or condensed in an analogous manner to that depicted by Mizuhara [18] for the pseudo base adduct. Breslow's proposal resembled that of Wiesner and Valenta [12] in that a carbanionic intermediate was involved. If these mechanisms are valid, the protons of the bridge methylene group of thiamin should exchange with deuterium when the reactions are performed in D_2O, but Ingraham and Westheimer [24] found no such exchange. Breslow [25] confirmed the lack of exchange at the bridge methylene group but found at the same time that the C-2 proton of the thiazolium ring does exchange readily with deuterium, and on this basis he proposed what is now the generally accepted mechanism of catalysis by thiamin.

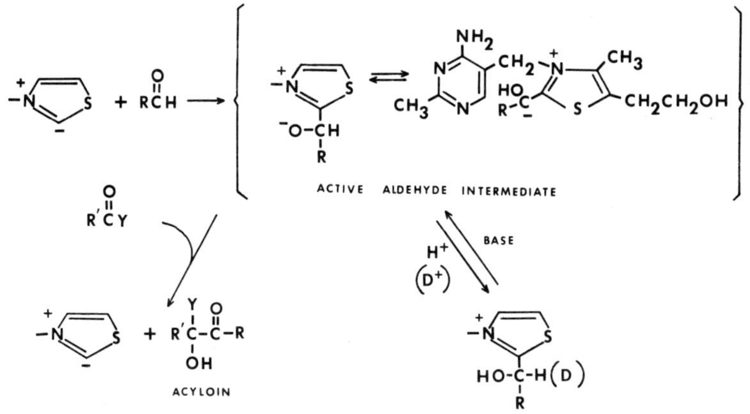

Fig. 2 The central role of the α-carbanion of HET in the acyloin condensation.

Breslow [26,27] recognized the analogy between cyanide ion, **B** (Scheme 2), and the thiamin C-2 carbanion, **C** (Scheme 2), and adapted the mechanism proposed by Lapworth [14,15] for catalysis of the benzoin condensation by CN^- to the case for thiamin catalysis. A general form of Breslow's mechanism is presented here for the acyloin condensation (Fig. 2). Much experimental evidence has been obtained for many aspects of this general mechanism. Breslow [25–27] provided evidence for the C-2 carbanion by means of hydrogen–deuterium exchange studies. The existence of the "active acetaldehyde" intermediate in biological systems was postulated long before this mechanism was proposed, and concrete data that the intermediate is actually an adduct of thiamin pyrophosphate at the C-2 position were quickly obtained. Krampitz and co-workers [28,29] synthesized 2-(α-hydroxyethyl)-thiamin-PP (HET-PP, Fig. 1) and showed that it was chemically competent in several enzymatic systems. For example, they found that acetolactate synthase produced [^{14}C]acetoin from [^{14}C]HET-PP and acetaldehyde. Goedde et al. [30] showed that the complete pyruvate dehydrogenase system from yeast mitochondria produced [^{14}C]acetyl-CoA from [^{14}C]HET-PP. In addition, HET-PP was isolated from preparations of pyruvate decarboxylase that were actively converting pyruvate to acetaldehyde and CO_2 [30–32]. In Fig. 2, the α-carbanionic form of the aldehyde adduct of thiamin is depicted as playing a central role in the various reactions. We obtained evidence in support of that role [33,34] by demonstrating H—D exchange at the α-carbon atoms of HET and HBT (Fig. 2); thiamin catalyzes the formation of benzoin only under conditions of pH and temperature in which such exchange occurs.

The generalized scheme presented in Fig. 2 does not delineate why C-2 of thiamin and C-α of HET or HBT should be activated positions on the respective molecules, nor does it represent a role for the aminopyrimidine moiety

. of thiamin in catalysis. Many investigators have approached these problems both theoretically and experimentally, and the results of the investigations are the subject matter of the rest of this review.

RELATIONSHIP OF THE STRUCTURE OF THIAMIN TO ITS CATALYTIC ACTIVITY

Electronic Considerations

The generally accepted mechanisms for pyruvate decarboxylase, transketolase, and other enzyme reactions in which TPP is a cofactor [35] suggest that, in large measure, the structure of thiamin has evolved so as to stabilize carbanions on C-2 and on the α-carbon of the α-hydroxylalkyl intermediates. One must determine, therefore, which electronic features of the thiazolium and aminopyrimidine rings of thiamin support the formation of these carbanions.

The pK_a of (C-2)-**H** of thiamin has been estimated variously between 14 and 20 [36–38], while a direct measurement gives a value of 12.7 [39]. The rate of exchange of (C-2)-**H** for solvent deuterium is a more readily measured indication of acidity, and furthermore it is fast relative to other carbon acids; e.g., at pH 7, (C-2)-**H** exchanges completely with solvent D_2O within a few seconds. These measurements indicate that thiamin is an unusually strong carbon acid [40]. By LCAO–MO calculations, Pullman [41,42] estimated an excess π-electron density of -0.13 on C-2; he suggested that the high electron density on C-2 is similar to that found on an acetylenic carbon atom and, as a result, the (C-2)-**H** would be similarly acidic. Subsequent observations that H—D exchange at C-2 was more rapid in oxazolium salts than in thiazolium salts (although a more negative charge was calculated to reside on the latter) stood in disagreement with the calculations [43]. Revised calculations then showed a $+0.19$ charge on C-2 [44] of 3,4-dimethylthiazolium salts. More recent extended Hückel and CNDO/2 calculations on thiamin give C-2 of thiamin a net atomic charge of $+0.15$ or $+0.11$, respectively [45]. Carbon-13 nmr spectroscopy also has been used to investigate charge distributions in molecules [46]. The **C-2** chemical shift, 155.3 ppm downfield from tetramethylsilane [47], also suggests a partial positive charge on **C-2** since a rough correlation of ^{13}C shielding with electron density has been found for heteroaromatic compounds [48]. On the other hand, from bond-lengths observed in X-ray crystallographic studies of thiamin [49] and TPP [50], various resonance forms have been proposed for the thiazolium ring, and inspection of these forms suggested that either a slight negative charge or no net charge exists on **C-2**. The X-ray method is rather indirect, however,

TABLE 1

Comparison of Rates of Exchange and Ring Opening at C-2 with J_{CH} and with Electronegativity of Heteroatoms Bonded to C-2

$$CH_3{-}\overset{+}{N}{-}CH_3$$

(structure with $H{-}\underset{2}{C}{-}X$ ring)

X	$J_{(C_2)-H}{}^a$	Electronegativity of heteroatom[b]	Rate of exchange $k_{obs}/[OD^-]$ (M^{-1} sec^{-1})	Rate of ring opening[c] (M^{-1} sec^{-1})
N—CH$_3$	220	3.07	1.3×10^2	2.2×10^{-4}
S	216	2.44	3.7×10^5	1.6×10^1
O	246	3.50	3.8×10^7	8.0×10^4

[a] Data from Haake et al. [52].
[b] Data from Allred [53].
[c] Data from Duclos and Haake [22].

and the agreement between the more recent molecular orbital calculations and cmr data favors a partial positive charge of uncertain magnitude on C-2. This conclusion also agrees with chemical intuition since a partial positive charge on C-2 will polarize the (C-2)-H bond, withdrawing electron density from the hydrogen atom and thus making it more susceptible to base-catalyzed ionization.

Two additional factors have been considered in explaining the acidity of (C-2—H of thiamin. One is the increased s character of the C—H bond, similar to acetylene; the other is the presence of electronegative atoms bonded near the ionizable C—H bond. Both factors are correlated and can be assessed from the magnitude of C—H coupling constants [51]. The value of J_{CH} for (C-2)—H of thiamin is 217 Hz [47], substantially smaller than that of acetylene (248 Hz) but much greater than expected for a carbon orbital whose formal hybridization is sp^2 [48]. Since thiamin is considerably more acidic than acetylene on both kinetic and thermodynamic grounds [47], the C—H couplings are not strictly correlated with acidity. On the other hand, comparison of J_{CH} for "azolium" ions (Table 1) shows that the magnitude of J_{CH} generally follows a pattern expected for the difference in electronegativity of the atoms in the ring. However, another discrepancy should be noted. Nitrogen is considerably more electronegative than sulfur, yet the values of J_{CH} indicate that thiazolium salts and imidazolium salts should have similar acidity. Actually, the rate of H–D exchange of the C-2 position is 3000 times greater in thiazolium salts than in imidazolium salts (Table 1) [22,52,53]. This difference may be due to some special ability of the sulfur atom to stabilize the incipient negative charge on C-2 in the transition state. Breslow [36] has suggested that the sulfur d_{xy} orbital overlaps with the filled sp_xp_y

orbital of C-2. If this factor is important, it is unique to sulfur since oxygen and nitrogen do not have *d* orbitals of low enough energy to interact favorably. Jordan's [45] MO calculations for thiamin indicate sizable *d*-orbital participation in the two bonds to sulfur in the thiazolium ring. Electronegativity effects must supercede stabilization by *d–σ* overlap since oxazoles exchange 100 times faster at C-2 than thiazoles, although stabilization by *d–σ* overlap is not possible in oxazoles.

Another type of sulfur participation has been proposed by Haake [52] (Scheme 3). The partial positive charge on sulfur might cause sufficient *d*-orbital contraction so that the benzyne-type structure **E** (Scheme 3) could

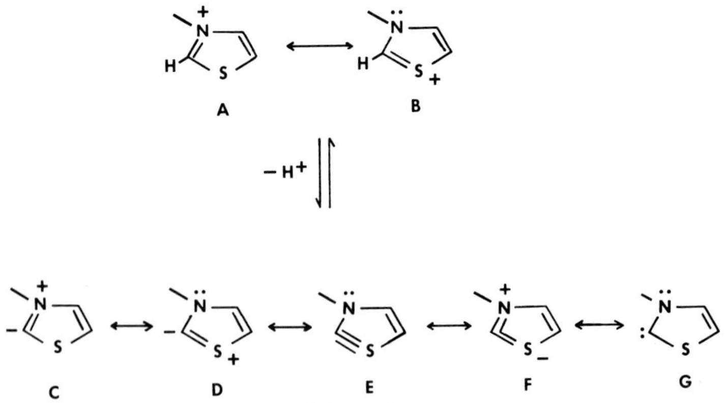

Scheme 3

make an important contribution to the transition state for ionization of thiazolium salts. Haake's MO calculations [52] using the ω technique place 65% of the formal positive charge of the thiazolium ring on the sulfur atom. Other calculations by Pullman [41] and Jordan [45] place more of the positive charge on the nitrogen atom. From crystallographic studies, Pletcher and Sax [50] conclude that the nitrogen atom carries $+0.75$ charge and the sulfur atom $+0.35$. Significant contributions of form **E** (Scheme 3) or of another form **F** (Scheme 3) suggested by Breslow [26,54] would lead to an increase in bond order of the (C-2)—S bond in the deprotonated (**C–G**, Scheme 3) relative to the protonated ring (**A**, **B**, Scheme 3). Such an increase has been predicted from MO calculations [45]. On the other hand, significant contribution by resonance form **G** (Scheme 3), also suggested by Breslow [36], would lead to a decrease in bond order for the (C-2)—S bond. Haake [52] has pointed out that in form **G** (Scheme 3), C-2 is electron deficient, and this form might therefore be of considerably higher energy than the other contributors. The carbenoid form (**G**, Scheme 3) had been suggested initially since ionization at C-2 of benzothiazolium salts leads to dimers (e.g., **3**), and these salts react with carbene trapping agents (e.g., azides) to give products characteristic of carbenes [55,56]. In addition, the carbanion of *N*,*N'*-diphenylimidazolium ion behaves as a nucleophilic carbene, adding to the electron-deficient tetracyanoethylene to form a cyclopropane derivative [43,57]. However, the five-membered rings of benzothiazoles and dipheny-

3

limidazolium salts appear not to be stabilized by aromaticity to the same extent as simpler thiazoles, since benzothiazolium salts undergo base-catalyzed ring opening about 10^4 times as readily as simple thiazolium salts [26]. Furthermore, benzaldehyde reacts with imidazolium salts to give a nonaromatic product [43], whereas aromaticity is maintained in the reaction of thiazolium salts. Since allenes and acetylenes must be linear, intuition suggests that resonance forms **E** and **F** (Scheme 3) would not be important; nevertheless, the calculation of bond orders in the C-2 anion favors these forms rather than **G** (Scheme 3).

Consideration of the aromaticity of the thiazolium ring is important in connection with the formation of the C-2 carbanion because thiazolium salts are in equilibrium with pseudo bases and with products of ring opening, which become more important as the pH is raised [22,23]. Duclos and Haake [22] reported that, in the case of oxazolium salts, ring-opened products are apparently favored at neutral pH, whereas these become important for thiazolium salts only above pH 9. The second-order rate constant for reaction of OH$^-$ with 3,4-dimethyloxazolium ion to form a pseudo base is almost 10^4 times greater than for 3,4-dimethylthiazolium ion (Table 1), whereas oxazolium salts form a carbanion at C-2 only 10^2 times more readily than thiazolium salts. Consequently, the concentration of C-2 carbanion at pH 7 may be greater for thiazolium salts than for oxazolium salts. Although imidazolium salts are 10^5 times more stable toward ring opening than thiazolium salts, they form C-2 carbanions much more slowly. Thus, of the azolium species, the thiazolium ion has the optimum combined ability to generate a C-2 carbanion while retaining the ring-closed form at physiological pH.

Substituents have a small but measurable effect on the acidity of the thiazolium C-2. The electron-donating groups at C-4 and C-5 of thiamin retard C-2 exchange, whereas the electron-withdrawing groups at N-3 accelerate the exchange. The rate-retarding effect of the substituents is shown by the rates of base-catalyzed exchange at C-2 for the 3-methyl-, the 3,4-dimethyl-, and the 3,4-dimethyl-5-(2-hydroxyethyl)thiazolium ions, which are

TABLE 2

Rates of (C-2)—H Exchange

Compound[a]	Relative rate	References
A[b]	100	
E	60	58
D	40	27,37
C	12	58
B	4	47

a

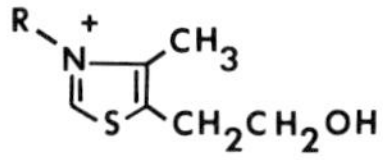

A: R=

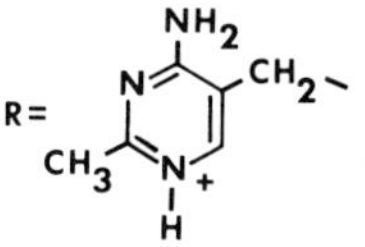

B: R=CH₃

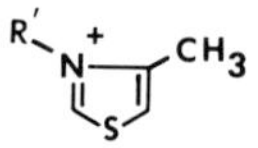

C: R'=CH₃

D: R'=

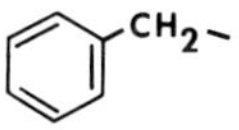

E:

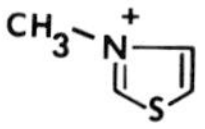

[b] The rate of exchange of (C-2)—H of N-1'-unprotonated thiamin is too rapid to measure by conventional nmr methods.

in the ratio of 15:3:1 (Table 2) [27,37,47,58]. The rate-enhancing effect of the aminopyrimidylmethyl group is shown by the 25-fold faster exchange of thiamin relative to 3,4-dimethyl-5-(2-hydroxyethyl)thiazolium ion. The inductive effect is transmitted through the bridge methylene group, since a methylene group transmits inductive effects about as well as a vinyl group [59]. Contrary to Chauvet-Monges *et al.* [60], we find that pyrophosphorylation of thiamin to give TPP has little effect on C-2 exchange. In the range pH 3–4, exchange at C-2 occurs more rapidly for thiamin than for TPP. However, when the thiamin solution contains a mixture of inorganic pyrophosphate and dimethylpyrophosphate to adjust the ionic strength to the same value as that of the TPP solution (I. L. Hansen and H. Z. Sable, unpublished observations), the rates of exchange at C-2 were essentially the same. We have reported previously that exchange at C-2 is markedly affected by the ionic strength of the solution [61]. Ullrich et al. [7], however, did find that changes in length of the C-5 substituent or replacement of the β-hydroxyl group by chlorine or hydrogen produces a change in the pK for pseudo base formation, and this might also affect C-2 exchange rates.

Hyperconjugative electronic structures have been proposed to account for various aspects of the interaction between the thiazolium ring and the aminopyrimidine ring [41,61]. Resonance forms such as **B** (Scheme 4) would

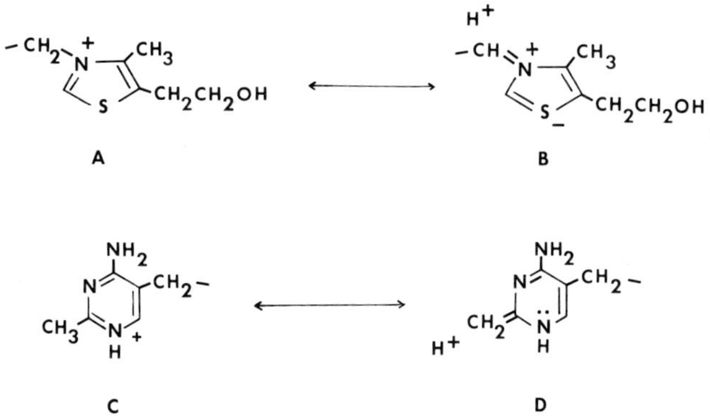

Scheme 4

allow a resonance interaction between the thiamin rings and might account for the change in frequency of the (C-4)—CH_3 nmr signal when N-1′ is protonated [61]. However, in HET, where such a resonance form can also be written, the corresponding shift in resonance frequency of the pmr signal of (C-4)—CH_3 is not observed (A. A. Gallo, unpublished observations). These spectral differences between thiamin and HET are probably related to conformational effects (see the section on conformation, below). Further- more, the bonds between the bridge carbon and N-3 in thiamin [49] and TPP [50] are of normal length, whereas an abnormally short C-2′–methyl distance of 1.477 Å in TPP has been ascribed to hyperconjugation [50]. This latter type of hyperconjugation (**D**, Scheme 4) is consistent with the H–D exchange of the C-2′ methyl group in acidic media [62], whereas the bridge methylene protons do not exchange under the same conditions.

As described above, stabilization of a second carbanion on the α-carbon of the α-hydroxyethyl intermediate also appears to be important for the catalytic mechanism. The (C-α)—**H** group of HET has a pK_a of 17 [63]; it is a weaker acid, therefore, than (C-2)—**H** of thiamin but is still quite acidic for a carbon acid. Breslow [36] has suggested that the anion is stabilized by resonance (**A ↔ B**, Scheme 5).

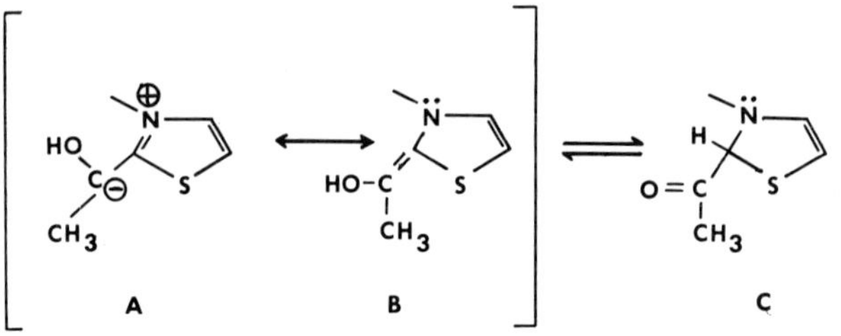

Scheme 5

Here again, the aromaticity of the thiazolium ring is important; otherwise, ketonization of the forms **A** ↔ **B** (Scheme 5) would lead to **C** (Scheme 5) and block further steps in the mechanism. Resonance of the type **A** ↔ **B** (Scheme 5) requires orientation of the plane containing the two α-substituents (alkyl and OH) parallel to the thiazolium ring in order that the orbital containing the unshared electron pair may align parallel to the π orbital at C-2. From an examination of space-filling models, Schellenberger proposed that the unshared pair of electrons of the (C-4′)—NH_2 group forms a hydrogen bond with the (C-α)—OH group (A, Scheme 6); such interaction would require the two substituents on C-α to be practically perpendicular to the plane of the thiazolium ring [8], thereby disfavoring **B** (Scheme 5). An alternative structure, **B** (Scheme 6), is discussed in the concluding section. Ullrich *et al.* [7] have

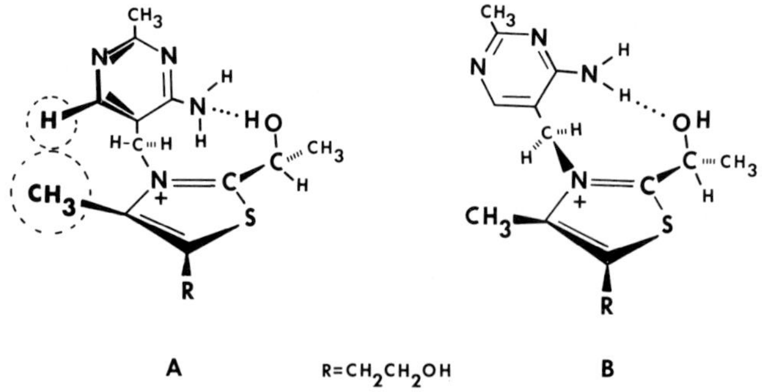

A $R=CH_2CH_2OH$ **B**

Scheme 6

suggested that such a conformation of HET (**A**, Scheme 6) may account for the slow formation of the α-carbanion from the protonated species, at least in the enzymatic process. However, an amino group-promoted destabilization of a conformation of HET required for resonance stabilization in the carbanion-forming process seems unlikely since HET analogs that do not have an aminopyrimidine substituent at N-3 exchange slightly slower than the parent compound [64] and the rates of H–D exchange at C-α are presumed to be a measure of the stability of the carbanionic form of an analog (see the section on acidic and basic sites in the aminopyrimidine ring, below). We have used hydroxybenzyl analogs to investigate the relative ease of formation of the C-α carbanion since (C-α)—H exchange is more rapid in these molecules than in hydroxyethyl derivatives [64]. The (C-α)—H exchanges twice as fast in HBT as in α-hydroxybenzyl-3-benzylthiazolium ion; this is consistent with the fact that a benzene ring is less electron withdrawing than an aminopyrimidine ring. An intramolecular N···H—O bond involving the (C-α)—OH of the latter compound is, of course, impossible in this case. In the crystal structure of HET, no evidence was found for an intramolecular hydrogen bond between the pyrimidine amino group and the (C-α)—OH [65]; instead the 2-(α-hydroxyethyl) group assumed a conformation that favors an intramolecular electrostatic interaction between its oxygen and the partial positive charge on the sulfur atom (see the section on conformation). A series of substituents at N-3 of thiazolium and hydroxybenzylthiazolium compounds had parallel effects on the rates of exchange at C-2 and at C-α, respectively, suggesting a common mechanism [64]. The relative rates of exchange at C-2 in the thiazolium compounds with an aminopyrimidylmethyl, a benzyl, or a methyl substituent was 25:3:1 [47]. For the same substituents, the relative rates at the C-α position of the hydroxybenzyl derivatives was 2.9:1.3:1, respectively [64]. Inductive differences between an aminopyrimidine ring, benzene ring, and a methyl group appear to be sufficient to account for the observed differences [64]. The attenuation of the substituent effect on the exchange rate at C-α relative to exchange at C-2 is understandable since the reaction center is one carbon farther away from the substituent. Resonance of the type A ↔ B (Scheme 5) requires delocalization of charge into the thiazolium ring, and MO calculations [42] indicate that this delocalization is energetically favorable. Partial negative charges of −0.44 and −0.34 were assigned to C-α of the HET and HBT α-carbanions, respectively, with the remaining part of the charge delocalized into the thiazolium ring and also into the phenyl ring in the case of HBT. These data, however, do not exclude conformations such as A or B (Scheme 6) (see the section on conformation) from being transiently involved in the interaction of substrates with C-2 (see the concluding section).

At least two factors contribute to the acidity of the (C-α)—H: (a) Resonance involving the canonical forms A and B (Scheme 5) stabilizes the α-carbanion. (b) Carbon magnetic resonance studies [64] show a downfield shift of 24 ppm for C-2 when thiamin is converted to HET. The chemical shift of C-2 of HET is in the region normally assigned to carbonyl carbons of esters and amides, which bear a partial positive charge. Such a charge would facilitate the breaking of the (C-α)—H bond. On the basis of the correlation between total charge densities and ^{13}C chemical shifts found for some heteroaromatic carbocations [66], charges of +0.19 and +0.35 can be calculated (A. A. Gallo, unpublished observations) for C-2 of thiamin and HET, respectively. In excellent agreement are the Hückel populational analyses of Jordan [45, 67], who calculated charges of +0.15 and +0.30 for C-2 of thiamin and HET, respectively. In contrast, X-ray crystallographic studies of HET [65] placed a partial negative charge on C-2 of HET. Because the crystallographic technique is rather indirect and since the results of cmr agree with theoretical calculations, we must conclude that a partial positive charge resides on C-2 of HET. The X-ray data for HET did provide an additional insight, however, which might be relevant. The intramolecular electrostatic interaction between the thiazolium sulfur atom and the (C-α)—OH group, which was found in the crystal structure, would tend to pull electrons away from the (C-α)—OH bond, making both the OH and (C-α)—H groups more acidic. There may be some evidence for this acidifying effect, since the pK_a of the α-OH group of HET is approximately 11 [68], making it about seven orders of magnitude more acidic than ethanol (pK_a 18); that factor would appear to be too great to be accounted for just by the substituent effect of the thiazolium ring.

Conformation

The conformation of thiamin appears to be of critical importance for its catalytic function, since small steric perturbations lead to much reduced catalytic activity. Breslow [26] found that, whereas both N-methyl- and N-ethyl-4-methylthiazolium ions catalyzed the benzoin condensation, N-isopropyl-4-methylthiazolium ion was inactive, presumably due to steric inhibition of C-2 reactivity. Biggs and Sykes [69] found that the analogs of thiamin in which the two rings are bridged by a dimethylene group (4) and an α-methylmethylene group (5) have 67 and 5%, respectively, of the activity of thiamin in catalyzing the acetoin condensation. The activity of 4 is close to what one would predict [59] for attenuation of an inductive effect by a second

4   5

methylene group (51%), and this result suggests that the distance between the pyrimidine and thiazolium rings may not be critical. On the other hand, the low activity of 5 may reflect not only decreased reactivity at C-2 due to hindrance by the added methyl group, but also some subtle change in the relative orientation of the two rings. This point of view is supported by

Schellenberger [8], who ascribes the lack of coenzymatic activity of 6'-methylthiamin and the partial activity of 6'-methyl-4-northiamin to a difference in relative orientation of the thiamin rings. Models show that the (C-4)—CH_3 group is almost in contact with (C-6')—H in one conformation (A, Scheme 6). The extra methyl group at C-6' of the 6'-methylthiamin derivative leads to twisting of the molecule about the CH_2 bridge, destabilizing that conformation, and this destabilization is eliminated when the 4-methyl group is removed. Coenzymatic activity, however, is a complex function of different steric and electronic factors for each of the steps of the enzymatic mechanism, and it also involves binding to the enzyme. Consequently, it might be more instructive to compare these compounds in simple non-enzymatic tests [e.g., rates of (C-2)—H exchange or adduct formation].

The catalytically active conformation of HET proposed by Schellenberger (A, Scheme 6) is different from the crystal conformation in that the crystal conformation [65] has the (C-2)—(C-α) bond rotated in such a way that the (C-α)—OH makes a close contact with the sulfur atom of the ring. This interaction appears to have a great influence on the conformation of the 2-(α-hydroxyethyl) side chain in the crystalline state, since apparently unhindered rotation* about the (C-2) —(C-α) bond ($\phi_{(C-2)-(C-\alpha)}$) would allow an intramolecular hydrogen bond between the pyrimidine amino group and the (C-2)—OH group (see Scheme 7), but this is not observed in the crystal [65].

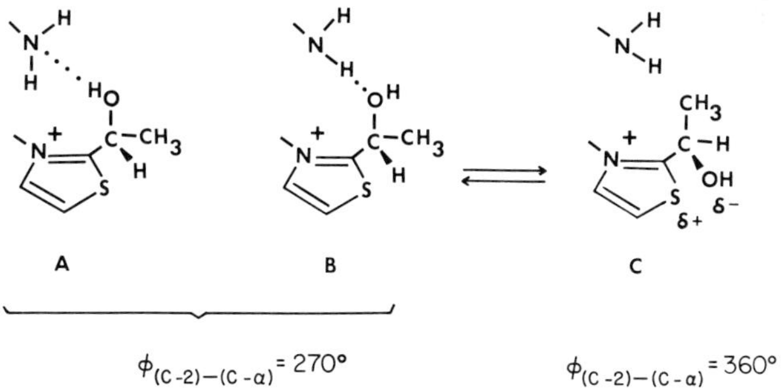

$$\phi_{(C-2)-(C-\alpha)} = 270° \qquad \phi_{(C-2)-(C-\alpha)} = 360°$$

Scheme 7

The strength of the first type of hydrogen bond (A, Scheme 7) is open to question in any case, since the amino group is, at best, a very weak base and is consistently a hydrogen bond donor rather than an acceptor as discussed below in the section on acidic and basic sites in the aminopyrimidine ring. The second type (B, Scheme 7) is an alternative that is discussed in the concluding

---

* $\phi_{(C-2)-(C-\alpha)}$, rotational angle about the (C-2)—(C-α) bond in 2-(α-hydroxyalkyl) derivatives of thiamin; defined in more detail in Jordan [67].

section. Steric factors would not preclude the formation of **A** or **B** (Scheme 7) in solution, however, since the conformational energy difference between **A** or **B** and **C** (Scheme 7) appears [67] to be negligible. Carbon magnetic resonance studies suggest that in HBT another (C-2)—(C-α) rotamer may exist in solution. The aminopyrimidine carbon atoms are shielded by 1–1.5 ppm [64] by the (C-α)–benzene ring, consistent with a significant population of a rotamer that places the (C-α)–benzene ring and the aminopyrimidine ring in a coplanar arrangement [corresponding to a (C-2)—(C-α) rotational angle of 150°]. This rotamer would be only 0.4 kcal/mole [67], less stable than the most stable rotamer **C** (Scheme 7).

The solution conformation of thiamin is probably different from that of HET since, in the proton nmr spectrum of thiamin, the (C-6′)—H and (C-4)—CH$_3$ signals are shifted downfield relative to the corresponding signals for HET. In the crystal structure of HET [65], (C-6′)—H is situated 2.16 Å above the plane of the thiazolium ring and (C-4)—CH$_3$ is above the pyrimidine ring so that each should be strongly shielded by the respective ring currents. Such shieldings are indeed observed [34,70]). A similar and somewhat larger effect is observed in the pmr and cmr spectra of HBT and other α-hydroxybenzyl derivatives compared to the respective unsubstituted compounds, due to additional shielding by the phenyl ring [64]. Simultaneous shielding of the proton signals of (C-6′)—H and (C-4)—CH$_3$ would be expected only in a narrow range of HET conformations approximating the crystal conformation. The data support the conclusion that the solution and crystal conformations of HET may be similar. The data do not define the solution conformation of thiamin, since there are many possible conformations of thiamin which have (C-6′)—H and (C-4)—CH$_3$ in less shielded positions than in HET. In one observed crystal conformation of thiamin [49], (C-2)—H is over the pyrimidine ring, where it should be strongly shielded by the ring current, in comparison with related compounds lacking an aminopyrimidine ring. The amount of this shielding, calculated according to the procedure developed by Johnson and Bovey [71,72] for the benzene ring but using a factor that can be estimated to relate the ring current of a heteroaromatic ring to that of benzene [73,74] (in this case an aminopyrimidine ring), shows that for the crystal conformation of thiamin, (C-2)—H should be shielded by approximately 1.6 ppm (A. A. Gallo, unpublished observations). In another conformation ("V" conformation) deduced from nmr data for the complex of thiamin and indoleacetate [75], this shielding is absent (Table 3) [49,76]. Since the chemical shifts of (C-2)—H in thiamin and some analogs are close to that found in the reference compound (Table 3), it is clear that in solution there is no ring current shielding of (C-2)—H in thiamin. Possibly weak lattice forces stabilize thiamin in a conformation in the crystal that is different from that found in solution. This suggestion is

TABLE 3

Predicted and Observed Proton Chemical Shifts of (C-2)—H

	Shielding effect			
	Predicted		Observed	
Compound[a]	Crystal conformation	"V" conformation	δ(ppm)	$\Delta\delta$(ppm)[b]
B	—	—	9.69	—
D	+0.2[c]	+0.2	9.75	+0.06
A	−1.6[d]	+0.1	9.72	+0.03
F	—	—	9.87	+0.18
G	—	—	9.67	−0.02

[a] See below and Table 2.

F:

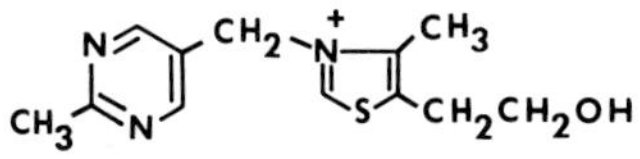

G:

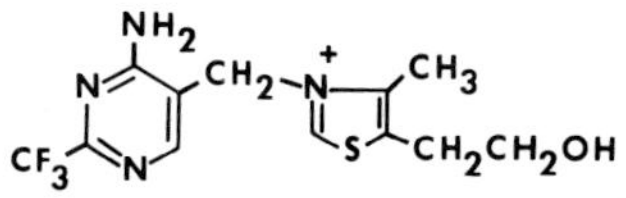

[b] Chemical shift difference, $\Delta\delta$, of (C-2)—H; δ (A, D, F, G) − δ (B).
[c] Power et al. [76].
[d] Kraut and Reed [49].

supported by recent theoretical calculations [45], which show that there is a wide range of relative orientations of the two rings of thiamin which have similar energy. The conformational energy differences among the crystal conformation of thiamin [49], TPP [50,77], and the "V" conformation [75] are all low enough that any one of these conformations or a mixture could predominate in solution. In further support of the latter argument, it was found that (thiamin H$^+$)·(CdCl$_4$$^-$) crystallizes [78] in a conformation similar to that of HET and different from that found for thiamin when the counterion was Cl$^-$. This HET-like conformation of thiamin also cannot be the predominant conformer in solution, since it does not agree with the observed deshielding of the (C-6′)—H and (C-4)—CH$_3$ of thiamin relative to HET. The conformational energy calculations [45] and examination of molecular models indicate that conformations resulting from changes in ϕ_{CN}* from the crystal conformation of thiamin [49] over a range of 180° have about the

* $\phi_{CN'}$ rotational angle about the (N-3)—CH$_2$ bond; defined in more detail in Jordan [45].

same energy. Since in this conformational range, (C-2)—H moves back and forth between the shielding and deshielding regions of the pyrimidine ring, a rapidly equilibrating mixture of these conformations could be responsible for the nearly similar chemical shifts of (C-2)—H in 3,4-dimethyl-5-(2-hydroxyethyl)thiazolium ion and thiamin (Table 3). In contrast, the conformational mobility of HET [67] is restrictied by the substituent at C-2; consequently, only a limited number of conformations, which differ mainly by small changes in ϕ_P,† may be important in solution. Studies of ^{13}C nmr longitudinal relaxation times (T_1) support this supposition (A. A. Gallo, unpublished observations). For thiamin, the T_1 values indicate substantial ring rotation relative to molecular tumbling, whereas the values for HET show that the two rings are much less mobile, being essentially "locked" into position by the C-2 substituent. This model can also explain the observation (A. A. Gallo, unpublished observations) that the pmr signal of (C-4)—CH_3 of HET is not affected by protonation or deprotonation of N-1', whereas in thiamin there is a change of 0.1 ppm [61]. Thiamin, being more flexible, can adopt a different set of conformations when N-1' becomes deprotonated, so a change in shielding of the (C-4)—CH_3 group can result; this flexibility is absent in HET. A change in conformation when N-1' of thiamin is deprotonated could result from a change in the electrostatic interaction between the positively charged N-3 and the changing charge at N-1'. On the other hand, the crystal conformation of thiamin in the N-1' deprotonated state [79] is nearly identical to that in the protonated state [49]. Crystal forces other than electrostatic interactions may therefore play a predominant role in determining the conformation in both crystals.

The crystal conformation of the 5-(2-hydroxyethyl) side chain is not strongly determined for thiamin and HET; in the crystal structure of HET [65] this group occupies a disordered position, and for both compounds lattice forces and hydrogen bonding appear to have a dominant influence on the conformation of this group. In all of the crystal structures of thiamin and its derivatives the torsional angle defined by S(1)—C(5)—C(5$_\alpha$)—C(5$_\beta$) is always near 90°. Also, in thiamin itself, the C(5)—C—C—OH ethanelike linkage is *gauche* rather than *anti*. The *gauche* arrangement is maintained in TPP, where the negatively charged pyrophosphate chain is folded over the positively charged thiazolium ring [77]. The conformation of this part of TPP is probably determined largely by the distribution of formal charges in the molecule, since, in this case, separation of unlike charges is minimized. In contrast, in the case of metal ion complexes of TPP in solution, in which the charge on the pyrophosphoryl moiety is effectively neutralized by the metal

† $\phi_{P'}$ dihedral or torsional angle for the bonds (N-3)—CH_2—(C-5')—(C-4'); defined in more detail in Sax *et al.* [65].

ion, pmr studies [80–82] support an extended conformation. Even in the absence of metal ions, the temperature dependence of proton T_1 values for TPP shows that in the range 9°–60°C there is no distinct conformational change; rather there is a continuous unfolding of the molecule as the temperature is raised [82].

Acidic and Basic Sites in the Aminopyrimidine Ring

In addition to (C-2)—H of the thiazolium ring, several groups in the aminopyrimidine moiety of thiamin are potential acidic and basic sites. These include the three nitrogen atoms and, less obviously, the (C-2')—CH$_3$ group [50,62]. The participation of the aminopyrimidine moiety in the steps of the catalytic mechanism, in binding of the divalent cation essential for the coenzymatic action of thiamin, and in apoenzyme–coenzyme binding is still only partially understood, and opposing theories remain to be resolved. Most notably, Schellenberger [8] assumes that the amino group is relatively basic (at least when TPP is enzyme bound) so that it may be considered an intramolecular base or possibly a nucleophile. More recently it has become clear that amino groups in purines and pyrimidines have considerable acidic character and are more akin to amides and amidines rather than to the aromatic amines, which they superficially resemble [61,83–89]; e.g., rotation about the C—N bond is restricted [89,90]. Breslow proposed [26,27] that an inductive effect of the pyrimidine moiety stabilizes the C-2 carbanion and other carbanionic intermediates in the synthesis of acyloins. He also proposed, very cautiously, that the amino group might participate as an intramolecular general base, but he pointed out that the amino group is relatively unreactive, presumably due to resonance donation of the unshared pair of electrons on the amino nitrogen atom into the pyrimidine ring. Indeed, since high yields of thiamin or its analogs are obtained when a primary halide (e.g., **6**) is condensed with a substituted thiazole (e.g., **7**) [91,92], all the nitrogens of the pyrimidine moiety must be poor nucleophiles in comparison with the N atom of the thiazole. Under certain circumstances the amino groups of aminopyrimidines do exhibit basic or nucleophilic properties; e.g., the amino group of cytidine can be acylated under relatively mild conditions [93,94].

Ordinary aromatic amines are protonated on the amino nitrogen, i.e.,

$$\text{ArNH}_2 + \text{H}^+ \rightleftharpoons \text{ArNH}_3{}^+$$

the pK for this process being about 4–5. Protonation of the corresponding purine and pyrimidine compounds occurs with similar pK values [21,95], but several lines of evidence show that the amino group is not the proton acceptor. Brown *et al.* [96] observed that the ultraviolet spectrum of an aqueous solution of protonated 4-aminopyrimidine resembles the spectra of pyrimidines methylated on a ring nitrogen atom but does not resemble spectra of such pyrimidines methylated on the amino nitrogen. Mizukami and Hirai [95] studied the variation of pK_a with σ_m (Hammett's meta substituent constants) for a series of substituted 2-methyl-4-aminopyrimidines (**8**) and concluded that protonation occurred on N-1. Moehrle and Tenczer [87] suggest that the protonation must occur preferentially on N-1' of

8   R = H, CN, CHO, COOH,
COOC$_2$H$_5$, CONH$_2$

(**1a**) on the basis of the pK_a values of 4-aminopyrimidine (5.7) and 5-aminopyrimidine (2.6) in comparison with thiamin (pK_a 5.0). Direct evidence for ring protonation was obtained by Jardetzky *et al.* [97] in an nmr study of pyrimidines and nucleosides dissolved in anhydrous trifluoroacetic acid. They concluded that the order of basicity is ring N > amino N > oxygen and that protonation of the amino group did not occur since the protonated states of the molecules still had only two protons in the amino group. Proton magnetic resonance spectroscopy of solutions of thiamin chloride hydrochloride, in liquid SO$_2$, between $-25°$ and $-70°$C, shows conclusively that the protonation occurs at N-1' [87]. Crystallographic studies [50,65] confirm this conclusion: In crystals of the protonated forms of thiamin-PP and HET, the proton is always associated with N-1'. In crystals of thiamin-PP hydrochloride one of the amino protons appears to participate in a hydrogen bond

with the slightly basic N-3'. Such behavior has been seen previously in nonaqueous solutions of unprotonated 2-aminopyrimidine (**9**) and also in 4-methylaminopyrimidine (**10**) [98]. Sax *et al.* [65] point out that in other crystals the amino protons are involved in H bonds with solvent water. From infrared spectral data Mason [98] has calculated the π-electron charge densities on the amino nitrogen in a series of aminopyridine, aminopyrimidines, and aniline. These are aniline, 1.802; 2-aminopyridine, 1.742; 2-aminopyrimidine, 1.682; 4-aminopyrimidine, 1.658. The last value is the smallest in 11 compounds he studied.

The hydrogen-bonding ability of the protons of the amino groups of heterocyclic primary amines is not restricted to the crystalline state, to nonaqueous solutions, or to ring-protonated forms. The well-known double-helical structure of deoxyribonucleic acid is stabilized by multiple hydrogen bonds between adenine–thymine pairs and between guanine–cytosine pairs. The amino groups of adenine, cytosine, and guanine are involved in these bonds [99], and in each case one of the amino protons is bonded to a ring nitrogen or to carbonyl oxygen of the pairing base. In no case is a proton of one base hydrogen bonded to the lone pair of electrons of any of the amino groups of the other. Furthermore, this H bonding is not confined to macromolecular structure nor to the hydrophobic core of the nucleic acid. Raszka and Kaplan [100] found intramolecular H bonding of the amino protons of adenine with the isoalloxazine moiety of flavin adenine dinucleotide, as well as related intermolecular H bonding of adenosine monophosphate with flavin mononucleotide, all in aqueous solutions. Pinnavaia *et al.* [88] report similar intermolecular H bonding in the self-assembly of planar tetramers of 5'-guanosine monophosphate in aqueous solution.

Other evidence that these amino groups are qualitatively different from aromatic primary amines was obtained by McConnell *et al.* [83–85,101] and Raszka and Kaplan [86,90] for purine and pyrimidine nucleotides and by Suchy *et al.* [61] for thiamin. In the studies cited, aqueous solutions of the compounds were examined by pmr at ambient temperature, and in all cases the amino protons gave discrete signals. Such behavior is extraordinary for amines, since exchange of the protons with those of solvent H_2O is usually so rapid that no separate signal can be observed. The integrated intensity of the amino proton signals of both the protonated and unprotonated species never exceeded two per molecule, again showing that the amino group is not the site of protonation. Chauvet-Monges *et al.* [102] recorded the spectrum of thiamin in concentrated H_2SO_4 at $-20°C$ and observed two additional signals at δ 12.5 and 13.0 ppm, which they ascribed to protons bound to N-3' and N-1', respectively. In contrast to results in nonaqueous acidic media [87,97,102], protons bound to ring nitrogens usually exchange so rapidly with solvent protons in dilute aqueous acid that they give no separate signals.

In some of these systems the exchange of NH_2 protons is so slow that simple cooling to 0–10°C may be enough to resolve the NH_2 signal into two signals [85,90], indicative of differing environments for the two protons caused by slow rotation around the C—N bond [103,104]. The second-order rate constants for catalysis of exchange of the protons by H^+, OH^-, and general acids and bases have been measured in several of the cases studied. McConnell *et al.* [83,85,101] noted that k_{H^+} and k_{OH^-} are several orders of magnitude lower than the diffusion-controlled catalytic rate constants, 10^{10}–10^{11} sec^{-1} M^{-1} [105], characteristic of aliphatic and aromatic amines. For adenine and guanine mononucleotides and polyadenylic acid at about 30°C [79,80], $k_{H^+} = 10^6$–10^8 and $k_{OH^-} = 10^7$ to 5×10^8 M^{-1} sec^{-1}; for cytidine nucleotides $k_{OH^-} = 10^9$ M^{-1} sec^{-1} [85]. The catalytic rate constants are determined from the dependence of the line width of the nmr signal on the pH. McConnell and Seawell's lowest values for k_{H^+} are approximately the same as those observed for N,N'-dimethylurea [106]. In the case of thiamin [61] k_{OH^-} is of the same order as that for the nucleotides, but $k_{H^+} = 7 \times 10^2$ M^{-1} sec^{-1}, i.e., 10^3 times smaller than the lowest values observed by McConnell. The extremely small value of k_{H^+} seems to be incompatible with any exchange mechanism that involves direct protonation of the NH_2 group. The smallness of k_{H^+} permits the observation of a third mode of catalysis of the exchange

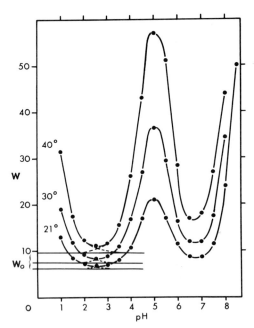

Fig. 3 The pH dependence of the line width of the (C-4′)—NH_2 signal; 60 MHz spectra of 1 M aqueous solutions of thiamin at the temperatures indicated. Line width (W) is expressed in hertz; W_o is the line width of the signal in the absence of exchange. [Reproduced from [61] by permission of the *Journal of Biological Chemistry*.]

of the amino protons, namely, general acidic and general basic catalysis by the protonated and unprotonated forms of thiamin, respectively (Figs. 3 and 4). Other general bases, e.g., acetate and phosphate, also catalyze the exchange with solvent; McConnell has reported similar general acidic–general basic catalysis by imidazole and 2-methylimidazole [101], and he points out that direct protonation of the amino group by H^+ would exclude the ability to observe catalysis by phosphate and amines.

The particular ring nitrogen atom, whether N-1' or N-3', that is the site of protonation in aqueous solutions of thiamin has been a source of controversy. Chauvet-Monges et al. [102] have suggested that a mixture of N-3'- and N-1'-protonated molecules is produced upon acidifying thiamin. We find [47], however, that the changes in the ^{13}C resonances upon protonation are consistent only with predominant N-1' protonation. In particular, C-6' is strongly shielded and C-4' deshielded upon protonation, whereas deshielding of C-6' and shielding of C-4' would be expected if N-3' were protonated

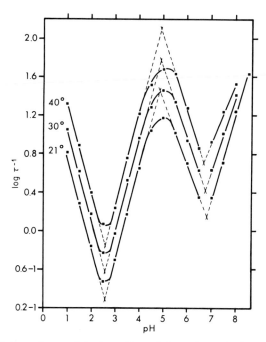

Fig. 4 The pH dependence of the logarithm of the reciprocal lifetime ($\log \tau^{-1}$) of the protons in the (C-4')—NH_2 group of thiamin; 60 MHz spectra of 1 M aqueous solution of thiamin at the temperatures indicated. The τ^{-1} expressed in seconds^{-1} refers to $2\pi (W - W_o)$. [Reproduced from [61] by permission of the *Journal of Biological Chemistry*.]

[107,108]. Also, the similarities of the ultraviolet spectra of protonated and methylated pyrimidines [96] and the analogy with the meta-substituted pyrimidines [95] and the pK_a values cited by Moehrle and Tenczer [87] are strong presumptive evidence that N-1' is the site of protonation. As there is no reason to assume that distribution of charge in a small molecule will differ markedly in solution and in the crystalline state, it is probably also safe to argue by analogy from the crystallographic data [50,65], which show protonation at N-1'. Although we have expressed reservations [64] about trying to assess small changes in charge distribution from X-ray data, there can be little doubt that crystallographic studies are valuable in assessing larger amounts of charge. Pletcher and Sax [50] have calculated the distribution of charge in the protonated species and found that about half of the positive charge resides on N-1' and most of the remainder is on the amino group. Bond lengths found in protonated thiamin [49] and some reference compounds are shown in Table 4. Clearly, the lengths of the C—N bonds linking the amino group to the ring in protonated thiamin and in unprotonated adenosine [109] are much shorter than that in acetanilide, reflecting the same partial double-bond character that is characteristic of amides, pyridine, etc. This is consistent with the decreased π-electron densities on the amino N atoms noted above [98], since formation of the multiple bond requires partial delocalization of these electrons. Among the ribonucleotides only CMP shows restricted rotation of the NH_2 group at neutral pH [90], and CMP has the greatest tendency for self-association via hydrogen bonding in aqueous solution. This may mean that the lone pair of electrons of the

TABLE 4

Carbon–Carbon and Carbon–Nitrogen Bond Lengths (Å)

Carbon–nitrogen bonds in representative Compounds[a]		Bond lengths in protonated thiamin[b]	
Aromatic in $C_6H_5NHCOCH_3$	1.426	(N-1')—(C-2')	1.333
Shortened, partial C=N		(C-2')—(N-3')	1.306
In C_5H_5N	1.352	(N-3')—(C-4')	1.367
In $HCONH_2$	1.322	(C-4')—(C-5')	1.434
RC≡N	1.158	(C-5')—(C-6')	1.354
		(C-6')—(N-1')	1.362
5'-Bromo-5'-deoxyadenosine[c]		(C-4')—NH_2	1.316
(C-6)—NH_2	1.29		

[a] "Handbook of Chemistry and Physics," 46th ed. p. F119.
[b] Kraut and Reed [49].
[c] Voet and Rich [109].

amino group is more extensively delocalized into the ring in aminopyrimidines than it is in aminopurines. Table 4 also shows that, in the protonated structure, the (C-2′)—(N-3′) and (C-5′)—(C-6′) bonds, respectively, are not only substantially shorter than the other C—N and C—C bonds in the ring, but are also shorter than the corresponding bonds in unprotonated thiamin [79]. We believe, therefore, that there is little contribution from a species protonated at N-3′, as proposed by other authors [102,110]. It seems clear that, in monoprotonated thiamin, the proton is associated with the ring nitrogen para to the amino group, in contrast to the case of aminopurines [111] which are protonated on the ortho nitrogen atom. It is worth noting that ring protonation is not unique to heterocyclic compounds; when 1,3,5-triaminobenzene is dissolved in dilute aqueous acid the benzenium ion (11) is formed in preference to an ammonium form [112].

11

CONCLUSIONS

It seems clear that formation of the C-2 carbanion, essential for thiamin action, proceeds much faster in thiazolium salts than in imidazolium salts, while oxazolium salts at neutral pH cannot give rise to carbonions since they exist predominantly in the ring-opened form. Further, the thiazolium ring is especially well suited for the role that thiamin plays, because it can also increase the acidity of the α-CH group of HET by virtue both of the adjacent partial positive charge on C-2 and of the resonance stabilization of the resulting α-carbanion. Another feature of the thiazolium ring that is essential for thiamin action is its aromaticity; this feature allows the integrity of the ring to be maintained throughout the catalytic process, against any influences that might divert the mechanism to the formation of nonaromatic products. On the basis of these electronic considerations it seems natural that the primary site of the catalytic activity is a thiazolium ring rather than some other moiety. The aminopyrimidine ring plays a secondary role in bolstering the acidity of (C-2)—H inductively and possibly may also act as a Lewis acid during the reaction of the C-2 carbanion with substrates (A, Scheme 8).

Nuclear magnetic resonance results support the concept that the solution conformations of thiamin and HET are different. Furthermore, both the

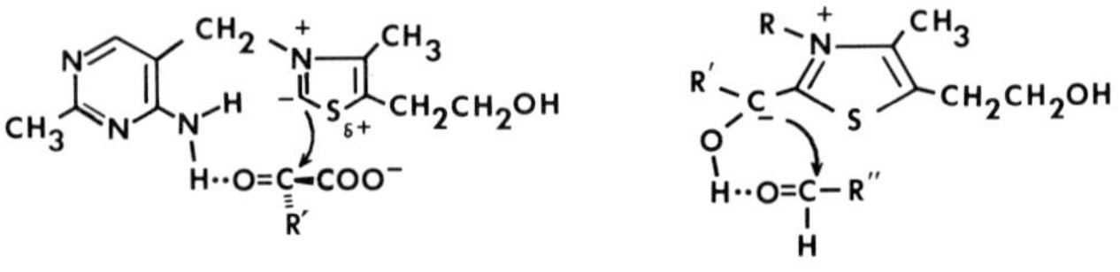

Scheme 8

calculations of Jordan [45,67] and inspection of molecular models show that thiamin is more conformationally mobile than HET with respect to the relative orientations of the pyrimidine and thiazolium rings; preliminary ^{13}C relaxation measurements support this tentative conclusion. Although the "V" conformation of thiamin is supported by a study of the complexes of indolyl derivatives and thiamin and satisfies the shielding criteria discussed in the section on conformation, this conformation may be specifically stabilized by the π–π interaction, so that in the absence of such an interaction, other conformations such as **B** (Scheme 6) (shown for HET) that differ slightly in the relative orientation of the pyrimidine and thiazolium rings are also possible. These conformations would permit the C-4′ amino group to participate in the catalytic mechanism leading to the formation of HET (**A**, Scheme 8), although it is unlikely that **B** (Scheme 6) is the predominant solution conformation of HET (see the section on conformation).

The participation of the C-4′ amino group as a proton donor (**B**, Scheme 6) rather than acceptor (**A**, Scheme 6) is based on the extensive and consistent evidence that the amino groups of aminopurines and aminopyrimidines, including thiamin, have properties which indicate that they should be considered Lewis acids rather than bases, although they may be weakly basic or weakly nucleophilic in special circumstances (e.g., in the formation of thiochrome) [21,113] and in reactions leading to their acylation [93,94]. On this basis, one may formulate a two-step mechanism for the acyloin condensation in which each transition state (**A** and **B**, Scheme 8) involves stabilization of a developing oxyanion by an intramolecular acidic function. Structure **A** (Scheme 8) also takes account of the partial positive charge assigned to the sulfur atom [49,50] and suggests that electrostatic interaction aids in the approach and positioning of the α-ketocarboxylate substrate. In agreement with such a concept, preliminary experiments (G. A. Bantle and H. Z. Sable, unpublished observations) show that the formation of HBTFT from benzaldehyde and TFT [92] occurs faster than the formation of HBT; the pK for protonation of the pyrimidine ring of TFT is almost 4 pK units lower than that of thiamin (J. E. Stuehr, H. Z. Sable, and W. Rosenberg, unpublished observations), and correspondingly the amino group of TFT would be more acidic.

Although thiamin is an efficiently designed catalyst for carbanion-forming processes, its activity is small relative to enzymes that utilize thiamin coenzymes. Factors that contribute to the differential in reaction rate between the inherent catalytic activity of thiamin and that of enzymes for which thiamin is a cofactor probably include the following:

1. Basic sites on the enzyme participate in the abstraction of the proton at C-2, C-α, or α-OH, depending on the particular reaction.

2. The enzyme itself may stabilize a particular conformation, possibly through a thiamin–tryptophan interaction [7], which might favor the participation of the amino group as proposed above (Scheme 8). The interaction of cationic groups on the enzyme, such as protonated lysine and arginine residues, or metal ions with N-1' as proposed by Schellenberger [8], would tend to further increase the Lewis acid character of the 4'-amino group required for this participation.

3. The enzyme environment might be very hydrophobic, as suggested by Lienhard et al. [63, 114], who showed that an ethanol environment accelerated specific steps in the mechanism. It is interesting, in this light, that the pK_a of (C-2)—H is actually increased slightly in methanol solution compared to aqueous solutions [39]. On the other hand, Ullrich et al. [7] have suggested that the lipophilic nature of the enzyme active site might shield very active intermediates such as the α-carbanion of HET from solvent molecules and prolong the lifetimes of these intermediates.

4. Another feature of the enzyme–coenzyme interaction may be specific recognition of the aminopyrimidine ring by the enzyme, representing a secondary binding function of this moiety; the aminopyrimidyl moiety is common to many coenzymes and a binding site for this group may have evolved from a common progenitor through the course of evolution [27,115]. This thesis would be supported by evidence for a common mode of binding of thiamin to thiamin-dependent enzymes [116]. One would need to know which face of enzyme-bound thiamin is exposed to solvent; although thiamin is achiral, it exists in mirror-image conformations with moderate energy barriers separating them, and enzymes might preferentially interact with one set of mirror-image conformations.

ACKNOWLEDGMENT

This chapter represents Paper XI in the series entitled "Coenzyme Interactions" from this laboratory; for Paper X see Gallo and Sable [64]. This work was supported by National Science Foundation Grants GB-41161 and PCM 73-02219 and by National Institutes of Health Grants 5T1 GM35 and AM-18888.

ADDENDUM

The literature search for this chapter was completed early in the spring of 1976. In this addendum we describe important observations that have come to our attention since that time.

We have referred to a suggestion by Breslow [36] that the anionic form of the thiamin–aldehyde adduct (Scheme 5) is stabilized by resonance (A ↔ B). Gutowski and Lienhard [117] have prepared thiamin thiazolone-PP and thiamin thiothiazolone-PP. These compounds are related to **B** (Scheme 5) but have an oxygen atom or a sulfur atom, respectively, in place of the α-hydroxyethyl moiety. The authors propose that both compounds are transition state analogs for TPP-dependent enzymes, and they provide evidence that, in accord with such a proposal, these substances bind to the TPP sites of the pyruvate dehydrogenase complex of *E. coli* much more strongly than TPP itself.

Pletcher, Sax and their associates [118–120] have extended their studies of the crystal structures of thiamin and its derivatives and analogs. The conformation of HBT in the crystalline lattice [118] agrees well with the conformation we have proposed above on the basis of our nmr studies. In the crystalline lattice of thiamin picrolonate [119] there is stacking interaction in solution between the neutral pyrimidine ring and the picrolonate anion, that is related to the interaction in solution between thiamin and indolyl compounds [75] from which we originally deduced the "V" conformation. However, in the crystalline lattice the thiamin molecule does not present a "V" conformation, but is found in the same conformation that these authors observed previously [49]. The same conformation is seen in crystals of the neutral zwitterion of TPP [120]. In those crystals the pyrophosphate side chain is extended away from the pyrimidine ring. Previously we found similar conformations in pmr studies of Ni^{2+} and Mn^{2+} complexes of TPP [80,81].

REFERENCES

1. C. Funk, *J. Physiol.* (*London*) **43**, 395 (1911).
2. R. R. Williams and T. D. Spies, "Vitamin B_1 and its Use in Medicine." Macmillan, New York, 1938.
3. K. Lohmann and P. Schuster, *Biochem. Z.* **294**, 188 (1937).
4. I. G. Leder, *Metab. Pathways*, 3rd Ed. **7**, 57–85 (1975)
5. F. A. Robinson, "The Vitamin Co-factors of Enzyme Systems," p. 6. Pergamon, Oxford, 1966.
6. A. V. Morey and E. Juni, *J. Biol. Chem.* **243**, 3009 (1968).
7. J. Ullrich, Y. M. Ostrovsky, J. Eysaguirre, and H. Holzer, *Vitam. Horm.* (*N.Y.*) **28**, 365–398 (1970).

8. A. Schellenberger, *Angew. Chem., Int. Ed. Engl.* **6**, 1024 (1967).
9. L. O. Krampitz, *Annu. Rev. Biochem.* **38**, 213 (1969).
10. W. Langenbeck and Z. Hutschenreuter, *Z. Anorg. Allg. Chem.* **188**, 1 (1930).
11. W. Langenbeck, *Ergeb. Enzymforsch.* **2**, 314 (1933).
12. K. Wiesner and Z. Valenta, *Experientia* **12**, 190 (1956).
13. K. G. Stern and J. L. Melnick, *J. Biol. Chem.* **131**, 597 (1939).
14. A. Lapworth, *J. Chem. Soc.* **83**, 995 (1903).
15. A. Lapworth, *J. Chem. Soc.* **85**, 1209 (1904).
16. T. Ugai, S. Tanaka, and S. Dokawa, *J. Pharm. Soc. Jpn.* **63**, 269 (1943).
17. T. Ugai, S. Dokawa, and S. Tsubokawa, *J. Pharm. Soc. Jpn.* **64**, 3 (1944).
18. S. Mizuhara and P. Handler, *J. Am. Chem. Soc.* **76**, 571 (1954).
19. S. Mizuhara, R. Tamura, and H. Arata, *Proc. Jpn. Acad.* **27**, 302 (1951).
20. R. Breslow, *Chem. Ind. (London)* B.I.F. Rev., R28 (1956).
21. D. E. Metzler, *in* "The Enzymes" (P. D. Boyer. H. Lardy, K. Myrbäck, eds.), 2nd ed., Vol. 2, Chapter 9, p. 295. Academic Press, New York, 1960.
22. J. M. Duclos and P. Haake, *Biochemistry* **13**, 5358 (1974).
23. J. M. Duclos, Ph.D. Thesis, Wesleyan University, Middletown, Connecticut (1973).
24. L. L. Ingraham and F. H. Westheimer, *Chem. Ind. (London)* p. 846 (1956).
25. R. Breslow, *Chem. Ind. (London)* p. 893 (1957).
26. R. Breslow, *J. Am. Chem. Soc.* **80**, 3719 (1958).
27. R. Breslow and E. McNelis, *J. Am. Chem. Soc.* **81**, 3080 (1959).
28. L. O. Krampitz, G. Gruell, C. S. Miller, J. B. Bicking, H. R. Skeggs, and J. M. Sprague, *J. Am. Chem. Soc.* **80**, 5893 (1957).
29. L. O. Krampitz, I. Suzuki, and G. Gruell, *Brookhaven Symp. Biol.* **15**, 282 (1962).
30. H. W. Goedde, H. Inouye, and H. Holzer, *Biochim. Biophys. Acta* **50**, 41 (1961).
31. H. Holzer and K. Beaucomp, *Angew. Chem.* **71**, 776 (1959).
32. G. L. Carlson and G. M. Brown, *J. Biol. Chem.* **236**, 2099 (1961).
33. J. J. Mieyal, R. G. Votaw, L. O. Krampitz, and H. Z. Sable, *Biochim. Biophys. Acta* **141**, 205 (1967).
34. J. J. Mieyal, G. Bantle, R. G. Votaw, I. A. Rosner, and H. Z. Sable, *J. Biol. Chem.* **246**, 5213 (1971).
35. E. Juni. *J. Biol. Chem.* **236**. 2302 (1961).
36. R. Breslow, *Ann. N.Y. Acad. Sci.* **98**, 445 (1962).
37. P. Haake, L. P. Bausher, and W. B. Miller, *J. Am. Chem. Soc.* **91**, 1113 (1969).
38. J. Crosby and G. E. Lienhard, *J. Am. Chem. Soc.* **92**, 5707 (1970).
39. R. F. W. Hopmann and G. P. Brugnoni, *Nature (London), New Biol.* **246**, 157 (1973).
40. D. J. Cram, "Fundamentals of Carbanion Chemistry." Academic Press, New York, 1965.
41. B. Pullman and C. Spanjaard, *Biochim. Biophys. Acta* **46**, 576 (1961).
42. B. Pullman and A. Pullman, "Quantum Biochemistry," pp. 636–821. Wiley (Interscience), New York, 1963.
43. T. C. Bruice and S. Benkovic, "Bioorganic Mechanisms," Vol. 2, pp. 204–226. Benjamin, New York, 1966.
44. R. L. Collin and B. Pullman, *Arch. Biochem. Biophys.* **108**, 535 (1964).
45. F. Jordan, *J. Am. Chem. Soc.* **96**, 3623 (1974)
46. G. C. Levy and G. L. Nelson, "Carbon-13 Nuclear Magnetic Resonance for Organic Chemists," pp. 136–148. Wiley (Interscience), New York, 1972.
47. A. A. Gallo and H. Z. Sable, *J. Biol. Chem.* **249**, 1382 (1974).
48. J. B. Stothers, "Carbon-13 NMR Spectroscopy," pp. 239–278. Academic Press, New York, 1972.

49. J. Kraut and H. J. Reed, *Acta Crystallogr.* **15**, 747 (1962).

50. J. Pletcher and M. Sax, *J. Am. Chem. Soc.* **94**, 3998 (1972).

51. J. B. Stothers, "Carbon-13 NMR Spectroscopy," pp. 331–334. Academic Press, New York, 1972.

52. P. Haake and W. B. Miller, *J. Am. Chem. Soc.* **85**, 4044 (1963).

53. A. L. Allred, *J. Inorg. Nucl. Chem.* **17**, 215 (1961).

54. R. Breslow, *J. Am. Chem. Soc.* **79**, 1762 (1957).

55. H. Quast and S. Hunig, *Angew. Chem., Int. Ed. Engl.* **3**, 800 (1964).

56. J. A. Elvidge, J. R. Jones, C. O'Brien, E. A. Evans, and H. C. Sheppard, *Adv. Heterocycl. Chem.* **16**, 1–31 (1974).

57. H. W. Wanzlick, *Angew. Chem.* **74**, 129 (1962).

58. W. Hafferl, R. Lundin, and L. L. Ingraham, *Biochemistry* **2**, 1298 (1963).

59. L. P. Hammett, "Physical Organic Chemistry," 2nd ed., pp. 366–368. McGraw-Hill, New York, 1970.

60. A. M. Chauvet-Monges, C. Rogeret, C. Briand, and A. Crevat, *Biochim. Biophys. Acta* **304**, 748 (1973).

61. J. Suchy, J. J. Mieyal, G. Bantle, and H. Z. Sable, *J. Biol. Chem.* **247**, 5905 (1972).

62. D. W. Hutchinson, *Biochemistry* **10**, 542 (1971).

63. J. Crosby, R. Stone, and G. E. Lienhard, *J. Am. Chem. Soc.* **92**, 2891 (1972).

64. A. A. Gallo and H. Z. Sable, *J. Biol. Chem.* **251**, 2564 (1976).

65. M. Sax, P. Pulsinelli, and J. Pletcher, *J. Am. Chem. Soc.* **96**, 155 (1974).

66. E. Dradi and G. Gatti, *J. Am. Chem. Soc.* **97**, 5472 (1975).

67. F. Jordan, *J. Am. Chem. Soc.* **98**, 808 (1976).

68. P. Pulsinelli, Ph.D. Thesis, University of Pittsburgh, Pittsburgh, Pennsylvania (1970).

69. J. Biggs and P. Sykes, *J. Chem. Soc.* p. 1848 (1959).

70. J. Ullrich and A. Mannschreck, *Eur. J. Biochem*, **1**, 110 (1967).

71. E. C. Johnson, Jr. and F. A. Bovey, *J. Chem. Phys.* **29**, 1012 (1958).

72. J. W. Emsley, J. Feeney, and L. H. Sutcliffe, "High Resolution Nuclear Magnetic Resonance Spectroscopy," Vol. 1. Pergamon, Oxford, 1966.

73. T. R. Janson, A. K. Kane, J. F. Sullivan, K. Knox, and M. E. Kenney, *J. Am. Chem. Soc.* **91**, 5210 (1969).

74. J. J. Mieyal, L. T. Webster, Jr., and U. A. Siddiqui, *J. Biol. Chem.* **249**, 2633 (1974).

75. J. E. Biaglow, J. J. Mieyal, J. Suchy, and H. Z. Sable, *J. Biol. Chem.* **244**, 4054 (1969).

76. L. Power, J. Pletcher, and M. Sax, *Acta Crystallogr., Sect. B* **26**, 143 (1970).

77. J. Pletcher and M. Sax, *Science* **154**, 1331 (1966).

78. M. F. Richardson, K. Franklin, and D. M. Thompson, *J. Am. Chem. Soc.* **97**, 3204 (1975).

79. J. Pletcher, M. Sax, S. Sengupta, J. Chu, and C. S. Yoo, *Acta Crystallogr., Sect. B* **28**, 2928 (1972).

80. A. A. Gallo, I. L. Hansen, H. Z. Sable, and T. J. Swift, *J. Biol. Chem.* **247**, 5913 (1972).

81. A. A. Gallo and H. Z. Sable, *J. Biol. Chem.* **250**, 4986 (1975).

82. H. J. Grande, R. L. Houghton, and C. Veeger, *Eur. J. Biochem.* **37**, 563 (1973).

83. B. McConnell, M. Raska, and M. Mandel, *Biochem. Biophys. Res. Commun.* **47**, 692 (1972).

84. B. McConnell and P. C. Seawell, *Biochemistry* **11**, 4382 (1972).

85. B. McConnell and P. C. Seawell, *Biochemistry* **12**, 4426 (1973).

86. M. Raszka, *Biochemistry* **13**, 4616 (1974).

87. H. Moehrle and J. Tenczer, *Pharm. Acta Helv.* **48**, 489 (1973).
88. T. J. Pinnavaia, H. T. Miles, and E. D. Becker, *J. Am. Chem. Soc.* **97**, 7198 (1975).
89. R. R. Shoup, H. T. Miles, and E. D. Becker, *J. Phys. Chem.* **76**, 64 (1972).
90. M. Raszka and N. O. Kaplan, *Proc. Natl. Acad. Sci. U.S.A.* **69**, 2025 (1972).
91. A. Schellenberger and K. Winter, *Hoppe-Seyler's Z. Physiol. Chem.* **344**, 16 (1966).
92. J. A. Barone, H. Tieckelmann, R. Guthrie, and J. F. Holland, *J. Org. Chem.* **25** 211 (1960).
93. H. G. Khorana, A. F. Turner, and J. P. Vizsolyi, *J. Am. Chem. Soc.* **83**, 686 (1961).
94. R. C. Bleaney, A. S. Jones, and R. T. Walker, *Tetrahedron* **31**, 2433 (1975).
95. S. Mizukami and E. Hirai, *J. Org. Chem.* **31**, 1199 (1966).
96. D. J. Brown, E. Hoerger, and S. F. Mason, *J. Chem. Soc.* p. 4035 (1955).
97. O. Jardetzky, P. Pappas, and N. G. Wade, *J. Am. Chem. Soc.* **85**, 1657 (1963).
98. S. F. Mason, *J. Chem. Soc.* p. 3619 (1958).
99. R. F. Steiner and R. F. Beers, Jr., "The Polynucleotides." Elsevier, Amsterdam, 1961.
100. M. Raszka and N. O. Kaplan, *Proc. Natl.. Acad. Sci. U.S.A.* **71**, 4546 (1974).
101. B. McConnell, *Biochemistry* **13**, 4516 (1974).
102. A. M. Chauvet-Monges, Y. Martin-Borret, A. Crevat, and J. Fournier, *Biochimie* **56**, 1269 (1974).
103. H. T. Miles, R. B. Bradley, and E. D. Becker, *Science* **142**, 1569 (1963).
104. R. R. Shoup, E. D. Becker, and H. T. Miles, *Biochem. Biophys. Res. Commun.* **43**, 1350 (1971).
105. E. Grunwald, A. Loewenstein, and S. Meiboom, *J. Chem. Phys.* **27**, 630 (1957).
106. J. F. Whidby and W. R. Morgan, *J. Phys. Chem.* **77**, 2999 (1973).
107. R. J. Pugmire and D. M. Grant, *J. Am. Chem. Soc.* **93**, 1880 (1971).
108. R. J. Pugmire and D. M. Grant, *J. Am. Chem. Soc.* **90**, 697 (1968).
109. D. Voet and A. Rich, *Proc. Natl. Acad. Sci. U.S.A.* **68**, 1151 (1971).
110. J. Fournier, E. J. Vincent, A. M. Chauvet, and A. Crevat, *Org. Magn. Reson.* **5**, 573 (1973).
111. J. J. Christensen, J. H. Rytting, and R. M. Izatt, *Biochemistry* **9**, 4907 (1970).
112. T. Yamaoka, H. Hosoya, and S. Nagakawa, *Tetrahedron* **24**, 6203 (1968).
113. J. Pletcher, M. Sax, C. S. Yoo, J. Chu, and L. Power, *Acta Crystallogr., Sect. B* **30**, 496 (1974).
114. J. Crosby and G. E. Lienhard, *J. Am. Chem. Soc.* **92**, 5707 (1970).
115. M. Buehner, G. C. Ford, D. Moras, K. W. Olsen, and M. G. Rossmann, *Proc. Natl. Acad. Sci. U.S.A.* **70**, 3052 (1973).
116. H. C. Dunathan and J. G. Voet, *Proc. Natl. Acad. Sci. U.S.A.* **71**, 3888 (1974).
117. J. A. Gutowski and G. E. Lienhard, *J. Biol. Chem.* **251**, 2863 (1976).
118. J. Pletcher, M. Sax, G. Blank, and M. Wood, *J. Am. Chem. Soc.* **99**, 1396 (1977).
119. W. Shin, J. Pletcher, G. Blank, and M. Sax, *J. Am. Chem. Soc.* **99**, 3491 (1977).
120. J. Pletcher, M. Wood, G. Blank, W. Shin, and M. Sax, *Acta Crystallogr.* (submitted for publication).

CHAPTER

6

Cytokinin Antagonists: Regulation of the Growth of Plant and Animal Cells

Sidney M. Hecht

INTRODUCTION

Cytokinins are a class of hormones that were first isolated from plant tissues. They regulate cell division and growth in plants and participate in the differentiation process through their interaction with other plant hormones [1]. Naturally occurring cytokinins are typically N^6-substituted adenine derivatives, and several highly active species have been isolated at the purine, ribonucleoside, and ribonucleotide levels [2–5]. Cytokinins have also been isolated as components of transfer RNA and are believed to be present in the tRNA's of most forms of life [2–6]. Sequencing of several tRNA's has shown cytokinin-active nucleosides to be present exclusively in the position adjacent to the 3' end of the anticodon triplet [7–18], but extensive study of these modified nucleosides as components of tRNA has failed to establish firmly either their function in tRNA or a possible relationship between their occurrence in the macromolecule and their role as growth regulators [19–24].

Certain cytokinins have also been shown to affect the behavior of mammalian cells. For example, they have been reported to regulate the growth of mammalian cells [25–27], to inhibit platelet aggregation [28], and to have immunosuppressive activity [29]. Cytokinins have also been employed as potential anticancer agents in clinical trials [30,31].

Although cytokinins are thought to be required for the growth of all plants, most plants grow without an exogenous source of cytokinin, presumably because they produce their own. This has limited the study of cytokinins to the relatively small number of plant tissues that require added cytokinin for optimal growth and has prompted the search for an antagonist that could block the utilization of endogenous cytokinin, thus extending the study of cytokinins to additional biological systems. A cytokinin antagonist would also be of use in determining the mechanism of cytokinin action in plants and in mammalian systems and might possibly indicate directly whether the mechanism of cytokinin action in plants is related to that in mammals.

PREPARATION AND TESTING OF
POTENTIAL ANTICYTOKININS

Potential cytokinin antagonists were designed on the assumptions that (1) expression of cytokinin activity involved the binding of cytokinin-active species to one or more cellular receptor sites and (2) it would be possible to prepare compounds that were very poor cytokinins but that were sufficiently similar in structure to the cytokinins to permit them to compete for available cytokinin receptor sites, thus diminishing utilization of the cytokinins. This is analogous to an approach that has been employed for the preparation of enzyme inhibitors. For example, allopurinol (4-hydroxypyrazolo[3,4-d]-pyrimidine) is a potent inhibitor of hypoxanthine utilization by xanthine oxidase [32,33], and formycin {7-amino-3-(β-D-ribofuranosyl)pyrazolo[4,3-d]pyrimidine} 5'-triphosphate inhibits the incorporation of the isomeric ATP into polyribonucleotides by DNA-dependent RNA polymerase [34]. Thus, heterocyclic modification of the purine nucleus in each case afforded a substrate analog capable of blocking utilization of the normal substrate.

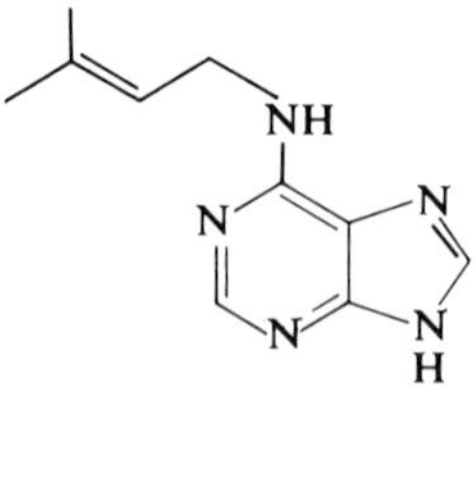

(1)
i⁶Ade

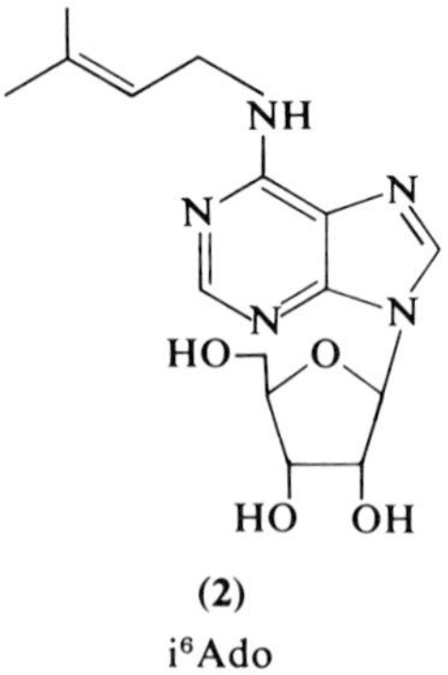

(2)
i⁶Ado

The species utilized as the basis for initial chemical modification was N^6-(3-methyl-2-butenyl)adenine [i⁶Ade (**1**)], which is a potent, naturally occurring cytokinin. Modification of this species in the heterocyclic moiety was known to afford cytokinin analogs having greatly diminished cytokinin activity [1,35], and alterations of this type were therefore employed in an attempt to prepare compounds with anticytokinin activity.

Substituted Pyrazolo[4,3-d]pyrimidines

Four substituted pyrazolo[4,3-d]pyrimidines [(**3**)–(**6**)] were prepared initially as potential cytokinin antagonists. These compounds were obtained by treatment of the known (3-methyl-)7-thiopyrazolo[4,3-d]pyrimidines [36,37] with methyl iodide to afford the respective 7-methylthio derivatives. The latter were heated in the presence of the appropriate amine at reflux temperature under nitrogen to afford (**3**)–(**6**) [38].

(**3**), R = H
(**4**), R = CH₃

(**5**), R = H
(**6**), R = CH₃

By extrapolation of structure–activity relationships determined in the purine series to cytokinin analogs (**3**)–(**6**), it was anticipated that the two analogs with the exocyclic isopentyl substituents [(**5**) and (**6**)] would have lower activity as cytokinins than those with isopentenyl substituents [(**3**) and (**4**)] and that the presence of a methyl group in the 3 position would also diminish the activity of analogs (**4**) and (**6**) relative to (**3**) and (**5**), respectively. It was hoped that assay of compounds with successively weaker cytokinin activities would reveal that the ability of these species to bind to cytokinin receptor sites did not diminish as quickly as their ability to function as cytokinins, so that certain of the species lacking or having very poor cytokinin activity would be capable of blocking the utilization of more potent cytokinins.

The cytokinin activity of compounds (**3**)–(**6**) was assayed in the tobacco bioassay. This involved the growth of tobacco callus in a standard medium containing the mineral salts and organic compounds (including 11.4 μM indole-3-acetic acid) specified by Linsmaier and Skoog [39]. The test compounds were included in replicate cultures at several concentrations, and the individual cultures were maintained in continuous, diffuse light for 4–5 weeks and then used for measurement of fresh weight yields.

As expected, cytokinin analogs (3)–(6) were all considerably less active than i^6Ade (1). Thus, while maximal growth was obtained in response to 9×10^{-3} μM i^6Ade, the isomeric 7-(3-methyl-2-butenylamino)pyrazolo[4,3-d]pyrimidine (3) was maximally active at about $2 \mu M$ concentration. Although analogs (3)–(6) were all much less active as cytokinins than the corresponding purines, they did show the expected order of activities among themselves. Compound (3) was about three times more active than compound (4) and at least ten times more active than (5), the latter of which did not elict maximal growth response at any tested concentration. It was anticipated that analog (6) would be the least active as a cytokinin, and the compound was, in fact, without growth-promoting activity.

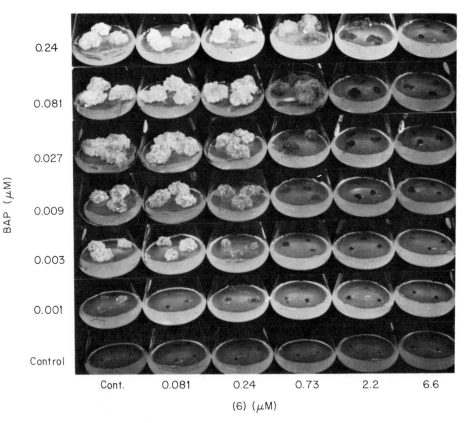

Fig. 1   Effect on the growth of tobacco tissue of various concentrations of 6-benzyl-aminopurine (BAP) and 3-methyl-7-(3-methylbutylamino)pyrazolo[4,3-d]pyrimidine (6).

Compound (6) was tested for its ability to inhibit utilization of potent cytokinins by tobacco callus. Figure 1 shows the results obtained growing tobacco on media containing various concentrations of N^6-benzyladenine and 3-methyl-7-(3-methylbutylamino)pyrazolo[4,3-d]pyrimidine (6). From this figure and the graph of fresh weight yields shown in Fig. 2, it is clear that compound (6) completely inhibited tobacco callus grown on an optimal $(2.7–8.1 \times 10^{-2} \ \mu M)$ concentration of N^6-benzyladenine when used in approximately 100-fold molar excess relative to the cytokinin. At concentrations of N^6-benzyladenine and compound (6) that gave incomplete inhibition of callus growth, the culture containing a higher concentration of cytokinin exhibited greater callus growth, while that having a higher concentration of (6) was further inhibited, which is consistent with the belief that compound (6) inhibited the growth of the tobacco tissue by competing with N^6-benzyladenine for cytokinin "receptor sites." Similar results were obtained when i⁶Ade was utilized as the cytokinin [40,41].

That the antagonist is not merely toxic to the plant tissue in an irreversible fashion can be judged by the observation (Fig. 2) that, at the highest tested concentration of N^6-benzyladenine (0.24 μM, which is past the growth optimum for this cytokinin), the addition of compound (6) resulted in growth enhancement, as though a smaller "effective" amount of the cytokinin were

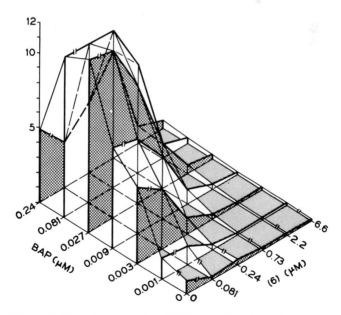

Fig. 2 Effect of 6-benzylaminopurine (BAP) and 3-methyl-7-(3-methylbutylamino)-pyrazolo[4,3-d]pyrimidine (6) on the fresh weight yield of tobacco callus.

TABLE 1

Biological Activity of Substituted 7-Aminopyrazolo[4,3-d]pyrimidines[a]

R"	R'	Range[b] of conc tested (μM)	Cytokinin activity min conc (μM) for Detection	Maximum growth	Growth inhibition, min conc (μM) for Detection	Complete inhibition[c]
H	HN—(isopentyl)	0.009–20	1.0	20	NA	—
H	HN—(isopentenyl)	0.001–20 (3)	0.08	1.0	NA	—
CH₃	OH	0.08–20	NA	—	NA	—
CH₃	S—	0.24–20	NA	—	NA	—
CH₃	HN—(isopentenyl)	0.08–20 (4)	0.24	6.6	NA	—
CH₃	(chain with N)	0.24–20	NA	—	NA	—
CH₃	HN—cyclohexyl	0.73–20	NA	—	NA	—
CH₃	HN—cyclopentyl	0.73–20	NA	—	6.6	NR
CH₃	HN—(isopropyl)	0.24–20	2.2	7[d]	6.6	NR
CH₃	HN—OH	0.73–20	NA	—	2.2	NR
CH₃	HN—(long chain)	0.24–20	NA	—	3.0	NR
CH₃	HN—(chain)	0.03–6.6	NA	—	0.2	2.2
CH₃	HN—(chain)	0.03–6.6	NA	—	0.1	0.73
CH₃	HN—(branched)	0.009–20 (4)	NA	—	0.1	0.73
CH₃	HN—(chain)	0.009–20	NA	—	0.03	0.5
CH₃	HN—(chain)	0.009–20 (3)	NA	—	0.03	0.2

[a] Abbreviations: NA, not active; NR, not reached.
[b] All values averages of two tests except as indicated by numbers in parentheses.
[c] Only slight growth stimulation.
[d] In presence of 0.003 μM i⁶Ade.

actually being used. The same conclusion can be reached from the fact that transfer of tobacco tissue on and off media containing a cytokinin antagonist (but always containing i^6Ade) resulted in reversible inhibition of the rate of growth [42]. Not surprisingly, the reversible nature of the growth inhibition by compound (6) was apparent only for moderate concentrations of cytokinin and (6). At concentrations of (6) greater than 1 μM (Fig. 2), for example, no amount of added N^6-benzyladenine could overcome the inhibition of tissue growth.

The preparation of a compound (6) having biological characteristics that might be expected of a true cytokinin antagonist prompted the preparation of several additional analogs. Of initial interest was the validity of an assumption used in the design of the potential anticytokinin, namely, that the anticytokinin need be sufficiently similar in structure to the cytokinin to permit participation in the same type of cellular "receptor complexes." Structure–activity relationships for cytokinins have been studied in detail, and it is known, e.g., that activity is optimal for purine derivatives having N^6-alkyl substituents with about five carbon atoms; much larger or smaller substituents result in compounds having lower activity. Therefore, several potential cytokinin antagonists were prepared which differed from (6) only with regard to the substituent on the exocyclic nitrogen atom [41]. As shown in Table 1, the n-pentyl derivative was the most active of those tested in this series. Just as those purines having a five-carbon substituent of N^6 were the most active as cytokinins, within this series of antagonists anticytokinin activity was optimal for those cytokinin analogs having an exocyclic substituent with four to six carbon atoms. The analogs without any alkyl substituent at the 7 position were without activity as cytokinins or anticytokinins. Thus, the assumption that an antagonist should resemble a cytokinin structurally was correct.

Substituted Pyrrolo[2,3-d]pyrimidines

The apparently successful preparation of a class of cytokinin antagonists by a rational procedure suggested that other heterocyclic classes of cytokinin antagonists should also be accessible by the same procedure. Therefore, a number of substituted pyrrolo[2,3-d]pyrimidines were tested as potential cytokinins and anticytokinins. As shown in Table 2, all of the 4-alkylamino-2-methylthiopyrrolo[2,3-d]pyrimidines tested were active as anticytokinins, and two of the compounds (the 4-cyclohexylamino and 4-cyclopentylamino derivatives) were exceptionally active [43]. Since substituents on the exocyclic nitrogen having about five carbon atoms had given optimal results in the first series of cytokinins tested, compounds having much larger and smaller substituents were not used in the pyrrolo[2,3-d]pyrimidine series. Nevertheless,

TABLE 2

Biological Activity of Substituted 4-Aminopyrrolo[2,3-d]pyrimidines[a]

R	R'	Range of conc tested (μM)	Cytokinin activity min conc (μM) for		Growth inhibition, min conc (μM) for	
			Detection	Maximum growth	Detection	Complete inhibition[b]
(structure)	H	0.24–20	0.62	6.6	NA	—
(structure)	H	0.24–20	5.8	>20	NA	—
H	SCH₃	0.24–20	NA	—	NA	—
(structure)	SCH₃	0.009–20	NA	—	0.24	2.2
(structure)	SCH₃	0.27–20	NA	—	0.40	2.0
(structure)	SCH₃	0.08–20	NA	—	0.1	1
(structure)	SCH₃	0.009–20	NA	—	0.009	0.05
(structure)	SCH₃	0.24–20	NA	—	6.6	>20
(structure)	SCH₃	0.003–20	NA	—	0.05	0.60
(structure)	SCH₃	0.24–20	NA	—	10	>20

[a] Abbreviations: NA, nonactive; NR, not reached.
[b] In presence of 0.003 μM i⁶Ade.

the need for an N^4 substituent of appropriate size was demonstrated in the sense that the analog lacking an alkyl substituent on N^4 (4-amino-2-methyl-thiopyrrolo[2,3-d]pyrimidine) was not active as a cytokinin or anticytokinin. Moreover, just as formal deletion of the 7-methyl substituent in the substituted pyrazolo[4,3-d]pyrimidine series effected conversion of those anticytokinins to species with cytokinin activity, two 4-alkylaminopyrrolo[2,3-d]pyrimidines lacking the methylthio group had cytokinin activity (Table 2) [43,44].* Thus, a close structural relationship between compounds with cytokinin and anticytokinin activity also existed for the second series of heterocycles.

Improved Syntheses of Potential Anticytokinins

The initial preparation of 7-alkylamino-3-methylpyrazolo[4,3-*d*]pyrimidines was carried out via an eight-step reaction sequence [38,40]. Chromatographic purification was necessary after three transformations, and the overall yield was approximately 1%. Since a number of technical problems were encountered which also made it difficult to scale up the synthesis, an alternate sequence has now been devised (Scheme 1). Although this sequence still involves eight steps starting from 3,5-dimethylpyrazole, no chromatographic

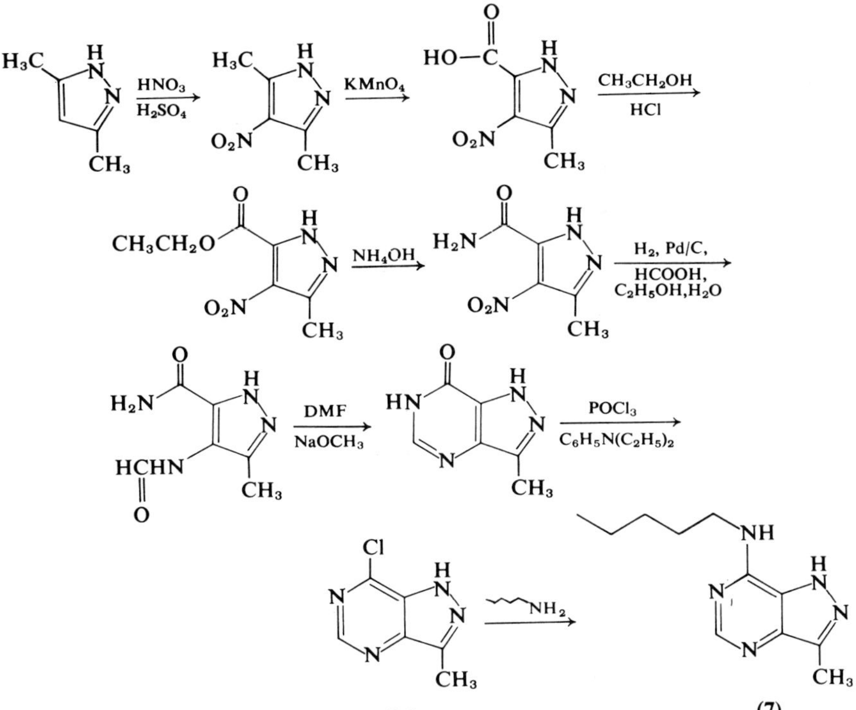

Scheme 1   (7)

* It should be noted, however, that two 7-alkylamino-3-methyl-5-methylthiopyrazolo-[4,3-*d*]pyrimidines were tested and found to have weak *cytokinin* activity. An apparent anomaly of this type has been noted once before [45], and it is probably correct to assert that for very weakly active cytokinins, the structure–activity relationships derived from single changes in highly active molecules may provide less predictive value.

purification is required and compound (**7**), e.g., can be obtained in substantial quantities in an overall yield of 20%.

   Certain types of heterocycles structurally related to the cytokinins can also be prepared by a novel and very concise procedure. This is illustrated in Scheme 2 for the synthesis of 2-methyl-4-(3-methylbutylamino)pyrazolo[3,4-*d*]pyrimidine-3-carbonitrile (**8**), which was accessible in 66% yield via the condensation of 3-amino-4,5-dicyano-1-methylpyrazole [46–48] and *N*-(3-methylbutyl)formamide. The formation of compound (**8**) presumably proceeded via intermediate (**i**), which is identical in structure with the intermediate species that would form during the Dimroth rearrangement [49,50] of 4-amino-2-methyl-5-(3-methylbutylamino)pyrazolo[3,4-*d*]pyrimidine-3-carbonitrile to give (**8**). Thus, the Dimroth product can be obtained in what is essentially an anhydrous medium, permitting the preparation of compounds [like (**8**)] having water-sensitive substituents. This procedure has been utilized for the preparation of a number of 4-alkylaminopyrazolo[3,4-*d*]pyrimidines [51] and is applicable to the synthesis of other types of heterocycles, e.g., 4-alkylaminopyrrolo[2,3-*d*]pyrimidines.

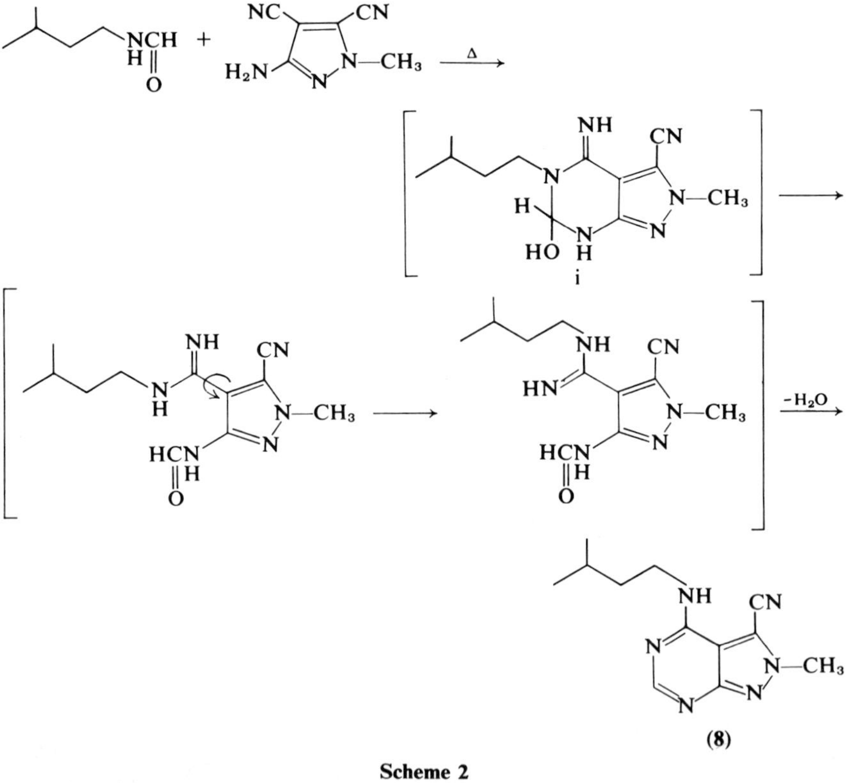

Scheme 2

CHARACTERIZATION OF HETEROCYCLES
AS SPECIFIC ANTICYTOKININS

The preparation of two types of heterocycles with anticytokinin activity prompted the synthesis of additional classes of potential cytokinin antagonists [52–55] and reports that certain 4-substituted 7-(β-D-ribofuranosyl)pyrrolo-[2,3-d]pyrimidines were anticytokinins [52,55]. Since inhibition of cytokinin-promoted growth may be effected by compounds that are not anticytokinins, it is obviously important to distinguish between those compounds that are merely inhibitory to growth and those that are specific cytokinin antagonists and function by competing with cytokinins for cytokinin receptor sites. Three criteria have been proposed [44]:

1. The apparent anticytokinin activity of compounds in a given heterocyclic series should reside with those compounds which resemble cytokinins structurally.
2. The antagonist must diminish the utilization of cytokinin in some system, the functioning of which is limited by the availability of cytokinins.
3. The inhibition of growth obtained at a given concentration of cytokinin antagonist and cytokinin must be "reversible," i.e. less inhibition must be apparent in cultures containing the same concentration of antagonist, but more cytokinin.

Since the molecular basis of cytokinin activity in plant tissues is not understood at present, individual inhibitors cannot be evaluated in terms of the second criterion, but conformity to the first and third can be tested experimentally.

Substituted Pyrazolo[4,3-d]pyrimidines

Inhibitors in the substituted pyrazolo[4,3-d]pyrimidine series were assayed for specificity as anticytokinins. For 3-methyl-7-(3-methylbutylamino)-pyrazolo[4,3-d]pyrimidine (6), e.g., it was found that inhibition of cytokinin-promoted growth could be reversed by the addition of greater concentrations of cytokinins (Figs. 1 and 2). Moreover, the growth-inhibitory effect of (6) could be counteracted by lower concentrations of the more active cytokinin i[6]Ade (1) and also by high concentrations of N,N'-diphenylurea, a weak cytokinin [41]. Quantitatively, reversal of inhibition by i[6]Ade was about 3 times as effective as that achieved with 6-benzylaminopurine and 500 times that obtained with diphenylurea. Thus, the relative effectiveness of the three cytokinins in reversing inhibition by compound (6) was in reasonable agreement with their relative potencies in promoting the growth of tobacco callus in the absence of the antagonist.

As discussed above, the activity of compounds in this series as antagonists correlated well with gross structural similarity to the most active cytokinins (e.g., i⁶Ade); compounds without the exocyclic 7-substituent were without activity as cytokinins or antagonists. Thus, the pyrazolo [4,3-*d*]pyrimidines prepared as potential anticytokinins fulfill two of the criteria outlined for cytokinin antagonists.

It should be noted, however, that while the inhibition caused by very high concentrations of (**6**) ($> 1 \mu M$) could not be reversed by any concentration of added cytokinin, exogenously supplied adenine at least doubled the range over which i⁶Ade would counteract (**6**). This effect of adenine suggested that at high concentrations the antagonist may function by more general interference in purine metabolism.

Substituted Pyrrolo[2,3-d]pyrimidines

The extent to which growth inhibition by compounds in the substituted pyrrolo[2,3-*d*]pyrimidine series could be reversed by cytokinins is illustrated for compound (**9**) in Fig. 3. In the presence of $9 \times 10^{-3} \mu M$ i⁶Ade, compound (**9**) effected complete inhibition of growth at $2.4 \times 10^{-2} \mu M$ concentration. As the concentration of added cytokinin was increased, the inhibitory effect of greater concentrations of compound (**9**) could be reversed. Thus, when tobacco callus was cultivated on a medium containing 0.73 μM i⁶Ade, at least partial reversal of inhibition could be obtained for concentrations of (**9**) as high as 2.2 μM. Furthermore, for any single inhibitory concentration of (**9**)

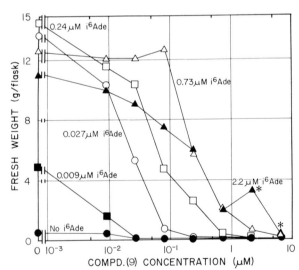

Fig. 3 Fresh weight yields of tobacco tissue resulting from growth on various concentrations of i⁶Ade (**1**) (0–2.2 μM) and 4-cyclopentylamino-2-methylthiopyrrolo[2,3-*d*]-pyrimidine (**9**) (0–6.6 μM). Asterisks denote cultures in which buds formed.

(9)   (10)

up to 2.2 μM, increased amounts of added cytokinin increasingly reversed inhibition of growth. In fact, for i[6]Ade concentrations from 9×10^{-3} to 2.2 μM, the amount of i[6]Ade required to begin to overcome inhibition by (9) was in constant proportion to the concentration of (9) utilized. Bioassay of the pyrrolo[2,3-d]pyrimidine analog without the exocyclic substituent (4-amino-2-methylthiopyrrolo[2,3-d]pyrimidine) revealed the compound to be without activity as a cytokinin or anticytokinin, which is consistent with the belief that the 4-alkylamino-2-methylthiopyrrolo[2,3-d]pyrimidines are specific anticytokinins [43].

To further characterize the nature of the growth-inhibitory effect obtained with the substituted pyrrolo[2,3-d]pyrimidines, experiments were carried out to determine whether inhibition was specific for cytokinins or could be reversed by added auxins. The results of experiments utilizing serial combinations of (10), indole-3-acetic acid, and i[6]Ade are shown in Fig. 4. At 0.08 μM i[6]Ade concentration (Fig. 4A), growth increased in proportion to added auxin, but growth inhibition by (10) was not affected by any concentration of added auxin. At higher concentrations of added i[6]Ade (e.g., 6.6 μM i[6]Ade, Fig. 4D) yields of tobacco tissue still increased as a function of added indole-3-acetic acid, but the further addition of compound (10) had no inhibitory effect. The observation that the inhibitory effects of (10) were dependent on the concentration of i[6]Ade but independent of indole-3-acetic acid concentration was consistent with the results that would be expected of a specific anticytokinin. The same experiment was carried out using various concentrations of (10) and gibberellic acid (up to 2.2 μM each) in the presence of 11.4 μM indole-3-acetic acid and 0.003 or 0.08 μM i[6]Ade. Gibberellic acid also failed to reverse the inhibitory effect of compound (10) [43].

Other Growth Inhibitors

A number of substituted 4-alkylaminopyrazolo[3,4-d]pyrimidines were tested for cytokinin and anticytokinin activity, and three analogs [(11)–(13)] were found to be inhibitory to the growth of tobacco tissue on 3×10^{-3}

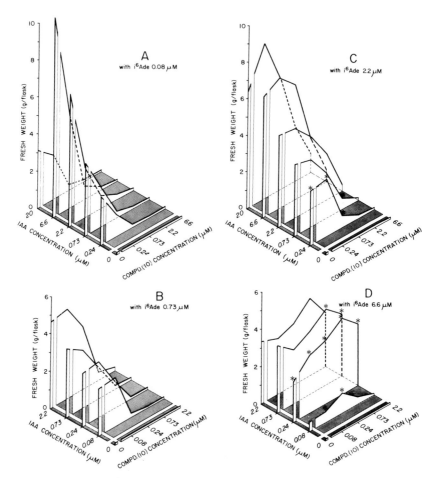

Fig. 4 Effect of various concentrations of i⁶Ade, 4-cyclohexylamino-2-methylthio-pyrrolo[2,3-d]pyrimidine (10), and indole-3-acetic acid on the fresh weight yields of tobacco callus. Asterisks denote cultures in which buds formed.

μM i⁶Ade. However, the structurally related heterocycle 4-aminopyrazolo-[3,4-d]pyrimidine (14) was equally as inhibitory to growth of the tobacco tissue, and high concentrations of i⁶Ade were ineffective in reversing the inhibition of growth caused by compounds (11)–(13), suggesting that these compounds had little specific cytokinin antagonist activity [44].

Four 7-alkylamino-3-(β-D-ribofuranosyl)pyrazolo[4,3-d]pyrimidines were also tested as potential anticytokinins, and three compounds [(15)–(17)] were found to effect significant inhibition of tobacco callus growth. As in the pyrazolo[3,4-d]pyrimidine series, the structurally related heterocycle lacking

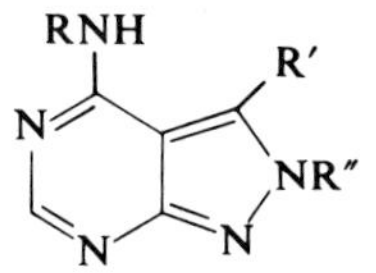

RNH structure

(**11**), R = [structure], R' = CN, R'' = CH₃

$(11), R = $ 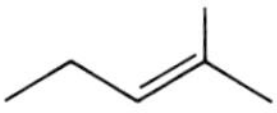 , $R' = CN,$ $R'' = CH_3$

$(12), R = $, $R' = CONH_2,$ $R'' = H$

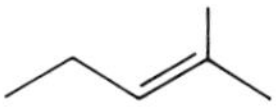

$(13), R = $, $R' = CONH_2,$ $R'' = H$

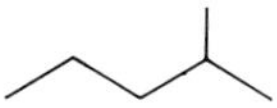

$(14), R = H,$ $R' = H,$ $R'' = H$

a substituent on the exocyclic amino group {7-amino-3-(β-D-ribofuranosyl)-pyrazolo[4,3-d]pyrimidine, formycin (**18**)} was also inhibitory to tobacco tissue grown on i^6Ade. Unlike the results obtained with the pyrazolo[3,4-d]-pyrimidines, while inhibition caused by formycin could not be counteracted by high concentrations of i^6Ade, the cytokinin did effect partial reversal of the growth inhibition caused by compounds (**15**)–(**17**). Thus, in spite of the lack of correlation between inhibition obtained with these analogs and their structural relationship to the cytokinins, compounds (**15**)–(**17**) may have some specific anticytokinin activity [44].

Iwamura and his co-workers have described the inhibitory activity obtained with several 4-alkylamino-7-(β-D-ribofuranosyl)pyrrolo[2,3-d]pyrimidines [52, 55]. The activity of the analogs correlated reasonably well with their structural

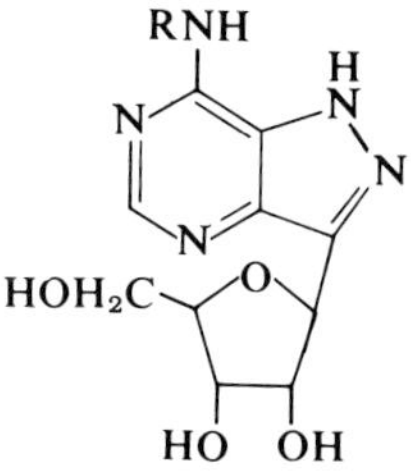

RNH structure with H, N, N, HOH₂C, O, HO OH

(**15**), R = [structure]

(**16**), R = [structure]

(**17**), R = [structure]

(**18**), R = H

relationship to the more active cytokinins. Interestingly, although the analog without a substituent on the exocyclic nitrogen atom {4-amino-7-(β-D-ribofuranosyl)pyrrolo[2,3-d]pyrimidine, tubercidin} was as inhibitory to the growth of the tissue as the best of the alkylated analogs, the 4-methylamino derivative was without inhibitory activity, suggesting that the more potent compounds (having larger alkyl substituents) may exert their inhibitory effects by a different mechanism than tubercidin itself. Iwamura et al. [52,55] did not report on the extent of reversal of inhibition by their analogs that could be obtained using high concentrations of a cytokinin. Since their assays were carried out using kinetin, which is only about 10% as active as i[6]Ade in the tobacco bioassay, it is also difficult to correlate the inhibitory activities of the 4-alkylamino-7-(β-D-ribofuranosyl)pyrrolo[2,3-d]pyrimidines with those of other inhibitors on the basis of published data. Bioassay of a sample of one analog {4-furfurylamino-7-(β-D-ribofuranosyl)pyrrolo[2,3-d]-pyrimidine} provided by Professor Iwamura indicated detectable inhibition of tissue growth at 0.73 μM concentration when the medium contained 3×10^{-3} μM i[6]Ade. Complete inhibition of growth by this analog was achieved at 5 μM concentration, so that this compound was about 100 times less inhibitory than compound (9) and 25 times less active than anticytokinin (7). Although the furfuryl derivative was found to be less inhibitory to the growth of tobacco callus than tubercidin itself (the latter of which gave inhibition detectable at 0.36 μM and complete at 0.88 μM), inhibition caused by the furfuryl derivative was more effectively countered by i[6]Ade [43].

It would seem that structural alteration of adenine afforded analogs some of which were inhibitory to the growth of tobacco callus. These compounds were not specific anticytokinins in that the inhibition obtained could not be reversed by high concentrations of added cytokinin. Nevertheless, the addition of alkyl substituents to the exocyclic nitrogen atoms afforded compounds which had anticytokinin or weak cytokinin activity, depending on the specific exocyclic substituent, and some of which also had more general activity as growth inhibitors at high concentration.

UTILIZATION OF ANTICYTOKININS IN PLANT BIOASSAYS

Inhibition of Cytokinin-Independent Tobacco Callus

One important purpose in preparing the cytokinin antagonists was to facilitate the study of cytokinins in species that produce their own. Therefore, 3-methyl-7-(3-methylbutylamino)pyrazolo[4,3-d]pyrimidine (6) was tested for activity on a strain of cytokinin-independent tobacco callus, which had

been shown to have endogenous cytokinin and to produce zeatin and i⁶Ade
from exogeneous [¹⁴C]adenine [56]. As expected, compound (6) inhibited
tissue growth in this strain in the same fashion as in the cytokinin-dependent
strain, and counteraction of inhibition could be achieved using somewhat
lower concentrations of added cytokinin. Strong evidence was thus provided
that this tissue utilized endogenous cytokinin [41].

Enhancement of Cytokinin-Promoted Budding

As shown in Figs. 3 and 4, compounds (9) and (10) promoted bud formation
in some assays carried out at high cytokinin concentrations. The effect of these
compounds on bud formation was especially apparent at an auxin (indole-3-
acetic acid) concentration of 11.4 μM, which is generally utilized in the
tobacco bioassay for cytokinins; at lower concentrations of indole-3-acetic
acid, (9) and (10) were less effective in augmenting the formation of buds by
i⁶Ade [43]. The observation that these two compounds reinforced i⁶Ade in
promoting bud formation while opposing its growth-promoting effects can
be interpreted to mean that there are different cytokinin "receptor sites"
which control the two functions and that compounds (9) and (10) are utilized
as cytokinin analogs in the promotion of bud formation but antagonize
cytokinin-promoted callus growth.

Determination of the "Active" Form of Cytokinins

The preparation and testing of heterocyclic analogs of cytokinins as potential cytokinins and anticytokinins have afforded considerable data pertinent
to the question of cytokinin "activation" prior to utilization in the promotion
of growth [52,57]. For example, (3-substituted) 4-alkylaminopyrrolo[2,3-*d*]-
pyrimidines and 4-alkylaminopyrazolo[3,4-*d*]pyrimidines contain carbon
rather than nitrogen atoms at the 3-position and should be metabolically
stable at this position. Since a number of these species have been found to
have cytokinin activity [44], it would seem that activation of such analogs
at the 3 position was not a prerequisite to their utilization as cytokinins. On
the assumption that cytokinin analogs that are purines can promote cell
division and growth by the same mechanism as these heterocyclic analogs,
one may conclude that activation of the corresponding position (7) in purines
is also unnecessary for the promotion of cell division and growth. Extension
of these principles to other heterocycles that are cytokinins has suggested
that exogenously supplied heterocycles can function as cytokinins without
undergoing ribosylation, separation of the exocyclic substituent from the
heterocyclic nucleus, or other types of modification [38,44].

Other Tests

Preliminary investigations have been made of the effects of 3-methyl-7-(3-methylbutylamino)pyrazolo[4,3-*d*]pyrimidine (**6**) on several plant materials. The compound did not affect seed germination, but high concentrations (180 μM) diminished the development of roots in wheat and radish seedlings and in *Coleus* cuttings. Similar inhibitory results were obtained using i[6]Ade, although some differences were noted in the color of the radish cotyledons after treatment. Young tomato and tobacco plants were unaffected by compound (**6**) (when the antagonist was supplied by spraying as an aqueous solution), but maintenance of tomato seedlings in nutrient media containing from 12 to 120 μM (**6**) resulted in severe wilting and death of the seedlings. The antagonist was also tested for its ability to enhance senescence, e.g., of discs excised from leaves of sweet corn, but no consistent effects were observed [41].

EFFECTS OF CYTOKININ ANALOGS
ON MAMMALIAN CELLS

Following the report of the initial clinical trial of N^6-(3-methyl-2-butenyl)-adenosine [i[6]Ado(**2**)] as an anticancer agent [30], Gallo and his co-workers reported that this cytokinin ribonucleoside was capable of regulating the transformation and growth of human lymphocytes which had been treated with phytohemagglutinin (PHA) [58]. The ribonucleoside effected stimulation of lymphocyte growth, relative to drug-free controls, when added at low concentration (0.8 μM) late after PHA addition and inhibition of growth when added at higher concentration (8 μM) soon after PHA treatment. Regulatory effects could be mediated only by cytokinin ribonucleosides, and stimulation of growth was achieved only with i[6]Ado. This was true in spite of the fact that several of the tested compounds were close structural analogs of i[6]Ado and had intense cytokinin activity in the tobacco bioassay, while others were naturally occurring cytokinins [59].

Gallo *et al.* studied the mechanism by which i[6]Ado affected the growth of PHA-transformed human lymphocytes. They showed that i[6]Ado did not compete with PHA for receptor sites on the lymphocyte cell surface and that incubation of PHA in the presence of the ribonucleoside for periods up to 16 hr had no effect on the mitogenic or transforming activity of the protein. Since i[6]Ado was also found not to affect the growth of lymphocytes that had entered S phase, it was concluded that the effect of the cytokinin was exerted prior to genome activation but after interaction of PHA with the cell membrane [59].

Transformation of mammalian cells is known to be associated with, and possibly caused by, changes in the intracellular concentration of cyclic AMP. To determine whether the primary mode of action of i^6Ado might be regulation of the intracellular concentration of cyclic AMP, PHA-transformed cells were treated with various concentrations of the cyclic AMP analog N^6, $O^{2'}$-dibutyryl cyclic AMP. The dose–response curves for stimulation and inhibition of DNA synthesis and lymphocyte transformation obtained with N^6, $O^{2'}$-dibutyryl cyclic AMP and with i^6Ado were remarkably similar in shape, suggesting strongly that the cytokinin might well function primarily via regulation of intracellular cyclic AMP concentration. Supportive evidence was derived from the observation that exogenous i^6Ado affected the ability of N^6, $O^{2'}$-dibutyryl cyclic AMP to regulate the growth of transformed lymphocytes [59].

If regulation of lymphocyte growth by i^6Ado were mediated via cyclic AMP metabolism, then the cytokinin should affect at least one of the enzymes involved in production and degradation of the cyclic nucleotide. Hecht et al. studied the effect of a number of cytokinins on the high K_m phosphodiesterase activity from beef heart [60]. All of the cytokinins were found to be reasonably good inhibitors of the high K_m activity (apparent K_m 70 μM) when measured at single concentrations of inhibitor (2 mM) and cyclic AMP (0.2 mM). Two of the cytokinins, i^6Ade and i^6Ado, were utilized in initial velocity studies and found to be competitive inhibitors with apparent K_i values of 129 and 109 μM, respectively. Also tested as potential inhibitors of the low-affinity enzyme were a number of cytokinin analogs in the pyrazolo[4,3-d]pyrimidine, pyrazolo[3,4-d]pyrimidine, and pyrrolo[2,3-d]pyrimidine series. The cytokinin analogs were also found to be inhibitors of the cyclic AMP phosphodiesterase activity and four of the analogs were used in kinetic studies and shown to be competitive inhibitors with apparent K_i values from 83 to 19 μM (Table 3).

TABLE 3

K_i Values for Cyclic AMP Phosphodiesterase Inhibitors [a]

Compound	Apparent K_i (μM)
N^6-(3-Methyl-2-butenyl)adenine (1)	129
N^6-(3-Methyl-2-butenyl)adenosine (2)	109
3-Methyl-7-(3-methylbutylamino)pyrazolo[4,3-d]pyrimidine (6)	35
3-Methyl-7-n-pentylaminopyrazolo[4,3-d]pyrimidine (7)	48
4-Cyclopentylamino-2-methylthiopyrrolo[2,3-d]pyrimidine (9)	83
4-(3-Methyl-2-butenylamino)pyrazolo[3,4-d]pyrimidine-3-carboxamide (12)	19

[a] Apparent K_m = 70 μM.

Significantly, related heterocycles such as adenosine and 7-amino-3-methyl-pyrazolo[4,3-d]pyrimidine, which lacked the exocyclic N-substituent, were without activity as inhibitors of the high K_m cyclic AMP phosphodiesterase activity, and heterocycles such as 7-(2-hydroxyethylamino)-3-methylpyrazolo-[4,3-d]pyrimidine and 4-ethylamino-2-methylpyrazolo[3,4-d]pyrimidine-3-carbonitrile having exocyclic N-substituents of nonoptimal length (as judged by activity in the tobacco bioassay) were found to be low in inhibitory activity.

Several of the compounds shown to be inhibitory to the high K_m phosphodiesterase activity from beef heart were tested as inhibitors of the associated low K_m activity. None of the compounds tested was at all inhibitory to the low K_m activity at concentrations comparable to the apparent K_m (0.2 μM), although one compound, 3-methyl-7-(3-methylbutylamino)pyrazolo[4,3-d]-pyrimidine (6), was tested at higher concentration and shown to be a weak inhibitor of this activity.

The observation that a number of cytokinin analogs were better inhibitors of the high K_m cyclic AMP phosphodiesterase activity than either i[6]Ade or i[6]Ado prompted the evaluation of several as regulators of the transformation and growth of PHA-treated lymphocytes. In spite of the fact that no tested cytokinin ribonucleoside other than i[6]Ado has been reported to effect both stimulation and inhibition of lymphocyte transformation and growth and that the corresponding purines were all completely inactive in the lymphocyte assay system [59], some of the cytokinin analogs in the substituted pyrazolo-[4,3-d]pyrimidine series were capable of eliciting both stimulatory and inhibitory responses. One compound, 3-methyl-7-n-pentylaminopyrazolo-[4,3-d]pyrimidine (7), was found to be several times more active than i[6]Ado in regulating the extent of transformation and growth of the PHA-treated lymphocytes with regard to both stimulatory and inhibitory effects [61]. Also surprising was the finding that the ribonucleoside analog of (7), 7-n-pentylamino-3-(β-D-ribofuranosyl)pyrazolo[4,3-d]pyrimidine, was slightly less active than i[6]Ado in regulating the extent of lymphocyte transformation and growth.

Although several cytokinin analogs that inhibited the cyclic AMP phosphodiesterase activity from beef heart were found to be capable of regulating the extent of transformation and growth of PHA-treated human lymphocytes, as might be expected if these compounds were also inhibitory to the cyclic AMP phosphodiesterase activity from human lymphocytes, the observed regulatory activity of the compounds on intact lymphocyctes obviously does not prove that they exert their effects via cyclic AMP metabolism. Direct analysis of the high and low K_m activities derived from PHA-transformed lymphocytes is relatively difficult since transformation is specific for the T-lymphocytes but the assay is ordinarily carried out using a complete leukocyte fraction that includes many other types of cells (B-lymphocytes, monocytes, etc.) that are not transformed by PHA but that nonetheless undoubtedly contain cyclic AMP phosphodiesterase activities. Thus, although both the high and low K_m activities derived from the PHA-treated leukocyte fraction were inhibited by i⁶Ado and both activities were inhibited to a greater extent by compound (7) (which is more potent than i⁶Ado in regulating transformation and growth of intact T-lymphocytes), direct proof that transformation and growth of T-lymphocytes are mediated via inhibition of cyclic AMP phosphodiesterase activity is still lacking.

Further progress in the investigation of the possible growth regulation of mammalian cells by cytokinins via inhibition of cyclic AMP phosphodiesterase activities clearly depended on the availability of a homogeneous population of mammalian cells which responded physiologically to exogenous cytokinins. Since the growth of mouse fibroblasts cells has been reported to be inversely proportional to the intracellular concentration of cyclic AMP [62–65], it was anticipated that compounds which functioned by inhibiting cyclic AMP phosphodiesterase activity would also inhibit fibroblast growth.

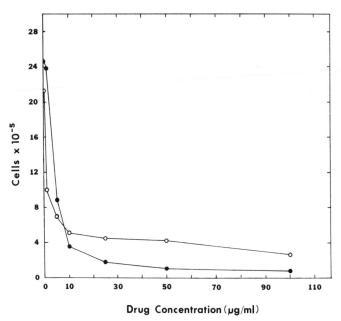

Drug Concentration (μg/ml)

Fig. 5 The inhibitory effect of various concentrations of i⁶Ado (2) and 3-methyl-7-*n*-pentylaminopyrazolo[4,3-*d*]pyrimidine (7) on mouse fibroblast (3T6) cells. The experiment was conducted as described previously [65a] except that 35 mm tissue culture dishes (containing 2.5 ml of medium) were each inoculated with 10⁵ cells. After incubation with a single concentration of (2) or (7) for 48 hr, the cells were harvested separately from triplicate cultures and counted [●—●, i⁶Ado; ○—○, (7)].

TABLE 4

Inhibition of Mouse Fibroblasts (3T6) by Cytokinins and Related Compounds

Compound	Inhibits (μg/ml)	Kills (μg/ml)
Adenosine	1.0	2.5
N^6-(3-Methyl-2-butenyl)adenosine [i^{6}Ado (2)]	2.5	10
5'-Deoxyadenosine	25	200
5'-Deoxy-i^{6}Ado	10	25
3-Methyl-7-(3-methyl-2-butenylamino)pyrazolo[4,3-d]-pyrimidine (4)	—	100
3-Methyl-7-(3-methylbutylamino) pyrazolo[4,3-d]pyrimidine (6)	—	10
3-Methyl-7-n-pentylaminopyrazolo[4,3-d]pyrimidine (7)	1	5
7-Amino-3-methylpyrazolo[4,3-d]pyrimidine	100	200
7-n-Pentyl-3-(β-D-ribofuranosyl)pyrazolo[4,3-d]pyrimidine (15)	10	> 200
7-Cyclopentyl-3-(β-D-ribofuranosyl)pyrazolo[4,3-d]-pyrimidine (16)	10	> 200
7-n-Hexyl-3-(β-D-ribofuranosyl)pyrazolo[4,3-d]pyrimidine (17)	> 200	> 200

In fact, this was consistent with reports that certain cytokinins were cytotoxic to cultured mouse cells [25,26].

As shown in Table 4 and Fig. 5 [65a], incubation of mouse fibroblast (3T6) cells in the presence of i^{6}Ado effected inhibition of the cells at relatively low concentration. However, adenosine itself is known to be highly cytotoxic to cultured mouse fibroblasts by virtue of its conversion to adenosine 5'-monophosphate, the latter of which inhibits uridine biosynthesis [66]. Since adenosine was found to be more cytotoxic to mouse fibroblasts than i^{6}Ado, it seemed reasonable to determine whether the toxicity obtained using i^{6}Ado might be due to its utilization as an analog of adenosine. Therefore, i^{6}Ado was tested in a cell line (3T6-TM) known to lack the (adenosine kinase) activity required to convert adenosine to the toxic adenosine 5'-monophosphate [67]. Although a high concentration (100 μg/ml) of i^{6}Ado was required to inhibit growth, the cytokinin was more active than adenosine against this cell line. The same order of toxicities was also observed in 3T6 cells for 6-amino-9-(5'-deoxy-β-D-ribofuranosyl)purine (5'-deoxyadenosine) and 6-(3-methyl-2-butenylamino)-9-(5'-deoxy-β-D-ribofuranosyl) purine (5'-deoxy-i^{6}Ado), which could not be converted to the respective 5'-monophosphates by adenosine kinase [68]. Thus, while part of the inhibition of 3T6 cells by i^{6}Ado did seem to be due to its utilization as an analog of adenosine, a second mechanism for cytokinin-mediated inhibition of cell growth was also apparent. The existence of an alternate mechanism was also confirmed by the observation that inhibition of cells cultured on i^6Ado (20 μg/ml) could not be reversed by added uridine, whereas cells cultured on toxic does of adenosine were spared by the further addition of uridine to the culture medium.

Certain substituted pyrazolo[4,3-d]pyrimidines were also inhibitory to the growth of 3T6 cells, and compound (7) was especially toxic (Table 4). In fact, a dose–response curve determined for this compound (Fig. 5) indicated that it was a more potent inhibitor of fibroblast growth than i^6Ado. That the inhibition caused by (6) and (7) cannot be attributed to their conversion to analogs of 5'-AMP can be appreciated from the much lower toxicity of three 7-alkylamino-3-(β-D-ribofuranosyl)pyrazolo[4,3-d]pyrimidine derivatives (Table 4) and from the inability of uridine to reverse the toxicity caused by (6) and (7).

As the dose–response curve for i^6Ado in Fig. 5 suggests, very low concentrations of cytokinins and substituted pyrazolo[4,3-d]pyrimidines (e.g., 7) have little effect on the growth of 3T6 cells, but small increments in the concentrations of these compounds beyond the maximum sublethal dose cause a dramatic increase in the extent of cell inhibition. To determine whether the cytokinins and substituted pyrazolo[4,3-d]pyrimidines functioned in the same fashion, precise dose–response curves were determined for low concentrations of i^6Ado and compound (6) and both compounds were then tested at maximum sublethal doses against 3T6 cells, both singly and in combination. It was anticipated that if the two compounds functioned in the same fashion, addition of both to the culture medium would have the same effect as slightly exceeding the maximum sublethal dose of a single compound, namely, a substantial increase in the number of cells killed. If, on the other hand, the two compounds inhibited 3T6 cells by unrelated mechanisms, combination of the two in the culture medium should not result in a synergistic effect. The i^6Ado was tested in replicate cultures of 3T6 cells at a concentration of 0.1 μg/ml and, after incubation for 4 days, 8% of the cells were killed (relative to a drug-free control). Similar incubation of compound (6) at a concentration of 1.0 μg/ml resulted in loss of 21% of the cells, while incubations with both compounds simultaneously gave 47% killing. Thus, the effect of the two compounds in combination was greater than that which would have been expected by simple addition of their individual effects, which is consistent with the interpretation that their mechanisms of action are related biochemically [68].

The ability of the cytokinins and cytokinin analogs to inhibit cyclic AMP phosphodiesterase activities in mouse fibroblast cells was assayed by incubation of [^{3}H]cyclic AMP in cell-free extracts of 3T6 cells in the presence of a cytokinin or cytokinin analog. The amount of [^{3}H]cyclic AMP remaining was

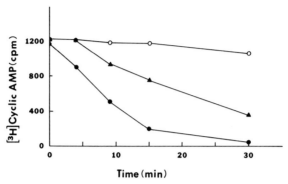

Fig. 6 Inhibition by i⁶Ado (**2**) of cyclic AMP degradation by the low K_m cyclic AMP phosphodiesterase activity from 3T6 cells. [³H]Cyclic AMP (2.6 μM) was incubated in the presence of a cell-free extract from 3T6 cells (containing 120 μg protein/250 μl incubation mixture), and aliquots were removed at predetermined time intervals and analyzed for cyclic AMP content; ●—●, curve obtained in the absence of inhibitor; ▲—▲, curve obtained in the presence of 400 μM i⁶Ado; ○—○, boiled enzyme control. The amount of the cyclic nucleotide was measured utilizing a cyclic AMP-dependent protein kinase, by a modification of the method of Gilman [69]. The amount of cyclic nucleotide tested was such that the binding of [³H]cyclic AMP to the kinase was proportional to the amount actually present in individual aliquots.

measured at predetermined time intervals using a cyclic AMP-dependent protein kinase activity, by a modification of the procedure of Gilman [69]. As shown for the low K_m activity in Fig. 6, i⁶Ado was inhibitory to both the high and low K_m phosphodiesterases. Compound (**7**) was tested as a potential inhibitor of the low K_m cyclic AMP phosphodiesterase activity and found to be more inhibitory than i⁶Ado, which is consistent with its greater cytotoxic activity against the intact cells. Although it was not possible to obtain a sample of the high K_m phosphodiesterase activity that was free from contamination with the low K_m activity, protein containing exclusively low K_m cyclic AMP phosphodiesterase activity was isolated by chromatography on Sephadex G-100. This material was utilized for measuring the initial velocity of cyclic AMP hydrolysis in the presence and absence of i⁶Ado. It was found that the apparent K_m for this conversion was 2 μM and that i⁶Ado was a competitive inhibitor of the transformation with an apparent K_i of 6 μM [61].

Cultures of 3T6 cells were maintained on a nutrient medium containing a sublethal dose of i⁶Ado and, after several passages, transferred to a similar medium containing a higher concentration of i⁶Ado. Repetition of this procedure using increasingly higher concentrations of i⁶Ado afforded cells that could be cultured in the presence of 200 μg/ml of i⁶Ado and that retained resistance to i⁶Ado after 15 passages in the absence of the cytokinin. Several clones were derived from these resistant cells, and one of these (3T6-CYT3), which had the same morphological characteristics as 3T6 cells and grew with the same generation time, was chosen for study of the mechanism by which resistance had developed.

Divekar *et al.* [70] previously reported the development of Sarcoma 180 cells resistant to adenosine analogs, including i[6]Ado. The resistant cells had somewhat greater adenosine deaminase activity, although i[6]Ado was not observed to be a substrate for this enzyme. The change reported to be responsible for resistance to adenosine analogs in the S180 cells was a 20,000-fold reduction in adenosine kinase activity. In the present case, the i[6]Ado-resistant 3T6 cells were analyzed for adenosine deaminase, adenosine kinase, and cyclic AMP phosphodiesterase activity. Adenosine deaminase activity in the resistant cells was determined using adenosine as the substrate at a concentration of 50 μM and shown to be about 75% higher than the same activity in the i[6]Ado-sensitive cells. Neither i[6]Ado nor compound (7) was found to have significant activity as an inhibitor of adensosine deaminase. Analysis of the ability of cell-free extracts of the normal and resistant cells to convert i[6]Ado to its corresponding 5'-monophosphate was carried out at at an i[6]Ado concentration of 100 μM, but no product formation was detected. It has been reported that i[6]Ado is a substrate for adenosine kinase from mouse cells [70–72], although the apparent lack of activity could be due to substrate inhibition of the enzyme [71] or to the choice of assay conditions that were otherwise inappropriate. Measurement of the high and low K_m cyclic AMP phosphodiesterase activities from the two lines revealed that the i[6]Ado-resistant line had approximately 30% more low K_m activity than the sensitive line and almost three times as much high K_m activity [61].

Although the presence of increased cyclic AMP phosphodiesterase activity in i[6]Ado-resistant 3T6 cells does not prove that i[6]Ado normally inhibits 3T6 cells by raising the intracellular concentration of cyclic AMP (an effect that could possibly be overcome by the greater amounts of cyclic AMP-degrading activity in the i[6]Ado-resistant cells), it is interesting that the i[6]Ado-resistant 3T6 cells were killed by (7). Compound (7) is a better inhibitor than i[6]Ado of the low K_m cyclic AMP phosphodiesterase activity from 3T6 cells. It is also worth noting that a compound (19) structurally related to those discussed

(19)

here has been reported to be a potent inhibitor of the cyclic AMP phospho-diesterase activities from rat and rabbit brain [73]. If the cytokinins and anticytokinins do control the growth of mammalians cells via regulation of cyclic AMP metabolism, the possibility should also be considered that the compounds may affect the functioning of adenylate cyclase and cyclic AMP-dependent protein kinase activities in such cells.*

ACKNOWLEDGMENTS

The plant bioassay results described in this report were obtained as part of a continuing collaboration with Professor Folke Skoog and Dr. Ruth Schmitz of the Institute of Plant Development, University of Wisconsin. I would also like to thank my co-workers at Massachusetts Institute of Technology, especially Dr. R. Bruce Frye, Dr. Hector Juarez, and Dr. Ulrich Jordis. This work was supported in part by research grants from the National Cancer Institute (CA 14896) and the donors of the Petroleum Research Fund, administered by the American Chemical Society.

REFERENCES

1. F. Skoog and D. J. Armstrong, *Annu. Rev. Plant Physiol.* **21**, 359 (1970).
2. D. S. Letham, *Bio-science* **19**, 309 (1969).
3. J. P. Helgeson, *Science* **161**, 974 (1968).
4. H. G. Zachau, *Angew. Chem., Int. Ed. Engl.* **8**, 711 (1969).
5. D. S. Letham, *Phytochemistry* **12**, 2445 (1973).
6. D. J. Armstrong, P. K. Evans, W. J. Burrows, F. Skoog, J.-F. Petit, J. L. Dahl, T. Steward, J. L. Strominger, N. J. Leonard, and S. M. Hecht, *J. Biol. Chem.* **245**, 2922 (1970).
7. H. G. Zachau, D. Dütting, and H. Feldmann, *Hoppe-Seyler's Z. Physiol. Chem.* **347**, 212 (1966).
8. J. T. Madison and H. Kung, *J. Biol. Chem.* **242**, 1324 (1967).
9. M. Staehelin, H. Rogg, B. C. Baguley, T. Ginsberg, and M. Wehrli, *Nature (London)* **219**, 1363 (1968).
10. B. G. Barrell and F. Sanger, *FEBS Lett.* **3**, 275 (1969).
11. M. Uziel and H. G. Gassen, *Biochemistry* **8**, 1643 (1969).
12. B. P. Doctor, J. E. Loebel, M. A. Sodd, and D. B. Winter, *Science* **163**, 693 (1969).
13. H. M. Goodman, J. Abelson, A. Landy, S. Zadrazil, and J. D. Smith, *Eur. J. Biochem.* **13**, 461 (1970).
14. H. Ishikura, Y. Yamada, and S. Nishimura, *FEBS Lett.* **76**, 68 (1971).
15. D. Hirsh, *J. Mol. Biol.* **58**, 439 (1971).
16. S. Hashimoto, S. Takemura, and M. Miyazaki, *J. Biochem. (Tokyo)* **72**, 123 (1972).

* The i⁶Ado had no effect, however, on the cyclic AMP-dependent protein kinase from beef heart used to assay changes in cyclic AMP concentration in our studies. Cytokinins have also been shown to affect the phosphorylation of proteins and activity of protein kinases in Chinese cabbage and tobacco leaves, but the protein kinase activities were found not to be cyclic AMP dependent [74,75].

17. T. C. Pinkerton, G. Paddock, and J. Abelson, *J. Biol. Chem.* **248**, 6348 (1973).
18. N. J. Holness and G. Atfield, *FEBS Lett.* **46**, 268 (1974).
19. F. Fittler and R. H. Hall, *Biochem. Biophys. Res. Commun.* **25**, 441 (1966).
20. M. L. Gefter and R. L. Russell, *J. Mol. Biol.* **39**, 145 (1969).
21. H. Hayashi, H. Fisher, and D. Söll, *Biochemistry* **8**, 3680 (1969).
22. M. D. Litwack and A. Peterkovsky, *Biochemistry* **10**, 994 (1971).
23. S. M. Hecht, L. H. Kirkegaard, and R. M. Bock, *Proc. Natl. Acad. Sci. U.S.A.* **68**, 48 (1971).
24. S. M. Hecht, B. J. B. Johnson, L. H. Kirkegaard, and R. M. Bock, *Fed. Proc., Fed. Am. Soc. Exp. Biol.* **30**, 1271 (1971).
25. A. Hampton, J. J. Biesele, A. E. Moore and G. B. Brown, *J. Am. Chem. Soc.* **78**, 5695 (1956).
26. M. H. Fleysher, *J. Med. Chem.* **15**, 187 (1972), and references therein.
27. D. Suk, C. L. Simpson, and E. Mihich, *Cancer Res.* **30**, 1429 (1970).
28. K. Kikugawa, K. Iizuka, and M. Ichino, *J. Med. Chem.* **16**, 358 (1973).
29. B. Hacker and T. L. Feldbush, *Biochem. Pharmacol.* **18**, 847 (1969).
30. R. Jones, Jr., J. T. Grace, Jr., A. Mittelman, and M. W. Woodruff, *Proc. Am. Assoc. Cancer Res.* **9**, 35 (1968).
31. A. Mittelman, J. T. Evans, and G. B. Chheda, *Ann. N.Y. Acad. Sci.* **255**, 225 (1975).
32. G. H. Hitchings, *Ann. Rheum. Dis.* **25**, 601 (1966).
33. G. B. Elion, *Ann. Rheum. Dis.* **25**, 608 (1966).
34. M. Ikehara, K. Murao, F. Harada, and S. Nishimura, *Biochim. Biophys. Acta* **155**, 82 (1968).
35. F. Skoog, H. Q. Hamzi, A. M. Szweykowska, N. J. Leonard, K. L. Carraway, T. Fujii, J. P. Helgeson, and R. N. Loeppky, *Phytochemistry* **6**, 1169 (1967).
36. R. K. Robins, F. W. Furcht, A. D. Grauer, and J. W. Jones, *J. Am. Chem. Soc.* **78**, 2418 (1956).
37. R. K. Robins, L. B. Holum, and F. W. Furcht, *J. Org. Chem.* **21**, 833 (1956).
38. S. M. Hecht, R. M. Bock, R. Y. Schmitz, F. Skoog, N. J. Leonard, and J. L. Occolowitz, *Biochemistry* **10**, 4224 (1971).
39. E. M. Linsmaier and F. Skoog, *Plant Physiol.* **18**, 100 (1965).
40. S. M. Hecht, R. M. Bock, R. Y. Schmitz, F. Skoog, and N. J. Leonard, *Proc. Natl. Acad. Sci. U.S.A.* **68**, 2608 (1971).
41. F. Skoog, R. Y. Schmitz, R. M. Bock, and S. M. Hecht, *Phytochemistry* **12**, 25 (1973).
42. J. P. Helgeson, G. T. Haberlach, and S. M. Hecht, *in* "Plant Growth Substances, 1973," pp. 485–493. Hirokawa Publ. Co., Tokyo, 1973.
43. F. Skoog, R. Y. Schmitz, S. M. Hecht, and R. B. Frye, *Proc. Natl. Acad. Sci. U.S.A.* **72**, 3508 (1975).
44. S. M. Hecht, R. B. Frye, D. Werner, S. D. Hawrelak, F. Skoog, and R. Y. Schmitz, *J. Biol. Chem.* **250**, 7343 (1975).
45. R. Y. Schmitz, F. Skoog, S. M. Hecht, R. M. Bock, and N. J. Leonard, *Phytochemistry* **11**, 1603 (1972).
46. C. L. Dickinson, J. K. Williams, and B. C. McKusick, *J. Org. Chem* **29**, 1919 (1964).
47. S. M. Hecht, D. Werner, D. D. Traficante, M. Sundaralingam, P. Prusiner, T. Ito, and T. Sakurai, *J. Org. Chem.* **40**, 1815 (1975).
48. R. A. Earl, R. J. Pugmire, G. R. Revankar, and L. B. Townsend, *J. Org. Chem.* **40**, 1822 (1975).
49. E. C. Taylor and P. K. Loeffler, *J. Am. Chem. Soc.* **82**, 3147 (1960).

50. N. J. Leonard, S. Achmatowicz, R. N. Loeppky, K. L. Carraway, W. A. H. Grimm, A. Szweykowska, H. Q. Hamzi, and F. Skoog, *Proc. Natl. Acad. Sci. U.S.A.* **56**, 709 (1966).

51. S. M. Hecht and D. Werner, *J. Chem. Soc., Perkin Trans.* 1 p. 1903 (1973).

52. H. Iwamura, T. Ito, Z. Kumazawa, and Y. Ogawa, *Biochem. Biophys. Res. Commun.* **57**, 412 (1974).

53. M. J. Robins, and E. M. Trip, *Biochemistry* **12**, 2179 (1973).

54. J. H. Rogozinska, C. Kroon, and C. A. Salemink, *Phytochemistry* **12**, 2087 (1973).

55. H. Iwamura, T. Ito, Z. Kumazawa, and Y. Ogawa, *Phytochemistry* **14**, 2317 (1975).

56. J. W. Einset and F. Skoog, *Proc. Natl. Acad. Sci. U.S.A.* **70**, 658 (1973).

57. G. G. Deleuze, J. D. McChesney, and J. E. Fox, *Biochem. Biophys. Res. Commun.* **48**, 1426 (1972).

58. R. C. Gallo, J. Whang-Peng, and S. Perry, *Science* **165**, 400 (1969).

59. R. C. Gallo, S. M. Hecht, J. Whang-Peng, and S. O'Hopp, *Biochim. Biophys. Acta* **281**, 488 (1972).

60. S. M. Hecht, R. D. Faulkner, and S. D. Hawrelak, *Proc. Natl. Acad. Sci. U.S.A.* **71**, 4670 (1974).

61. S. M. Hecht and H. Juarez, in preparation.

62. M. L. Heidrick and W. L. Ryan, *Cancer Res.* **31**, 1313 (1971).

63. J. R. Sheppard, *Nature (London), New Biol.* **236**, 14 (1972).

64. J. Otten, G. S. Johnson, and I. Pastan, *Biochem. Biophys. Res. Commun.* **44**, 1192 (1971).

65. W. Seifert and D. Paul, *Nature (London), New Biol.* **240**, 281 (1972).

65a. S. M. Hecht, R. B. Frye, D. Werner, T. Fukui, and S. D. Hawrelak, *Biochemistry* **15**, 1005 (1976).

66. K. Ishii and H. Green, *J. Cell Sci.* **13**, 429 (1973).

67. T-S. Chan, K. Ishii, C. Long, and H. Green, *J. Cell. Physiol.* **81**, 315 (1973).

68. S. M. Hecht and R. B. Frye, in preparation.

69. H. Gilman, *Proc. Natl. Acad. Sci. U.S.A.* **67**, 305 (1970).

70. A. Y. Divekar, M. H. Fleysher, H. K. Slocum, L. N. Kenny, and M. T. Hakala *Cancer Res.* **32**, 2530 (1972).

71. A. Y. Divekar and M. T. Hakala, *Mol. Pharmacol.* **7**, 663 (1971).

72. B. Hacker, *Biochim. Biophys. Acta* **102**, 198 (1965).

73. J. Schultz, *Arch. Biochem. Biophys.* **163**, 15 (1974).

74. R. K. Ralph, P. J. A. McCombs, G. Tener, and S. J. Wojcik, *Biochem. J.* **130**, 901 (1972).

75. R. K. Ralph, S. Bullivant, and S. J. Wojcik, *Biochim. Biophys. Acta* **421**, 319 (1976).

Specific Chemical Probes for Elucidating the Mechanism of Steroid Hormone Action: Progress Using Estrogen Photoaffinity Labeling Agents

John A. Katzenellenbogen, Howard J. Johnson, Jr.,
Harvey N. Myers, Kathryn E. Carlson, and
Robert J. Kempton

INTRODUCTION

The initial step in the action of steroid hormones (S, Scheme 1) in target tissues is thought to be the binding of hormone with specific, high-affinity cytoplasmic binding proteins, which are generally referred to as receptors (R, Scheme 1). The formation of the steroid–receptor complex (R·S) is reversible, and the binding is characterized both by high affinity (K_d in the range of 10^{-9} M] and great stereospecificity. Subsequent to steroid binding, the complex R·S undergoes an "activation" (conversion to R'·S). This process, which is not well understood, converts the complex to a form that has high affinity for certain nuclear constituents, so that the complex then moves into the nucleus and becomes bound to chromatin. While the chromatin binding sites have not yet been characterized, nuclear binding of the steroid–receptor

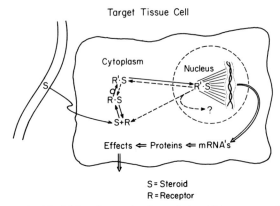

Scheme 1   Model for the mechanism of steroid hormone action.

complex is known to cause specific changes in gene expression. The altered pattern of messenger RNA synthesis is soon reflected by changes in cellular proteins. Ultimately, the characteristic biochemical and physiological effects elicited by the hormone in its target cells are manifested. The basic pattern for steroid hormone action at the molecular level holds true with little alteration for all classes of steroids: estrogens, androgens, progestins, and corticosteroids. In each case, it is the receptor protein whose presence establishes that the cell will be steroid responsive and whose binding specificity determines what type of steroid the cell will be responsive to.

Because of the dual role that steroid receptors play (determining tissue and steroid specificity and providing intracellular communication) it is not surprising that a major thrust of research in steroid hormone action has focused on these species (for recent reviews, see O'Malley and Schrader [1], King and Mainwaring [2], and Westphal [3]). In general, however, steroid receptor proteins have proved recalcitrant to the standard approach of biochemical study, namely, the isolation and purification to homogeneity of native proteins prior to detailed physiochemical analysis. This is attested to by the fact that, while 15 years have elapsed since the presence of steroid receptors was first demonstrated, substantial progress in steroid receptor purification has been recorded only in the past few years [4–9].

There are significant reasons why this situation exists. The receptors are only trace components; even in target tissues, purifications of the order of 1:100,000-fold would be required to achieve homogeneity. The thermal stability of receptors is low, particularly in the absence of steroidal ligand, and they often have a proclivity for irreversible aggregation. Another difficulty stems from the fact that the action of steroids in target tissue cells is dynamic, involving several different forms of the steroid–receptor complex, movement between subcellular compartments, and binding to other cellular constituents. This dynamic character cannot be preserved during isolation; in fact, the physical properties of the receptors often appear to change as they become purified. Alternative approaches to receptor characterization, in particular the chemical approach of affinity labeling, in which reactive steroid derivatives or analogs are used to selectively label the receptor binding site, thus have particular merit. This approach does not depend on rigorously purified receptor preparations in order to produce stable, covalently labeled receptors for further characterization.

One can imagine a number of uses to which the covalently linked hormone–receptor species produced by this affinity labeling process could be put to advantage. (1) Specific hormone binding components can be identified in heterogeneous binding preparations. This is especially important in the development of reliable clinical assays for receptors. (2) The covalently labeled binding proteins can be purified under disaggregating or denaturing conditions, in which the binding activity of the native protein would be lost. This approach would be extremely useful in situations in which receptor purification is hampered by receptor lability or aggregation. (3) Rigorous physiochemical analysis of the purified, labeled species can be undertaken. (4) Chemical analysis of labeled peptides that result from degradation of the purified protein can yield information on the binding site composition and topography. A particularly intriguing application in this regard would be the comparison of receptors for the same hormone but from different target cells (e.g., estrogen receptors from uterus vs. hypothalamus). Any differences that were apparent could suggest the potential for selective pharmacological effects. (5) The covalent interaction of these reactive steroids with the hormone receptor sites could result in unusual pharmacological activity (prolonged agonistic and antagonistic effects). (6) It is possible that certain dynamic features of steroid hormone action may eventually be studied by the affinity labeling, particularly the photoaffinity labeling, technique. (7) Finally, it might be possible to study the ternary interaction of the hormone–receptor complex with chromatin, which is a crucial yet poorly understood process in the mechanism of steroid hormone action.

Despite the attractiveness and promise of the technique of affinity labeling for studying hormone mechanisms at the receptor level, intensive efforts to exploit this approach have begun only recently. The progress of studies in the heterogeneous systems has been slow, so most of the results achieved to date are preliminary and suggestive rather than conclusive. In this chapter we describe our efforts to achieve selective labeling of the estrogen receptor protein from uterine preparations of immature rats and lambs using the technique of photoaffinity labeling. (For reviews of affinity labeling of hormone binding sites, see Katzenellenbogen [10,11].)

FUNDAMENTAL CONSIDERATIONS RELATING
TO THE DESIGN OF AFFINITY LABELING
REAGENTS FOR THE ESTROGEN RECEPTOR

The salient feature of an affinity labeling process is that covalent bond formation takes place within a complex of protein and labeling reagent $(R \cdot E - x)$. The high, localized concentration of reagent $(E - x)$ in the complex results in an enhanced reaction between the reagent and nearby amino acid residues (proximity effect), which is the basis for the anticipated selectivity of the affinity labeling process. These considerations are summarized in Scheme 2.

Several reports [10–14] have outlined certain features that a binding-preceded labeling process should display; these can be summarized as follows: (1) There should be evidence for a reversible equilibrium between reagent and binding site; (2) the labeling reaction should display a rate or binding saturating effect; (3) nonreactive ligands that compete for the same binding site should exert a protective effect, reducing the rate and in certain cases the extent of labeling; and (4) in situations in which a selective labeling can be achieved, there should be a stoichiometric relationship between the extent of labeling and the extent of inactivation. These properties, in total or in part, are generally used as the criteria for establishing that a labeling process is indeed occurring by an affinity labeling mechanism as opposed, in particular, to a simple bimolecular mechanism.

One can distinguish at this stage between two types of affinity labeling: conventional and photoaffinity labeling. In the former, the covalent bond formation involves the reaction between a nucleophilic residue on the protein and a reactive electrophilic function on the labeling reagent; the types of functions used in this case are typical alkylating (haloketones and esters) and acylating (imidates) groups. In photoaffinity labeling, the reactive function is photosensitive. In the dark it is inert; thus, the reagent behaves like a normal reversible binding ligand. Upon irradiation, a highly reactive species (e.g., carbene or nitrene) or an electronically excited state

Reversible interaction:

$$R + E \xrightleftharpoons{K_d} R \cdot E$$

Affinity labeling:

$$R + E - x \xrightleftharpoons{K_d'} R \cdot E - x \longrightarrow \overparen{R \cdot E - x}$$

Labeling = "binding affinity" × "reaction efficiency"

Scheme 2   Kinetic model for affinity labeling process (R, estrogen receptor; E, estradiol E—x, estrogen affinity label).

(ketone $n\pi^*$ triplet) is produced, which can then react rather indiscriminately with a variety of residues in the binding site.

The differences between these two modes of affinity labeling go beyond the obvious fact that photoaffinity labeling provides an opportunity for external control over the covalent attachment process, permitting the selection of conditions optimal for labeling selectivity. A complete discussion of these differences and a rationale for our selection of the technique of photoaffinity labeling over conventional labeling in our present studies can be found elsewhere [10,11].

The rather simplistic kinetic model shown in Scheme 2 holds for the labeling of homogeneous preparations of binding proteins. However, the situation with labeling in heterogeneous binding preparations is considerably more complex. For example, a typical preparation of estrogen receptor from rat or lamb uterus, even after two stages of purification (ammonium sulfate precipitation and Sephadex G-200 column chromatography), may contain only 1 part receptor to 10,000 parts of other proteins. Thus, the labeling reagent can interact not only with the receptor binding site, but with binding sites on many other proteins. Although these sites may bind hormone with lower affinity, their greater number can severely compromise the selectivity of the labeling process. A full discussion of the factors involved in the affinity labeling of heterogeneous binding systems is given by Katzenellenbogen [10,11].

One can conceive of two independent aspects of the labeling process: binding affinity and labeling efficiency (Scheme 2). The first aspect, binding affinity, is related to the degree to which the modified ligand (affinity labeling reagent) has retained its affinity for the binding site. Labeling efficiency is a measure of how successful the covalent-bond-forming reaction operates once the labeling reagent is within the binding site complexes. In order for the affinity labeling process to be successful (and, in particular, for the labeling of a receptor in a heterogeneous preparation to be selective), it is essential that both the binding affinity and the labeling efficiency of an affinity labeling reagent be as high as possible.

These general considerations led us to devise the following empirical protocol for the development of affinity labeling reagents for the selective covalent labeling of the uterine estrogen receptor:

1. On the basis of the best available information concerning the tolerance of the receptor for binding various substituted estrogenic ligands and its sensitivity toward the covalent attachment reactions, candidate affinity labeling compounds are designed and synthesized in nonradiolabeled form. These materials are fully characterized by chemical and spectroscopic means to verify their structure.

2. By means of indirect competition assays, the binding affinity and inactivation efficiency (a measure of labeling efficiency) are determined.

3. Those derivatives that appear to be promising at this stage are then prepared in radioactive form, with a high specific activity tritium label.

4. The reversible binding affinity and binding specificity of the reagents in radiolabeled form are assayed directly. This is followed by direct determination of the efficiency and selectivity of their covalent attachment to the receptor in the binding preparation.

On the basis of the considerations outlined above, we selected as the first compounds to prepare as potential affinity labeling reagents for the uterine estrogen receptor a set of relatively simple photosensitive derivatives of steroidal estrogens (estradiol and estrone) and the nonsteroid estrogen hexestrol. A number of factors affected our selection of these compounds.

First, at the time we initiated this project, the only photoattaching function that had been studied was the diazocarbonyl group; shortly thereafter, Knowles [15] introduced the aryl azide function. We selected these two groups for investigation. Later, we enlarged our purview to aromatic ketones and other, more unusual chromophoric functions.

The second consideration relates to binding affinity. There is a considerable amount of data in the literature concerning the relationship between estrogen structure and binding affinity that are of relevance to the design of affinity labeling reagents. The most complete studies have been done by Korenman [16], Geynet *et al.* [17], Terenius [18], Katzenellenbogen *et al.* [19], and Ellis.* The work of Ellis, although unfortunately not yet published, is the most complete, providing the relative binding affinities of some 140 estrogens and related derivatives.

Briefly, the binding specificity of the uterine estrogen receptor is such that among steroids it binds well only to estrogens, yet it also binds to a variety of nonsteroidal compounds, many having di- or triarylethylene structures. Within the estrogen skeleton, hydroxyl groups at 3 and 17β are needed for high binding affinity; substituents are well tolerated at positions 6 (ketone), 7α (methyl and alkyl), 11β (alkyl, methoxy, but not hydroxy), 16 (polar better than nonpolar), and 17α (ethynyl). Derivatives substituted in the A ring often bind, but with considerably reduced affinities. In the stilbestrol–hexestrol series, large internal substituents such as two ethyl groups are needed for activity, and ring substituents are well tolerated. Some of the triarylethylene

---

* Ellis has determined the binding affinity of 140 steroid derivatives for a partially purified estrogen-binding protein of mature rat uterus using a competition assay with [^{3}H]estradiol and charcoal–dextran adsorption. These data are available from Dr. David J. Ellis, Institute of Biological Sciences, Syntex Research Center, Stanford Industrial Park, Palo Alto, California 94304.

antiestrogens, which are considerably more bulky than estradiol, have surprisingly high binding affinities [20].

The third consideration deals with attachment efficiency. Little is known about the chemical constitution of the estrogen binding site of the uterine estrogen receptor protein other than its sensitivity to sulfhydryl reagents [21], although recently even this has been questioned [22]. Thus, there is little to guide one in selecting attaching functions and positioning them about the ligand molecule in order to produce a labeling reagent that will have high attachment efficiency.

The final consideration in the design of this initial set of reagents was synthetic accessibility. While it appeared from binding studies that the compounds derivatized in the B and C rings might be potentially higher binders, no definitive or quantitative predictions of the effectiveness of such derivatives could be made. Therefore, we chose to functionalize the most accessible sites, that is, those on or near the 3 and 17β hydroxyl groups in the steroid and the two phenolic hydroxyls in the hexestrol molecule. We felt that such a set of compounds would enable us to test the different assay procedures in our reagent development protocol. Even if these first reagents ultimately proved not to be successful in achieving an efficient and selective covalent labeling of the receptor, they would at least point the way for more effective (but possibly chemically more complex) second-generation labeling reagents.

RESULTS

Chemical Synthesis of Estrogen Photoaffinity Labeling Reagents

We have described the preparation of the estrogen derivatives and analogs shown in Scheme 3 by relatively standard routes [23]. These compounds contain either a diazocarbonyl or an azide photoreactive function or, in the case of 8, an aryl ketone. In many cases, analogous derivatives in the steroidal and nonsteroidal series were prepared.

Recently, we have improved significantly the preparation of the diazo-ketopropyl or diazoacetonyl ethers 3 and 4 [24]. Previously these compounds were synthesized in six steps by alkylation with ethyl bromoacetate followed by hydrolysis, hydroxyl protection, conversion to the acid chloride, reaction with diazomethane, and hydroxyl deprotection. However, we have found that, under carefully controlled conditions, the phenolate ion reacts with bromodiazoacetone to form the diazoacetonyl ether directly in one step.

Studies Using Nonradiolabeled Photosensitive Estrogen Derivatives

DETERMINATION OF THE BINDING AFFINITY BY COMPETITION ASSAY

Standard protein competitive binding techniques using dextran-coated charcoal to remove free steroid were employed to determine the reversible

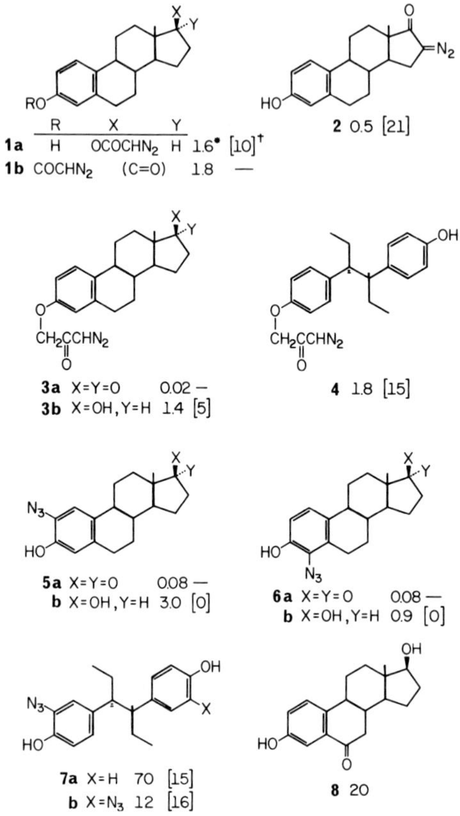

	R	X	Y	
1a	H	OCOCHN$_2$	H	1.6* [10]†
1b	COCHN$_2$	(C=O)		1.8 —

2 0.5 [21]

3a X=Y=O 0.02 —
3b X=OH,Y=H 1.4 [5]

4 1.8 [15]

5a X=Y=O 0.08 —
 b X=OH,Y=H 3.0 [0]

6a X=Y=O 0.08 —
 b X=OH,Y=H 0.9 [0]

7a X=H 70 [15]
 b X=N$_3$ 12 [16]

8 20

Scheme 3 Photosensitive estrogen derivatives and analogs. [* Numbers are binding affinities (RAC × 100) relative to estradiol = 100; † numbers in brackets are percent inactivation efficiencies. See text for details.]

binding affinity of these derivatives with the estrogen receptor from rat and lamb uterus. Generally, cytosol preparations were utilized in these assays without purification. The methodology we use has been described in full [19].

A particularly convenient feature of our assay procedure is the use of blood microtitration assay hardware. Up to 96 incubations (maximum volume 125 μl) can be carried out simultaneously on disposable polystyrene blood microtitration plates. These plates can be cooled on ice, and all the samples can be vortexed simultaneously. An adhesive film cover protects them during centrifugation.

A typical set of competition curves is shown in Fig. 1. By comparing the midpoints of competition on the sigmoidal curves from the plot of bound [³H]estradiol vs. log concentration of competitor, one can determine the

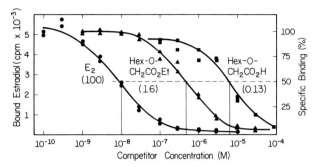

Fig. 1 Binding affinity assay. Various concentrations of competitor, E_2 (estradiol), Hex-O-CH$_2$CO$_2$Et [4-O-(carbethoxymethyl)hexestrol], and Hex-O-CH$_2$CO$_2$H [4-O-(carboxymethyl)hexestrol] were incubated with 10^{-8} M [^{3}H]estradiol in rat uterine cytosol (1.4 uterine equivalents/ml). Charcoal–dextran was used to adsorb unbound steroid. The RAC × 100 values are given in parentheses. (For details, see [19].) [Reprinted with permission from Katzenellenbogen et al., Biochemistry **12**, 4085 (1973). Copyright by the American Chemical Society.]

relative binding ability. This can be converted to the ratio of association constants (RAC) by a simple formula derived by Korenman [25] and Rodbard [26]. These data are generally expressed as RAC × 100 values, so that the competitor binding can be considered as a percentage of the estradiol binding affinity. The RAC × 100 values for the photosensitive estrogen derivatives that we have prepared are given under their structures in Scheme 3.

A general feature of these data that can be noted readily is that the binding affinity of a particular derivative reflects to a degree the binding affinity of its parent ligand. Thus, the hexestrol derivatives in all cases bind with higher affinity than the estradiol derivatives, which show better binding than the estrones. The derivatives of hexestrol deserve additional comment.

Compared to the steroids, hexestrol is both conformationally mobile and symmetrical (*meso*) (Scheme 4). Introduction of a single ring substituent destroys the symmetry of the ligand, producing a pair of enantiomers. If one maintains congruency between the configuration of the hexane chain and the stereochemistry of the BC ring region of the natural steriods, then one of these enantiomers will bind with its substituent in the A ring binding region of the receptor, and the other will bind with the substituted ring in place of the steroid D ring. Furthermore, with each enantiomer, two binding conformers will be possible because of the free rotation about the phenylhexane bond. While this symmetry and conformational mobility introduces some ambiguity in using hexestrol derivatives to evaluate the bulk tolerance of the estrogen binding site of the receptor, it provides a bonus in that one has essentially four alternative positions for accommodating a hexestrol ring substituent in the estrogen binding site, while the same group on the steroid will be

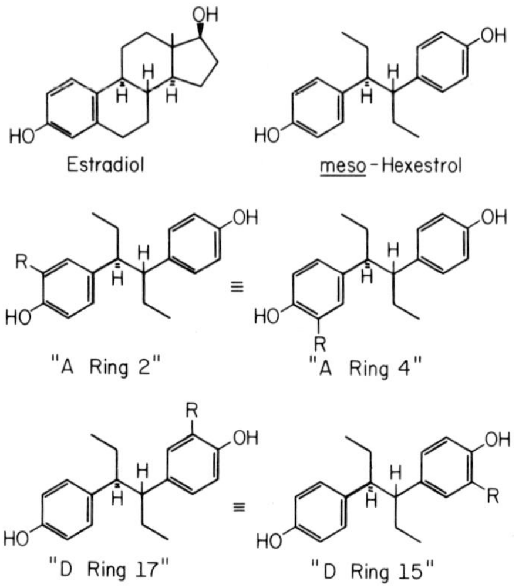

Estradiol

meso – Hexestrol

"A Ring 2"

≡

"A Ring 4"

"D Ring 17"

≡

"D Ring 15"

Scheme 4 Stereochemical and conformational relationships between estradiol and *meso*-hexestrol.

positioned unambiguously, but at one site only. The higher affinity of the hexestrol derivatives (the azides in particular) relative to the A-ring-substituted steroids suggests, in fact, that their substituent is being accommodated in the D ring binding region [19].

DETERMINATION OF LABELING EFFICIENCY (INACTIVATION EFFICIENCY) BY A PHOTOLYSIS-EXCHANGE COMPETITION ASSAY

The next stage of development of the photoaffinity labeling reagents involves establishing that there is a specific photointeraction of these compounds with the estrogen binding site. This information can be determined indirectly with the nonradiolabeled compounds using a photoinactivation (photolysis-exchange) assay (Scheme 5) [20]: Binding sites are filled with a minimum

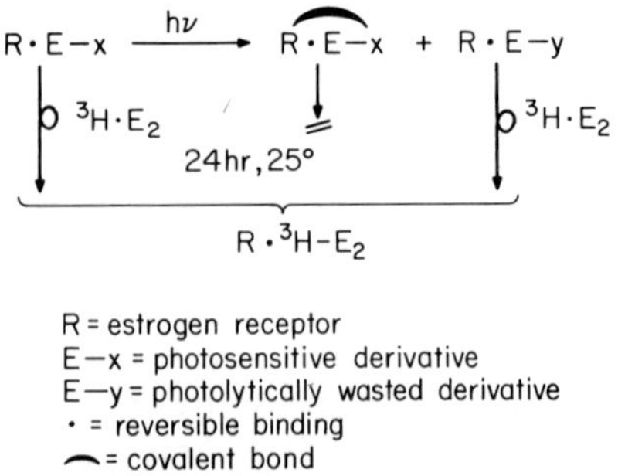

R = estrogen receptor
E—x = photosensitive derivative
E—y = photolytically wasted derivative
· = reversible binding
⌒ = covalent bond

Scheme 5 Scheme for determining inactivation efficiencies by the photolysis-exchange assay. (See text for details.)

saturating concentration of unlabeled photosensitive derivatives (saturating concentrations can be determined from the RAC × 100 values) and are then irradiated at either 254 or > 315 nm. The amount of estrogen-specific binding that remains after a period of photolysis is determined by quantitative exchange (24 hr, 25°C) against an excess of [^{3}H]estradiol using a procedure that we have developed [27]. The results of photolysis of the photosensitive compound bound to the receptor (R·E—x) can be either photoinactivation (for instance, by covalent attachment of the derivative to the receptor to generate a nonexchangeable species $\overgroup{\text{R·E—x}}$) or simply photodegradation or wastage of the compound (by reaction with solvent) to give a different reversible complex, now with a photoinert ligand (R·E—y). The exchange assay monitors total reversible binding capacity, R·E—x plus R·E—y. Thus, a decrease in this quantity from the initial binding capacity can be taken as evidence of a successful photoinactivation event.

In order to demonstrate the estrogen binding site specificity of this inactivation event, two photoinactivation time courses are run in parallel, one in which only the photosensitive reagent is present (possible inactivation by both site-specific and nonspecific processes), and the other in which the binding sites are first protected by filling them with unlabeled estradiol prior to addition of the photosensitive compound (inactivation possible only by nonspecific processes). The difference between the inactivation curves for the two experiments is taken as the site-specific photoinactivation process.

Different types of photoreactive behavior are observed. These can best be observed by comparing semilog plots of estrogen-specific binding (monitored by exchange) as a function of irradiation time (Figs. 2 and 3). When the ligand in the estrogen binding site is not photoreactive, e.g., estradiol, the rate of inactivation is slow and first order (Fig. 2, top curves). That this represents direct irradiative damage of the protein at 254 nm is established by the fact that empty sites and sites filled with photoinert estrogens are degraded at the same rate [28]. When sites are filled with a photosensitive derivative, the rate of inactivation is in most cases much more rapid. For example, the photoinactivation curves for hexestrol diazoketopropyl ether (4) (Fig. 2) show that sites filled with the derivative (bottom curves) are degraded several times faster than estrogen-filled sites (top curve), the site specificity being demonstrated by the protective effect of estradiol (middle curve). The saturability of the site-specific inactivation process is evident from the fact that the same site-specific inactivation efficiency (*vide infra*) is found at two saturating concentrations of the derivative 4 (compare the left and right sides of Fig. 2).

The biphasic nature of the inactivation curves is not surprising. The rate of the initial, rapid portion is the same as that seen for the loss of the

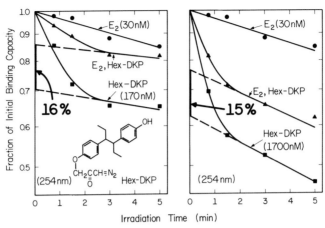

Fig. 2 Effect of concentration on photoreactivity of hexestrol diazoketopropyl ether (Hex-DKP, **4**) at 254 nm. This compound (■) at two concentrations [left, 170 nM (0.1 × 30/RAC nM); right, 1700 nM (30/RAC nM)] was irradiated in the presence of rat uterine cytosol at 2 uterine equivalents/ml. The irradiation time courses of binding sites filled only with estradiol (30 nM, ●) or protected with estradiol prior to the addition of **4** (▲) are also shown. The reaction efficiency is determined by extrapolation of the linear portion of the lower two curves back to zero time. Binding capacity was determined by exchange with [³H]estradiol and is corrected for nonspecific binding. Aliquots contained 0.082 uterine equivalent, and initial binding capacity was 22,000 dpm, or 2.2 pmoles/uterine equivalent. The degree of estrogen receptor site saturation is 90% (left) and 99% (right). (For details, see [20].) [Reprinted with permission from Katzenellenbogen *et al.*, *Biochemistry* **13**, 2986 (1974). Copyright by the American Chemical Society.]

photoreactive chromophore when the derivative is irradiated under the same conditions in organic solvents. The second portion of the curve represents a slower process that parallels the inactivation rate of estradiol-filled sites. This probably represents merely the protein photodegradation process observed with photoinert ligands which still proceeds after the photoreactive ligand has been completely consumed. (In this case the inert ligand is the discharged species E—y.) By extrapolating the linear portions of the protected and unprotected curves back to zero time, a fraction of the binding capacity is subtended. This fraction is considered to be the estrogen site-specific inactivation efficiency of the photosensitive ligand; that is, it represents the percentage of the estrogen binding sites that can no longer bind estrogen reversibly after 1 equivalent of bound photosensitive derivative has been completely photolyzed. The photoinactivation efficiency of the photosensitive derivatives is listed in brackets below their structures in Scheme 3.

A second type of photoinactivation behavior is illustrated by 6-oxoestradiol (**8**) (Fig. 3). Here, irradiation can be done at longer wavelengths because of the

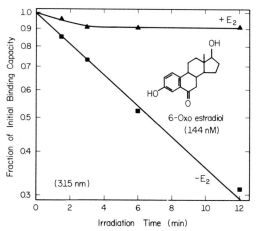

Fig. 3 Time course of irradiation of 6-oxoestradiol (**8**) at 315 nm. This compound at 144 nM was irradiated in the presence of rat uterine cytosol, both with estradiol (30 nM) preincubation (▲) and without estradiol preincubation (■). Determination of binding capacity is as in Fig. 2. (For details, see [20].) [Reprinted with permission from Katzenellenbogen *et al.*, *Biochemistry* **13**, 2986 (1974). Copyright by the American Chemical Society.]

more accessible high-intensity chromophore. The inactivation rate appears to be first order and is site specific.

It is not difficult to rationalize the difference between the biphasic and linear type of photoinactivation kinetics (Scheme 6). In the former case, irradiation of the photosensitive derivative (diazo or azide) causes an irreversible photoactivation that involves loss of nitrogen to produce a carbene or nitrene; subsequent reaction causes either receptor inactivation or reagent discharge. Reagents of this type can be termed "photolabile reagents"; once they are consumed, the rate of photoinactivation returns to that of sites filled with photoinert (photodischarged) ligand.

The second type of behavior is consistent with a process in which the photoactivation event is reversible. The reactive species is presumably in an electronically excited state, which has not suffered irreversible bond scission.

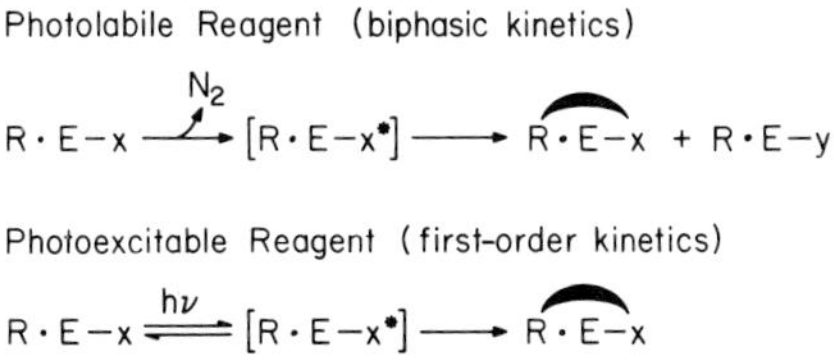

Photolabile Reagent (biphasic kinetics)

$$R \cdot E-x \xrightarrow{N_2} [R \cdot E-x^\bullet] \longrightarrow \overset{\frown}{R \cdot E-x} + R \cdot E-y$$

Photoexcitable Reagent (first-order kinetics)

$$R \cdot E-x \underset{h\nu}{\overset{h\nu}{\rightleftharpoons}} [R \cdot E-x^\bullet] \longrightarrow \overset{\frown}{R \cdot E-x}$$

Scheme 6 Two modes for the kinetics of receptor inactivation by photosensitive estrogen derivatives.

Reaction can then proceed to cause binding inactivation, or the excited species can decay back to the original molecule. Reagents of this type can be termed "photoexcitable" or "reversibly photoactivatable" ligands; the inactivation rate is expected to be first order, as the unaffected binding sites should always be filled with the photosensitive reagent.

It is important to remember that the photolysis-exchange assay determines the site-specific inactivation efficiency of a compound. Without the use of radiolabeled derivatives, however, it is not possible to determine whether the inactivation that is being measured actually represents covalent attachment, the alternative being a site-specific photodestruction by an atom or energy transfer process or redox reactions that do not, in fact, covalently link the ligand and protein. Thus, one must make a distinction between the "photo-inactivation efficiency" of a compound, which is being measured by such an assay, and the "photocovalent attachment or labeling efficiency," which is actually the important factor in the labeling process and whose determination requires the use of radiolabeled derivatives. Nevertheless, this assay is exceedingly useful because it can determine for a candidate photoaffinity labeling reagent, while still in nonradiolabeled form, the upper limit of its labeling efficiency.

Several compounds that have no obvious means of undergoing photocovalent attachment but have an intense and accessible ultraviolet or visible chromophore were investigated. In this regard the Parke-Davis antiestrogens CI-628, CI-680, and 9411 x 27 (Scheme 7) are particularly intriguing. The first two are potent photoinactivators at 315 nm (see Fig. 4 for CI-628); the last, although the highest affinity binder of the three, is the least efficient inactivator. This was presumed to be the result of a new mechanism for excited-state deactivation that is permitted when the phenolic oxygen is unsubstituted [20].

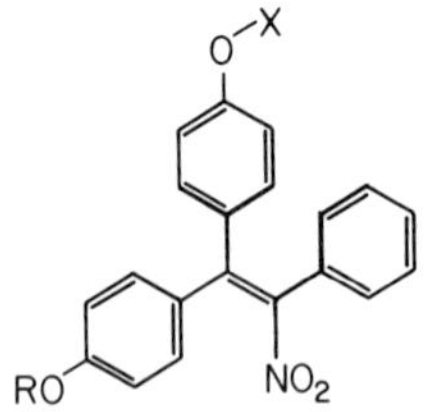

CI-628	R = CH$_3$, X = CH$_2$CH$_2$·N◯	4[*]	[50][†]
CI-680	R = CH$_3$, X = CH$_2$CH$_2$CH$_2$·NMe$_2$	34	[25]
9411x27	R = H , X = CH$_2$CH$_2$CH$_2$·NMe$_2$	222	[0]

Scheme 7 Photosensitive antiestrogens. [* Numbers are binding affinities (RAC × 100) relative to estradiol = 100; † numbers in brackets are percent inactivation efficiencies. See text for details.]

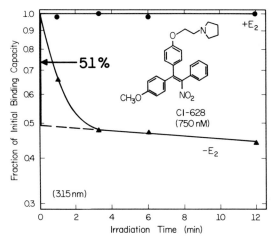

Fig. 4 Time course irradiation of Parke-Davis antiestrogen CI-628 at 315 nm. This compound at 750 nM was irradiated in the presence of rat uterine cytosol, both with estradiol (30 nM) preincubation (●) and without estradiol preincubation (▲). Determination of binding capacity and reaction efficiency are as in Fig. 2. Sites irradiated in the absence of either derivative (empty sites) gave data superimposable with the protected line (●). (For details, see [20].) [Reprinted with permission from Katzenellenbogen *et al.*, *Biochemistry* **13**, 2986 (1974). Copyright by the American Chemical Society.]

The inactivation efficiencies of the other derivatives show that in most cases there is no clear relationship between binding affinity and inactivation efficiency (cf. **2** and **5b**). Also, comparison of the inactivation efficiencies of the A-ring-substituted steroidal vs. hexestrol derivatives (**3, 5,** and **6** vs. **4** and **7**) shows the latter to be more efficient inactivators. This may indicate that the A ring binding site of the receptor is relatively unsusceptible to reaction with the photoreactive intermediates, as the steroidal derivatives must present their substituents in this region of the receptor and are relatively unreactive. On the other hand, the hexestrol derivatives, which probably have their substituent accommodated near the D ring site (*vide ante*, Scheme 3), react considerably more efficiently (Scheme 8). Further evidence for the differential susceptibility of the receptor site to photoreaction comes from the relatively high inactivation efficiency of the 16-diazoestrone (**2**) and the equivalent inactivation efficiencies seen for the mono- and disubstituted hexestrol derivatives **7a** and **7b**.

Studies Using Radiolabeled Photosensitive Estrogen Derivatives in Unpurified Rat Uterine Receptor Preparations

REVERSIBLE BINDING TO RECEPTOR PREPARATIONS

On the basis of their relatively good binding affinities and inactivation efficiencies, photosensitive estrogen derivatives **2, 4, 7a,** and **8** have been

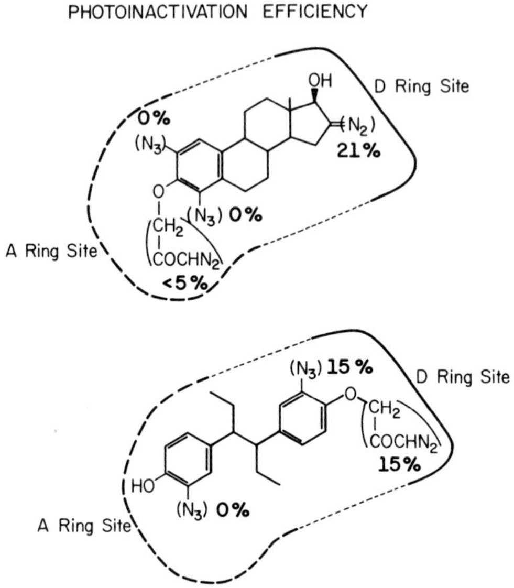

Scheme 8　Summary of photoactivation efficiencies of selected photoactive estrogens.

prepared in tritium-labeled form. Studies with the derivatives **4** and **7a** are presently most advanced. The extent of nonspecific binding observed with these derivatives is much higher than that of either estradiol or hexestrol, and, although in both cases estrogen-specific binding can be demonstrated in uterine cytosol preparations, protein fractionation techniques must be used.

Sucrose gradient sedimentation is a convenient way to distinguish between estrogen receptor binding and nonspecific binding [29]. Under conditions of low ionic strength, the rat uterine estrogen receptor sediments as a high molecular weight aggregate (ca. 8 S); most of the nonspecific binding proteins sediment as 4 S species. Figure 5 shows a sucrose gradient profile with [³H]-hexestrol azide (**7a**).

While estradiol binding is found only in the 8 S region of the gradient (dashed line), most of the [³H]hexestrol azide (**7a**) is bound by nonreceptor proteins sedimenting in the 4 S region. However, hexestrol azide binding in the 8 S region is evident, and it appears to be estrogen specific. It is quantitatively equivalent to the binding observed with [³H]estradiol, and it is subject to competition by excess unlabeled estradiol. The 4 S binding is not estrogen specific (no competition by unlabeled estradiol), but it is of quite high affinity, as evidenced by the fact that excess unlabeled hexestrol azide causes nearly complete exclusion of binding from the 4 S region.

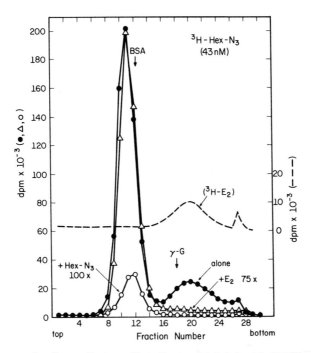

Fig. 5 Sucrose density gradient profiles of cytosol labeled with [³H]hexestrol azide (Hex-N₃, **7a**). Rat cytosol was incubated with 43 nM [³H]Hex-N₃ alone (●) or after preincubation with 75-fold excess unlabeled estradiol (E₂, △) or 100-fold excess unlabeled Hex-N₃ (○). The cytosol complex was charcoal treated and centrifuged for 13 hr at 246,000 g through gradients of 5–20% sucrose. Bovine serum albumin (4.7 S) and γ-globulin (7.0 S) were added as internal markers. The dashed line, with scale on right vertical axis, is binding profile for 30 nM [³H]estradiol alone [35].

PHOTOCOVALENT ATTACHMENT TIME COURSE STUDIES

Despite the high level of nonspecific binding observed with these compounds in unpurified rat uterine cytosol, we proceeded to investigate their capacity for photocovalent attachment. Figure 6 shows the time course of photocovalent attachment of [³H]hexestrol diazoketopropyl ether (**4**) to the rat uterine cytosol preparations. Similar results were obtained with [³H]hexestrol azide (**7a**). In this experiment, the extent of covalent attachment was determined by boiling ethanol extraction of aliquots applied to filter paper discs. In the dark, covalent attachment of these reagents to proteins proceeds only very slowly (see Fig. 6, legend), but upon irradiation, rapid attachment ensues (top curve). It is evident that the labeling is not highly specific for the receptor, however, since the amount of attachment that is obtained is in excess of the amount of labeling expected (Top curve vs. inset values *a* and *b*; see Fig. 6, legend).

In order to establish whether some of the incorporation in the curve was estrogen site specific, an experiment was run simultaneously with a sample in which the receptor sites had been blocked by the addition of unlabeled estradiol prior to the addition of [³H]hexestrol diazoketopropyl ether (Fig. 6, dashed curve). Protection by estradiol did cause a depression in the incorporation of [³H]hexestrol diazoketopropyl ether that was of the order predicted from the inactivation efficiency of this compound (cf. Scheme 3; see inset value *b*).

The data in Fig. 6 also indicate that not all the covalent attachment is dependent upon the photochemically reactive chromophore, as continued irradiation for times beyond which the chromophore has been completely consumed (ca. 1.5 min), still leads to additional covalent attachment, though at a slower rate. Furthermore, slow incorporation was also observed in the

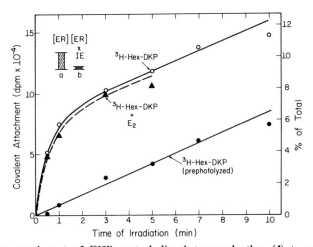

Fig. 6 Photoattachment of [³H]hexestrol diazoketopropyl ether (**4**) to rat uterine cytosol proteins. Rat cytosol was incubated with 167 n*M* [³H]hexestrol diazoketopropyl ether (○) or [³H]hexestrol diazoketopropyl ether prephotolyzed for 3 min at 254 nm (●). The protected sample [³H]Hex-DKP + E₂ (▲) was first incubated with unlabeled estradiol (3000 n*M*) to fill the receptor sites and then with [³H]hexestrol diazoketopropyl ether. Following photolysis at 254 nm for the times indicated, the covalent attachment was assayed as follows: Duplicate 75 μl samples were spotted onto 2.5 cm Whatman 3MM filter paper discs. The discs were dried for 5 min, washed two times for 15 min each in boiling 100% EtOH, rinsed two times at room temperature in 1:1 ether:80% EtOH, rinsed two times in ether, and dried. Radioactivity was determined by liquid scintillation counting. The incorporation into dark controls (0.7–1 × 10⁴ dpm–direct; 1–3 × 10⁴ dpm–prephotolyzed) has been subtracted from each point. The amount of estrogen receptor in the cytosol ([ER]) is indicated as the striped bar *a*. As the inactivation efficiency (IE) of hexestrol diazoketopropyl ether is 15% (Scheme 3), a maximum of only 15% of the receptor-bound reagent can be expected to become covalently attached ([ER] × IE, indicated as hatched bar *b*.) [35].

case where the compound was photolyzed prior to the addition of cytosol (Fig. 6, bottom curve). Brunswick and Cooperman [30] have also noted such secondary covalent attachment in their studies with diazomalonyl derivatives of cAMP, and we have described a slow, nonspecific photocovalent attachment of estrogens and other phenolic compounds to proteins [28].

It was apparent from these studies that it would be essential to fractionate the radiolabeled protein components in order to determine whether the estrogen receptor had become labeled.

ELECTROPHORETIC ANALYSIS ON DENATURING UREA–POLYACRYLAMIDE GELS

At first, fractionation on gels containing 8 M urea was investigated because under these conditions only covalently labeled proteins would be observed. A typical electrophoretic profile after labeling with hexestrol azide (**7a**) is seen in Fig. 7.

In this case, unfractionated rat uterine cytosol was saturated with [³H]-hexestrol azide (**7a**) (both in the absence and presence of a high concentration of unlabeled **7a**) and irradiated for 1 min at 254 nm. This is sufficient time to consume nearly all the photosensitive compound with minimal photolytic damage to the receptor. The proteins were then stripped of unattached compounds by charcoal adsorption, added to buffer containing 8 M urea, and subjected to polyacrylamide gel electrophoresis in the presence of 8 M urea.

Electrophoretic analysis of the total labeled material (Fig. 7A, top curve) shows at least five peaks of radiolabeled material: a peak at the top of the gel, which may represent species that are still aggregated; a large peak at slice 12–13, which coelectrophoreses with ¹⁴C-labeled albumin; a mobile band in slice 34–36; and material coincident with the tracer dye (ran off the gel in the top case).

Covalent labeling of the estrogen receptor should be blocked by prior addition of an excess of unlabeled estradiol or photosensitive derivative (**7a**). The middle trace of Fig. 7A is an example of the latter experiment. Here, substantial protection of the labeling of the first and fifth components is evident. It is unlikely that the fifth component corresponds to the receptor, however, because when estradiol is used as the protector the labeling of this component persists (not shown); also, it migrates as a 4S species on low salt sucrose density gradient sedimentation (*vide infra*). The behavior of the first component is consistent with that of the estrogen receptor (*vide infra*).

The same sample of labeled material was subjected to sucrose gradient centrifugation, and the electrophoretic profiles of the 4 S and 8 S regions of the gradient are shown in Figs. 7B and 7C, respectively. The principal labeled components in the 4 S region are albumin and the rapidly moving peak. Again, the latter species remains unlabeled in the presence of excess hexestrol azide but not in the presence of excess unlabeled estradiol (separate experiment, not shown), so it is not the estrogen receptor. Material obtained from the 8 S region of the gradient (Fig. 7C) shows label in the aggregated region and in the regions directly ahead of the albumin peak. Excess hexestrol azide, as well as estradiol (experiment not shown), does afford substantial protection for the aggregated material and suggests that the estrogen receptor may be migrating as a disperse collection of aggregates in this region of the gel.

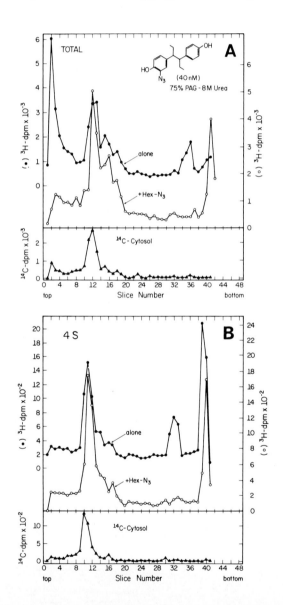

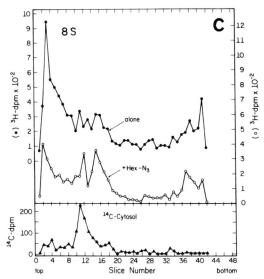

Fig. 7 Acrylamide gel electrophoresis of unfractionated and sucrose gradient-fractionated cytosol, photocovalently labeled with [³H]hexestrol azide ([³H]Hex-N₃, **7a**). Fresh rat cytosol was incubated with 40 nM [³H]Hex-N₃, alone (●) or after preincubation with 100-fold excess of unlabeled Hex-N₃, irradiated for 1 min at 254 nm, and charcoal treated. A portion was sedimented through a 5–20% sucrose gradient and the 4 S and 8 S regions were collected separately. Electrophoresis of the total sample (A) or the 4 S (B) or 8 S (C) gradient region was performed in alkaline pH gels, cross-linked with N,N'-diallytartardiamide containing 8 M urea and 7.5% acrylamide. The gels were sliced into 2.3 mm sections, dissolved in 2% periodic acid, and assayed by liquid scintillation counting. Lower panels show the profile of total uterine cytosol proteins (▲). A small aliquot of total cytosol, ¹⁴C-labeled by the formaldehyde–borohydride procedure, was added just prior to electrophoresis. [Reprinted with permission from Katzenellenbogen *et al.*, *Journal of Toxicology and Environmental Health*. Suppl. 1, 205 (1976). Copyright by the Hemisphere Publishing Corporation.]

Additional studies have been done with radiolabeled hexestrol diazoketopropyl ether (**4**). This derivative has a lower receptor binding affinity, and its dissociation rate from the binding site is more rapid. Therefore, its estrogen-specific binding is difficult to determine by direct measurement, and no 8 S peak is evident upon sucrose density gradient analysis. Measurement of covalently incorporated radioactivity produced by irradiation of this derivative with whole cytosol preparations also indicates that much nonspecific labeling is occurring. Electrophoretic analysis of total cytosol preparations, photolytically labeled in the absence and presence of excess unlabeled estradiol (to determine the site specificity of the labeling process), gave erratic and ambiguous results.

Analysis of the covalently labeled rat uterine cytosol proteins in the 8 M urea gel system was hindered by two factors. First, it appeared that the components labeled in an estrogen-specific fashion were not being adequately disaggregated by the 8 M urea treatment. A further indication of this was that count recovery from these gels was only in the range of 30%; large amounts of protein and radioactivity appeared caked at the top of the stacking gel after some runs. Second, while we initially had thought that it would be advantageous to work in a urea denaturing system, as only the covalently labeled proteins would be observed, a major shortcoming of such systems is that the mobility of the urea-denatured estrogen receptor is not known. In the face of these difficulties, we sought ways to increase the selectivity and reduce the ambiguity of the electrophoretic assay of covalent attachment.

Studies Utilizing Partially-purified Estrogen Receptor Preparations

PARTIAL PURIFICATION OF LAMB CYTOSOL

We have recently developed a procedure for the partial purification and disaggregation of the uterine estrogen receptor [31] based on the limited trypsinization procedure first reported by Rat [32]. Lamb uteri have proven to be a convenient and plentiful source of receptor for these purification studies, and, since hexestrol azide and hexestrol diazoketopropyl ether showed similar binding affinity and inactivation efficiency with receptor from either rat or lamb uterus, these partially purified lamb receptor preparations seemed to be most suitable to continue our affinity labeling studies.

The two-stage purification procedure (ammonium sulfate precipitation, Sephadex G-200 column chromatography) effects a 10–30-fold purification of receptor and removes a major portion of the proteins that contribute to the nonspecific binding. The trypsinization converts the lamb uterine receptor to a form that has an apparent molecular weight of ca. 50,000 and tolerates some further manipulation without reaggregation [31].

ELECTROPHORETIC FRACTIONATION OF ESTROGEN RECEPTOR PREPARATIONS: ANALYSIS OF COVALENT BINDING

We [31] and others [33,34] have shown that the trypsinized forms of the estrogen receptor can be successfully subjected to polyacrylamide gel electrophoresis in the Ornstein-Davis buffer system while retaining their binding activity. Using this electrophoretic system (together with a solvent extraction assay, see below), we have been able to analyze the reversible and covalent binding of the photoreactive estrogen analogs to proteins [35]. While we have found that hexestrol diazoketopropyl ether (4) does not bind to proteins in the receptor preparations irreversibly, at least a portion of hexestrol azide (7a) does become irreversibly associated with receptor through irradiation.

The behavior of hexestrol diazoketopropyl ether (4) was first studied. Initially, we were concerned by the rapidity with which this compound dissociated from the receptor, and we assumed that it would not be feasible to subject the receptor—hexestrol diazoketopropyl ether complex to the Sephadex G-200 column chromatography. We were pleased to find, however, that we could apply the receptor purification–disaggregation sequence to this complex despite its rapid rate of ligand exchange. The material elutes as a single peak in the middle of the included volume of a Sephadex G-200 column (MW ca. 50,000); the quantity of bound radioactivity and the elution volume are consistent with that observed for the estradiol–receptor complex. Furthermore, the estrogen specificity of the peak material could be ascertained by exchanging fractions against an excess of labeled estradiol, in both the presence and absence of an excess of unlabeled estradiol. In this way, the estrogen-specific binding capacity was shown to coincide with the peak of hexestrol diazoketopropyl ether binding activity.

Figure 8 shows the electrophoretic profiles of [^{3}H]estradiol ([^{3}H]E$_2$) and [^{3}H]hexestrol diazoketopropyl ether ([^{3}H]Hex-DKP) bound to the partially purified lamb uterine cytosol protein preparation. Figure 8A shows that estradiol migrates as a single species with high mobility. The specificity of estradiol binding in this preparation is evident from the low level of nonspecific binding (Fig. 8A, lower curve). Figure 8B is a control that shows the electrophoretic profile for hexestrol diazoketopropyl ether after irradiation in the absence of proteins; a portion of the photoproduct migrates with the dye band (and is most likely the carboxylic acid arising from the photochemical Wolff rearrangement of the diazoketone [24]), while the rest of the photoproduct (hexestrol [24]) barely penetrates the separation gel. Free estradiol and free, unphotolyzed hexestrol diazoketopropyl ether also migrate only to the top of the separation gel (not shown). None of the peaks in Fig. 8B is in a position that will interfere with the assay of estrogen receptors.

After ammonium sulfate precipitation and Sephadex G-200 chromatography, the binding of hexestrol diazoketopropyl ether to the estrogen receptor can be observed on gels (Fig. 8C). The quantity of estrogen-specific binding that is found with this compound, however, is considerably less than that seen with estradiol on the same sample of cytosol, which suggests that a substantial fraction of receptor complexes has undergone dissociation during the fractionation procedures.

Figure 8D shows the profile of radioactivity obtained by electrophoresis of the same sample of semipurified receptor complex after photolysis. The intriguing feature of this profile is that there is more radioactivity in the estrogen receptor region than prior to photolysis.

In order to determine whether the radioactivity found in the gels after irradiation is covalently bound to protein, the individual gel slices were extracted in an organic solvent, and the extract (noncovalent binding) and gel slice (covalent binding) were counted separately (see Fig. 8). By using this technique, it was apparent that all of the radioactivity in the gel

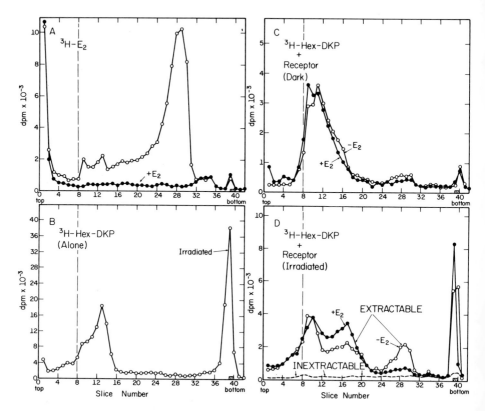

Fig. 8 Lamb uterine cytosol was incubated with either [³H]estradiol (panel A) or [³H]hexestrol diazoketopropyl ether (**4**) and then purified through the Sephadex G-200 column chromatography step. An aliquot of the [³H]hexestrol diazoketopropyl ether peak was exchanged for 3 hr at 0°C against a 1000-fold excess of unlabeled estradiol. Portions of both [³H]hexestrol diazoketopropyl ether samples (with and without unlabeled estradiol) were irradiated at 254 nm for 1 min. All samples were analyzed by electrophoresis on alkaline pH, 6% polyacrylamide gels cross-linked with *N,N'*-diallyl-tartardiamide. The slices from gel D were first extracted with 0.5 ml ethyl acetate at 25°C for 2 hr to remove noncovalently bound counts and then all gel slices were prepared for scintillation counting as described in the legend of Fig. 7. Panel B shows the profile of [³H]hexestrol diazoketopropyl ether irradiated and electrophoresed in the absence of proteins. The vertical dashed line at slice 8 in all panels indicates the interface between the stacking and separation gels, and the small shaded rectangle at slice 38 indicates the position of the tracking dye band (bromphenol blue) [35].

shown in Fig. 8D was extractable and, therefore, that [³H]hexestrol diazoketo-propyl ether (**4**) had not bound to the receptor covalently. Furthermore, the material that was reversibly bound to the receptor after photolysis was shown to be hexestrol by thin layer chromatographic comparison with an authentic sample. We have previously shown that hexestrol is one of the major photo-products of hexestrol diazoketopropyl ether in water [24]. Thus, the higher (reversible) binding to receptor seen after photolysis of hexestrol diazoketo-propyl ether (Fig. 8D) can be explained by exchange in favor of the higher affinity ligand, hexestrol, which has been generated photolytically.

Hexestrol azide (Hex-N$_3$, 7a) has a much higher binding affinity for the estrogen receptor than the diazo compound 4; therefore, fewer problems with complex dissociation and nonspecific binding were encountered. A large amount of estrogen-specific binding of [^{3}H]hexestrol azide is evident in the profile from cytosol preparations that were only ammonium sulfate precipitated and trypsinized (Fig. 9B). The profiles of free derivative, both before and after photolysis, are shown in Fig. 9A.

Prior to photolysis, nearly all the radioactivity associated with the receptor region is solvent extractable and is chromatographically identical with an authentic sample of unlabeled hexestrol azide. After photolysis, however, nearly half of the receptor-associated material is no longer solvent extractable (Figs. 9C and D). The material that is extractable is hexestrol and not unconsumed starting material.

Simple calculations show that the quantity of radioactivity in the receptor region of the gel that is nonextractable after photolysis corresponds to approximately 15–20% of the material bound at the time of photolysis. This is the maximum incorporation that would be expected on the basis of the 15% inactivation efficiency of hexestrol azide as determined by the photolysis exchange assay [20].

At this point, we are in the process of characterizing the nature of this solvent-inextractable material that runs in the estrogen receptor region. These further investigations should establish whether this material is actually the estrogen receptor. Nevertheless, the following points can be made to substantiate that the species that has been labeled with [^{3}H]hexestrol azide (7a) is the estrogen receptor. (1) Both the reversible and the irreversible binding of this derivative is subject to competition with unlabeled estradiol; (2) the electrophoretic mobility of the species labeled with [^{3}H]hexestrol azide (7a) both reversibly and irreversibly, is the same as that of the receptor–estradiol complex; and (3) the extent of irreversible labeling is consistent with that determined to be an upper limit by our inactivation studies.

The following points can be made to substantiate that the linkage connecting a portion of the hexestrol derivative to the receptor after photolysis is, in fact, covalent in nature. (1) The unphotolyzed derivative or the derivative photolyzed alone, prior to the addition of protein, can be quantitatively extracted from the receptor, while a portion of the derivative is not extractable when it is bound to the receptor at the time of irradiation. In further studies to be described elsewhere, we have found that (2) the material that

is not extractable with organic solvents can be extracted with buffer and that
(3) the radioactivity in this buffer extract can be precipitated with trichloro-
acetic acid or hot ethanol.

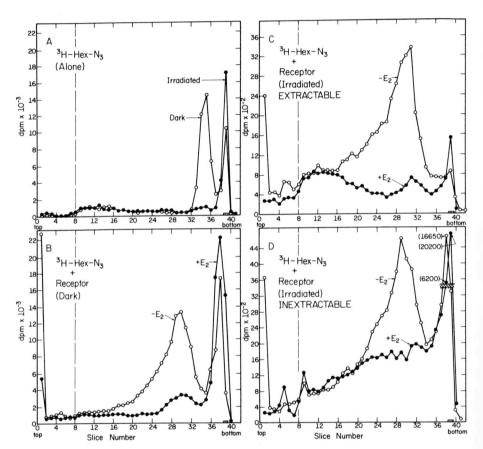

Fig. 9 Acrylamide gel electrophoresis of [³H]hexestrol azide (**7a**) performed as described in the legend of Figure 8. Panel A shows the profile of [³H]hexestrol azide,
before and after irradiation, electrophoresed in the absence of protein. Panels B, C, and
D represent partially purified lamb uterine cytosol incubated with [³H]hexestrol azide
in the absence and presence of 1000-fold excess of unlabeled estradiol. Samples for C
and D were irradiated for 5 min at 254 nm before electrophoresis. Toluene-extractable
radioactivity (C) represents [³H]hexestrol azide which is not covalently attached to the
protein while the inextractable radioactivity (D) may indicate covalent binding. The
vertical line at slice 8 in all panels indicates the interface between the stacking and
separation gels, and the small shaded rectangle at slice 38 indicates the position of the
tracking dye band (Bromphenol Blue) [35].

DISCUSSION AND CONCLUSION

In our initial consideration of photoaffinity labeling reagents for the estrogen receptor, we were concerned that the available information on structure–binding affinity relationships might not be a reliable guide to the design of new reagents of high binding affinity; furthermore, there was essentially no information that could guide in the design of reagents with high labeling efficiency. Therefore, the concern that governed our selection of compounds for initial study was chemical simplicity, and we chose derivatives that for the most part could be prepared by derivatization of known steroidal and nonsteroidal estrogens.

Utilizing these photosensitive estrogen derivatives first in nonradiolabeled form, we are able to determine by indirect competition techniques the receptor binding affinity and inactivation efficiency of these compounds. Our approach of simply adding a photoreactive function to an estrogenic molecule resulted in compounds that generally had their binding affinity reduced considerably from that of the parent estrogen, and they had only modest inactivation efficiencies. Still, these data, which were easily obtained, gave us a basis for the selection of those derivatives that were most promising for further study by direct methods, after preparation of the reagents in radiolabeled form.

While the competitive binding studies gave us information only about the interaction of the photosensitive estrogens with the receptor site, direct binding studies revealed their nonspecific (low affinity) as well as their estrogen-specific binding. At this stage, it became apparent that we had neglected an important factor in our efforts to design a selective labeling reagent. Our principal concern in terms of binding had been to maximize estrogen receptor binding affinity; we had not considered the effects of structural modification and substitution on binding to other, nonreceptor proteins.

It is well known that low affinity ("nonspecific") binding of lipophilic ligands to proteins is generally relatively nonspecific with regard to structure and stereochemistry, being governed principally by ligand lipophilicity. As most of the derivatization we have done in preparing our first set of photosensitive estrogens involves either the masking of hydrophilic substituents or the addition of hydrophobic ones, it is not surprising that the nonspecific binding of these more lipophilic derivatives is much higher than that of estradiol.

The quantitative structure–activity relationship approach developed by Corwin Hansch is particularly relevant to these considerations. Hansch has found that the biological activity or the binding affinity of a series of derivatives can often be related in a quantitative fashion with their lipophilicity, measured most conveniently as their 1-octanol: water partition coefficient [36]. The following relationship was shown to hold for the binding of lipophilic molecules to serum albumin:

$$\log K_a = 0.75 \log P + C$$

where K_a is the equilibrium association constant, P is the l-octanol: water partition coefficient, and C is an experimentally determined constant that depends upon the degree of binding saturation. This equation places on a quantitative basis, the relationship between lipophilicity and protein binding: It predicts that for each 10-fold increase in partition coefficient, protein binding affinity will increase by approximately 7-fold. As serum albumin binding of steroids is generally considered to be a good model for their nonspecific binding, this equation can be used to predict the degree of nonspecific binding expected for new steroid derivatives and analogs. It is particularly convenient to use this approach because Hansch has developed empirically (on the basis of measurements of P on a large number of model compounds) a set of substituent parameters that can be summed to predict the partition coefficients of new compounds. Using the substituent parameters and the equation above, one can calculate the relative nonspecific binding expected for the photoaffinity labeling reagents that we have been studying. These values are shown under the structures in Scheme 9. It is evident that, relative

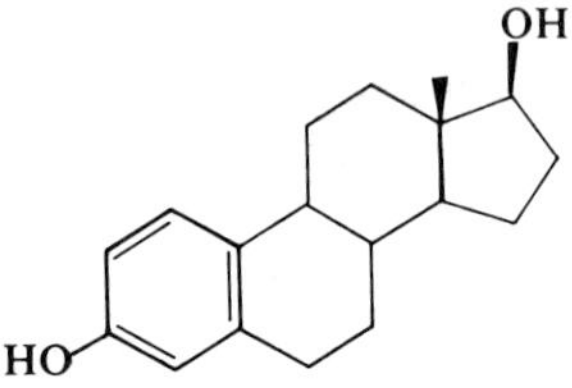

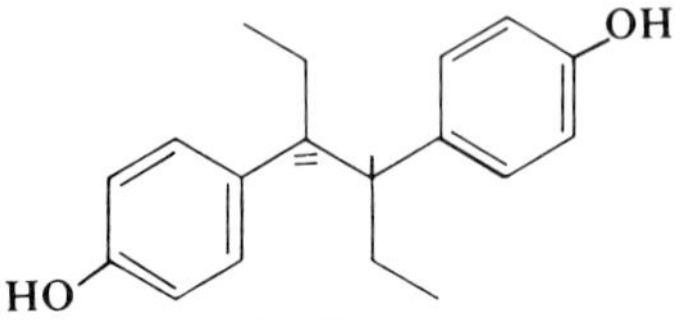

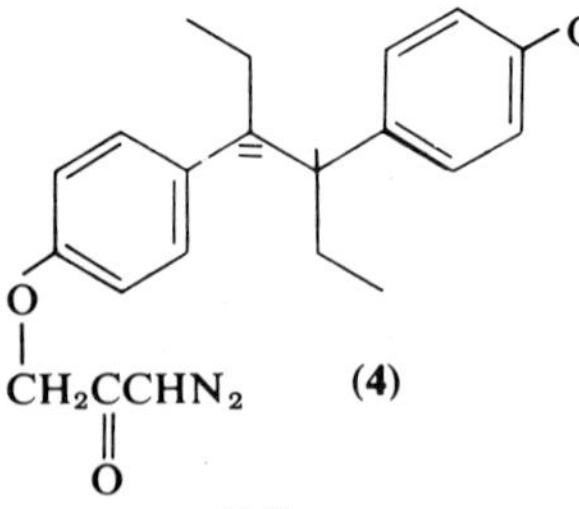

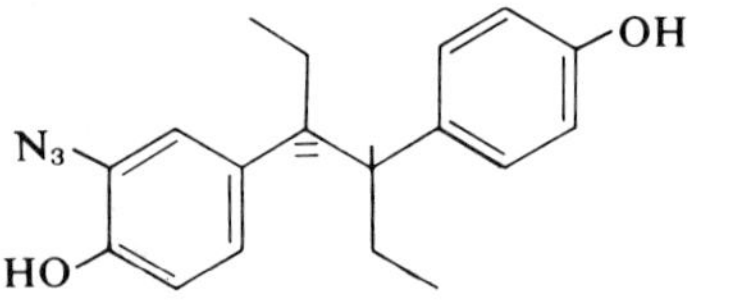

Scheme 9 Nonspecific binding affinity of photosensitive estrogen derivatives calculated according to Hansch [36]. (The bracketed numbers below the structures represent the predicted nonspecific binding affinity of the derivatives relative to that for estradiol. For details, see text.)

to estradiol, both hexestrol and its azide and diazoketopropyl ether derivatives have greatly elevated levels of nonspecific binding. Obviously, in the design of new photosensitive estrogen derivatives, we will be concerned with the implication of structural modification not only in terms of estrogen-receptor-specific binding, but in terms of nonspecific binding affinity as well.

Another limitation of the derivatives that we have studied thus far is their relatively modest inactivation efficiencies, which are measured by the photolysis-exchange assay procedure. Furthermore, there is no assurance that the labeling efficiencies measured in direct assay with the radiolabeled derivatives will even be as high as the inactivation efficiencies measured indirectly.

The low labeling efficiency of some of the photosensitive derivatives and the low receptor purity in cytosol preparations from whole rat or lamb uterus have placed severe constraints on the success of selective labeling experiments conducted directly in cytosol preparations. While electrophoretic fractionation of labeled cytosol preparations on denaturing (8 M urea) gels may suffice to resolve components that correspond to the estrogen receptor labeled with hexestrol azide (Fig. 7), it appears that, with most of these first-generation photoaffinity labeling reagents that we have prepared, partial purification of the receptor prior to photolytic labeling is essential to any successful identification and characterization of labeled receptor species. Electrophoretic fractionation of (trypsinized) receptor in an active binding state, coupled with solvent extraction of the gel slices, provides a convenient and unambiguous assay for the covalent labeling of receptor. The results of using this assay with hexestrol azide (Fig. 9) have thus far been very encouraging. Still, it is clear that major advances in reagent selectivity and efficiency are required before the technique of affinity labeling can fulfill its promise as a tool in elucidating the molecular details of steroid hormone action.

Currently, our efforts are being directed toward the preparation of a second generation of estrogen affinity labeling reagents. We have designed these compounds on the basis of our experiences with the relatively simple diazocarbonyl and azide first-generation derivatives, and we hope that they will display all the features (high receptor binding affinity with acceptably low nonreceptor binding, and high labeling efficiency) that are required to make the labeling of the estrogen receptor a simple and effective process.

ACKNOWLEDGMENTS

We are grateful for the support of this research by grants from the National Institutes of Health (AM 15556) and the Ford Foundation (700–0333 and graduate fellowship to H.N.M.) J.A.K. is an Alfred P. Sloan Fellow (1974–1976) and a Camille and Henry Dreyfus Teacher-Scholar (1974–1979).

REFERENCES

1. B. W. O'Malley and W. R. Schrader, *Sci. Am.* **234**, 32 (1976).
2. R. B. J. King and W. I. P. Mainwaring, "Steroid-Cell Interactions." Univ. Park Press, Baltimore, Maryland, 1974.
3. U. Westphal, "Steroid-Protein Interactions." Springer-Verlag, Berlin and New York, 1971.
4. G. A. Puca, E. Nola, V. Sica, and F. Bresciani, *in* "Methods in Enzymology" (B. W. O'Malley and J. G. Hardman, eds.) Vol. 36, Part A, pp. 331–349. Academic Press, New York, 1975.
5. E. R. DeSombre and T. A. Gorell, *in* "Methods in Enzymology" (B. W. O'Malley and J. G. Hardman, eds.), Vol. 36, Part A, pp. 349–374. Academic Press, New York, 1975.
6. V. Sica, I. Parikh, E. Nola, G. A. Puca, and P. Cuatrecasas, *J. Biol. Chem.* **248**, 6543 (1974).
7. H. Truong and E. E. Baulieu, *FEBS Lett.* **46**, 321 (1974).
8. W. T. Schrader, R. W. Buller, R. W. Kuhn, and B. W. O'Malley, *J. Steroid Biochem.* **5**, 989 (1974).
9. R. W. Kuhn, W. T. Schrader, R. G. Smith, and B. W. O'Malley, *J. Biol. Chem.* **250**, 4220 (1975).
10. J. A. Katzenellenbogen, *Annu. Rep. Med. Chem.* **9**, 222 (1974).
11. J. A. Katzenellenbogen, *Biochem. Actions Horm.* (in press).
12. B. R. Baker, "Design of Active Site-Directed Irreversible Enzyme Inhibitors." Wiley, New York, 1967.
13. F. Davidoff, S. Carr, M. Lanner, and J. Loeffler, *Biochemistry* **12**, 3107 (1973).
14. A. R. Main, *Essays Toxicol.* **4**, 59–105 (1974).
15. J. R. Knowles, *Acc. Chem. Res.* **5**, 155 (1972).
16. S. G. Korenman, *Steroids* **13**, 163 (1969).
17. C. Geynet, C. Millet, H. Truong, and E. E. Baulieu, *Gynecol. Invest.* **3**, 2 (1972).
18. L. Terenius, *Acta Pharmacol. Toxicol.* **31**, 441 and 449 (1972).
19. J. A. Katzenellenbogen, H. J. Johnson, Jr., and H. N. Myers, *Biochemistry* **12**, 4085 (1973b).
20. J. A. Katzenellenbogen, H. J. Johnson, Jr., K. E. Carlson, and H. N. Myers, *Biochemistry* **13**, 2986 (1974).
21. E. V. Jensen, D. J. Hurst, E. R. DeSombre, and P. W. Jungblut, *Science* **158**, 385 (1967).
22. G. A. Puca, E. Nola, U. Hibner, G. Cicala, and V. Sica, *J. Biol. Chem.* **250**, 6452 (1975).
23. J. A. Katzenellenbogen, H. N. Myers, and H. J. Johnson, Jr., *J. Org. Chem.* **38**, 3525 (1973).
24. J. A. Katzenellenbogen, H. J. Johnson, Jr., H. N. Myers, and R. J. Kempton, *Biochemistry* **16**, 1964 (1977).
25. S. G. Korenman, *Endocrinology* **87**, 1119 (1970).
26. D. Rodbard, *in* "Receptors for Reproductive Hormones" (B. W. O'Malley and A. R. Means, eds.), pp. 289–326. Plenum, New York, 1973.
27. J. A. Katzenellenbogen, H. J. Johnson, Jr., and K. E. Carlson, *Biochemistry* **12**, 4092 (1973).
28. J. A. Katzenellenbogen, T. S. Ruh, K. E. Carlson, H. S. Iwamoto, and J. Gorski, *Biochemistry* **14**, 2310 (1975).
29. D. Toft, G. Shyamala, and J. Gorski, *Proc. Natl. Acad. Sci. U.S.A.* **57**, 1740 (1967).

30. D. J. Brunswick and B. S. Cooperman, *Biochemistry* **12**, 4074 (1973).
31. K. E. Carlson, L.-H. K. Sun, and J. A. Katzenellenbogen, *Biochemistry* **16**, 4288 (1977).
32. R. L. Rat, C. Vallet-Strouve, and T. Erdos, *Biochimie* **56**, 1387 (1974).
33. G. Vallet-Strouve, L. Rat, and J. M. Sala-Trepat, *Eur. J. Biochem.* **66**, 327 (1976).
34. C. Secco-Millet, O. Soulignac, P. Rocher, E. E. Baulieu, and H. Richard-Foy, *C. R. Acad. Sci.* (*Paris*) **284**, 137 (1977).
35. J. A. Katzenellenbogen, K. E. Carlson, H. J. Johnson, Jr., and H. N. Myers, *Biochemistry* **16**, 1970 (1977).
36. C. Hansch, *in* "Drug Design" (J. Ariens, ed.), Vol. I, Chapter 2. Academic Press, New York, 1971.

8

The Redox Chemistry of 1,4-Dihydronicotinic Acid Derivatives

R. J. Kill and D. A. Widdowson

INTRODUCTION

Recent years have seen a renaissance of interest in the chemistry of 1,4-dihydronicotinic acid derivatives, in terms of both the fundamental redox chemistry of the systems and the coenzyme roles of NAD(P)H. Early studies concentrated on the hydride nature of the redox process, but much of the recent work has emphasized the facility of radical and electron transfer reactions. Some enzymatic reactions involving NAD(P)H are likely to involve electron transfer processes, and the assumption of hydride reduction by NAD(P)H enzymes is questionable.

While there is some repetition of data from earlier reviews [1–8], this survey concentrates on the more recent developments in the chemistry of 1,4-dihydronicotinic acid derivatives and, wherever appropriate, results of enzyme studies are introduced for comparison.

THE NATURAL SYSTEMS

In order to design suitable model reactions and to interpret the fundamental chemistry of dihydronicotinamides in relation to the enzymatic process, it is

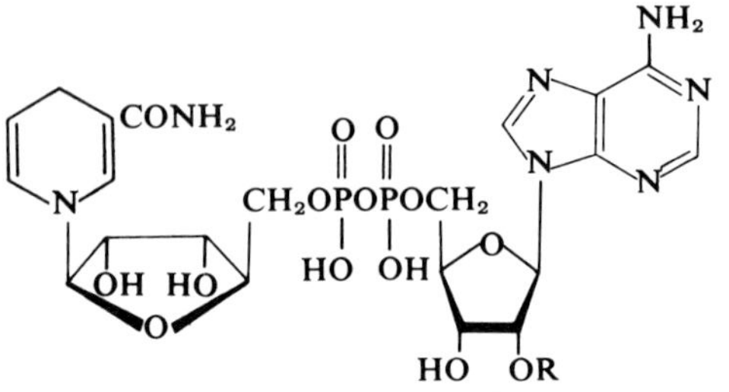

(1)

$NAD^+, R = H$

$NADP^+, R = PO_3H_2$

necessary to understand something of the natural system. There are some 250 known enzymatic reactions that depend on the pryidine nucleotide coenzymes NAD(P)H (1) [9]. Some of the enzymes involved are specific for either NADH or NADPH, while the others accept either. The general process is the reversible reduction/oxidation of a suitable substrate (S) in the presence of the appropriate enzyme (Scheme 1).

NAD(P)H-Dependent enzymes include the dehydrogenases, transhydrogenases, diaphorases, phosphorylases, and oxidases. The reactions catalyzed by the enzymes take place within the ternary (enzyme–coenzyme–substrate) complexes. The transfer of hydrogen is direct [10] and involves no exchange with solvent or apparently with any part of the enzyme. Previous reports [11,12] to the contrary have since been disproved [13]. The transfer is stereospecific with respect to both the coenzyme and the substrate [14]. The enzyme controls the stereospecificity [2].

The enzymatic mechanism is complex and, for example, for liver alcohol dehydrogenase (LADH), seven steps are currently recognized [14a], with product desorption as the rate-determining step. The determination of the exact nature of the hydrogen transfer *in vivo* is hampered by the complexity of the ternary system. It is, however, generally accepted that the lack of exchange of the hydrogen atom with the aqueous medium precludes its transfer as a proton, unless it be, in a formal sense, during a concerted process [15]. Kinetic substituent effects have shown [16–18] that in the transition state there is a partial negative charge on the substrate. Whether this arises from the transfer of a hydride ion or an electron remains to be satisfactorily resolved.

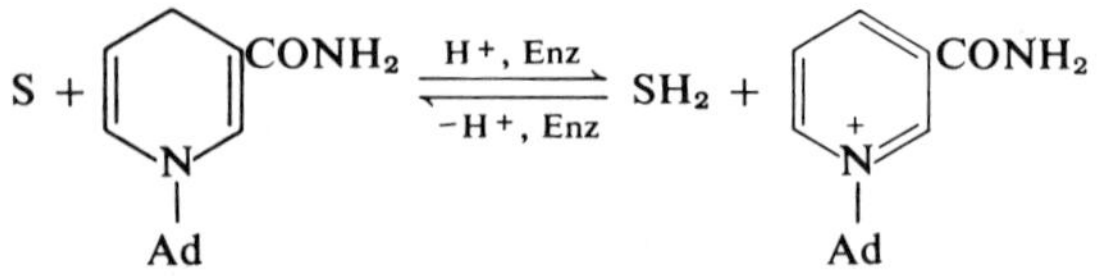

Scheme 1

The role of the enzyme in the reaction is not clear. Possibilities include the provision of an environment of the correct dielectric, the orientation by physical constraints of the coenzyme and substrate in a favorable manner, and the activation by chemical means of one or both of the reactants. It is known that the coenzyme is bound in extended form [19] with a modified electronic structure [20,21]. Many of the reactions require metal ions, the role of which is becoming evident from X-ray structural studies of the dehydrogenases [22]. The enzymatic structures have been determined at resolutions down to 2.4 Å [22]. The structural data for alcohol [22], lactate [23], malate [24], and glutamate [25] dehydrogenases have been reviewed. These data show that, at least for liver alcohol dehydrogenase, the metal cation (Zn^{2+}) is complexed in the active site. It is involved in binding the substrate in juxtaposition with the coenzyme and may act as an electrophilic catalyst for the reaction [17,17a,26].

COENZYME MODELS

NAD(P)H Models

Two types of dihydropyridine have been widely used as models of NAD(P)H: the N-substituted 1,4-dihydronicotinamides (2), which strongly resemble the natural system (1), and the Hantzsch esters (3). Both classes of compound are readily synthesized [8], and they have distinct advantages as models of NAD(P)H. Unlike the natural systems, they are stable and sufficiently simple for ready spectroscopic assay. Their solubilities in most organic solvents facilitate handling, but the use of such solvents creates problems in the drawing of comparisons with the natural aqueous system.

Hadju and Sigman [27] have shown that a carboxylate ion adjacent to the pyridine nitrogen, as in (2d), accelerates the reduction of the acridinium ion by a factor of 100 over the unsubstituted compound (2a). Stabilization of a

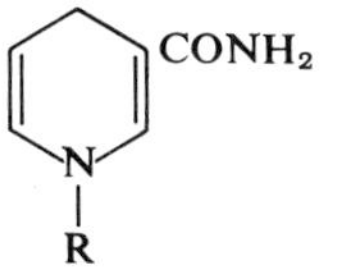

(2a), R = benzyl
(2b), R = Me
(2c), R = n-Pr
(2d), R = o-carboxy benzyl
(2e), R = 2,6-dichlorobenzyl

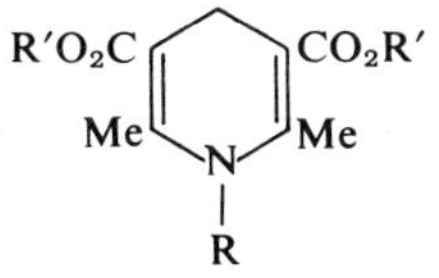

(3a), R = H
(3b), R = Me

partial positive charge on the pyridine nitrogen in an intermediate charge-transfer complex is the suggested explanation. In lactate [23,28] and malate [29] dehydrogenases there is a glutamate carboxyl group in the vicinity of the bound coenzyme dihydronicotinamide ring.

$NAD(P)^+$ Models

Few models of $NAD(P)^+$ have been shown to oxidize alcohols, since thermodynamically the redox equilibrium favors the formation of the pyridinium ion. However, N-methyl-3,4,5-tricyanopyridinium perchlorate (20) has been used to oxidize fluorenol to fluorenone [30] (see section on oxidation of alcohols). Suckling [31] has studied the oxidation of benzyl alkoxides with a variety of pyridinium salts (see section on oxidation of alcohols).

ELECTROCHEMISTRY OF $NAD(P)^+$-RELATED SYSTEMS

While a direct correlation cannot be drawn, a comparison of the electrochemical properties of the $NAD(P)^+$-related systems and their chemical

TABLE 1

Polarographic Reduction Potentials of $NAD(P)^+$-Related Compounds

Compound	Solvent	pH	Potential, $E_{1/2}$ (w.r.t. SCE)[a]	Reference
α-NAD$^+$	Water	3–9.6	ca. -0.96	33
	DMSO	—	-0.98	32
β-NAD$^+$	Water	1–11.8	ca. -1.00	33
NADP$^+$	Water	3–10	ca. -1.00	32
	DMSO	—	-1.06	32
DNAD$^{+\,b}$	Water	3–9 6	ca. -0.98	33
	DMSO	—	-1.00	32
DNADP$^+$	Water	3–9.4	ca. -0.92	33
NMN$^{+\,c}$	Water	1–11.8	ca. -0.90	35
	DMSO	—	-0.99	32
N^1-Methylnicotinamide	Water	4–13	ca. -1.11	36
	Water	3.7–12	ca. -1.06	35
	Acetonitrile	—	-1.04	32
	DMSO	—	-1.01	32
Nicotinamide	Acetonitrile	—	-2.00	32
	DMSO	—	-2.01	32

[a] SCE, standard calomel electrode.
[b] DNAD$^+$, deamino-NAD$^+$.
[c] NMN$^+$, nicotinamide mononucleotide.

redox reactions is useful. The susceptibility to one- or two-electron processes under particular conditions has a bearing on the chemical properties of the compounds. A study of the redox potentials (Tables 1 and 2) relative to those of the substrates allows an estimate to be made of the feasibility of an *in vitro* reaction.

Electrochemical Reduction of NAD(P)⁺ Analogs

The electrochemical reduction of NAD(P)$^+$ and its model compounds has been the subject of several recent studies [32–36]. A wide range of compounds has been shown to undergo an initial reversible one-electron oxidation to

TABLE 2

Voltametric Oxidation Potentials of NAD(P)H and Related Compounds[a]

Compound	Working electrode[b]	Support electrolyte[c]	Solvent	Potential[d] (w.r.t. to SCE)	Reference
NADH	GCE	pH 7.1 buffer	Water	0.67	38
	Pt	TBAP	DMSO	0.89	38
NADPH	GCE	pH 7.1 buffer	Water	0.72	38
	Pt	TBAP	DMSO	0.84	38
NMNH	GCE	pH 7.1 buffer	Water	0.67	38
	Pt	TBAP	DMSO	0.76	38
(2a)	GCE	TBAB	Acetonitrile	0.46 $(E_{1/2})$[e]	40
	GCE	TBAB	Acetonitrile	0.57	40
	GCE	TEAP	Acetonitrile	0.60	37
	GCE	—	Ethanol	0.45 $(E_{1/2})$[e]	40
	GCE	—	Ethanol	0.55	40
	GCE	pH 7–13 buffer	Water	0.24	41
	Pt	pH 7–13 buffer	Water	0.42	41
	Au	pH 7–13 buffer	Water	0.36	41
(2b)	GCE	TEAP	Acetonitrile	0.53	37
(2c)	GCE	TBAB	Acetonitrile	0.39 $(E_{1/2})$[e]	40
	GCE	TBAB	Acetonitrile	0.51	40
	GCE	TEAP	Acetonitrile	0.87	37
	GCE	—	Ethanol	0.43 $(E_{1/2})$[e]	40
	GCE	—	Ethanol	0.52	40
(2e)	GCE	TEAP	Acetonitrile	0.60	37

[a] The reader is referred to four other compilations of dihydropyridine oxidation potentials [39–42].

[b] GCE, glassy carbon electrode.

[c] Abbreviations: TBAP, tetra-*n*-butylammonium phosphate; TEAP, tetraethylammonium phosphate; TBAB, Tetra-*n*-butylammonium tetrafluoroborate.

[d] Voltametric peak potential unless otherwise indicated.

[e] Voltametric half-peak potential.

(4) [32–34]. This is followed by a coupling to either 4,4'- or 6,6'-dimers, depending on the steric effects of the *N*-substituents [33]. In aprotic media the dimers are the sole products of the reduction. Electrochemical oxidation of these products regenerates the original cations.

In protic media the radical (4) is observed both to dimerize and to undergo further reduction to the anion (5). This is rapidly protonated to give a

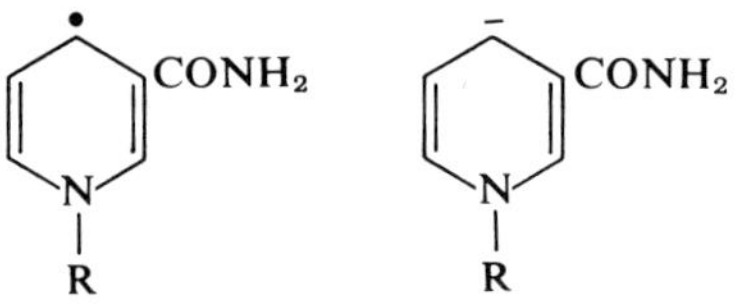

dihydro species, the exact nature of which is not clear. Elving claims that although NAD^+ is reduced largely to 1,4-NADH [33], 1-methylnicotinamidium salts are reduced to the 1,6-isomers [34]. Burnett and Underwood [36], however, report that the model compounds also afforded the 1,4-isomer. In protic media the dimers may be further reduced.

Electrochemical Oxidation of NAD(P)H Analogs

The electrochemical oxidation of the dihydronicotinamides (2) in acetonitrile solution (Scheme 2) gave [37] the 1-alkylnicotinamidium ions (6) in yields dependent on the conditions. The initial step is the formation of protonated pyridinyl radical (7) by a one-electron process. This species is not electroactive at the oxidation potential of the parent (2). In the absence of

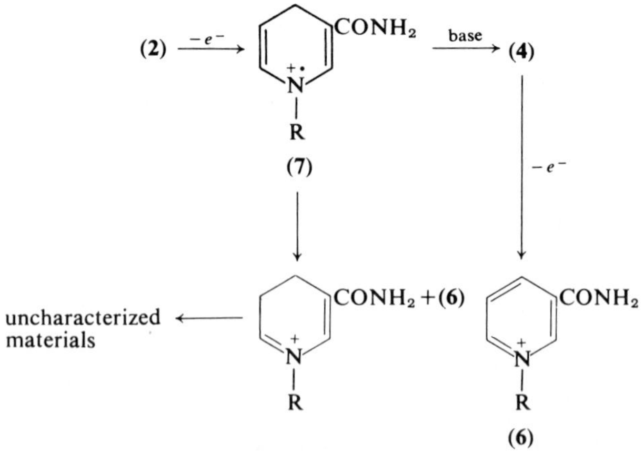

Scheme 2

base the radical (7) disproportionates to afford the pyridinium salt (6) (40%) and a number of uncharacterized products.

In basic solutions the radical cation (7) is deprotonated to the pyridinyl radical (4), which undergoes rapid anodic oxidation to the cation (6). Under these conditions the reaction is quantitative. Oxygen also blocks the disproportionation and acid decomposition reactions of the primary product (7).

The natural species (1) are not soluble enough in acetonitrile for a direct comparison to be made with these results. However, Elving and his co-workers [38] found anodic oxidation of NMNH, NADH, and NADPH in dimethyl sulfoxide to occur via a two-electron process (Scheme 3). In contrast

$$RNH \longrightarrow RN^+ + H^+ + 2e^-$$

Scheme 3

Kuthan *et al.* [39] have observed one-electron oxidation of a variety of dihydropyridines at a dropping mercury electrode. These variations could be due to solvent effects, to differing experimental conditions, or to intrinsic differences between the natural and model systems. In the absence of experiments carried out under comparable conditions the question remains unanswered. The reduction potentials of a large number of dihydropyridines have recently been reported [40–42].

ALCOHOL DEHYDROGENASE MODELS

The publication [43] in 1951 of the Theorell mechanism for the liver alcohol dehydrogenase reaction led to a search for *in vitro* hydride reductions as dehydrogenase models. This generated much new chemistry of the dihydronicotinamides, and many hydride and radical chain redox reactions have since been reported (Table 3) [30,31,44–65]. Recently, more sophisticated studies of the redox reactions have been made and the possibility of electron transfer processes considered. *In vitro* asymmetric reductions have been observed in metal complexes. The involvement of complexation at the zinc ion has assumed greater importance since metal ion catalysis has been shown *in vitro* (see Table 3) and there is strong evidence [17,17a,22,26] for the *in vivo* participation. The precise mechanisms of the enzymic reactions, however, are still unproved.

Reduction of Carbonyl Functions

Isolated carbonyl groups are not reduced by NAD(P)H (1), 1,4-dihydronicotinamides (2), or Hantzsch esters (3) alone. Ultraviolet irradiation of a

TABLE 3

Alcohol Dehydrogenase Models

Substrate	Reductant	Product[a]	Reference
Reduction of carbonyl compounds			
Acetaldehyde	(1)/hv	Ethanol	44
Alloxan	(2a)	ns	45
Benzaldehydes	(23)	Benzyl alcohols	31
Benzil	(3a)	Benzoin	46
	(2a)/Mg^{2+}	Benzoin	47
Benzoin	(2a)/Mg^{2+}	1,2-Diphenylethane-1,2-diol	47
Benzoyl formate ester	(2a)/Mg^{3+}	Mandelate ester	48
Benzoylformic acid	(3)	Mandelic acid	49
Chloral	(2a)	ns	50
Cyclobutanone	(2a)/O_2	Butyric acid plus butyrolactone	51
N-(2,6-Dichlorobenzyl)-3-(o-formylbenzoyl)-1,4-dihydropyridine	hv	N-(2,6-Dichlorobenzyl)-3-(o-hyroxymethylbenzoyl)-pyridinium ion	52
Fluorenone	(3a)/Zn^{2+}	Fluorenol	53
Hexachloroacetone	(2a)	Hexachloroisopropanol	50,54
3-Hydroxypyridine-4-carboxaldehyde	(3a),(2c)	3-Hydroxy-4-hydroxymethyl pyridine	53,56
α-Ketoglutarate ester	(1)/hv	α-Hydroxyglutarate ester	53
2-Mercaptobenzophenone	(3)	α-Phenyl-o-mercapto-benzyl alcohol	57
5-Nitrosalicylaldehyde	(3)	3-Hydroxy-5-nitrobenzyl alcohol	58
1,10-Phenanthroline-2-carboxaldehyde	(2c)/Zn^{2+}	1,10-Phenanthroline formyl alcohol	59
Pyridine-2-carboxaldehyde	(2a)/M^{n+}	2-Hydroxymethylpyridine	60
Pyridine-4-carboxaldehyde	(2a)/M^{n+}	4-Hydroxymethylpyridine	60
Pyridoxal phosphate	(2c), (3a)	Pyridoxin phosphate	55,56,56
Pyruvate ester	(2a)/Mg^{2+}	Lactate ester	48
	(1)	Lactate ester	61
Pyruvic acid	(3)	Lactic acid	62
Salicylaldehyde	(3)	o-Hydroxybenzyl alcohol	58
Trifluoroacetophenone	(2a)	Trifluoromethyl phenyl-carbinol	63
	(2a)/Mg^{2+}	Trifluoromethyl phenyl carbinol	64
2-Benzoylpyridine	(2a)/Mg^{2+}	Phenyl-2-pyridylcarbinol	64a
Oxidation of alcohols			
Benzyl alcohols	(21)	Benzaldehydes	31,65
Fluorenol	(20)	Fluorenone	30

[a] ns, not specified.

(8)   (10)

(9)   (11)

mixture of NADH and acetaldehyde [44], however, leads to the production of ethanol and NAD^+. Since NADH may be converted to NAD^+ by irradiation in the absence of substrate the reaction most likely proceeds via photoexcitation of the coenzyme rather than the substrate. A photochemical intramolecular reduction of the aldehyde function by a 1,4-dihydropyridine has been reported [52].

A wide variety of activated substrates have been reduced by NAD(P)H models (Table 3). Increasing the electrophilicity of the carbonyl carbon atom by attaching electronegative functions is a technique that has been widely employed to facilitate these reductions. Thus, although acetone and acetophenone are inert, their halogenated analogs hexachloroacetone (8) [50,54] and trifluoroacetophenone (9) [63,64] are readily reduced to the carbinols (10) and (11) by the model coenzymes (2) or (3). For hexachloroacetone (8) it is necessary to use polar solvents [e.g., dimethylsulfoxide (DMSO)] to obtain the carbinol cleanly. Less polar solvents (even nitromethane) or the addition of radical initiators or photolysis lead to the production of dechlorinated materials. These effects can be rationalized as a change of mechanism from a hydride process to a radical chain reaction. The potential for ionic or radical reaction in 1,4-dihydronicotinamides is clearly finely balanced and very dependent on conditions.

In parallel with the involvement of zinc ions in the dehydrogenase systems [17,17b,22,26], metal ions, notably those of zinc and magnesium, have been observed to catalyze the reduction of a number of substrates (see Table 3) [48,60,64,64a]. Trifluoroacetophenone is reduced much more rapidly by N-benzyl-1,4-dihydronicotinamide (2a) when magnesium ions are present [64]. In compounds such as 1,10-phenanthroline-2-carboxaldehyde, where chelation (as 12) of the added metal ion is possible, the catalysis is especially pronounced [59]. The relative efficiency of a series of metal ions as catalysts for the reduction of pyridine-2-carboxaldehyde (13) by the dihydronicotinamide (2a) has been determined [60] as $Cu^{2+} > Zn^{2+} > Pb^{2+} > Cd^{2+}$. This coincides with the order of the stability of the complexes (14) [66]. The counterion also affects the efficiency of catalysis, which decreases in the

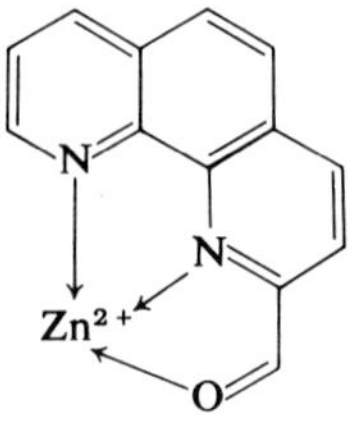

(12) 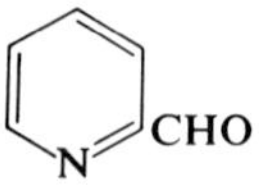   (13) 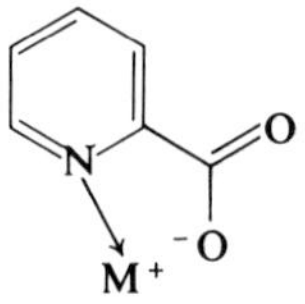   (14)

order $NO_3^- > Br^- > Cl^- > AcO^-$. This would seem to indicate competitive binding to the cation by counterion. Salicylaldehyde, thiophene-2-carboxaldehyde, and furfural are reported [60] not to be reduced by the NAD(P)H model (2a) even in the presence of zinc salts. The authors have suggested that the metal ion is not strongly enough coordinated by oxygen and sulfur to be held near the aldehyde functions. Earlier workers [55], however, had shown that oxygen at least is well able to coordinate a range of metal ions, including zinc, all of which catalyze the reduction of pyridoxal phosphate analogs by NAD(P)H models.

Asymmetric reduction of the carbonyl function by a model system has only recently been achieved. The dihydronicotinamides (15), in the presence of zinc or magnesium ions, were found [48] to reduce ethyl benzoylformate to ethyl mandelate of optical purity up to 38%. In the absence of metal ions no reduction of substrate occurred.

The reduction of trifluoroacetophenone (9) by the dihydronicotinamide (2a) (see above) is catalyzed by, but not dependent on, metal ions. It is therefore interesting [64] that treatment of the phenone (9) with the chiral dihydronicotinamide (R)-(−)-(15c) alone affords optically inactive carbinol (11), whereas the addition of magnesium perchlorate to the reaction leads to the production of (S)-trifluoromethylphenylcarbinol of 16% optical purity. The mechanism by which the cation induces asymmetry is not clear. It may be [48] that it coordinates the dihydropyridine, fixing its conformation in such a way as to induce chiral reduction and, at the same time, increasing its reactivity. Alternatively, the metal ion may coordinate both reactants [64] with the chirality of the reducing agent, giving rise to a preferred orientation for the carbonyl group in the complex.

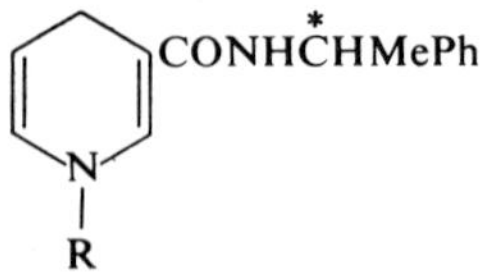

(15a), R = benzyl
(15b), R = 2,6-dichlorobenzyl
(15c), R = n-Pr

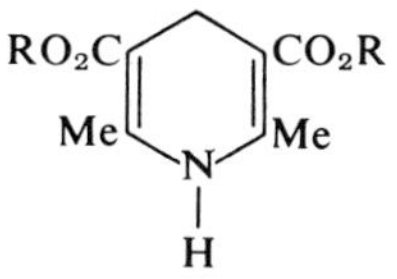

(16), R = (−)-menthyl

The modified Hantzsch ester (16) has been used [67] to effect chiral reductions of pyruvate and benzoylformate esters with zinc ion catalysis. The asymmetric induction was particularly pronounced (78%) when the substrates used were the (−)-menthyl esters.

In dogfish lactate dehydrogenase the metal ion of the alcohol dehydrogenases is replaced by the imidazole proton of a histidine residue (17) [23].

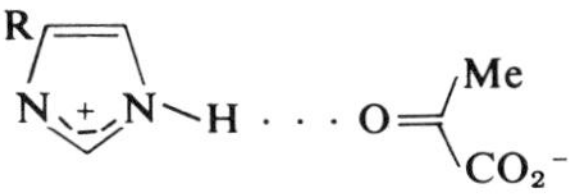

(17)

Attempts to mimic this in model systems have included intramolecular proton catalysis as in 3-hydroxypyridine-4-carboxaldehyde (18) [56,67]. The importance of the hydrogen bond is shown by the failure of NAD(P)H models to efficiently reduce pyridine-4-carboxaldehyde or the phenolate of the aldehyde (18). A polar mechanism for the reduction of (18) by NAD(P)H models is indicated by the increase in the rate of reaction with solvent polarity [56]. This process may also be catalyzed by transition-metal ions, whose effectiveness increases in the same order, $Mg^{2+} < Mn^{2+} < Zn^{2+} < Co^{2+} < Ni^{2+}$, as the stability of their complexes with salicylaldehyde [56].

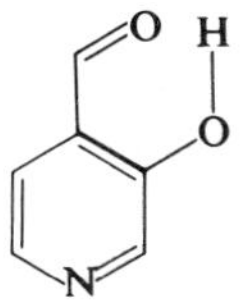

(18) 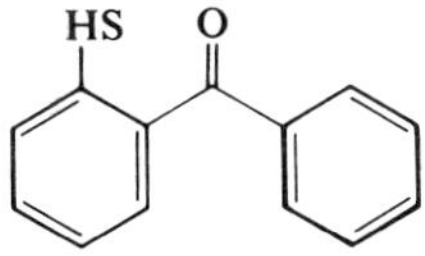   (19)

The reduction [57] of 2-mercaptobenzophenone (19) with the Hantzsch ester (3a) was apparently designed as an intramolecular proton-assisted process. However, not only does the 4-mercapto isomer fail to react, but the 2-hydroxy and 2-amino compounds are also inert. The reaction is enhanced by radical initiators and does not involve direct transfer of hydrogen. The involvement of an intermediate thiyl radical was suggested [57].

Oxidation of the Alcohol Function

In 1965 Wallenfels [30] reported the oxidation of fluorenol to fluorenone by N-methyl-3,4,5-tricyanopyridinium perchlorate (20). This unusual NAD$^+$ model has a very high electron affinity by virtue of the three cyano functions.

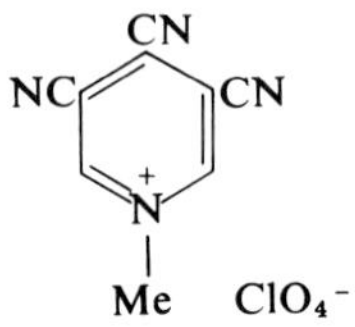

(20)

The yield of fluorenone was only 8% and the corresponding reduction product of (20) was probably not a 1,4-dihydropyridine. High-temperature oxidation of benzyl alcohol by pyridine, nicotine, and nicotinic acid methiodides had been reported by Kadis [65].

Shirra and Suckling [31] have studied the equilibria of Scheme 4. Plots of log K_{eq} vs. σ^* (σ for m-Cl) for the reactions were linear, with slopes of

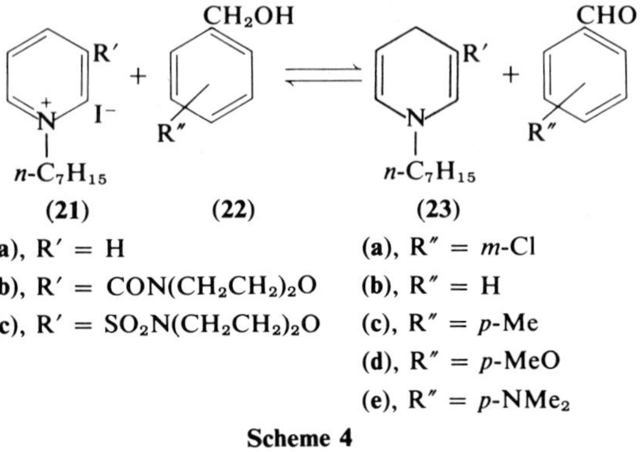

(a), R' = H
(b), R' = CON(CH$_2$CH$_2$)$_2$O
(c), R' = SO$_2$N(CH$_2$CH$_2$)$_2$O

(a), R" = m-Cl
(b), R" = H
(c), R" = p-Me
(d), R" = p-MeO
(e), R" = p-NMe$_2$

Scheme 4

$\rho = 0$, -0.8 and -2.0 for salts (21a)–(21c), respectively. These values reflect the control of the equilibria involving (21b) and (21c) by electron density on the benzyl carbon atom and in the aromatic ring of (22). The dihydropyridine (23a) is so reactive that the effect of R" is unobserved since the equilibrium of Scheme 4 lies overwhelmingly to the left for this compound. In the cases of the pyridinium salts (21b) and (21c), it is not clear whether the negative ρ values are indicative of an intermediate charge-transfer complex, cation radical, or cation. The only conclusion is that there is a transfer of negative charge to the benzyl alcohol in the transition state.

TRANSHYDROGENASE MODELS

The transhydrogenase enzymes catalyze the equilibrium of Scheme 5. The *in vivo* hydrogen transfer is direct and stereospecific [68]. The reaction

$$NAD^+ + NADPH \rightleftharpoons NADH + NADP^+$$

Scheme 5

also occurs nonenzymatically [69], without stereospecificity (see Table 4). It has been suggested [69] that this nonenzymatic contribution may be responsible for the lack of absolute stereospecificity in enzymatic reactions requiring long incubation times [70].

A number of transhydrogenase models have now been examined (Table 4) [27,69,71–78]. Cilento [71] has reported approximate kinetic data on the

TABLE 4

Transhydrogenase Models

Substrate	Reductant	Product[a]	Reference
3-Acetylpyridine adenine dinucleotide	(1)	3-Acetyldihydropyridine adenine dinucleotide	71–73
Acridine	(3a)	9,10-Dihydroacridine	74
N-Benzyl-3-acetylpyridinium[+]	(1)	N-Benzyl-1,4-dihydro-3-acetylpyridine	71
N-Benzylnicotinamidium[+]	(1), (2a)	N-Benzyl-1,4-dihydro-nicotinamide	71
Isoquinoline	(3a)	1,2,3,4-Tetrahydroiso-quinoline	73
N-Methylacridinium[+]	(2a)	N-Methyl-9,10-dihydro-acridine	75
N-Methylphenazinium[+]	(2a)	N^9-Methyl-9,10-dihydro-phenazine radical cation	76
Methyl viologen	(2a)/catalyst	Methyl dihydroviologen radical cation	76
NAD[+]	(1)	NADH	69
	(2c)	NADH	77
Phenanthridine	(3a)	1,2-Dihydrophenanthridine	74
Phenazine	(2)/O₂/H[+]	ns	78
N-Propylnicotinamidium[+]	(2c)	N-Propyl-1,4-dihydroni-cotinamide	69
Quinoline	(3a)	1,2,3,4-Tetrahydroquinoline	74
3,5-Dicarbethoxy-1,2,6-trimethylpyridinium[+]	(3b)	3,5-Dicarbethoxy-1,2,6-trimethyl-1,2-dihydro-pyridine	71a

[a] ns, not specified.

(24a), R′ = PhCH₂, R″ = NH₂, Me
(24b), R′ = n-Pr, R″ = NH₂

reduction of the nicotinamides (24a) by NAD(P)H, which indicates a bimolecular reaction pathway. Spectrophotometric examination of the couples (24b)–(2c) and NAD⁺–NADH led Ludoweig and Levy [69] to conclude that the reaction proceeds via a charge-transfer complex. Such complexes had previously been reported [79,80] in other model systems. Kinetic isotope measurements [75] on the reduction of the N-methylacridinium ion by (2a) also indicated the existence of an intermediate complex.

While Ludoweig and Levy [69] could not detect any electron spin resonance (esr) signals in the reactions that they studied, another model system [76], a mixture of the dihydronicotinamide (2a) and N-methylphenazinium methosulfate (25), did show esr signals due to the radical cation (26). Although (26) is formally the product of hydrogen atom transfer from (2a)

(25)   (26)

to (25), it could also be the result of electron donation by (2a) followed by solvent (ethanol) proton abstraction. The fate of the (presumed) radical or radical cation of the dihydronicotinamide (2a) was not reported.

OLEFIN HYDROGENASE MODELS

NAD(P)H is the cofactor in a number of enzymatic saturations of olefinic bonds [81–84]. Labeling studies [81] have shown that the C-4 hydrogen of NAD(P)H adds to the more electrophilic end of the double bond in a number of structurally quite different steroidal olefins. Watkinson *et al.* [81] considered three basic mechanisms for these reductions (Scheme 6). Their investigations showed that only mechanism (a) accurately accounted for the

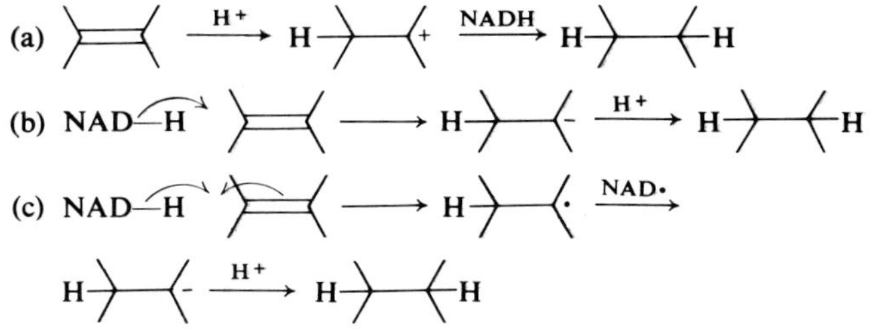

Scheme 6

hydrogen distribution in the products of the enzymatic reduction. The second step of this mechanism, however, was unresolved between a hydride ion transfer or an electron transfer followed by hydrogen atom abstraction.

The orientation of addition of all the model systems (Table 5) [46,85–92]

TABLE 5

Olefin Hydrogenase Models

Substrate	Reductant	Product[a]	Reference
2-Benzalindane-1,3-diones	(2)	2-Benzylindane-1,3-diones	85
3-Benzoylpropenoic acid	(3)	3-Benzoylpropanoic acid	86,87
1-Benzoyl-3,3,3-trifluoro-propene	(3)	1-Benzoyl-3,3,3-trifluoro-propene	86
Benzylidene acetophenone	(3a)	3-Phenylpropiophenone	46
1,1-Dicyanoalkenes	(2)	1,1-Dicyanoalkanes	88,89
Enacyl chlorides	(3a)	Ketenes	87
Fumarate ester	(3a)	Succinate ester	46
	(1), (2), or (3)/$h\nu$	Succinate ester	90
Fumaric acid	(3a)	Succinic acid	46
	(1), (2), or (3)/$h\nu$	Succinic acid	90
Maleate ester	(3a)	Succinate ester	46
Maleic acid	(3a)	Succinic acid	46
Maleic anhydrides	(3a)	Succinic anhydrides	46,91
tert,tert-Muconic acid	(3a)	ns	46
ω-Nitrostyrene	(2)	ω-Nitroethylbenzene	88
Tetracyanoethylene	(2)	Pentacyanopropenide anion	92
3,4,5,6-Tetrahydrophthalic anhydride	(3a)	cis-Cyclohexane-1,2-dicarboxylic acid	46

[a] ns, not specified.

for which this datum has been recorded is the same as that found by Watkinson. The reduction of 1-benzoyl-2-trifluoromethylethylene with 4,4-dideutero-(3a), for instance, affords 2-deutero-3,3,3-trifluoropropiophenone [86]. The reaction was observed to be polar and possibly proceeds via the enolate (27). The reductions of benzoylacrylic acid [86], ω-nitrostyrene [88], and 2-benzalindane-1,3-dione [85] probably follow similar pathways.

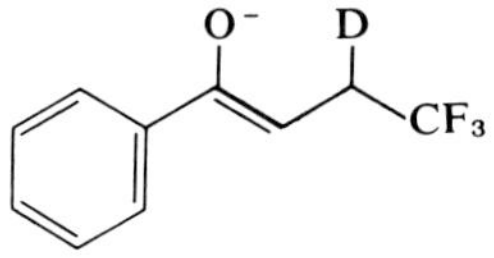

(27)

Wallenfells [92] has reported that the reduction of tetracyanoethylene (TCNE) (28) by the dihydronicotinamide (2c) affords the pentacyanopropenide ion (29).

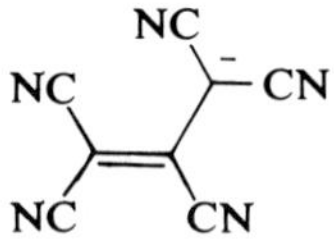

(29)

The TCNE radical anion was detected fleetingly by its characteristic ultraviolet absorption. The authors postulate a mechanism involving hydride ion transfer and disproportionation [Scheme 7, route (a)] but, in view of the

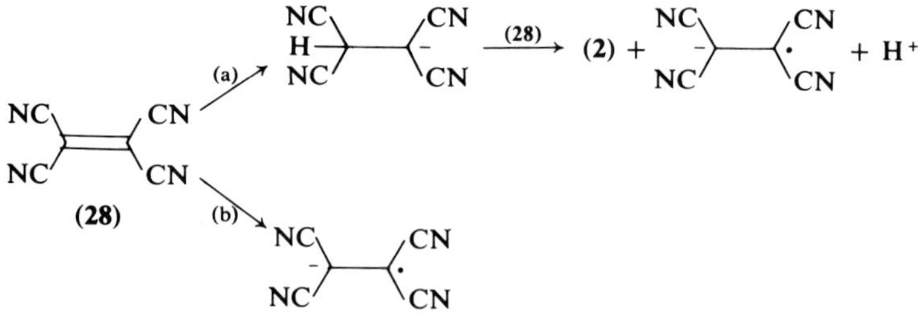

Scheme 7

known power of TCNE as an electron acceptor [93], route (b) cannot be ignored.

The reduction [89] of the *gem*-dicyanoalkene (30a) by the dihydronicotinamide (2a) in ethanol/acetic acid is much more efficient than that of (30b) (Scheme 8). Again, both hydride and electron transfer processes could be involved.

$$RCH{=}C(CN)_2 \xrightarrow[\text{EtOH/ACOH}]{(2a)} RCH_2CH(CN)_2$$

(30a), R = Ph   9:1   15 min, 20°C, 83%

(30b), R = Me   12 hr, 20°C, 27%

Scheme 8

REDUCTION OF QUINONES

The ability of dihydropyridine derivatives to reduce quinones to dihydroquinones has been known for many years (Table 6) [17,17a,27,45,46,76,94–99]. The intermediacy of semiquinone anion radicals in the reduction of quinones is also well known [100].

The esr quintets recorded during dihydronicotinamide reduction of 1,4-benzoquinone and 1,4-naphthoquinone have been assigned [94] to the semiquinone radical anions. No signals due to the (presumed) dihydronicotinamide radical cation were observed. This may be due to rapid disproportionation. No esr signals were detected during the reduction [94] of chloranil or 2,6-di-tert-butylquinone under similar conditions.

Electron transfer catalysts such as pyocyanine (PC) have been reported [76] to enhance the rates of reduction of acenaphthoquinone and anthraquinone by the dihydronicotinamide (2a). A 30-fold increase in the semiquinone

TABLE 6

Quinones[a]

Substrate	Reductant	Reference
Acenaphthoquinone	(2)	76
Anthraquinone	(2)	76
p-Benzoquinone	(2a)	17,17a,45,94–96
Chloranil	(2a)	94–96
	(2d)	27
	(3a)	46,97
2-Chloro-p-benzoquinone	(1)	17
2,6-Di-tert-butyl-p-benzoquinone	(2a)	94
2,6-Dichloroindophenol	(2)	95
2-Methyl-p-benzoquinone	(2a)	95,98
o-Naphthoquinone	(2a)	96,99
p-Naphthoquinone	(2a)	94
Tetramethyl-p-benzoquinone	(2a)	95

[a] In all cases the product is the hydroquinone.

radical esr signals in these reactions was recorded in the presence of PC over that with (2a) alone.

Hadju and Sigman [27] have observed a large increase in the rate of reduction of chloranil by N-o-carboxybenzyl-1,4-dihydronicotinamide (2d) over its unsubstituted analog (2a). Since a 4′-carboxylate substituent shows no effect over the unsubstituted material, the authors ascribed this phenomenon to the stabilization of an intermediate complex in which the N-1 nitrogen bears a partial positive charge. By analogy with the reduction of the methylacridinium cation, which shows a similar effect [27,75], the intermediacy of a charge-transfer complex in the reduction of quinones by NAD(P)H and its models can be postulated. Charge-transfer complexes are known [101] between hydroquinones and pyridinium cations but have yet to be reported between quinones and dihydropyridines.

FLAVINS

The reduction of flavins by an NAD(P)H is an important step in the respiratory electron transport chain (Scheme 9). Ohno and co-workers [102]

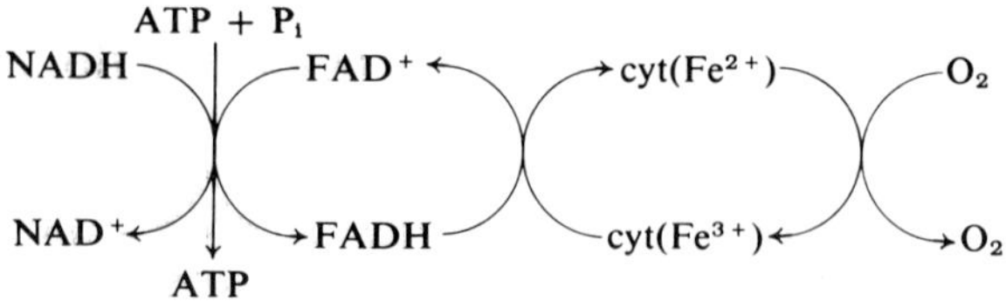

$$\text{ATP} + \text{P}_1$$

NADH — FAD$^+$ ← → cyt(Fe^{2+}) — O$_2$

NAD$^+$ ← → FADH — cyt(Fe^{3+}) ← → O$_2^-$

ATP

Scheme 9   A simplified respiratory chain.

have recently reported a biomimetic respiratory chain in which dihydronico-tinamide (2a), N-methylphenazium methosulfate (PMS), and hemin serve as models for NAD(P)H, flavin, and cyctochromes respectively. When lumi-flavin was used in place of PMS, at a concentration some 10^{-5} of those of (2a) and hemin, it was found to turn over 10^5 cycles in the 10 min that the reaction took to complete. Such a high efficiency of catalysis is unprecedented. When ADP and tetra- n-butylammonium phosphate were added to the reaction mixture, biomimetic ATP synthesis was observed. An analogous model of the photosynthetic electron transport process has also been re-ported [98].

The nature of the hydrogen transfer in the reduction of flavins has been the subject of many studies (Table 7) [98,99,102–109]. The flavin site to which the coenzyme transfers hydrogen is not easily determined because the basicity of the two possible receptor nitrogen atoms (1 and 5) leads to facile exchange processes. Bruice [103] has attempted to overcome this problem by reducing

TABLE 7

Flavins[a]

Substrate	Reductant	Reference
5-Deazaflavins	(2a)	103
Flavin mononucleotide	(2a)	99,107
Flavin nicotinamide bis coenzymes	—	105–107
Lumiflavin	(2)	102,108
Proflavin hemisulfate	(2)	98
Others	(2), (3)	109

[a] In all cases the product is the dihydroflavin.

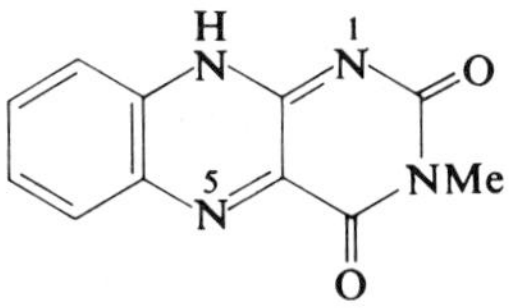

(31)

the 5-deazaflavin (31) with NADH in D_2O. The reduced deazaflavin contained no deuterium at C-5, which shows that position to be the hydrogen delivery site. The general chemical properties of the deazaflavin (31) are similar to the corresponding flavin, and the rates of reduction by NADH can be approximated to those of the flavin. This evidence is supported by theoretical calculations, which show N-5 to be the most electrophilic center of the flavin nucleus [110], and by a report [111] that the sulfite anion adds to flavins at the 5 position.

The reduction of flavin mononucleotide (FMN) by NADH proceeds via the FMN anion radical, which has been observed by esr spectroscopy [104]. No resonance due to NADH[+] was recorded. The electron transfer was thought to occur via a donor–acceptor complex. Donor–acceptor complexes between oxidized nicotinamide systems and reduced flavins have been described both in flavoproteins and in model systems. These donor–acceptor complexes have been well characterized [112] by correlation of the long-wavelength absorption with the energies of the lowest vacant orbitals of the NAD(P) analogs. The complexes have been shown [113,114] to be catalytic in some enzymes but not in others.

Long-wavelength-absorbing complexes of reduced nicotinamides and oxidized flavins have been known for some time [115,116] and have been shown to be catalytic intermediates in a number of enzymatic processes [108].

These complexes, which have also been observed in models [105,117], have been described as biradical [108] or covalent [106]. However, all the recent evidence [105,106,114,117,118] supports a donor–acceptor structure. An increase in the electron deficiency of the oxidized flavin by protonation enhances its rate of reduction by NAD(P)H [119]. Conversely, the rate is reduced by a decrease of electron density on the dihydronicotinamide ring or a decrease of the electron deficiency of the flavin.

It has been demonstrated [117] that the overall rate of reduction of lumiflavin by the dihydronicotinamide (2b) is equal to the rate of disappearance of an initial, rapidly formed complex. Further, the rate of reduction of oxygen to hydrogen peroxide by a mixture of (2b) and oxidized lumiflavin was greater than the rate of oxygen by the dihydroflavin [117]. Since (2b) reacts only slowly with oxygen, it is apparent that the intermediate complex is the effective reductant under aerobic conditions.

Blankenhorn [105–107] has studied the bis coenzymes (32) and (33). These species possess long-wavelength absorptions of the same type as are seen

(32), $n = 2-4$ (33), $n = 2-4$

in the intermolecular interaction between flavins and nicotinamides. They show linear variation of absorbance with concentration, demonstrating the intramolecular nature of the absorption bands. The dependence of the oxidation–reduction potential and the proton magnetic resonance shift of the flavin C-6 and C-9 protons on the chain length (n) indicates that as the two ring systems become more able to adopt an orientation parallel to each other the charge-transfer interaction becomes more intense.

The intermediacy of charge-transfer complex would permit rationalization of the known [118,119] two orders of magnitude greater rate of reduction of flavins by dihydronicotinamide (2c) over NADH. Relatively strong inter- [109] and intramolecular [120] complexing of adenine with flavins is known and would be expected to lead to competitive, nonproductive complexation in the case of NADH.

TABLE 8

Imines and Iminium Salts

Substrate	Reductant	Product[a]	Reference
Benzylideneacetophenone	(3)	Benzylacetophenone	46
Benzylideneaniline	(3)	ns	46
Cyclic imines	(2a)	Cyclic amines	88
5-Deazaflavin	(2a)	Dihydro-5-deazaflavin	103
1,2-Dihydroquinoline	(2a)	1,2,3,4-Tetrahydroquinoline	74,97
Iminium salts	(2)	Amines	122
Indolenines	(2), (3)	Indoles	122–126
Malachite green	(2a)	Leuco malachite green	45,86
3-Methylbenzothiazolium chloride	(2)	ns	121
Thiamine hydrochloride	(2)	ns	121
α,β-Unsaturated iminium salts	(2), (3)	Enamines	127

[a] ns, not specified.

IMINES AND IMINIUM SALTS

In vivo reductions of iminium cations are known [121]. There are scant data on the enzymatic reactions, although model reductions have been studied for some time (Table 8) [45,46,74,86,88,97,103,121–127]. In 1955, Mauzerall and Westheimer [45] reported the reduction of malachite green (34) to its leuco base by the NAD(P)H model (2a). Apart from a value of k_H/k_D of 4.5, no kinetic data were recorded.

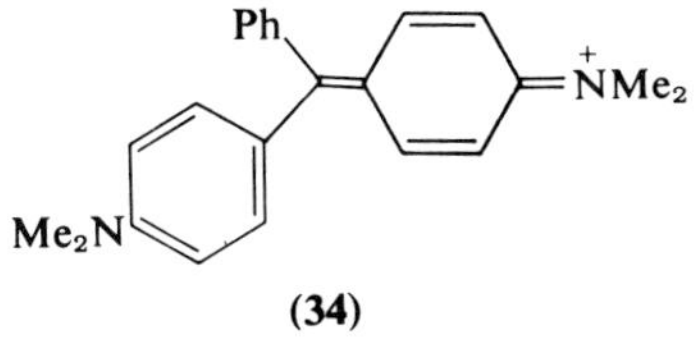

(34)

More recently a wide variety of imines and iminium salts have been reduced by NAD(P)H-related systems (see Table 8). Much interest in the reduction of indolenines was stimulated by a claim [123] that tryptophan residues acted as hydrogen carriers in reactions mediated by yeast and rabbit muscle dehydrogenases. However, single turnover reactions have since shown [13] this claim to be based on the observation of a small-scale side reaction.

There is some evidence for charge-transfer complexes as intermediates in the reductions of indolenines.

The reduction of substituted phenyl-2-methylindolenine methohydrogen sulfates (35) with the Hantzsch ester (3a) shows [124] a rate enhancement of

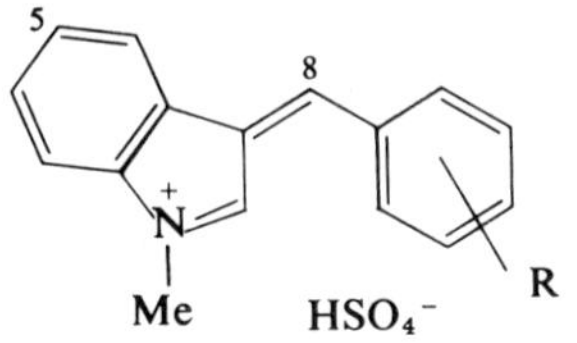

(35)

500-fold on changing the solvent from ethanol to acetonitrile. The latter is known to promote the formation of donor–acceptor complexes, whereas the greater solvation of the salt by ethanol could be expected to retard such a process. Hammett plots for a range of R substituents in the indolenine (35) show slightly better correlation with σ than σ^+. The derived value of ρ is $+1.5$. This indicates considerable quenching of the cationic character of (35) in the transition state.

Pandit et al. have described the reduction of saturated (36) [122] and α,β-unsaturated (37) [127] steroidal iminium salts as being under steric approach

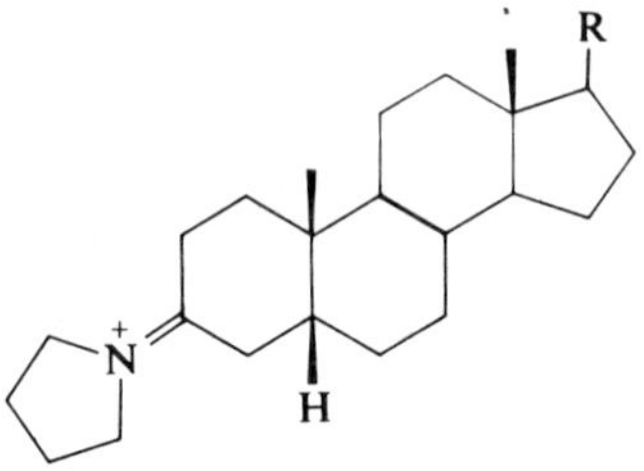

(36) 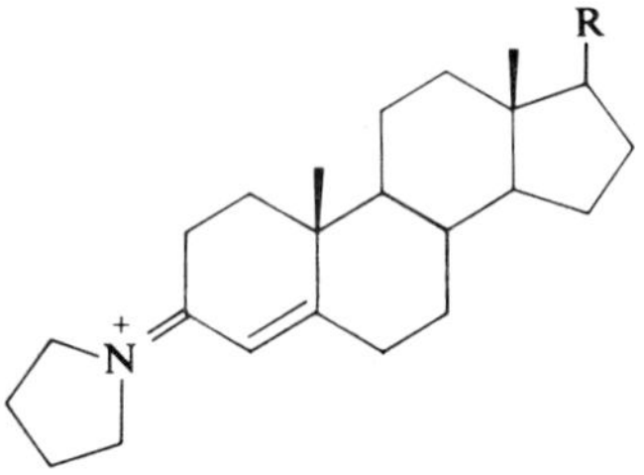   (37)

control. The authors suggested that the pyridine nucleotide-mediated micro-biological reduction of Δ^4-3-oxosteroids may proceed via the sequence of

$$C{=}C{-}C{=}O \longrightarrow C{=}C{-}C{=}\overset{+}{N} \longrightarrow C{-}C{=}C{-}N \longrightarrow C{-}C{-}C{=}O$$

Scheme 10

Scheme 10, in which the nitrogen atom is that of an amino acid residue. There is, however, good evidence [17,17a,26] that ketones are reduced by processes involving Lewis acid behavior by metal ions as described earlier.

Wallenfels [88] has reported the reduction in good yield (50–85%) of the aldimine double bond in the powerful lachrymators (38a)–(38d) by the dihydronicotinamide (2a). In contrast, compounds (38e) and (38f) possess only slight and no lachrymatory power, respectively, and while (38e) was

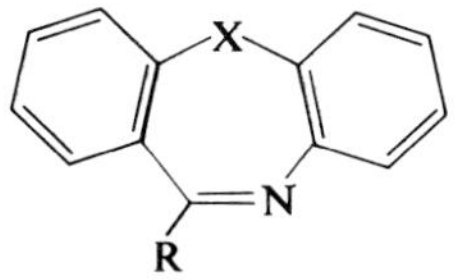

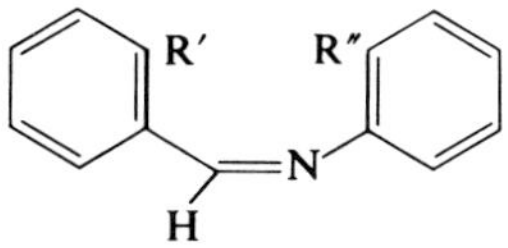

(38a), X = O, R = H   (39a), R′ = R″ = H
(38b), X = S, R = H   (39b), R′ = R″ = OH
(38c), X = NMe, R = H   (39c), R′ = H, R″ = OH
(38d), X = CH₂, R = H   (39d), R′ = OH, R″ = H
(38e), X = O, R = Me   (39e), R′ = Cl, R″ = OH
(38f), X = O, R = Ph   (39f), R′ = Cl, R″ = H

reduced by (2a) in 15–29% yield, (38f) was unaffected. Compounds (39a)–(39f) are not lachrymators and are again unreactive toward (2a). The explanation of these observations is not immediately clear. The lessening of ring strain that accompanies the reduction of (38a)–(38e) and presumably also accompanies their lachrymatory action may provide a part of the driving force. The polarity of the C=N bond must also be of importance to explain the observed low reactivities of (38e) and (38f).

OTHER NITROGEN COMPOUNDS

Because of their electrophilic nature, oxygenated nitrogen functions have been used as substrates for several *in vitro* reductions (Table 9) [46,99,128,

TABLE 9

Other Nitrogen Compounds

Substrate	Reductant	Product	Reference
Azobenzene	(3a)	Hydrazobenzene	46
Azoxybenzene	(2a)	Aniline, hydrazobenzene	128
Nitrobenzenes	(2a), (b)	Aniline, phenylhydroxylamine, hydrazobenzene	128
o-Nitrophenol	(2a)	o-Aminophenol (>30%)	128
p-Nitrophenol	(2a)	p-Aminophenol (<3%)	128
4-(o-Nitrophenyl)-3a	hν	4-(o-Nitrosophenyl)-2,6-lutidine 3,5-dicarboxylate	129
Nitrosobenzenes	(2a)	Aniline, azoxybenzene	99
	(2a)	Aniline, phenylhydroxylamine, hydrazobenzene	128
p-Nitrostilbene	(3a)	p-Aminostilbene	46

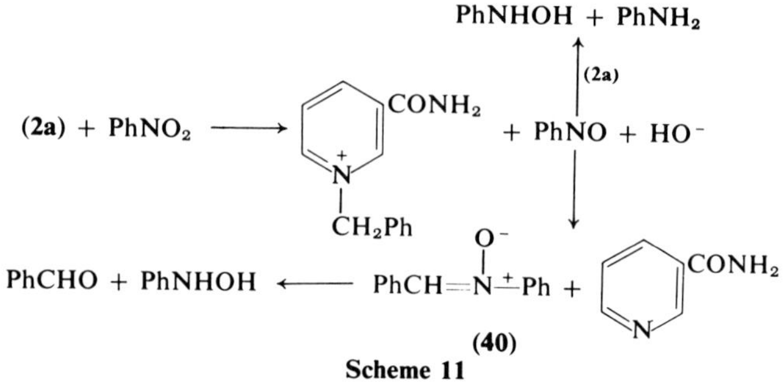

$$\text{PhNHOH} + \text{PhNH}_2$$

Scheme 11

129]. The reduction of nitrobenzene by dihydronicotinamides **(2a)** and **(2e)** at elevated temperatures affords [128] aniline, phenylhydroxylamine, and hydrazobenzene (Scheme 11). Quantities of nicotinamide and benzaldehyde obtained after hydrolysis of the reaction mixture were thought to derive from the formation of the nitrone **(40)** and its subsequent acid cleavage. *o*-Nitrophenol was more readily reduced than *p*-nitrophenol, a reflection of the additional polarization of the N—O linkage induced by hydrogen bonding. In general, electron-withdrawing substituents increased the rate of reduction, while electron-donating substituents decreased it. However, *o*-nitroanisole was reduced [128] almost as quickly as *o*-nitrophenol. The authors ascribe this effect to steric inhibition of resonance in the initial ground state.

A photochemical intramolecular reduction of a nitro group has been reported [129]. Irradiation of the 4-(2-nitrophenyl) Hantzsch esters **(41)** in the 360–392 nm band cleanly produces the nitroso compound **(42)**. The

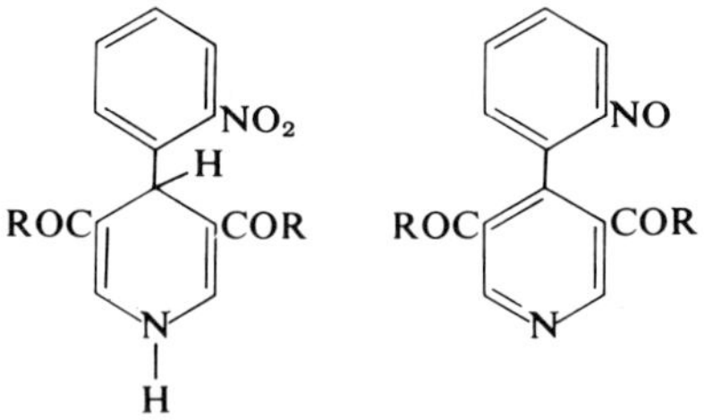

4-nitro isomer of **(41)** is unreactive. The process exhibits first-order kinetics for the initial 60% of reaction. After this point the rate decreases, perhaps because of donor–acceptor interactions between product and starting material.

Nitroso compounds are more easily reduced than the corresponding nitro compounds. The nitrone **(43)** has been isolated [128] in 89% yield from the

$$Me_2N{-}\langle\!\!\bigcirc\!\!\rangle{-}\overset{\overset{O^-}{\underset{|}{N^+}}}{=}CHPh$$

(43)

reaction of *p*-nitroso-*N*,*N*-dimethylaniline and **(2a)** in refluxing ethanol. Other products were nicotinamide and *p*-amino-*N*,*N*-dimethylaniline.

Azobenzene may be reduced to hydrazobenzene [128], and azoxybenzene to aniline and hydrazobenzene [128]. Hydrazones, oximes, nitroalkanes, inorganic nitrates, and nitrites are inert to dihydronicotinamide **(2a)** [128].

PEROXIDES, SULFIDES, AND DISULFIDES

Acyl peroxides react rapidly with the Hantzsch ester **(3a)** at temperatures too low for their unimolecular decomposition to act as the initiation reaction [130]. In contrast, alkyl peroxides react with **(3a)** only after initial thermal homolysis of the peroxide bond [131]. The alkyl peroxides show [132] a kinetic isotope effect of $k_H/k_D = 1.58$ for the reactions with **(3a)**. The acyl peroxides show [132] no isotope effect. This difference reflects the different rate-determining steps in the reactions. For the alkyl peroxides the rate-determining step involves hydrogen atom abstraction from the secondary amine moiety of **(3a)** by the alkoxy radical. In the case of the acyl peroxides the reaction mechanism has been suggested [132] to involve electron rather than hydrogen atom transfer.

The room-temperature reduction of tetramethylthiuram disulfide **(44)** by **(2a)** has been reported [133] to yield two major products, the nicotinamidium salts of the per- and dithiocarbamate anions **(45)** and **(46)**. A small amount

$$\overset{S}{\underset{\|}{(Me_2NCS)_2}} \qquad \overset{S}{\underset{\|}{Me_2NCSS^-}} \qquad \overset{S}{\underset{\|}{Me_2NCS^-}}$$

(44)   **(45)**   **(46)**

(1%) of carbon disulfide was detected by gas–liquid chromatography. Tetramethylthiuram monosulfide under similar conditions afforded only the nicotinamidium salt of **(46)**. No carbon disulfide was detected. In neither case was dimethylamine observed, even at trace levels. Nor were the reactions affected by radical traps. The anions **(45)** and **(46)** showed no tendency to add to the nicotinamidium cation, although the ultraviolet spectra of the salts were indicative of a donor–acceptor interaction similar to that of *N*-methyl-pyridinium iodide [134].

TABLE 10

Peroxides, Sulfides, and Disulfides

Substrate	Reductant	Product	Reference
Diacetyl peroxide	(3a)	Acetic acid, methane, carbon dioxide	130,132
Dibenzoyl peroxide	(3a)	Benzoic acid, carbon dioxide	130,132
Di-*tert*-butyl peroxide	(3a), (b)	*tert*-Butanol	130–132
Tetramethylthiuram disulfide	(2a)	Di- and perthiocarbonate anions	133
Tetramethylthiuram monosulfide	(2a)	Dithiocarbamate anion	133

Dimethyl sulfoxide is reported [128] to be inert toward dihydronicotinamides in refluxing ethanol. A list of the reactions pertaining to this section is given in Table 10 [130–133].

THIONES AND RELATED COMPOUNDS

Thiobenzophenone was one of the substrates to be reduced by NAD(P)H models [135] (Table 11) [88,135–139] in early studies. The rate of reaction was found [138] to be unaffected by the presence or absence of oxygen or by radical chain inhibitors. Polar solvents and electron-withdrawing substituents in the thioketone enhanced the rate, whereas less polar solvents and, with one exception, electron-donating substituents decreased the rate of reduction. The presence of an *o*-hydroxyl group in the substrate enhanced the reaction, probably via hydrogen bonding. The infrared spectrum of *o*-hydroxythiobenzophenone shows only a very weak C=S absorption.

TABLE 11

Thiones and Related Compounds

Substrate	Reductant	Product	Reference
Sulfite esters	(2a)	Alcohols plus H_2S	88,136
Sulfonium salts (R_3S^+)	(3b)	Alkane plus dialkyl sulfide	137
Thiobenzophenone	(2a)	α-Phenylbenzylmercaptan	135,138, 139
Thiopropanal *S*-oxide	(2a)	Propylmercaptan	88

Westheimer *et al.* [138] proposed a hydride ion transfer mechanism for this reaction. However, sodium borohydride does not readily reduce thiobenzophenone [140] and, more recently [139] esr measurements have implicated a radical ion pair in the mechanism. When dihydronicotinamide (**2a**) and thiobenzophenone are mixed in 2-methyltetrahydrofuran at 77 K, an intense esr signal, identical with that of the radical anion of thiobenzophenone, was recorded. When the mixture was warmed to room temperature a new signal, possibly due to a thiyl radical, was observed. Perturbations of the resonance at 77 K have been ascribed [139] to the proximity (at about 6 Å) of the counter-cation radical (**7**).

The ease of reduction of thiobenzophenone as opposed to benzophenone may be due in part to the greater polarization of the $C=S$ bond. The dipole moment of thiobenzophenone is 3.4 [141], while that of benzophenone is 2.95 [142].

Thiopropanal *S*-oxide (**47**), the lachrymatory agent of the onion is one of a

$$CH_3CH_2CH=S=O$$
(**47**)

number [88] of powerful lachrymators that are reduced by NAD(P)H models. The product of reduction of (**47**) is thiopropanol.

AROMATICS OTHER THAN PYRIDINES

Hydrogen addition to electron-deficient benzene nuclei has been observed in only a few instances (Table 12) [18,76,143,144].

The reduction of a series of dinitrobenzene sulfonates (**48**, Scheme 12) by NADH has been studied [18]. The reaction of (**48**) (R = NO_2) is catalyzed

TABLE 12
Aromatics Other than Pyridines

Substrate	Reductant	Product	Reference
3,5-Dinitrobenzaldehyde	(**2a**)	3-Formyl-5-nitrocyclohexa-2,5-diene nitronate anion	143
2,6-Dinitro-4-*X*-benzene sulfonate	(**1**)	2,6-Dinitro-4-*X*-benzene	18
Picryl chloride	(**2a**)	Trinitrobenzene	143
1,3,5-Trinitrobenzene	(**2a**)	3,5-Dinitrocyclohexa-2,5-diene nitronate anion	143
2,4,6-Trinitrobenzene sulfonate	(**1**)	1,3,5-Trinitrobenzene	144
2,3,5-Tetraphenyltetrazolium	(**2a**)	Triphenylformazene	76

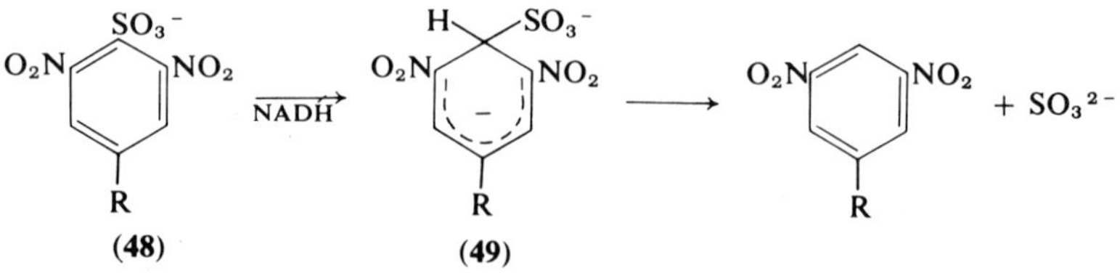

Scheme 12

by glutamate dehydrogenase, apparently at the active site [144]. The characteristic red color of the complex (49) is observed during these reactions (Scheme 12). Hammett plots for a range of substituents R in (48) clearly demonstrate the transfer of negative charge from NADH to (48) in the transition state. Displacement of chlorine from picryl chloride by (2a) has also been reported [143].

The nicotinamidium salt of the hydrido Meisenheimer anion (50), analogous to the intermediate (49) in Scheme 12, has been isolated [143] in high yield as a stable crystalline solid from the reduction of trinitrobenzene with (2a). Similarly, good yields of the salt of anion (51) have been obtained [143].

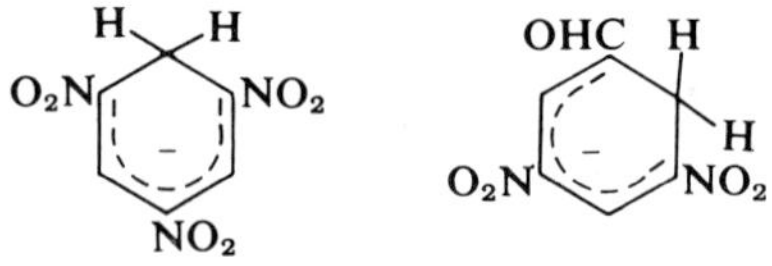

(50)   (51)

In this case there was no trace of reduction at the 4 position or at the carbonyl group.

REDUCTIVE DEHALOGENATION

The reductions of trityl [143] and tropylium [146] halides to their respective hydrocarbons can be readily conceived of as hydride ion processes. The kinetics [143] of the reduction of trityl bromide with the dihydronicotinamide (2a) are in agreement with a rate-limiting ionization of the halide followed by hydride abstraction. The addition of lithium bromide suppresses the ionization and inhibits the reaction.

Most other reported reductive halogenations (Table 13) [50,54,88,89,99, 137,143–150] are probably not hydride processes. The photochemical reduction [146] of bromotrichloromethane by (2a), (3a), and (3b) is a radical chain reaction with a quantum yield of 7–80 under the reported conditions. The addition of dihydroanthracene, a radical scavenger [151], causes a reduction in the reaction rate and a lowering of the quantum yield almost to unity. Dittmer and his co-workers [50,54] have reported a similar reaction between (2a) and hexachloroacetone. Under radical conditions (i.e., in cyclohexane with peroxides or irradiation) the products of reaction are a mixture of

TABLE 13
Reductive Dehalogenation

Substrate	Reductant	Product[a]	Reference
N-Bromoacetamide	(2a)	Acetamide	145
2-Bromo-1,1-dicyanoalkanes	(2)	1,1-Dicyanoalkanes	88,89
Bromomalononitrile	(2a)	Enamine, addition compound	137
	(3a)	Malononitrile	137
gem-Bromonitroalkanes	(2a)	Nitroalkanes	147
α-Bromophenones	(2a)	Phenones	88,137,148
	(3a)	Phenones	137
N-Bromosuccinimide	(2a)	Succinimide	99,145
Bromotrichloromethane	(3a), (b)	Chloroform	146
α-Chlorodeoxybenzoin	(2a)	Deoxybenzoin	149
10-Chloro-5,10-dihydrophen-arsazine	ns	As,As bonded dimer	88
2,6-Dichlorobenzyl bromide	(2a)	2,6-Dichlorotoluene	88
sym-Difluorotetrachloroacetone	(2a)	ns	50
Ethyl α-bromoacetate	ns	Ethyl acetate	88
Hexachloroacetone	(2a)	Penta- and tetrachloro-acetone	50,54
Iodobenzene chloride	(2)	ns	99
4-Nitrobenzyl bromide	ns	1,2-Di-(4-nitrophenyl)-ethane	88
Picryl chloride	(2a)	Trinitrobenzene	173
sym-Tetrachloroacetone	(2a)	ns	50
Trityl bromide	(2a)	Triphenylmethane	145
Tropylium halides	(2)	Cycloheptatriene	150

[a] ns, not specified.

penta-, sym-tetra-, and unsym-tetrachloroacetone and N-benzylnicotinamidium chloride (see section on alcohol dehydrogenase models).

Nonchain radical dehalogenations are also known. The reduction of gem-bromonitroalkanes has been shown [147] to proceed via the pathway of Scheme 13. The initial electron transfer may take place in a donor–acceptor complex. Subsequent elimination of bromide ion generates the radical (52),

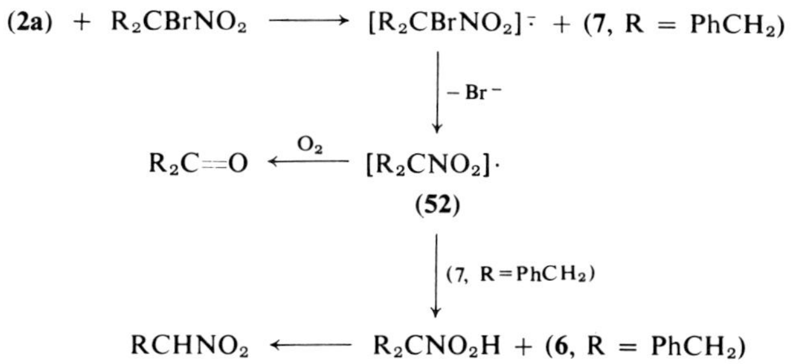

$$(2a) \ + \ R_2CBrNO_2 \ \longrightarrow \ [R_2CBrNO_2]^{\bar{\cdot}} \ + \ (7, \ R = PhCH_2)$$

$$\Big\downarrow -Br^-$$

$$R_2C{=}O \ \xleftarrow{\ O_2\ } \ [R_2CNO_2]\cdot$$
$$(52)$$

$$\Big\downarrow (7, \ R = PhCH_2)$$

$$RCHNO_2 \ \longleftarrow \ R_2CNO_2H \ + \ (6, \ R = PhCH_2)$$

Scheme 13

which may be trapped with oxygen, leading to the ketone. In the absence of oxygen the sole product is the nitroalkane, and radical species have been detected in the reaction by esr spectroscopy.

In a similar manner the reduction of the α-bromophenones (53) has been shown [148,152] to be an electron transfer process. In the presence of oxygen the intermediate radical species is trapped to afford a mixture of the α-diketone, α-hydroxyphenone, and the ester (54). The rates of reaction of the

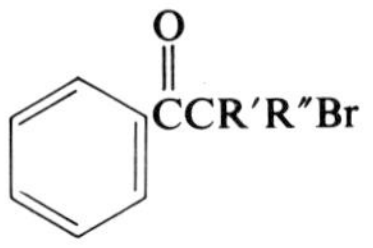

$$\text{C}_6\text{H}_5\text{-}\overset{\overset{\textstyle O}{\|}}{\text{C}}\text{CR}'\text{R}''\text{Br} \qquad \text{C}_6\text{H}_5\text{-}\overset{\overset{\textstyle O}{\|}}{\text{C}}\text{CR}'\text{R}''\text{O}_2\text{CPh}$$

(53a) R′=R″=H   (54)
(53b) R′=H, R″=Me
(53c) R′=R″=Me

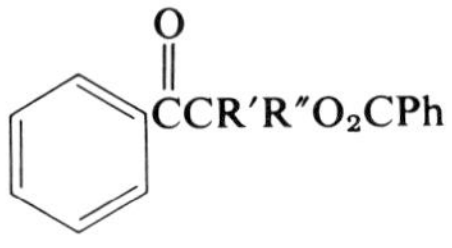

bromophenones (53) with (2a) decrease in the order (53b) > (53a) > (53c) [152]. This contrasts with their order of reactivity to nucleophiles, which is (53a) > (53b) > (53c). Presumably, the relative reactivity towards nucleophiles is governed by steric factors at the α-carbon, whereas reaction rate with electron donors is also affected by the stability of the α-radical generated by bromide ion elimination from the initial radical anion.

Further evidence for one-electron reduction of α-haloketones comes from the ability of iron polyphthalocyanine to catalyze the dehalogenation of α-chlorodeoxybenzoin by dihydronicotinamides [149]. A nonchain radical process has been proposed [149] for this reaction.

Although no mechanistic data were published, the reductive dehalogenation of 2-bromo-1,1-dicyanoolefins [89] could also involve an initial electron transfer step. The affinity of cyanoolefins for electrons is well known [93].

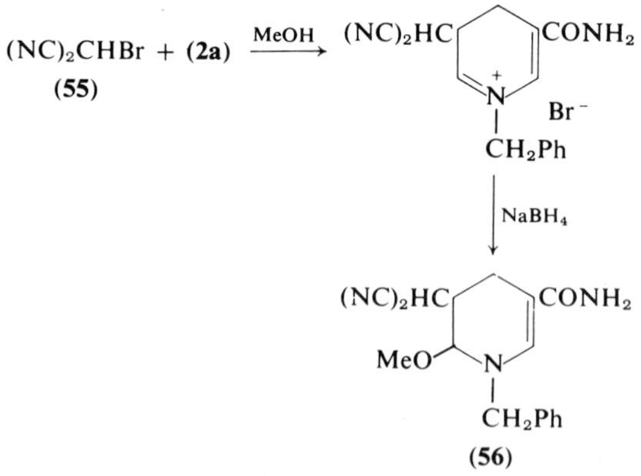

$$(\text{NC})_2\text{CHBr} + (2a) \xrightarrow{\text{MeOH}} (\text{NC})_2\text{HC}\underset{\overset{\textstyle |}{\underset{\textstyle CH_2Ph}{\overset{+}{N}}}}{\bigcirc}\text{CONH}_2 \quad \text{Br}^-$$

(55)

$$\Big\downarrow \text{NaBH}_4$$

$$(\text{NC})_2\text{HC}\underset{\overset{\textstyle |}{\underset{\textstyle CH_2Ph}{N}}}{\underset{\textstyle MeO}{\bigcirc}}\text{CONH}_2$$

(56)

Scheme 14

Kellogg [137] has reported a novel reaction of dihydronicotinamides with bromomalononitrile (55) (Scheme 14) in methanol; (2a) adds to (55) in enamine fashion. Reductive workup affords (56). This is the only known example of (2a) behaving as a enamine toward any electrophile other than a proton. In contrast, the Hantzsch ester (3b) reduces bromomalononitrile to malononitrile in 70% yield at room temperature.

TABLE 14

Radicals

Substrate	Reductant[a]	Reference
α,α-Diphenyl-β-picrylhydrazyl	(3)	158
	(2a)	45
Fenton's reagent	(1)	153
Hydrazyls	ns	88
β-Hydroxysuccinimyl	(1)	154
Ketyls	(2), (3)	155,156
Porphyrindine	(1)	153
Spirocyclohexylporpohyrexide	(1)	153
2,2,6,6-Tetramethyl-4-piperidone 1-oxide	(3)	96

[a] ns, not specified.

TABLE 15

Metal Ions and Metal Complexes

Substrate[a]	Reductant	Reference
1. Hemin	(2)	98
2. $Fe(CN)_6^{3-}$	NADH	72,157
3. Co(acac)$_3$	(2)	99
Fe(acac)$_3$	(2)	99
Mo(acac)$_3$	(2)	99
V(acac)$_3$	(2)	99
Ti(π-cp)$_2$Cl$_2$	(2)	99
V(π-cp)$_2$Cl$_2$	(2)	99
4. CuSO$_4$	(2a)	128
5. AgNO$_3$	(2)	145
6. Iron polythiocyanine	(2)	149

[a] acac, acetylacetone; π-cp, π-cyclopentadienide ligand.

RADICALS AND INORGANIC COMPOUNDS

A number of reductions of radical (Table 14) [45,88,96,153–156,158] and inorganic (Table 15) [72,98,99,128,145,149,157] species by NAD(P)H models have been reported. In many cases the reduced product was not reported, and in even more cases the fate of the dihydropyridine was not determined.

CONCLUSION

This survey has shown that the 1,4-dihydropyridine nucleus of the natural and model compounds is capable of a mechanistically diverse series of redox reactions. A change from an ionic to a radical process can sometimes be brought about by an apparently minor change in conditions. The distinction between hydride and electron transfer reactions, although they are very different in principle, is often difficult to establish in practice. This difficulty is even more evident in the enzymatic reactions in which a more complex set of processes is involved. The chemistry of the 1,4-dihydronicotinamide system suggests that the mechanisms of the enzymatic reactions could be very substrate dependent.

The use of 1,4-dihydronicotinic acid derivatives as reducing agents in synthesis is not yet developed. Their immediate potential would seem to be as mild homogeneous reagents for reductive dehalogenation of α-haloketones and as chiral reducing agents for carbonyl functions. In either function, the development of metal ion catalysis holds the greatest potential.

REFERENCES

1. N. O. Kaplan, in "The Enzymes" (P. D. Boyer, H. Lardy, and K. Myrbäck, eds.), 2nd ed., Vol. 3, p. 105. Academic Press, New York, 1960.
2. H. Sund, in "Biological Oxidations" (T. P. Singer, ed.), pp. 603 and 641. Wiley (Interscience), New York, 1968.
3. T. Bruci, in "Biorganic Mechanisms," p. 343. Benjamin, New York, 1966.
4. H. Sund, H. Diekmann, and K. Wallenfels, Adv. Enzymol. 26, 115 (1964).
5. F. H. Westheimer, Adv. Enzymol. 24, 441 (1962).
6. A. S. Mildvan, in "The Enzymes" (P. D. Boyer, ed.), 3rd ed., Vol. 2, p. 446. Academic Press, New York, 1970.
7. H. Sund, ed., "Pyridine Nucleotide Dependent Dehydrogenases," p. 39. Springer-Verlag, Berlin and New York, 1969.
8. U. Eisner and J. Kuthan, Chem. Rev. 72, 1 (1972).
9. K. Dalziel, in "The Enzymes" (P. D. Boyer, ed.), 3rd ed., Vol. 11, p. 2. Academic Press, New York, 1975.
10. F. H. Westheimer, H. F. Fisher, E. E. Conn, and B. Vennesland, J. Am. Chem. Soc. 73, 2403 (1951); J. Biol. Chem. 202, 687 (1953).

11. K. A. Schellenberg, *J. Biol. Chem.* **241**, 2446 (1966); **242**, 1815 (1967).
12. T. L. Chan and K. A. Schellenberg, *J. Biol. Chem.* **243**, 6284 (1968).
13. W. S. Allison, H. B. White, and M. J. Connors, *Biochemistry* **10**, 2290 (1970).
14. A. San Pietro, N. O. Kaplan, and S. P. Colowick, *J. Biol. Chem.* **212**, 941 (1955).
14a. K. Dalziel, *in* "The Enzymes" (P. D. Boyer, ed.), 3rd ed., Vol. 11, p. 10. Academic Press, New York, 1975.
15. G. A. Hamilton, *Prog. Bioorg. Chem.* **1**, 83 (1971).
16. J. P. Klinman, *J. Biol. Chem.* **247**, 7977 (1972).
17. J. W. Jacobs, J. T. McFarland, I. Wainer, D. Jeanmaier, C. Ham, K. Hamm, M. Wnuk, and M. Lam, *Biochemistry* **13**, 60 (1974).
17a. M. F. Dunn and J. S. Hutchison, *Biochemistry* **12**, 4882 (1973); M. J. Dunn, J. F. Biellmann, and G. Branlant, *ibid.* **14**, 3176 (1975).
18. L. Kurz and C. F. Frieden, *J. Am. Chem. Soc.* **97**, 677 (1975).
19. M. J. Adams, A. McPherson, M. G. Rossmann, R. W. Schevitz, and A. J. Wonacott, *J. Mol. Biol.* **51**, 31 (1970).
20. P. D. Boyer and H. Theorell, *Acta Chem. Scand.* **10**, 447 (1955).
21. N. O. Kaplan and H. S. Ramaswami, *in* "Pyridine Nucleotide Dependent Dehydrogenases" (H. Sund, ed.), p. 39. Springer-Verlag, Berlin and New York, 1969.
22. C-I. Branden, H. Jornavall, H. Eklund, and B. Furugren, *in* "The Enzymes" (P. D. Boyer, ed.), 3rd ed., Vol. II, pp. 107–190. Academic Press, New York, 1975.
23. J. Holbrook, A. Liljas, S. J. Steindel, and M. G. Rossman, *in* "The Enzymes" (P. D. Boyer, ed.), 3rd ed., Vol. 11, pp. 191–293. Academic Press, New York, 1975.
24. L. J. Banaszak and R. A. Bradshaw, *in* "The Enzymes" (P. D. Boyer, ed.), 3rd ed. vol. 11, pp. 369–396. Academic Press, New York, 1975.
25. E. L. Smith, B. M. Austen, K. M. Blumenthal, and J. F. Nyc, *in* "The Enzymes" (P. D. Boyer, ed.), 3rd ed., Vol. 11, pp. 294–368. Academic Press, New York, 1975.
26. H. Eklund, B. Nordström, E. Zeppezauer, G. Söderlund, I. Ohlsson, T. Bowie, and C.-I. Brändén, *FEBS Lett.* **44**, 200 (1974).
27. J. Hadju and D. S. Sigman, *J. Am. Chem. Soc.* **97**, 3524 (1975).
28. M. J. Adams, M. Buehner, K. Chandrasekhar, G. C. Ford, M. L. Hackert, A. Liljas, M. G. Rossmann, I. E. Smiley, W. S. Allison, J. Everse, N. O. Kaplan, and S. S. Taylor, *Proc. Natl. Acad. Sci. U.S.A.* **70**, 1968 (1973).
29. L. E. Webb, E. J. Hill, and L. J. Banaszak, *Biochemistry* **12**, 5101 (1973).
30. K. Wallenfels and W. Hanstein, *Angew. Chem., Int. Ed. Engl.* **4**, 869 (1965).
31. A. Shira and C. J. Suckling, *Tetrahedron Lett.* p. 3323 (1975).
32. K. S. V. Santhanam and P. J. Elving, *J. Am. Chem. Soc.* **95**, 5482 (1973).
33. C. O. Schmakel, K. S. V. Santhanam, and P. J. Elving, *J. Am. Chem. Soc.* **97**, 5083 (1975).
34. U. Bruhlmann and E. Hayon, *J. Am. Chem. Soc.* **96**, 6169 (1974).
35. C. O. Schmakel, K. S. V. Santhanam, and P. J. Elving, *J. Electrochem. Soc.* **121**, 345 (1974).
36. J. N. Burnett and A. L. Underwood, *J. Org. Chem.* **30**, 1154 (1965).
37. W. J. Blaedel and R. G. Haas, *Anal. Chem.* **42**, 918 (1970).
38. R. D. Braun, K. S. V. Santhanam, and P. J. Elving, *J. Am. Chem. Soc.* **97**, 2591 (1975).
39. J. Kuthan, J. Volke, V. Volkeová, and V. Šimonek, *Collect. Czech. Chem. Commun.* **39**, 3438 (1974).

40. J. Andrews, Ph.D. Thesis, Imperial College, London (1976).
41. P. Leduc and D. Thevenot, *J. Electroanal. Chem.* **46**, 543 (1973).
42. J. Stradins, G. Duburs, J. Beilis, J. Uldrikis, and A. F. Korotkova, *Khim. Geterotsikl. Soedin.* p. 84 (1972); J. Stradins, J. Beilis, J. Uldrikis, G. Duburs, A. E. Sausins, and B. Cekavicius, *ibid.* p. 1525 (1975).; J. Stradins, G. Duburs, J. Beilis, J. Uldrikis, A. E. Sausins, and B. Cekavicius, *ibid.* p. 1530.
43. H. Theorell and B. Chance, *Acta Chem. Scand.* **5**, 1127 (1951).
44. D. Abelson, E. Parthé, K. W. Lee, and A. Boyle, *Biochem. J.* **96**, 840 (1965).
45. D. Mauzerall and F. H. Westheimer, *J. Am. Chem. Soc.* **77**, 2261 (1955).
46. E. A. Braude, J. Hannah, and R. P. Linstead, *J. Chem. Soc.* p. 3257 (1960).
47. Y. Onnishi, M. Kagami, and A. Ohno, *Tetrahedron Lett.* p. 2437 (1975).
48. Y. Ohnishi, M. Kagami, and A. Ohno, *J. Am. Chem. Soc.* **97**, 4766 (1976).
49. R. Abeles and F. H. Westheimer, *J. Am. Chem. Soc.* **80**, 5459 (1958).
50. D. C. Dittmer and R. A. Fouty, *J. Am. Chem. Soc.* **86**, 91 (1964).
51. D. C. Dittmer, R. A. Fouty, and J. R. Potoski, *Chem. Ind.* (*London*) p. 152 (1964).
52. J. D. Sammes and D. A. Widdowson, *J. Chem. Soc., Chem. Commun.* p. 1023 (1972).
53. W. R. Frisell and C. G. Mackenzie, *Proc. Natl. Acad. Sci. U.S.A.* **45**, 1568 (1959).
54. D. C. Dittmer, L. J. Steffa, J. R. Potoski, and R. A. Fouty, *Tetrahedron Lett.* p. 827 (1961).
55. S. Shinkai and T. C. Bruice, *J. Am. Chem. Soc.* **94**, 8258 (1972).
56. S. Shinkai and T. C. Bruice, *Biochemistry* **12**, 1750 (1973); see also P. van Eikeren and D. L. Grier, *J. Am. Chem. Soc.* **98**, 4655 (1976).
57. K. A. Schellenberg and F. H. Westheimer, *J. Org. Chem.* **40**, 1859 (1965).
58. U. K. Pandit and F. R. Mas Cabré, *J. Chem. Soc. Chem. Commun.* p. 552 (1971).
59. D. J. Creighton and D. S. Sigman, *J. Am. Chem. Soc.* **93**, 6314 (1971); D. J. Creighton, J. Hajdu, and D. S. Sigman, *ibid.* **98**, 4619 (1976).
60. M. Shirai, T. Chrishina, and M. Tanaka, *Bull. Chem. Soc. Jpn.* **48**, 1079 (1975).
61. G. Di Sabato, *Biochemistry* **9**, 4594 (1970).
62. K. Wallenfels and D. Hofmann, *Tetrahedron Lett.* No. 15, p. 10 (1959).
63. J. J. Steffens and D. M. Chipman, *J. Am. Chem. Soc.* **93**, 6694 (1971).
64. Y. Ohnishi, T. Numakunai, and A. Ohno, *Tetrahedron Lett.* p. 3813 (1975).
64a. R. A. Gase, G. Boxhoorn, and U. K. Pandit, *Tetrahedron Lett.* p. 2889 (1976).
65. B. Kadis, *Abstr., 135th Meet. Am. Chem. Soc., Boston, 1959* p. 24 (1959).
66. K. Suzuki, M. Yasuda, and K. Yamasaki, *J. Phys. Chem.* **61**, 229 (1957).
67. K. Nishigama, N. Babu, J. Oda, and Y. Inouge, *J. Chem. Soc., Chem. Commun.* p. 101 (1976).
68. A. San Pietro, N. O. Kaplan, and S. P. Colwick, *J. Biol. Chem.* **212**, 941 (1955).
69. J. Ludowieg and A. Levy, *Biochemistry* **3**, 373 (1964).
70. G. Krakow, J. Ludowieg, J. H. Mather, W. M. Normore, L. Tosi, S. Udaka, and B. Vennesland, *Biochemistry* **2**, 1009 (1963).
71. G. Cilento, *Arch. Biochem. Biophys.* **88**, 357 (1960).
71a. T. J. van Bergen, T. Mulder, and R. M. Kellogg, *J. Am. Chem. Soc.* **98**, 1960 (1976).
72. M. J. Spiegel and G. R. Drysdale, *J. Biol. Chem.* **235**, 2498 (1960).
73. G. R. Drysdale, M. J. Spiegel, and P. Strittmatter, *J. Biol. Chem.* **236**, 2323 (1961).
74. E. A. Braude, J. Hannah, and R. P. Linstead, *J. Chem. Soc.* p. 3268 (1960).
75. D. J. Creighton, J. Hadju, G. Mooser, and D. S. Sigman, *J. Am. Chem. Soc.* **95**, 6855 (1973); J. Hajdu and D. Sigman, *ibid.* **98**, 6066 (1976).
76. N. Kito, Y. Ohnishi, M. Kagmai, and A. Ohno, *Chem. Lett.* p. 353 (1974).
77. J. Ludowieg and A. Levy, *Fed. Proc., Fed. Am. Soc. Exp. Biol.* **21**, 239 (1962).

78. L. A. Negievich, O. M. Grishin, and A. A. Yasnikov, *Ukr. Khim. Zh. (Russ. Ed.)*, **34**, 381 (1968); *Chem. Abstr.* **69**, 76221t (1968).
79. G. Cilento and P. Giusti, *J. Am. Chem. Soc.* **81**, 3801 (1959).
80. S. G. A. Alivisatos, F. Ungar, A. Jibril, and G. A. Mourkides, *Biochim. Biophys. Acta* **51**, 361 (1961).
81. A. Watkinson, D. C. Wilton, A. D. Rahimtula, and M. M. Akhtar, *Eur. J. Biochem.* **23**, 1 (1971).
82. M. A. Wilson and J. Cascarino, *Biochem. Biophys. Acta* **216**, 54 (1970).
82. T. P. Singer and E. B. Kearney, *in* "The Enzymes" (P. D. Boyer, H. Lardy, and K. Myrbäck, eds.), 2nd ed., Vol. 7, pp. 384–445. Academic Press, New York, 1963.
84. J. S. McGuire and G. Tompkins, *Fed. Proc., Fed. Am. Soc. Exp. Biol.* **19**, 29 (1960).
85. G. Duburs and J. Uldrikis, *Khim. Geterotsikl. Soedin.* No. 6, p. 83 (1970).
86. B. E. Norcross, P. E. Klinedinst, and F. H. Westheimer, *J. Am. Chem. Soc.* **84**, 797 (1962).
87. R. R. Roesler, *Diss. Abstr. Int. B* **31**, 594 (1970).
88. K. Wallenfels, W. Ertel, A. Hockendorf, J. Rieser, and K. H. Uberschor, *Naturwissenschaften* **62**, 459 (1975).
89. K. Wallenfels, W. Ertel, and K. Friedrich, *Justus Liebigs Ann. Chem.* **1973**, 1663 (1973).
90. Y. Ohnishi and M. Kagami, *Chem. Lett.* p. 125 (1975).
91. U. K. Pandit, J. B. Steevens, and F. R. Mas Cabré, *Bioorg. Chem.* **2**, 293 (1973).
92. H. Sund, H. Diekmann, and K. Wallenfels, *Adv. Enzymol.* **26**, 144 (1964).
93. M. Rabinovitz, *in* "The Chemistry of the Cyano Group" (Z. Rappoport, ed.), pp. 307–341. Wiley (Interscience), New York, 1970.
94. L. A. Negievich, O. M. Grishin, V. D. Pokhodenko, and A. A. Yasnikov, *Ukr. Khim. Zh. (Russ. Ed.)* **33**, 756 (1937); *Chem. Abstr.* **67** 107922n (1967).
95. K. Wallenfels and M. Gellrich, *Justus Liebigs Ann. Chem.* **621**, 149 (1959).
96. S. N. Zelenin, M. L. Khidekel', D. D. Mozzhukhin, E. N. Sal'nikova, and P. Kaikaris, *J. Gen. Chem. USSR (Engl. Transl.)* **37**, 1423 (1967).
97. E. A. Braude, J. Hannah, and R. P. Linstead, *J. Chem. Soc.* p. 3249 (1960).
98. K. Kano, T. Shibata, M. Kajiyara, and T. Matsuo, *Tetrahedron Lett.* p. 3693 (1975).
99. D. D. Mozzhukkin, M. L. Khidekel', O. N. Eremenko, E. N. Sal'nikova, and A. S. Astakhova, *J. Gen. Chem. USSR (Engl. Transl.)* **37**, 1416 (1967).
100. D. Bijl, H. Kainer, and A. C. Rose-Innes, *Naturwissenschaften* **41**, 303 (1954).
101. M. Shira, T. Koizumu, Y. Iguchi, and M. Tanaka, *Chem. Lett.* p. 915 (1975).
102. A. Ohno, T. Kimura, S. Oka, Y. Ohnishi, and M. Kagami, *Tetrahedron Lett.* p. 2371 (1975).
103. M. Brüstlein and T. C. Bruice, *J. Am. Chem. Soc.* **94**, 6548 (1972).
104. I. Isenberg, S. L. Baird, and A. Szent-Györgyi, *Proc. Natl. Acad. Sci. U.S.A.* **47**, (1961).
105. G. Blankenhorn, *Eur. J. Biochem* **50**, 351 (1975).
106. G. Blankenhorn, *Biochemistry* **14**, 3172 (1975); P. Hemmerich, *in* "Pyridine Nucleotide Dependent Hydrogenases" (H. Sund, ed.), p. 410. Springer-Verlag, Berlin and New York, 1970.
107. R. T. Proffitt, L. L. Ingraham, and G. Blankenhorn, *Biochim. Biophys. Acta* **362**, 534 (1974).
108. V. Massey, R. G. Matthews, G. P. Foust, L. G. Howell, C. H. Williams, G. Zanetti, and S. Ronchi *in* "Pyridine Nucleotide Dependent Hydrogenases" (H. Sund, ed.), p. 393. Springer-Verlag, Berlin and New York, 1970.
109. G. R. Penzer and G. K. Radda, *Q. Rev., Chem. Soc.* **21**, 43 (1967).

110. P.-S. Song, in Brustleim and Bruice [103, reference 3].
111. F. Muller and V. Massey, *J. Biol. Chem.* **214**, 4007 (1969).
112. T. Sakurai and H. Hosoya, *Biochem. Biophys Acta* **112**, 459 (1966).
113. V. Massey and G. Palmer, *J. Biol. Chem.* **237**, 2347 (1962).
114. V. Massey, and S. Ghisla, *Ann. N.Y. Acad. Sci* **227**, 446 (1974).
115. G. Haas, *Biochem. Z.* **290**, 291 (1937).
116. M. Dolin, *J. Biol. Chem.* **225**, 559 (1957).
117. D. J. T. Porter, G. Blankenhorn, and L. L. Ingraham, *Biochem. Biophys. Res. Commun.* **52**, 447 (1973).
118. T. C. Bruice, L. Main, S. Smith, and P. Y. Bruice, *J. Am. Chem. Soc.* **93**, 7327 (1971).
119. C. H. Suelter and D. E. Metzler, *Biochim. Biophys. Acta* **44**, 23 (1960).
120. D. B. McCormick, *in* "Molecular Associations in Biology" (B. Pullman, ed.), p. 337. Academic Press, New York, 1968.
121. S. Shinkai and T. Kunitake, *Chem. Lett.* p. 1113 (1974).
122. U. K. Pandit, R. A. Gase, F. R. Mas Cabré, and M. J. de Nie-Sarink, *J. Chem. Soc., Chem. Commun.* p. 211 (1975).
123. K. A. Schellenberg and G. W. McLean, *J. Am. Chem. Soc.* **8**, 1077 (1966).
124. R. W. Huffman and T. C. Bruice, *J. Am. Chem. Soc.* **89**, 6243 (1967).
125. T. Hino and M. Nakagawa. *J. Am. Chem. Soc.* **91**, 4598 (1969).
126. K. A. Schellenberg, G. W. McLean, H. L. Lipton, and P. S. Lietman, *J. Anm. Chem. Soc.* **89**, 1948 (1967).
127. U. K. Pandit, R. A. Gase, F. R. Mas Cabré, and M. J. de Nie Sarink, *J. Chem. Soc., Chem. Commun.* p. 627 (1974).
128. D. C. Dittmer and J. M. Kolyer, *J. Org. Chem.* **27**, 56 (1962).
129. J. A. Berson and E. Brown, *J. Am. Chem. Soc.* **77**, 447 (1955).
130. E. S. Huyser, J. A. K. Harmony, and F. L. McMillian, *J. Am. Chem. Soc.* **94**, 3176 (1972).
131. E. S. Huyser, C. J. Bredeweg, and R. M. VanScoy, *J. Am. Chem. Soc.* **86**, 4148 (1964).
132. E. S. Huyser and A. A. Kahl, *J. Org. Chem.* **35**, 3742 (1970).
133. C.-H. Wang, S. M. Linnel, and N. Wang, *J. Org. Chem.* **36**, 525 (1971).
134. E. M. Kosower, *J. Am. Chem. Soc.* **78**, 3497 (1956).
135. R. H. Abeles and F. H. Westheimer, *Fed. Proc., Fed. Am. Soc. Exp. Biol.* **15**, 675 (1956).
136. H. Diekmann, D. Hofmann, and K. Wallenfels, *Justus Liebigs Ann. Chem.* **79**, 697 (1964).
137. T. J. van Bergen and R. M. Kellogg, *J. Am. Chem. Soc.* **98**, 1962 (1976).
138. R. H. Abeles, R. F. Hutton, and F. H. Westheimer, *J. Am. Chem. Soc.* **79**, 712 (1957).
139. A. Ohno and N. Kito, *Chem. Lett.* p. 369 (1972).
140. J. C. Powers, Ph.D. Thesis, Harvard University, Cambridge, Massachusetts, (1958).
141. E. C. E. Hunter and J. R. Partington, *J. Chem. Soc.*, **87** (1933).
142. H. L. Donle and G. Volkert, *Z. Phys. Chem., Abt. B* **8**, 60 (1930).
143. R. J. Kill and D. A. Widdowson, unpublished results.
144. D. J. Bates, B. R. Goldin, and C. Frieden, *Biochem. Biophys. Res. Commun.* **39**, 502 (1970).
145. A. Brown and H. F. Fisher, *J. Am. Chem. Soc.* **98**, 5682 (1976).
146. J. L. Kurz, R. Hutton, and F. H. Westheimer, *J. Am. Chem. Soc.* **83**, 584 (1961).
147. R. J. Kill and D. A. Widdowson, *J. Chem. Soc., Chem. Commun.* p. 755 (1976).
148. A. H. Ingall, R. J. Kill, and D. A. Widdowson, unpublished results.

149. H. Inone, R. Aoki, and E. Imoto, *Chem. Lett.* p. 1157 (1974).
150. O. M. Grishin, Z. N. Parnes, and A. A. Yasnikov, *Izv. Akad., Nauk SSSR, Ser. Khim.* p. 1564 (1966).
151. E. C. Kooymann, *Discuss. Faraday Soc.* **19**, 163 (1951).
152. A. H. Ingall, Ph.D. Thesis, Imperial College, London (1973).
153. K. A. Schellenberg and L. Hellerman, *J. Biol. Chem.* **231**, 547 (1958).
154. P. C. Chan and B. H. J. Bielski, *J. Biol. Chem.* **250**, 7267 (1975).
155. M. L. Khidekel', A. S. Astakhova, M. F. Dmitrieva, S. N. Zelenin, G. I. Kozub, P. A. Kaikaris, and Y. A. Shvetsov, *J. Gen. Chem. USSR* (*Engl. Transl.*) **37**, 1407 (1967).
156. A. S. Astakhova and M. L. Khidekel', *Izv. Akad, Nauk SSSR, Ser. Khim.* p. 1909 (1964).
157. J. H. Quastel and A. H. M. Wheatley, *Biochem. J.* **32**, 936 (1938).
158. E. A. Braude, A. G. Brook, and R. P. Linstead, *J. Chem. Soc.* p. 3574 (1954).

CHAPTER

9

Models for the Role of Magnesium Ion in Enzymatic Catalysis of Phosphate Transfer and Enolate Formation

Ronald Kluger

INTRODUCTION

The study of the mechanism of organic reactions can be of particular biochemical significance. An area of special interest involves catalysis of reactions by metabolically important metal ions in aqueous solution. Although many metal ions are essential for metabolism, their functions are not clearly established in terms of organic reaction mechanisms. By contrast, many metal ions that are important catalysts in organic chemistry are not biochemically important. In this review, I deal primarily with catalytic patterns of magnesium ion in aqueous solution. This ion is metabolically significant [1,*2], and there are many cases in which it appears to catalyze organic reactions that can be related to enzymatic processes. Since the direct study of the function of magnesium ion in enzymatic catalysis has been considerably hindered by technical problems (primarily due to the complexity of the enzymatic environment and the necessarily low concentrations

* This book describes the dietary significance of magnesium in alleviating such diverse conditions as polio, epilepsy, weak teeth, cancer, alcoholism, and body odors. For a physiological but less entertaining perspective see Dowben [2].

of enzyme–magnesium complexes that are formed under normal circumstances), much of what is "known" about magnesium ion in enzymatic catalysis comes from extensions of organic model reactions [3†]. In order to limit the extent of this review, I have chosen to concentrate on two types of organic model reactions that appear to bear close analogy with enzymatic processes. These are (1) phosphate transfer and (2) enolate formation in decarboxylation and condensation reactions.

PHOSPHATE TRANSFERS

The general reaction equation for phosphate transfer involving magnesium ion is given in Eq. (1). The donor phosphate compound (D) loses the phosphate moiety to the acceptor (A) with the aid of enzyme (or another organic

$$
D—O—\overset{\overset{\displaystyle O}{\|}}{P}\overset{O^{\ominus}}{\underset{OH}{\diagdown}} + A—OH \underset{}{\overset{(Enz)}{\underset{Mg^{\textcircled{2+}}}{\rightleftharpoons}}} D—OH + A—O—\overset{\overset{\displaystyle O}{\|}}{P}\overset{O^{\ominus}}{\underset{OH}{\diagdown}} \tag{1}
$$

molecule) and magnesium ion. A common enzymatic reaction is that of kinase enzymes that utilize ATP as a donor [3]. Considerable study has been given to the question of where metal ions are coordinated to ATP at equilibrium [4–5,7]. Since magnesium ion is diamagnetic, no direct broadening of nuclear magnetic resonance (nmr) spectra of the substrate can be observed, making direct observation difficult. Virtually all potential binding sites have been shown to be significant at one time or other. The difficulty of working with ATP is further complicated by what must be a general consideration when dealing with any reactive assembly of species: Where does association of the metal ion occur *during* the course of reaction? Obviously, while magnesium ion may occupy a thermodynamically favorable position when coordinated to ATP in the absence of enzyme and acceptor, reaction may proceed through a complex or series of complexes whose structure is not the thermodynamically determined one. Therefore, any studies of phosphate transfer from ATP in the absence of enzyme may give complex results that might not bear on the enzymatic case. Alternative approaches to the problem have included modification of the structure of ATP [8] to encourage binding of metal ions in the absence of enzyme in a manner that will promote reaction and the study of structurally simpler phosphate donors in which the number of potential coordination sites is reduced.

Molecules simpler than ATP from which phosphate transfer can occur include acetyl phosphate, carbamyl phosphate, phosphoenolpyruvate, and

† This is a recent review, which encompasses part of the topic.

arginine phosphate. All have some metabolic significance as phosphate donors. Since my group's work has concentrated on acetyl phosphate, I have chosen to emphasize the role of magnesium with respect to this phosphate donor for this review. Supportive work on ATP and other donors is also noted.

Acetyl phosphate is important as an acyl donor in reactions such as in Eq. (2) [9]. These reactions proceed in the presence of enzymes that do not

$$
\underset{\substack{\parallel \ \parallel \\ CH_3COP}}{\overset{O \ \ O}{}} \underset{O^\ominus}{\overset{O^\ominus}{\diagdown}} + CoASH \; \rightleftharpoons \; CoA-S\overset{O}{\overset{\parallel}{C}}CH_3 + P_i \tag{2}
$$

require metal ions for activity. The alternative mode of reaction of acetyl phosphate involves its properties as a phosphate donor. For example, the reaction (3) catalyzed by acetate kinase [10] leads to the transfer of phosphate

$$
\underset{\substack{\parallel \ \parallel \\ CH_3COP}}{\overset{O \ \ O}{}} \underset{O^\ominus}{\overset{O^\ominus}{\diagdown}} + ADP \; \underset{}{\overset{Enz,\,(Mg^{(2+)})}{\rightleftharpoons}} \; CH_3C\overset{O}{\overset{\parallel}{}}O^\ominus + ATP \tag{3}
$$

from acetyl phosphate to ADP. The reaction in the presence of enzyme requires 1 equivalent of metal ion per equivalent of substrate. Furthermore, the enzymatic reaction that leads to hydrolysis of acetyl phosphate by transfer of phosphate to water involving acyl phosphatase as catalyst also utilizes magnesium ion [11]. As I noted earlier, ATP has many sites of coordination for magnesium as well as many potential and real phosphate cleavage points. By comparison, acetyl phosphate has only two modes of cleavage to be contended with, and there are relatively few potential coordination sites. Studies of the nonenzymatic reaction patterns of acetyl phosphate can be utilized to consider the functions that magnesium ion might fulfil in enzymatic reactions.

In order to elucidate the catalytic role of magnesium ion in the reactions of acetyl phosphate, a variety of approaches have been employed. A study of the kinetic effects of magnesium ion in the hydrolysis of acetyl phosphate was first performed by Koshland [12], extending Lipmann's pioneering observation that calcium ion promoted hydrolysis [13]. Koshland found that magnesium ion facilitated the hydrolysis of the dianionic form of acetyl phosphate but had no effect on the hydrolysis of the monoanion. On the basis of this observation, he proposed that, in the transition state of the magnesium ion-catalyzed decomposition of the dianion, magnesium finds acetyl phosphate as a bidentate ligand [reaction (4)] with the metal in a position analogous to that occupied by a proton in the hydrolysis of the monoanion. He postulated the role of magnesium to be in assisting the formation of a reactive

$$H_3C-\overset{\overset{\displaystyle O}{\parallel}}{C}\underset{\underset{\displaystyle O}{}}{\overset{\overset{\displaystyle Mg^{2+}}{}}{\cdots}}\overset{O^{\ominus}}{\underset{\underset{\displaystyle O}{\parallel}}{P}}-O^{\ominus} \longrightarrow H_3C-\overset{\overset{\displaystyle O}{\parallel}}{C}O^{\ominus}Mg^{2+}+ PO_3^{\ominus} \xrightarrow{\ H_2O\ } P_i \qquad (4)$$

metaphosphate species (PO_3^-) by promoting the departure of acetate as magnesium acetate. No additional evidence for the mechanism was presented, and Koshland noted that this was only a preliminary explanation.

Oestreich and Jones engaged in an extensive and thorough series of studies on the role of metal ions in the hydrolysis of acetyl phosphate. They studied the magnesium ion-catalyzed hydrolysis of acetyl phenyl phosphate [14]. This substrate is normally much less reactive than acetyl phosphate. Oestreich and Jones observed that magnesium ion significantly promoted the reaction leading to hydrolysis catalyzed by hydroxide. They also concluded that magnesium ion promotes attack of hydroxide on the anionic substrate by decreasing the electrostatic barrier to approach of these species. Later work by Cooperman [15] and Sigman [16] (*vide infra*) on other molecules that involve zinc ion-catalyzed phosphate transfer in more rigidly defined systems led to essentially the same conclusion. An interesting point raised by Oestreich and Jones concerns their observation that the rate constant for magnesium ion catalysis is less than twice that of proton catalysis, suggesting that no special polarization or chelation by magnesium ion is occurring.

Oestreich and Jones [17] also investigated the effect of metal ions on the hydrolysis of acetyl phosphate itself over the pH range 5.8–8.8. For the case of magnesium ion, they report that Eq. (5) describes the observed results,

$$k_o = k_u + k_c[\text{Mg}^{2+}] \qquad (5)$$

where k_c is the second-order rate constant first reported by Koshland [12]. However, these workers found that there is also a factor dependent on whether hydroxide or water is the attacking species. Thus, they suggest the rate expression (6). They utilize kinetic results to determine stability constants

$$\text{Rate} = k_u[\text{ACP}^{2-}]_{\text{free}} + k_3[\text{MgACP}] + k_4[\text{MgACP}][\text{OH}^-] \qquad (6)$$

for the metal–substrate complexes. On the basis of spectral data, Oestreich and Jones conclude that, in the equilibrium state, the metal bonds only to the phosphate portion of the substrate. They also conclude that their results are consistent with the metal catalyzing either C—O or P—O cleavage of the substrate.

Satchell and co-workers reinvestigated the magnesium ion-catalyzed hydrolysis of acetyl phosphate [18]. They showed that the pH dependence of the reaction rate constants is such that an additional term involving magnesium-coordinated hydroxide and metal-coordinated substrate is necessary. They suggest mechanisms in which magnesium promotes C—O cleavage of the anhydride by coordinating to the phosphate moiety. These workers measured the binding constant of magnesium ion to acetyl phosphate by the method that Burton developed for determining binding constants of magnesium and ATP [19]. Essentially, since 8-hydroxyquinoline is known to have a high affinity for magnesium, yielding a colored complex, by introduction of a competing substrate the equilibrium constant for the substrate can be determined from its effect on the magnesium–hydroxyquinoline complex concentration. Unfortunately, this method is not very accurate when the substrate binds magnesium ion weakly since 8-hydroxyquinoline has a very high affinity for magnesium, and it is difficult to observe significant changes. Satchell et al. found fault with the kinetic data of Oestreich and Jones [17]. Oestreich and Jones had extrapolated the binding constant for magnesium ion and acetyl phosphate dianion from kinetic data, obtaining a value of ~ 7 M^{-1} at 39°C. Satchell et al. report a directly determined value for K_{assoc} of more than 100 M^{-1}. They conclude that since Oestreich and Jones had severely underestimated K_{assoc} some incorrect approximations had resulted which led to erroneous kinetic parameters. When Kurt Nakaoka was studying a related problem in my laboratory, he noted that, if Satchell et al. are correct, then acetyl phosphate has a much higher affinity for magnesium than one could expect on the basis of other monophosphates of similar basicity (over an order of magnitude higher). This would imply that the carbonyl group of acetyl phosphate may play an important role in binding magnesium, leading to a higher affinity. However, Oestreich and Jones had spectroscopic evidence [14] that went against this conclusion. Therefore, Nakaoka determined the affinity constant for magnesium ion and a series of phosphates and phosphonates using the same method described by Burton [19,20] that Satchell et al. had employed. His results for other phosphates were in agreement with published values, but the value obtained for acetyl phosphate was close to the extrapolated by Oestreich and Jones [17] and differed by a large factor from the value of Satchell et al. Magnesium affinity correlated with basicity and appears to be independent of the presence of carbonyl groups. Furthermore, careful infrared and nmr studies confirmed [20,21] the conclusion that acetyl phosphate is likely not to undergo special coordination and therefore not to be involved in the ground state in a bidentate mode. Remarkably, in spite of their finding an abnormally high value for the binding constant of magnesium and acetyl phosphate, Satchell et al. [18] chose not to invoke a special binding mode and in considering possible mechanisms did not include bidentate structures.

Klinman and Samuel [22] studied the effect of magnesium ion on the site of cleavage of acetyl phosphate utilizing water enriched in ^{18}O. Koshland's original report [12] had proposed that P—O cleavage of the anhydride was occurring in the magnesium ion-catalyzed reaction. Oestreich and Jones pointed out that this was not necessarily the case [17] but provided no test to see if C—O cleavage occurs. Satchell *et al.* [18] had suggested that C—O cleavage rather than P—O cleavage was accelerated but provided no data to support this. Klinman and Samuel showed that the reaction in the presence of magnesium ion is complex and incompatible with the results of Satchell *et al.* However, their analysis reveals that magnesium ion accelerates the reaction of acetyl phosphate with water, which leads to C—O cleavage. Therefore, *in the absence of enzyme*, magnesium ion catalysis of the hydrolysis of acetyl phosphate does not provide information as to the mechanism of phosphorylation except in the indirect sense that whatever mode is found in the nonenzymatic case is not likely to be that in the enzymatic case involving phosphate transfer.

It was shown by Philip Wasserstein in our laboratory that if acetyl phosphate bound magnesium ion in the course of reaction solely to the phosphate portion of the molecule and this only neutralized charge, then acetylation rather than phosphorylation would be accelerated [23]. Wasserstein prepared dimethylacetyl phosphate, the fully neutral ester, and determined its reactivity patterns. He found that its rate of hydrolysis is considerably higher than any anionic form of acetyl phosphate. The rapid hydrolysis of the neutral compound occurs because reaction at the carboxyl moiety is facilitated in reactions with hydroxide and with water. This can be explained because, in the hydrolysis of acetates, the reaction is proportional in rate to the acidity of the

$$CH_3O-\underset{\underset{H_3CO}{|}}{\overset{\overset{O}{\|}}{P}}-O\overset{\overset{O}{\|}}{C}CH_3 \underset{k_{-1}}{\overset{k_1}{\rightleftharpoons}} CH_3-O-\underset{\underset{H_3CO}{|}}{\overset{\overset{O}{\|}}{P}}-O-\underset{\underset{OH}{|}}{\overset{\overset{HO}{|}}{C}}-CH_3 \overset{k_2}{\longrightarrow} \text{products} \quad (7)$$

conjugate acid of the leaving group [24]. In Eq. (7) $k_{obs} = k_1 k_2/(k_{-1} + k_2)$. Normally, $k_{-1} \gg k_2$ so that $k_{obs} = k_1 k_2/k_{-1}$. If k_1/k_{-1} is independent of the leaving group, then k_2 is rate determining. Jencks [24] showed that $\log k_2 = -\alpha p K_a$ so that, by conversion of phosphate to the much more weakly basic leaving group dimethyl phosphate, the rate of hydrolysis is increased. In the case of acetyl phosphate monoanion or dianion, addition of water to the carboxyl group is still as favorable to formation of a tetrahedral intermediate but is small compared to the rate constant for metaphosphate elimination mechanism, resulting in P—O cleavage due to the departure of acetate directly. Thus, charge neutralization by a metal ion in the case of a reaction proceeding via metaphosphate-type elimination is counterproductive. The results are readily interpretable on the basis of dimethyl phosphate being a better leaving group than monomethyl phosphate. Therefore, a role that magnesium ion is likely to have in this case is complexation to lower the pK_a of the leaving groups.* This decreases the enthalpic barrier to expulsion by weakening the C—O bond. The electrostatic barrier could not apply directly to the case of attack by water. Interestingly no acid-catalyzed reaction was observed. This is consistent with the function of acid catalysis is the unesterified compounds being improvement of the leaving ability of the phosphate group. The results obtained with acetyldimethyl phosphate show the utility of esterification as a model for metal ion coordination.

OTHER SUBSTRATES AND OTHER METALS

Selwyn observed that the ATPase of mitochondia requires metal ions for activation [25]. A plot of the ionic radius of those ions against relative activity of the enzyme showed that the radius is a key factor in controlling activity and that "fit" of the metal ion must therefore be important, suggesting bidentate coordinate. Tetas and Lowenstein [26] observed that Cu^{2+}, Zn^{2+}, and Mn^{2+} catalyzed the nonenzymatic hydrolysis of ATP at pH 5. At pH 8, these and other metals promoted an even faster hydrolysis, whereas in the absence of metals the rates at pH 5 and 8 are the same. This led Selwyn to suggest that the metal ions bind hydroxide and that this coordinated hydroxide serves as a nucleophile in the metal-ion-catalyzed hydrolysis of ATP. It should be noted that, in neutral solution, magnesium ion actually slows down the hydrolysis of ATP [26] whereas, in enzymatic cases, magnesium ion is a cosubstrate. At pH 9, complex and diverse reaction patterns are observed which can be explained by the metal ion binding to several sites of the complex substrate [25]. The choice of site is highly dependent on the size of the ion. Again, the most effective form occurs when the metal ion coordinates hydroxide so that it is set to act as a nucleophile.

Sigman and co-workers [16] found that the zinc complex of 1,10-phenanthroline carbinol is phosphorylated nonenzymatically by ATP [Eq. (8)]. On

$$\text{(8)}$$

* Hsu and Cooperman [24a] recently suggested that a role of zinc ion in promoting phosphate hydrolysis involves improving the leaving ability of phosphate. This is analogous to the suggestion we proposed to explain the reactivity of neutral acyl phosphates [23].

the basis of an examination of molecular models, it was concluded that, in this noncatalytic system (there is no turnover), reaction proceeds through a ternary complex in which neutralization of charge, enhancement of nucleophilicity, and alignment of reactants facilitate reaction [16]. These authors suggest that the metal ion serves as a template for reaction and may also improve the quality of the leaving group. However, the site of coordination of ATP to the complex could not be identified, and the relevance to catalytic systems is not certain.

Lloyd and Cooperman found that another stable zinc complex, pyridine-2-carbaldoxime, reacts with phosphorylimidazole to form the phosphorylated derivative of the metal complex and imidazole [27]. In the absence of the metal ion, no reaction between the two substrates occurs. Their conclusion is that, in this noncatalytic system, electrostatic barriers are decreased by the zinc ion, which also acts as a template for reaction. Inspection of molecular models indicated to these authors that the complex they propose would be strained but an enzymatic complex might not be. However, no direct evidence bearing on the question is presented.

Cooperman et al. also found that water and amines attack phosphorylimidazole anion but that carboxylates do not [15]. The metal ions Ca^{2+}, Mg^{2+}, and Zn^{2+} slowed the rate of attack by water and amines. Although these metal ions are cofactors for phosphate transfer enzymes, no catalysis is observed. Cooperman et al. attribute the lack of catalysis to the relative insensitivity of phosphate ester cleavage to charge polarization [28]. Since metal ions are not as good at polarizing charge as are protons [29], any effect must be negligible. Since there is no electrostatic barrier to be overcome, the function of the metal ion as a catalyst is lost. The lack of specificity of coordination in this model makes the results difficult to interpret. The general problem of where the metal ion is in the transition state of the reaction of interest is dealt with later in this review.

RECENT THEORIES OF CATALYSIS AND EXPLANATION OF P—O CLEAVAGE

Strain Activation

Spiro et al. proposed that metal ions could coordinate to a phosphate group in a manner in which the phosphate is a strained bidentate ligand, thus promoting reaction at phosphorus [30]. In an enzymatic reaction, binding energy from formation of the enzyme–substrate–metal complex allows formation of the strained chelate, in which reaction is now specifically directed to relieve strain. Binding energy is utilized to induce phosphorylation by a strain-relief pathway. In the case of acetyl phosphate, the "strained" mode is similar to that proposed by other workers [14–16] in suggesting that the function of the metal ion may be primarily electrostatic facilitation of nucleophilic attack. Our earlier work on acetyldimethyl phosphate indicated

$$
\underset{\substack{\| \\ \text{CH}_3\text{COP}}}{\overset{\text{O} \ \ \text{O}}{\underset{\ \ \ \|}{}}}\begin{array}{c} -\text{O}^{\ominus}\cdots \\ -\text{O}^{\ominus}\cdots \end{array}\text{Mg}^{\,2+}
$$

that a major effect of this type of coordination would be to make phosphate a better leaving group by decreasing its basicity [23]. Therefore, reaction at the carbonyl group rather than at phosphorus is likely to be enhanced, which is consistent with the observations of Klinman and Samuel [22]. We prepared a compound that is strained at phosphorus and whose charge is fully neutralized by esterification, acetylethylene phosphate, to see if reactivity could be

$$
\begin{array}{ccc}
\underset{\text{O}}{\overset{\text{O}}{\bigsqcup}}\!\!\!\text{P}\!\!-\!\!\text{Cl} + \underset{}{\overset{\oplus\ \ominus \ \ \|}{\text{H}_4\text{NOCCH}_3}} & \longrightarrow & \underset{\text{O}}{\overset{\text{O}}{\bigsqcup}}\!\!\!\text{P}\!\!-\!\!\overset{\text{O}}{\overset{\|}{\text{OCCH}_3}} & \overset{\text{O}_3}{\longrightarrow} & \underset{\text{O}}{\overset{\text{O}}{\bigsqcup}}\!\!\!\overset{\overset{\text{O}}{\|}}{\text{P}}\!\!-\!\!\overset{\text{O}}{\overset{\|}{\text{OCCH}_3}} & \quad (9)
\end{array}
$$

increased at phosphorus [Eq. (9)] [31]. This compound is much more reactive to hydrolysis than any other acyl phosphate in which the anhydride is not part of a ring. In acetylethylene phosphate, strain *external* to the anhydride linkage causes extremely rapid reaction to occur since the transition state for attack is expected to be less severely strained [32]. Thus, facilitated reaction occurs at phosphorus. This hypothesis was verified by noting that the pK_a of ethylenephosphoric acid is the same as dimethylphosphoric acid. Since reaction at the carbonyl group is sensitive only to leaving group basicity [23,24], the enhancement must arise from reaction at phosphorus. Therefore, Spiro's hypothesis [30] is reasonable if the energetics of the system permit the attainment of the proper mode of chelation without sacrifices that cancel any gain that is made; that is, the binding energy must be utilized directly along the reaction coordinate.

Stabilization of Pentacovalency

Benkovic and co-workers [33–35] proposed an explanation for magnesium ion catalysis of phosphate transfer in intramolecular reactions which utilizes portions of Spiro's hypothesis [30] but which removes some of the restrictive and unlikely energetic conditions placed on the nature of the binding mode. They studied reactions in which an internal carboxylate nucleophile reacts at phosphorus with great facility to displace an exocyclic phenoxide leaving group. This reaction is orders of magnitude faster than bimolecular processes. Any metal ion catalysis that promotes departure of phenoxide in the intramolecular reaction is assumed by necessity to occur by modification of the usual intramolecular path. The kinetic patterns observed were consistent with preequilibrium complexation of the metal ion followed by slow expulsion of the leaving group. The reaction is insensitive to electrostatic effects, so the metal ion catalysis that is observed must arise from another source. Benkovic et al. [33–35] propose that the pentacovalent intermediate expected to form on the normal reaction coordinate (or the corresponding transition state; an intermediate need not exist) is stabilized by coordination to the metal. Strain as such is not relieved or induced, but transition states are decreased in energy. (The original arguments are stated in terms of stabilization of an intermediate that of itself would not provide acceleration.) Benkovic et al. [33] recall Selwyn's consideration [25] of the importance of ionic radius in determining catalytic efficacy and note that it is important in the cases they studied.

MAGNESIUM ION CATALYSIS OF ENOLATE FORMATION

Another important catalytic function attributed to magnesium ion is in the formation of enolate species. A variety of enzymes that catalyze apparent "carbanion" formation in aldol condensations and decarboxylations have metal ion requirements that are either uniquely or best satisfied by magnesium ion [7,36,37]. Spector, for example, has reviewed the requirement for magnesium ion in the function of isocitrate lyase [38]. An extensive study of the role of magnesium ion in the function of malate synthase was reported by Eggerer [39,40] (recent reviews have been presented by Rose [41] and by Higgins et al. [42]). The malate synthase reaction involving condensation of acetyl-CoA onto glyoxalate appears to occur via carbanion formation catalyzed by the enzyme with the assistance of magnesium ion and the second substrate. Eggerer and Klette showed that proton exchange from acetyl-CoA did not occur in the absence of glyoxalate [40]. However, 0.1% exchange did accompany condensation. Pyruvate did not act as a cosubstrate, but it promoted the exchange reaction significantly. Eggerer and Klette proposed that combined acid–base catalysis was occurring. That is, the pyruvate or glyoxalate functioned effectively as a general base when acetyl-CoA was complexed to metal and enzyme.

Kosicki reported that magnesium ion appeared to function in oxaloacetate decarboxylase [Eq. (10)] by stabilization of the transition state leading to decarboxylation [43].* This was substantiated by his observation that deuteration at position 3 of pyruvate in D_2O was enhanced by this enzyme [Eg. (10,11)] In addition, a significant increase in rate of deuteration was found when magnesium ion without enzyme was added to a solution of pyruvate in D_2O-containing phosphate buffer [Eq. (11)].

$$^{\ominus}OCH_2CCO^- + Mg^{\textcircled{2+}} \rightleftharpoons {}^{\ominus}OC-CH_2C\overset{Mg^{2+}}{\underset{}{}}C=O \longrightarrow$$

$$H_2C=C\overset{Mg^{2+}}{\underset{}{}}C=O + CO_2 \quad (10)$$

$$H_3CCCO^{\ominus} + Mg^{\textcircled{2+}} \longrightarrow H_2C=C\overset{Mg^{\textcircled{2+}}}{\underset{}{}}C=O \overset{D_2O}{\longrightarrow} H_2\underset{D}{\overset{Mg^{\textcircled{2+}}}{C-C}}C=O \quad (11)$$

Pedersen demonstrated in 1948 that cupric ion catalyzed the bromination of neutral keto esters in the presence of acetate [44], although no complex could be demonstrated to exist directly between starting ester and the metal ion. Schellenberger *et al.* found that metal ions catalyze the enolization of α-keto acids [45], which is consistent with Kosicki's finding. These results suggested to us that if magnesium could facilitate enolization exchange by a well-understood pathway, the results could be applied to other enzymatic systems as a probe of the site of coordination. In order to study this process we chose a molecule that contains exchangeable hydrogens α to a carbonyl group and a phosphonate moiety that provides a binding site for magnesium ion:

$$CH_3CCH_2-P\overset{O^{\ominus}}{\underset{O^{\ominus}}{}} \quad (Mg^{\textcircled{2+}})$$

---

* It appears that this enzyme is in fact pyruvate kinase [43a]. Based on reported mechanistic information [43b] it appears that similar binding of oxaloacetate and phosphoenolpyruvate could promote both reactions:

The deuteration of this compound was found to be markedly enhanced in the presence of magnesium ion [20,46]. The hydroxide ion-catalyzed process appears to be accelerated by a factor of about 2000. The study of this molecule was complicated by the fact that it is itself a base ($pK \sim 7$) and can catalyze enolization of a second molecule in a bimolecular-type process. The rate of exchange reaches an apparent local maximum at the pK_a of the conjugate monoacid, suggesting that the most rapid reaction occurs in the cross-reaction of dianion and monoanion. This cross-reaction was also markedly enhanced by the addition of magnesium ion. A straightforward explanation of our observations is that magnesium ion stabilizes the incipient enolate and permits electrostatic repulsion between the base and substrate to be overcome.

As I noted earlier, many workers have suggested that a major function of metal ions in enzymes may involve reduction of electrostatic barriers that are present in reactions between anions. The enolization reactions of acetonylphosphonate provide interesting possibilities as means of probing electrostatic effects. General-base-catalyzed enolization involves initial proton abstraction from carbon to form an enolate ion. The encounter between a neutral substrate and base catalyst is not subject to an obvious electrostatic encounter effect. However, since an enolate is being formed, stabilization of the transition state leading to its formation may occur. If the substrate is anionic and the base is also anionic, then electrostatic effects of the metal ion could be important. In order to assess the relative importance of electrostatic effects compared to "Lewis acid" catalysis, Andrea Wayda investigated the rates of iodination of the monomethyl ester of acetonylphosphonate with a series of neutral and anionic base catalysts in the presence and absence of magnesium ion. She found that magnesium enhanced the rate of enolization of reactions catalyzed by charged and by uncharged bases [47]. However, the anionic bases were affected by a rate factor about 2-fold greater than the neutral bases. Thus, the catalytic effects of magnesium ion are divided approximately equally between electrostatic effects and others in this system. The situation with respect to nucleophilic attack by an anion on an anion may be quite different since, in order to add to an anionic center rather than abstract a proton from an adjacent carbon center, a considerably greater electrostatic barrier must be overcome.

Further data on the magnesium ion-catalyzed deuteration of methyl acetonylphosphonate has been obtained by Joseph Chan in our laboratory. If the function of the metal ion is stabilization of formation of the enolate ion, then formation of enolates resulting from proton abstractions at the 4 position and at the 2 position should both be catalyzed, consistent with the relative difference in acidities of the two positions. This was in fact what was observed [Eq. (12)].

$$\begin{array}{c} \text{Mg}^{2+} \\ \text{O}^{\ominus}\quad{}^{\ominus}\text{O} \\ \diagdown\;\diagup \\ \text{H}_2\text{C}{=}\text{C}\quad\text{P}{=}\text{O} \\ \text{CH}_2\quad\text{OCH}_3 \end{array}$$

$$\text{H}_3\text{CCCH}_2\text{P}\underset{\text{OCH}_3}{\overset{\text{O}^{\ominus}}{}} \qquad (12)$$

$$\begin{array}{c} \text{Mg}^{2+} \\ \text{O}^{\ominus}\quad{}^{\ominus}\text{O} \\ \text{H}_3\text{C}{-}\text{C}{=}\text{C}{-}\text{P}{=}\text{O} \\ |\qquad\quad\text{OCH}_3 \\ \text{H} \end{array}$$

CONCLUSIONS

Magnesium has been demonstrated to have a variety of catalytic effects on phosphate transfer and enolization. No evidence has been presented that makes any one mode an overwhelmingly clear-cut explanation for all cases. However, the functions of the ion as a Lewis acid, neutralizer of anions, improver of leaving groups, template for reactions, and stabilizer of pentacovalent transition states have received considerable experimental support. We have said that the most efficient and controlled use of the ion by an enzyme would involve an equilibrium substrate–metal binding mode that does not promote reaction (or inhibits reaction) but can be easily perturbed by an enzymatic environment to yield a binding mode that is catalytically productive [20]. Model systems providing a site for coordination and a site for catalysis that is nearby should continue to be suitable for studies of the role of the metal ion in analogy to enzymatic systems if turnover can occur.

The problems of understanding the catalytic role of magnesium ion in enzymatic reactions appear to have been amplified by the difficulties entailed in the study of the reactions of "simple" molecules. Although a good number of examples of nonenzymatic catalysis have been recorded, very little conclusive evidence has been obtained which actually details the mode of catalysis. In systems where magnesium is incorporated as a stable portion of the substrate via chelation, a noncatalytic reaction must result since the ion is a stoichiometric participant in the reaction and thus no turnover is observed. In systems where magnesium dissociates and associates freely, it is difficult to ascertain the location of the ion during the catalytic reaction by kinetic observations or by spectroscopic methods. However, if the reaction being studied is one whose mechanism is well understood *and* magnesium ion fits easily into these catalytic patterns, then reasonable estimates of the function of magnesium can be made. If the reaction occurs via a particularly enhanced internal pathway that is further promoted by magnesium, it is necessary to consider what function the metal ion may serve in this pathway [33–35]. A test of postulated modes of catalysis can be made by modifying the substrate via the introduction of ester groups and the bridging of ligands that mimic the proposed binding. A minimal criterion for some proposed modes of catalysis is that the function be alternatively accomplished by synthetic modification of the substrate. Thorough kinetic, spectroscopic, and synthetic studies must be combined to give a maximum amount of information. In my work, I hope to develop experimental procedures that will continue to distinguish among the many possibilities for catalysis.

ACKNOWLEDGMENTS

This review was written with the aid of a grant from the National Research Council of Canada. The work described from my laboratory was accomplished with the valuable collaboration of Philip Wasserstein, Kurt Nakaoka, Andrea Wayda, and Joseph Chan.

REFERENCES

1. J. I. Rodale, "Magnesium, the Nutrient that Could Change Your Life," Pyramid Books, New York, 1971.
2. R. M. Dowben, "General Physiology," pp. 67–70. Harper, New York, 1969.
3. S. J. Benkovic and K. J. Schray, in "The Enzymes" (P. D. Boyer, ed.), 3rd ed., Vol. 8, p. 201. Academic Press, New York, 1973.
4. R. Phillips, Chem. Rev. 66, 501 (1966).
5. J. H. Rytting, Chem. Rev. 71, 439 (1971).
6. C. M. Frey and J. E. Stuehr, J. Am. Chem. Soc. 94, 8898 (1972).
7. A. S. Mildvan, in "The Enzymes" (P. D. Boyer, ed.), 3rd ed., Vol. 2, p. 445. Academic Press, New York, 1970.
8. For example, B. S. Cooperman, Biochemistry 8, 5005 (1969).
9. F. Lipmann, Adv. Enzymol. 6, 242 (1946).
10. R. S. Anthony and L. B. Spector, J. Biol. Chem. 247, 2120 (1972).
11. A. L. Lehninger, J. Biol. Chem. 162, 340 (1946).
12. D. E. Koshland, Jr., J. Am. Chem. Soc. 74, 2286 (1952).
13. F. Lipmann and L. C. Tuttle, J. Biol. Chem. 153, 571 (1944).
14. C. H. Oestreich and M. M. Jones, Biochemistry 6, 1515 (1967).
15. G. J. Lloyd, C.-M. Hsu, and B. S. Cooperman, J. Am. Chem. Soc. 93, 4889 (1971).
16. D. S. Sigman, G. M. Wahl, and D. J. Creighton, Biochemistry 11, 2236 (1972).
17. C. H. Oestreich and M. M. Jones, Biochemistry 5, 2926 (1966).
18. P. J. Briggs, D. P. N. Satchell, and G. F. White, J. Chem. Soc. B p. 1008 (1970).
19. K. Burton, Biochem. J. 71, 388 (1959).
20. R. Kluger, P. Wasserstein, and K. Nakaoka, J. Am. Chem. Soc. 97, 4298 (1975).
21. K. Nakaoka, Undergraduate Honors Thesis, University of Chicago, Chicago, Illinois (1974).
22. J. P. Klinman and D. Samuel, Biochemistry 10, 2126 (1971).
23. R. Kluger and P. Wasserstein, Biochemistry 11, 1544 (1972).

24. J. F. Kirsch and W. P. Jencks, *J. Am. Chem. Soc.* **86**, 837 (1964).
24a. C.-M. Hsu and B. S. Cooperman, *J. Am. Chem. Soc.* **98**, 5652 (1976).
25. M. J. Selwyn, *Nature (London)* **219**, 490 (1968).
26. M. Tetas and J. M. Lowenstein, *Biochemistry* **2**, 350 (1963).
27. G. J. Lloyd and B. S. Cooperman, *J. Am. Chem. Soc.* **93**, 4883 (1971).
28. J. D. Chanley and E. Feageson, *J. Am. Chem. Soc.* **85**, 1181 (1963).
29. J. B. Breinig and M. M. Jones, *J. Org. Chem.* **28**, 252 (1963).
30. F. J. Farrell, W. A. Kjellstrom, and T. G. Spiro, *Science* **164**, 320 (1969).
31. R. Kluger and P. Wasserstein, *Tetrahedron Lett.* p. 3451 (1974).
32. F. H. Westheimer *Acc. Chem. Res.* **1**, 70 (1968).
33. J. J. Steffens, I. J. Siewers, and S. J. Benkovic, *Biochemistry* **14**, 2431 (1975).
34. J. J. Steffens, E. J. Sampson, I. J. Siewers, and S. J. Benkovic, *J. Am. Chem. Soc.* **95**, 936 (1973).
35. E. J. Sampson, J. Fedor, P. A. Benkovic, and S. J. Benkovic, *J. Org. Chem.* **38**, 1301 (1973).
36. I. A. Rose, *Adv. Enzymol.* **43**, 491 (1975).
37. L. B. Spector, *in* "The Enzymes" (P. D. Boyer, ed.), 3rd ed., Vol. 7, p. 357. Academic Press, New York, 1972.
38. J. A. Olson, *J. Biol. Chem.* **234**, 5 (1959).
39. H. Eggerer, *Biochem. Z.* **343**, 11 (1965).
40. H. Eggerer and A. Klette, *Eur. J. Biochem.* **1**, 447 (1967).
41. I. A. Rose, *in* "The Enzymes" (P. D. Boyer, ed.), 3rd ed., Vol. 2, p. 281. Academic Press, New York, 1970.
42. M. J. P. Higgins, J. A. Kornblatt, and H. Rodney, *in* "The Enzymes" (P. D. Boyer, ed.), 3rd ed., Vol. 7, p. 407. Academic Press, New York, 1972.
43a. D. J. Creighton and I. A. Rose, *J. Biol. Chem.* **251**, 69 (1976).
43b. D. J. Creighton and I. A. Rose, *J. Biol. Chem.* **251**, 61 (1976).
43. G. W. Kosicki, *Biochemistry* **7**, 4310 (1968).
44. K. J. Pedersen, *Acta Chem. Scand.* **2**, 252 (1948).
45. A. Schellenberger, G. Oehme, and G. Hubner, *Chem. Ber.* **98**, 1938 (1965).
46. R. Kluger and P. Wasserstein, *J. Am. Chem. Soc.* **95**, 1017 (1973).
47. R. Kluger and A. Wayda, *Can. J. Chem.* **53**, 2354 (1975).

10

Some Problems in Biophysical Organic Chemistry

Edward M. Kosower

INTRODUCTION

The development of physical organic chemistry over the past half-century has been so great as to create the illusion that all of the important facts are known and a certain malaise about what challenges remain for organic chemists in the future. However, for anyone who approaches complex systems or important questions on short time scales, the situation is that many high peaks remain to be scaled, not so much because they are there, but because of the interdependence of progress in one area on success in another. Without pretending that questions of high philosophy pervaded our work, but with the conviction that we should always probe as deeply as possible into any problem we encounter, I shall review briefly four questions that we tried to answer and what we learned in doing the work.

CAN A SUITABLE MODEL REACTION FOR NICOTINAMIDE ADENINE NUCLEOTIDE ENZYMATIC REACTIONS BE DEVISED?

In 1953–1954, when I was an NIH postdoctoral fellow with Professor F. H. Westheimer at Harvard, we sought to make thiol adducts of quaternary nicotinamide salts. Some of the spectroscopic changes noted were traced to a reaction of base with the ring (basic conditions were required). In order to understand the changes, the reaction of base with 1-methylpyridinium ion was examined. The only conveniently prepared 1-methylpyridinium salt was the iodide, and it was quickly found that the spectrum changed in absorption intensity with the square of the 1-methylpyridinium iodide concentration. The problem of the reaction of base with the ring was not further pursued. The new species was first interpreted as an adduct [1], but subsequent work carried out while I was on the staff of Lehigh University caused me to adopt the formulation of a charge-transfer complex for the new light-absorbing species [2] [Eq. (1)].

$$MPy^+ + I^- \rightleftharpoons MPy^+, I^- \xrightarrow{hv} MPy\cdot, I\cdot \tag{1}$$

The initial publication on the possibilities for charge-transfer complexation in biochemistry [3] following so soon after Mulliken's brilliant papers [4,5] caused numerous scientists to adopt this type of explanation without adequate evidence, a practice that luckily has not caused any permanent harm to the fields of either organic chemistry or biochemistry. In an effort to make the charge-transfer transition more easily observable, I prepared 1-ethyl-4-carbomethoxypyridinium iodide, and my first efforts at the University of Wisconsin were devoted to the discovery of the enormous solvent sensitivity of the charge-transfer absorption band. The positions of the maximum were adopted as measures of solvent polarity, and the Z values thus generated are now reasonably popular and useful parameters of solvent polarity [6,7]. In an attempt to correlate the position of the absorption band of 1-alkylpyridinium iodides with the one-electron reduction potential, a clever and skillful graduate student (Bill Schwarz) discovered the stability of 4-substituted 1-alkylpyridinyl radicals [8]. After I had moved to State University of New York at Stony Brook, Ed Poziomek succeeded in preparing useful quantities of 1-ethyl-4-carbomethoxypyridinyl radical and in studying its properties [9]. Irving Schwager made the first kinetic studies of the reaction and demonstrated directly the high reactivity of the pyridinyl radical toward tetrachloromethane [10] that was previously inferred by Westheimer in his work with

$$Py\cdot + CCl_4 \longrightarrow PyCl + \cdot CCl_3$$

$$Py\cdot + CCl_3 \longrightarrow PyCCl_3 \quad \text{(two isomers)} \tag{3}$$

$$PyCl \longrightarrow Py^+Cl^-$$

Hutton on the unexpected conversion of a dihydropyridine to a pyridinium salt in tetrachlormethane [Eqs. (2) and (3)] [11].

Harold Waits developed the procedure we currently use for the preparation of 1-ethyl-4-carbomethoxypyridinyl [Eq. (2)] [12]. Mohammad discovered the high reactivity of 4-nitrobenzyl chloride toward 1-ethyl-4-carbomethoxy-pyridinyl radical and was able to prove by use of the Z-value criterion that this was an electron transfer reaction [13] [Eq. (4)]. The usual reaction of the

$$
\begin{array}{ll}
\text{Py·} + \text{4-NBCl} \longrightarrow \text{Py}^+, \text{4-NBCl}^{\bar{\cdot}} & \\
\text{Py}^+, \text{4-NBCl}^{\bar{\cdot}} \longrightarrow \text{Py}^+\text{Cl}^- + \text{4-NB·} & \text{(4)} \\
\text{Py·} + \text{4-NB·} \longrightarrow \text{Py-4-NB} & \text{(two isomers)}
\end{array}
$$

radical with a halide (e.g., dibromomethane) is much less sensitive to the polarity of the solvent and is therefore an atom transfer reaction [Eq. (3)].

Ikegami and Itoh [14,15] prepared a number of pyridinyl diradicals, some of which show intramolecular complexation, as might have been expected

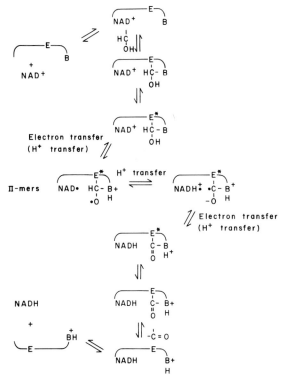

Fig. 1 Scheme illustrating a plausible one-electron pathway for the oxidation and reduction catalyzed by dehydrogenases and nicotinamide adeneine dinucleotides.

from the exciplexes discovered by Hirayama [16]. Hajdu showed that metal ions, particularly magnesium, promoted the intramolecular spectroscopic interaction [17] [Eq. (5)] and could be observed for as many as five or six methylene groups between the pyridinyl radicals.

$$(5)$$

Recent work by Chipman [18], Ohno [19], and Sigman [20] makes the intervention of pyridinyl radicals and/or charge-transfer complexes a reasonable possibility for enzymatic reactions of NADH. A possible scheme for the enzymatic reaction is shown in Fig. 1 [21,22]. Thus, although our efforts to approach the question of an appropriate model reaction led us quite far afield—our most recent work with Avraham Teuerstein concerns complexes of bispyridinium ions and pyridinyl radicals that exhibit an intervalence (charge-transfer) transition [23]—we believe that these studies will ultimately contribute to the solution of the original problem.

CAN ACETYLPHENYLHYDRAZINE BE AN INTRACELLULAR OXIDIZING AGENT FOR GLUTATHIONE?

One day in early 1965, my wife, Nechama S. Kosower, then on the faculty of the Albert Einstein College of Medicine in New York, asked how acetylphenylhydrazine could oxidize glutathione (GSH) to the disulfide GSSG. She had been inspired by the great Israeli physician Chaim Sheba to investigate questions related to the role of glutathione in certain genetic diseases. It was known that individuals who were unable to regenerate GSH from GSSG in their red blood cells were susceptible to hemolytic anemias after exposure to certain drugs (primaquine, for example) or certain foods (fava beans). The danger of fava beans had been recognized by Pythagoras, who warned his followers to abstain from (fava) beans. The drugs or foods apparently promoted the conversion of GSH into GSSG, a reaction that could not be reversed if the enzyme glucose-6-phosphate dehydrogenase were lacking. The question then became, Why is GSH important to the cell? To answer the question it was necessary to perturb the concentration of GSH within the cell (e.g., oxidize it to GSSG) and find out how the cell responded. Acetylphenylhydrazine had been used for this purpose as a "solid" form of the classic hemolytic agent phenylhydrazine. Thus, the question of how acetylphenyl-hydrazine could oxidize GSH to GSSG arose. The hypothetical reaction is shown in Eq. (6). After we noted that oxygen was required for the reaction

$$C_6H_5NHNHCOCH_3 + GSH \longrightarrow GSSG + ? \qquad (6)$$

to proceed, it became apparent that the true reaction was not as shown in Eq. (6) but perhaps that between a diazene and GSH [Eq. (7)].

$$C_6H_5N{=}NCOCH_3 + 2GSH \longrightarrow GSSG + C_6H_5NHNHCOCH_3 \qquad (7)$$

In connection with a problem that ultimately led to our work on mono-substituted diazenes [24], we had at hand methyl phenyldiazenecarboxylate. This reagent was tried and was an instant success as an intracellular GSH-oxidizing agent [25]. The ester, in a side reaction that my wife has put to good use in controlling the site of membrane damage caused by free radicals [26], hydrolyzes to the acid, which decarboxylates to air-sensitive phenyldiazene [Eq. (8)]. We have therefore made various amides as intracellular GSH-oxidizing agents, including "diamide," "DIP," and "DIP+2" [27,28]. The mechanism of the two-step oxidation has now been carefully elucidated by Walter Correa and Hanna Kanety-Londner [29,30] [Eq. (9)].

$$C_6H_5N{=}NCOOCH_3 \xrightarrow{H_2O} C_6H_5N{=}NCOOH \longrightarrow C_6H_5N{=}NH \qquad (8)$$

$$(CH_3)_2NCON{=}NCON(CH_3)_2 + GSH \longrightarrow (CH_3)_2NCON(SG)NHCON(CH_3)_2$$
$$\text{intermediate} \qquad (9)$$

$$\text{Intermediate} + GSH \longrightarrow GSSG + (CH_3)_2NCONHNHCON(CH_3)_2$$

The ability to oxidize mainly the GSH within cells has led us into many different biological and biochemical areas, including protein synthesis [31], the nature of interferon [32], neurotransmitter release in collaboration with my neurophysiologist friend Bob Werman [33,34], and even a theory of the molecular basis for learning and memory [35]. The latter suggests that disulfide bonds are the short-term form of memory in nervous systems. We have recently applied the GSH oxidation to measurement of the permeability of agents like DIP in the red blood cell membrane and have been able to find substantial differences among different species [36,37].

Although our answer to the original question is still uncertain, the application of straightforward physical organic principles has been helpful in creating a useful new way of probing certain biological questions about glutathione.

CAN IMPROVED POTENTIAL-INDICATING
DYES BE PREPARED ON THE BASIS OF
PHYSICAL ORGANIC PRINCIPLES?

Tasaki and his collaborators [38] introduced the idea of probing the dynamics of nerve action by means of a dye that changed in fluorescence with the potential change across the membrane. Tanizawa and I began an investigation of one of the classes of dyes that had given some hope that Tasaki's approach might become more useful if the fluorescence changes could be made larger. We examined the fluorescence behavior of 6-N-arylamino-2-naphthalenesulfonates (**1**) as a function of solvent polarity (the potential change might after all produce a change in the polarity of the dye binding site) and discovered that there were two types of fluorescence, one that could be assigned as arising from a naphthalene-localized state, $S_{1,np}$, and a second that came from a charge-transfer state, $S_{1,ct}$ [39]. Eventually, in collaboration with Michael Ottolenghi of Hebrew University, a student of his, Naomi Orbach, and an intelligent, hard-working student of mine, Hanna Dodiuk, we sorted out many of the problems to produce the scheme for the excited state shown in Fig. 2 [40]. We have more recently discovered that there are two isomeric charge-transfer states, $S_{1ct,(C)}$ and $S_{1,ct(U)}$, for a system like (**1**) [41], and Hanna Kanety [42] has good evidence that a similar pair of charge-transfer emitting states is responsible for some of the fluorescence observed for 8-N-arylamino-1-naphthalenesulfonate (**2**).

We still have hope that careful control of the substitution in an arylamino-naphthalenesulfonate or application of the principles we have learned about the relationship between quantum yield and the environment of the emitting

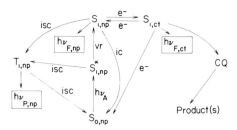

Fig. 2 Scheme illustrating the radiative and nonradiative processes for intramolecular donor–acceptor systems as exemplified by 2-N-arylamino-6-naphthalenesulfonates(ANS) (**1**). Symbols: S, singlet; T, triplet; np, nonplanar relationship of N-aryl group to naphthalene ring; ct. charge transfer; *, nonrelaxed state, used only for vibrational excitation in the scheme; $h\nu$, light absorbed; $h\nu_{F,np}$, light emitted by S_{1np}; $h\nu_{F,ct}$, light emitted by $S_{1,ct}$; isc, intersystem crossing between singlet and triplet; ic, internal conversion e⁻, electron transfer reaction; $h\nu_{p,np}$, light emitted by $T_{1,tp}$; CQ, product formed by chemical reaction, thus chemical quenching; vr., vibrational relaxation.

molecule might lead us to the design of a superior molecule for work in neurophysiology. In the meantime, we have developed some new and significant insight into the behavior of excited states, especially with respect to the occurrence of charge transfer.

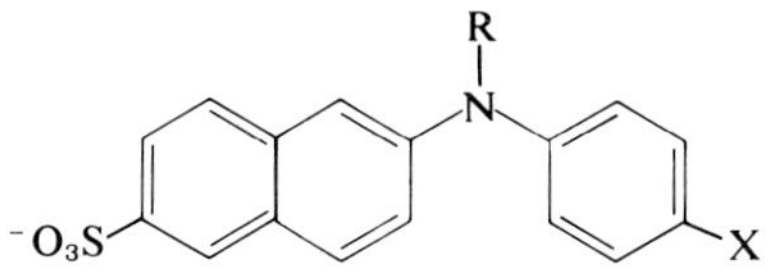

CAN NEUROTRANSMITTER RECEPTORS BE MADE TO MOVE?

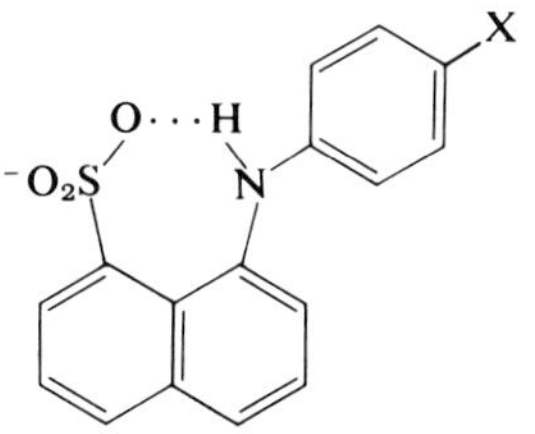

In a series of exciting conferences between myself and two neurophysiologists from Hebrew University (Bob Werman and Micha Spera), we discussed various ways in which physical organic chemistry could be useful to neurophysiology. One of the problems concerned the polymolecular requirement for certain neurotransmitters like acetylcholine and γ-aminobutyric acid [43]. We felt that it might be valuable to have a reagent that entered the membrane and helped the polymeric receptor (presumed to be responsible for the polymolecular requirement) to dissociate. At the time, we referred to the proposed agent as "membrane grease."

The design of the agent was based on fairly straightforward physical organic principles and conformed to the idea that there should be a hydrophobic part, a hydrophilic portion, and a connecting link. The hydrophobic part should contain a disordering element to favor mobility. A cyclopropane ring was selected rather than a double bond to avoid problems of sensitivity to oxygen. The hydrophilic portion was designed to be as small as possible, and the connecting link was an ester. The first compound thus written came to be called A_2C, as shown below. A_2C is still the most useful compound in

$$CH_3(CH_2)_7\overset{\displaystyle CH_2}{\overset{\diagup\diagdown}{CH-CH}}(CH_2)_7COOCH_2CH_2OCH_2CH_2OCH_3$$

$$A_2C$$

its class, thus extending the list of examples of biological agents in which the first one found is the best.

My wife Nechama, who had become associate professor of human genetics at Tel-Aviv University by that time, and I discussed at some length how we could best demonstrate the biological activity of A_2C. Eventually we chose the newly discovered phenomenon of cap formation [44]. (The reader should remember that our work began in the year that Singer and Nicolson brought out their model of the fluid mosaic membrane [45].)

The experiments were a success, and the agents were classified as membrane mobility agents [46]. We have demonstrated a difference in response to membrane mobility agent by malignant and normal cells [47] and have shown the usefulness of the agents in the process of cell fusion [48]. A comprehensive treatment of the cell fusion process has been prepared [49].

ACKNOWLEDGMENTS

I am grateful to the various agencies that supported much of the work mentioned, including the National Institutes of Health, the National Science Foundation, the Army Research Office (Durham, North Carolina), the U.S. Air Force Office of Scientific Research, the Petroleum Research Fund, the United States–Israel Binational Science Foundation (BSF), the Alfred P. Sloan Foundation, and Mr. Arthur Blumenfeld, New York.

Many students and postdoctorals carried out the work described. Their names have been mentioned in the text. Collaboration with A. J. Swallow of the Paterson Laboratories, Holt Radium Institute, Manchester, England, has been a pleasant part of the work on pyridinyl radicals but not cited specifically in the text.

Attempts to develop drugs on the basis of our glutathione work were carried out by an excellent Japanese co-worker, Tetsuto Miyadera [50,51], and have been continued by B. Pazenchevsky.

REFERENCES

1. E. M. Kosower, *J. Am. Chem. Soc.* **77**, 3883 (1955).
2. E. M. Kosower and P. E. Klinedinst, Jr., *J. Am. Chem. Soc.* **78**, 3497 (1956).
3. E. M. Kosower, *J. Am. Chem. Soc.* **78**, 3493 (1956).
4. R. S. Mulliken, *J. Am. Chem. Soc.* **74**, 811 (1952).
5. R. S. Mulliken, *J. Phys. Chem.* **56**, 801 (1952).
6. E. M. Kosower, *J. Am. Chem. Soc.* **80**, 3253, 3621, and 3267 (1958).
7. E. M. Kosower, "An Introduction to Physical Organic Chemistry." Wiley, New York, 1968.
8. W. M. Schwarz, E. M. Kosower, and I. Shain, *J. Am. Chem. Soc.* **83**, 3164 (1961).
9. E. M. Kosower and E. J. Poziomek, *J. Am. Chem. Soc.* **86**, 5515 (1964).
10. E. M. Kosower and I. Schwager, *J. Am. Chem. Soc.* **86**, 5528 (1964).
11. J. L. Kurz, R. Hutton, and F. H. Westheimer, *J. Am. Chem. Soc.* **83**, 584 (1961).
12. E. M. Kosower and H. P. Waits, *Org. Prep. Proced. Int.* **3**, 261 (1971).
13. M. Mohammad and E. M. Kosower, *J. Am. Chem. Soc.* **93**, 2713 (1971).
14. E. M. Kosower and Y. Ikegami, *J. Am. Chem. Soc.* **89**, 461 (1967).
15. M. Itoh and E. M. Kosower, *J. Am. Chem. Soc.* **90**, 1843 (1968).
16. F. Hirayama, *J. Chem. Phys.* **42**, 3163 (1965).
17. E. M. Kosower and J. Hajdu, *J. Am. Chem. Soc.* **93**, 2534 (1971).

18. J. J. Steffens and D. M. Chipman, *J. Am. Chem. Soc.* **93**, 6694 (1971).
19. A. Ohno and N. Kito, *Chem. Lett.* p. 369 (1972).
20. J. Hajdu and D. S. Sigman, *J. Am. Chem. Soc.* **97**, 3524 (1975).
21. E. M. Kosower, *Prog. Phys. Org. Chem.* **3**, 81 (1965).
22. E. M. Kosower, *in* "Free Radicals in Biology" (W. A. Pryor, ed.), Vol. 2, Chapter 1. Academic Press, New York, 1976.
23. E. M. Kosower and A. Teuerstein, *J. Am. Chem. Soc.* **98**, 1586 (1976).
24. E. M. Kosower, *Acc. Chem. Res.* **4**, 193 (1971).
25. N. S. Kosower, G. A. Vanderhoff, E. M. Kosower, and P.-k.C. Huang, *Biochem. Biophys. Res. Commun.* **20**, 469 (1965).
26. N. S. Kosower, Y. Marikowsky, B. Wertheim, and D. Danon, *J. Lab. Clin. Med.* **78**, 533 (1971).
27. N. S. Kosower, E. M. Kosower, B. Wertheim, and W. Correa, *Biochem, Biophys. Res. Commun.* **37**, 593 (1969).
28. E. M. Kosower, N. S. Kosower, H. Kanety-Londner, and L. Levy, *Biochem. Biophys. Res. Commun.* **59**, 347 (1974).
29. N. S. Kosower, K. R. Song, E. M. Kosower, and W. Correa, *Biochim. Biophys. Acta* **192**, 8 (1969).
30. E. M. Kosower and H. Kanety-Londner, *J. Am. Chem. Soc.* **98**, 3001 (1976).
31. N. S. Kosower, G. A. Vanderhoff, and E. M. Kosower, *Biochim. Biophys. Acta* **272**, 623 (1972).
32. N. S. Kosower and E. M. Kosower, *J. Mol. Med.* **1**, 11 (1975).
33. R. Werman, P. L. Carlen, M. Kushnir, and E. M. Kosower, *Nature (London), New Biol.* **233**, 120 (1971).
34. E. M. Kosower and R. Werman, *Nature (London), New Biol.* **233**, 121 (1971).
35. E. M. Kosower, *Proc. Natl. Acad. Sci. U.S.A.* **69**, 3292 (1972).
36. N. S. Kosower, G. Saltoun, and L. Levi, *Biochem. Biophys. Res. Commun.* **62**, 98 (1975).
37. N. S. Kosower, E. M. Kosower, and L. Levi, *Biochem. Biophys. Res. Commun.* **65**, 901 (1975).
38. I. Tasaki, E. Carbone, K. Sisco, and I. Singer, *Biochim. Biophys. Acta* **323**, 220 (1973).
39. E. M. Kosower and K. Tanizawa, *Chem. Phys. Lett.* **16**, 419 (1972).
40. E. M. Kosower, H. Dodiuk, K. Tanizawa, M. Ottolenghi, and N. Orbach, *J. Am. Chem. Soc.* **97**, 2167 (1975).
41. E. M. Kosower and H. Dodiuk, submitted for publication.
42. E. M. Kosower, H. Kanety, and H. Kanety, submitted for publication.
43. N. Brookes and R. Werman, *Mol. Pharmacol.* **9**, 571 (1973).
44. S. de Petris and M. C. Raff, *Nature (London), New Biol.* **241**, 257 (1973).
45. S. J. Singer and G. L. Nicolson, *Science* **175**, 720 (1972).
46. E. M. Kosower, N. S. Kosower, Z. Faltin, A. Diver, G. Saltoun, and A. Frensdorff, *Biochim. Biophys. Acta* **363**, 261 (1974).
47. S. Lustig, D. H. Pluznik, N. S. Kosower, and E. M. Kosower, *Biochim. Biophys. Acta* **401**, 458 (1975).
48. N. S. Kosower, E. M. Kosower, and P. Wegman, *Biochim. Biophys. Acta* **401**, 530 (1975).
49. E. M. Kosower, N. S. Kosower, and P. Wegman, *Biochim. Biophys. Acta* in press (1977).
50. E. M. Kosower and T. Miyadera, *J. Med. Chem.* **15**, 307 (1972).
51. T. Miyadera and E. M. Kosower, *J. Med. Chem.* **15**, 534 (1972).

11

Photoredox Reactions of Porphyrins and the Origins of Photosynthesis

D. Mauzerall

INTRODUCTION

The relation of structure to properties or function has always been a central problem of organic chemistry. It is the hope of molecular biologists to extend this approach to the complex structures of living matter. In this chapter I begin such an analysis for the process of photosynthesis. I overstress the work and ideas of my students and myself—not that they are necessarily the best; I simply know them best. A unified view of photosynthesis and an interpretation of the biosynthetic path of chlorophyll are presented. The photochemical properties of porphyrins are correlated with the proposed photosynthetic function. A simple theory to explain the structural requirements of the photon conversion unit is presented, followed by an indication of the future research in this field.

FUNCTIONAL DEFINITION OF PHOTOSYNTHESIS

Definition

Photosynthesis is that system which supplies the gradient of free energy necessary for all life on the earth. As opposed to the passive structure of crystals, the organization of living matter is that of active structures. Their organization is maintained in a steady state by a continual flow of matter and energy through them. The decrease in free energy during this process supplies the flow of entropy needed to stabilize these structures [1]. Although the emphasis of modern biological research has been on the flow of information necessary for the reproduction of these structures, it is the more general flow of entropy that is required for their existence.

Modern Gradient

At present, the free-energy gradient is composed of oxygen and reduced organic matter, i.e., foods. The process of respiration in most cells converts this gradient to adenosine triphosphate, the form of chemical energy useful to the cell. Photosynthesis continually replenishes this gradient by using the energy of solar photons to convert carbon dioxide and water, the products of respiration, to oxygen and reduced organic matter. It is interesting that, among the eight levels of oxidation–reduction between carbon dioxide and methane, nature works approximately midway at the fourth level, that of formaldehyde. Presumably this is a balance between the largest gradient of free energy and the chemical reactivity of foods.

Original Gradient

In the beginning, i.e., 3 billion years ago at the time of the origin of life, the environment differed in two ways from the present. First, the atmosphere was reducing [2] and, second, the protobiological environment was more aqueous. The latter difference arises from the reasonable assumption that life originated in the ocean and is discussed later. The reducing environment means that the useful free-energy gradient would have consisted of hydrogen-oxidized organic matter. Primeval photosynthesis would therefore have had to form this gradient from reduced organic compounds. It is usually argued [3] that the early forms of life were heterotrophic or chemosynthetic; i.e., the gradient of free energy lay in the disproportionation of the semioxidized organic material. This necessary material was assumed to be formed by the more esoteric, and much smaller (by a factor of $\sim 10^5$), sources of energy such as electrical discharges and solar ultraviolet (UV) light. If we allow that the level of self-replication was reached in this era, as is usually assumed, then the Malthusian argument can be applied. The inexorable power of exponential growth and its equally inexorable limitation by energy supply have been reaffirmed in recent times. I therefore argue that primeval photosynthesis was required at very early times for life processes to reach a useful steady state. Professor Granick has already taken the argument one step further and hypothesized a photoactive iron sulfide mineral as the photochemical element acting as an organizer for prebiotic reactions [4]. The suggestion is particularly striking since it was made before the iron–sulfur proteins, ferrodoxins, were discovered. Professor Krasnovskii has also argued for an early origin of photosynthesis [5].

BIOSYNTHETIC PATHWAY AND THE EVOLUTION OF PHOTOSYNTHESIS

The Sequence: Porphyrins → Metalloporphyrins → Chlorophylls

The extreme constancy of the biosynthetic pathway of hemes and chlorophylls is an excellent argument for the evolutionary unity of all living things. From the first committed precursor, δ-aminolevulinic acid, through the various steps of condensation to produce the natural isomer of the porphyrins, the pathway is identical in bacteria, plants, and man. Several good reviews of this knowledge are available [6–8]. Following the path of Professor Granick, we view this biosynthetic pathway as a window looking back into evolution. The first products fulfill a function in their time and are successively replaced, through mutation and selection, by molecules having more useful functions.

Various aspects of this argument have been presented elsewhere [9–9c], and I will give only a brief outline. The first porphyrin formed in the biosynthetic chain, uroporphyrin (Fig. 1), is very water soluble as befits its name. It readily

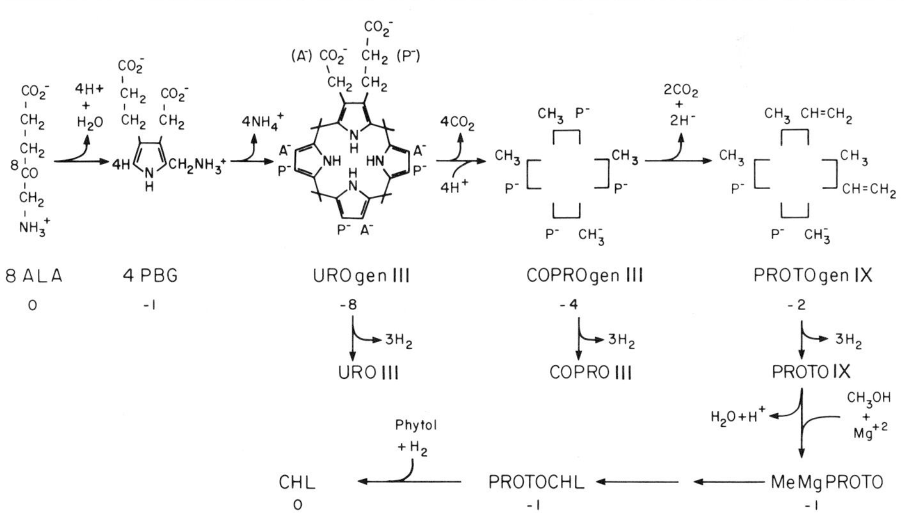

Fig. 1 Outline of the biosynthetic pathway of chlorophyll. The numbers below the abbreviations are the ionic charges per molecule in neutral solution (URO, uroporphyrin; COPRO, coproporphyrin; PROTO, protoporphyrin; gen refers to hexahydroporphyrin or porphyrinogen; PROTOCHL, protochlorophyllide; CHL, chlorophyll; PBG, porphobilinogen; ALA,γ-aminolevulinate). The Roman numerals refer to isomers, see e.g. ref. 6. (Reproduced with permission from Mauzerall [9c].)

oxidizes reduced organic compounds in anaerobic, neutral, aqueous solution when illuminated [9b,10]. For example, tertiary amines are oxidized to the carbonyl derivative and the corresponding secondary amine. The former would serve the function of oxidized organic material and of useful biogenetic molecules. The reduced porphyrin, urophlorin, thus formed is postulated to form hydrogen and porphyrin. Insufficient data on the thermodynamic values of porphyrins are available to know the exact position of this equilibrium. Experiments are planned to test this postulate.

The Prevalence of Uroporphyrin at Early Times

The particular isomer of uroporphyrin (uroporphyrinogen) used in biosynthesis, III (as shown in Fig. 1), is the most favored in a random cyclization of the pyrolic units, and thus its prebiotic origin is reasonable. Evidence that the closed macrocyclic porphyrinogen is the thermodynamically favored polymer [11] also supports this view. The thermodynamic data assembled by George [12] can be interpreted as showing that the formation of uroporphyrinogen from succinate and glycine at pH 7 is thermodynamically neutral. The aromatization to the porphyrin is highly favored, particularly for the reaction with oxygen. The fact that uroporphyrin is the basis of the structure of the ubiquitous vitamin B_{12} [6] and the finding of a tetrahydrouroheme as the prosthetic group of a sulfite reductase from a primitive bacterium [13] support the contention that uroporphyrin was prevalent in early biogenesis.

Evolution of Organized Structure

The movement from solution to organized structure is deduced from the steady increase in hydrophobicity along the pathway (Fig. 1). The ionic charge on the sequence of intermediates is uroporphyrin, -8; coproporphyrin, -4; protoporphyrin, -2; protochlorophyllide, -1; and chlorophyll, 0. The driving force for organized structures includes (a) the increased efficiency by optimizing the location of the reactants; (b) the possibility of photo electromotive forces if the structure ordered is in a membrane; and (c) the increased rate of the primary reaction, thus avoiding quenching by the concurrently discovered route to oxygen. I believe the momentous step to organized structure occurred between the level of magnesium porphyrins and chlorophyll (see below).

One can pinpoint the position on the biosynthetic pathway (Fig. 1) where test-tube chemistry is replaced by a more specific reaction; it is at the coproporphyrin to protoporphyrin stage. The thermodynamically favorable condensation and decarboxylation reactions of previous steps are replaced by an oxidative decarboxylation of a specific two out of four propionic acid residues. Enzymatic activity presumably entered at this step. It is just after this point that the first metalloporphyrin forms, a postulated turning point in the evolution of photosynthetic systems. It is interesting that the overall oxidation level of the complete pigment molecule rises as one goes along the chain to chlorophyll and bacteriochlorophyll. The fact that the latter macrocycles are reduced porphyrins has often been taken as indicating earlier biogenesis. The implied view that they are not is defended elsewhere [14].

THE MECHANISM OF PHOTOSYNTHESIS

Energy Conversion

The basic mechanism of photosynthesis is an electron transfer reaction between the chlorophyll excited state and the primary acceptor (or donor) [15]. Efficient energy conversion requires a barrier to the reverse electron transfer to the ground states which wastes the photon energy as heat. In the case of photosynthetic bacteria, this barrier amounts to 30 msec [15,16], allowing ample time for secondary electron transfers to occur. This barrier is explained by a simple theory of electron transfer between organic molecules based on quantum mechanical tunneling [9a,17]. It stresses that tunneling occurs at long distances and that the lifetime of the photoexcited state is finite. The electron tunneling is pictured as occurring from the excited state of the donor to a (virtual) level of the the acceptor. The electron essentially tunnels into a black hole since it and the corresponding positive hole are trapped by very rapid ($\leq 10^{-12}$ sec) relaxation processes. These rapid relaxations are caused by the complexity of the molecules under consideration. The 30 or so nuclei and the similar number of π electrons guarantee a large number of paths to the relaxed states. The probability of electron tunneling is given by $AK^2 \exp(-d/a)$. Here, A is a frequency related to the electron energy; for the present problem it is $\sim 10^{14}$ sec^{-1}. Because we are considering tunneling between π orbitals, the orientation factor K^2 enters. It and the exponential are essentially the overlap of the wave functions involved. The problem has some similarity to that of energy transfer [18] as stated by Hopfield [19]. However, I assume that the effect of "reorganization energy" is minimized in π systems. Experimentally, a very large number of organic aromatic ions undergo electron transfer reactions in solution with rates within a factor of 30 of the encounter-limited rate. Since the rotational relaxation time of a Stokes–Einstein molecule is the time required for it to diffuse a distance equal to its radius, the angular dependence of the electron transfer (K^2) may already account for one-tenth to one-third of this factor. The characteristic distance for the electron transfer, a, depends on the energy of the electron in the orbitals involved and thus weakly on the molecular environment. For carbon π orbitals and solvents with no empty levels close to those of the reactants, a is ~ 1 Å. For the reverse electron transfer to the ground state, a similar equation holds, but with a smaller characteristic distance, a'. The size of a' is determined by the energy levels of the stabilized charge-transfer state and thus by the trapping energy E_t. The trapping energy is the difference between the excited state of the system E and the thermalized charge-transfer state T (Fig. 2). Thus, the lifetime of the charge-transfer state of the system decreases as a strongly exponential function of distance between the donor and acceptor. However, the rate of formation of this state is an even more rapidly increasing function of the distance. This means that, at distances less than that where the formation of the charge-transfer state equals the natural lifetime of the excited state, the yield of the charge-transfer state rapidly rises to unity (Fig. 2). Thus, the yield–lifetime of the charge-transfer state will be greater than that of the original excited state only over a limited window along the distance axis. At short distances, although the yield of the charge-transfer state is unity, its lifetime rapidly becomes less than that of the original excited state. In other words, the excited state is

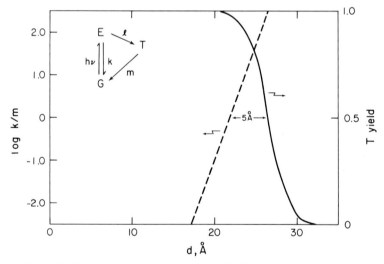

Fig. 2 The yield of the charge-transfer state T (solid line) and the logarithm of the lifetime of the T state normalized to that of the excited state E (dashed line) are plotted as a function of the distance d between the donor and acceptor (G denotes grand state). The orientation factor is assumed to be maximal; misorientation will move both curves toward smaller d. The numerical calculation uses an effective a (for l) of 1 Å and a' (for m) of 0.8 Å, and the lifetime is that of a triplet state, $k = 10^3$ sec^{-1}. Details are given in Mauzerall [17].

quenched, an all too common occurrence in solution-phase photochemistry and fully predicted by this simple theory. We have implicitly assumed that the electron spin decorrelation time is less than the charge recombination time. If this is not so, the lifetime of the triplet charge-transfer state may become limiting [17].

Evolutionary selection has optimized the parameters of distance and orientation in the reaction centers of photosynthesis. Because of the complex structure of this center, it is possible that the lifetime of the charge-transfer state at room temperature could be prolonged by a conformational change that decreases the coupling, e.g., by rotation of a molecule. The energy levels of the reactants are fixed by the molecules, and the trapping energy is optimized by the opposing requirements of longest lifetime (deeper trap) with maximum energy storage (shallower trap).

Photoredox Properties of Porphyrins and Metalloporphyrins

The redox properties of the ground state of metalloporphyrins have been well studied, and several excellent reviews are available [20]. The essential property is that both porphyrins and metalloporphyrins are amphoteric and undergo reversible one-electron oxidation and reduction to the corresponding cations and anions. We restrict ourselves to the π system, setting aside the variable-valence transition-metal chelates. Whereas the cations are quite stable, the anions require protection from oxygen and proton sources. The difference between the one-electron oxidation and reduction potential is constant at about 2.2 eV [21]. This is very close to the separation of the ground and first excited state, supporting the view that little reorganization energy occurs in these redox reactions. The metalloporphyrins are more easily oxidized and the porphyrins more easily reduced. This effect can be explained by considering that the metalloporphyrins carry a more negative charge, proportional to the electronegativity of the central substituent, than the unsaturated porphyrin free base [22].

The excited states of these pigments follow the same distinction in redox reactions [17]. But since one begins about 2 eV above the ground state, excited metalloporphyrins (specifically the closed-shell metal ions magnesium and zinc) are powerful reducing agents, and, conversely, excited porphyrins are strong oxidants. Porphyrins directly photooxidize many organic compounds [9b,10,23]. The initially formed porphyrin radical rapidly disproportionates to porphyrins and phlorin [10] in protonaceous solvents. The phlorins in turn can disproportionate to porphyrins and tetrahydroporphyrins and possibly on to the hexahydroporphyrins or porphyrinogens. These compounds are the first products of the biosynthetic pathway to the porphyrins (see the section on the sequence: Porphyrins $\rightarrow$ metalloporphyrins $\rightarrow$ chlorophylls). It is postulated that the reduced porphyrins, in the presence of a catalyst and/or light, could form hydrogen and the parent porphyrin.

The fact that the triplet state of zinc uroporphyrin reduces nicotinamide adenine dinucleotide phosphate to the free radical ($E_0 = -0.7$ V) with the same rate constant ($\sim 2 \times 10^8$ M^{-1} sec^{-1}) as it reduces ferricyanide ion ($E_0 = +0.4$ V) is a clear demonstration that this excited state is a powerful reductant. We estimate its redox potential to be at least -1.4 eV [24]. Simple, positively charged ring-N-substituted nicotinamides at 10^{-6} M oxidize all the triplet state of negatively charged zinc uroporphyrin. Thus, concentrated solutions are not required. Since the triplet state is formed in high yields from the singlet state of these porphyrins the overall quantum yield is high. A detailed study by Pat Carapellucci [24] of the coulombic effects on the reaction between the triplet state of zinc uroporphyrin and various acceptors led to the conclusion that the electron transfer occurs through about 10 Å of solvent water and has led to the electron tunneling mechanism described above (see the section on energy conversion).

OVERALL SCHEME

The scheme proposed for the evolutionary origin of photosynthesis can now be summarized. The first formed porphyrins photooxidized prevalent reduced organic and inorganic compounds and were recycled through the emission of hydrogen. The oxidized organic compounds may have been used directly in biogenesis. The formation of the magnesium (or zinc?) porphyrins allowed the direct photoreduction of intermediates to hydrogen. The resultant metalloporphyrin cations could also oxidize organic compounds, but possibly through intermediary (manganese) enzymes, because of the need of mechanistic matching of electron transfer to hydrogen transfer reactions. In this context it is interesting that Gaffron and Rubin [25] observed many years ago that algae can be induced to form and use hydrogen under anaerobic conditions. It is possible that the stability of the cations allowed a summation of the four electron transfers needed to form oxygen from water. With this literally epoch-making step, the modern era of photosynthesis was born. The larger free energy gradient eased the evolutionary path of multicellular organisms and led to modern mammalian chauvinism.

The reaction center of bacterial photosynthesis [26] contains no fewer than four bacteriochlorophyll and two bacteriopheophytin molecules. This rather large number of pigment molecules may be required for the highly efficient electron transfer. It is known that the electron spin left after transfer from the excited donor is spread over two of the bacteriochlorophylls [27,28]. The function of the bacteriopheophytins could be to provide slightly downhill way stations for the electron tunneling, thus providing for a further total transfer and favoring the forward path. Some evidence exists from picosecond spectroscopy for a transient bacteriopheophytin anion [29]. The bacteriopheophytin could also be a remnant of the primeval photosynthesis described above, directly using the magnesium-free pigment. Cellarius and Peters [30] observed that a small amount of bacteriopheophytin, $\sim 1\%$ of the bacteriochlorophyll, is made at the same time as the first bacteriochlorophyll formed in developing photosynthetic bacteria. Our understanding of this amazing product of three billion years of evolution is, it is hoped, at hand.

PRESENT AND FUTURE RESEARCH

The analytical stage of the study of photosynthesis is now ready to support the synthetic period. Several major discoveries have been made [31]: the isolation of bacterial reaction centers, the oscillations in the initial yield of oxygen per flash after dark adaption, and the importance of the electric field and ionic gradients across the thylakoid membranes. Similarly, the analysis of the photochemistry and mechanism of redox reactions of the porphyrin pigments is approaching a satisfactory state [9a,17]. We can now contemplate the synthetic stage of these studies, i.e., the attempt to assemble our knowledge into a coherent whole, to synthesize a photosynthetic unit. The first, stumbling steps along this path have already been taken. Our first attempt made use of the hydrophobic properties of chlorophyll. Pheophytin a (magnesium-free chlorophyll) was absorbed to polystyrene spheres of 50 nm radius in amounts ranging from isolated molecules to several layers [32]. A quantitative analysis of the two-dimensional energy transfer, of the aggregation using a one-dimensional Ising model, and of the sensitized photoreduction of an azo dye as a function of surface coverage led to the conclusion that energy transfer to reactive sites which included small aggregates or chain ends occurred. The quantum yield of the surface reaction was about the same as in free solution. Thus, this simple model showed the defining property of the *in vivo* photosynthetic unit: A large number of chlorophyll molecules cooperate to increase the optical cross section to solar photons and efficiently transmit the energy to the photoreactive site.

It is interesting that considerations similar to those advanced in the section on energy conversion also favor an optimum (finite) distance and orientation for the energy transfer to be efficient. Too weak an interaction, i.e., at too great a distance, leads to inefficient energy transfer. Too strong electronic interaction, i.e., too close, can lead to trapping and degradation of the energy. It seems that most if not all of the antenna pigments of photosynthetic systems are held in a protein matrix, presumably close to the optimum average interaction.

Another good model for the photosynthetic apparatus is the chlorophyll-containing lipid bilayer. Tien [33] and co-workers have studied the photo-electrical effects of the bilayer formed with extracts of chloroplasts. The bilayer structure is thought to be the basic element of the photosynthetic thylakoid and of other biological membranes. These workers and others have observed a wide variety of effects which depend on the redox properties and pH of the solutions on either side of the bilayer. Hong and I [9a,34] have used simple magnesium or zinc porphyrins in addition to chlorophylls to study the light-induced electron transfer across the lipid–water interface of a lipid bilayer. Our complete electrical equivalent circuit of the system allows the rate constants for the interfacial electron transfer reactions to be isolated. Two conclusions have been reached: (a) The bilayer structure together with asymmetric location of charge donors and acceptors allows vectorial photo-chemistry and thus the development of photopotentials as in the natural system. (b) The rise time of the electron transfer from the excited state is less than 100 nsec. Thus, this reaction successfully competes with the quenching by oxygen and so the experiment works in air, as does the photosynthetic system. Most solution-phase photoreactions occur through the triplet state and so, although useful for the primeval photosynthesis, are totally quenched by the present atmosphere.

These models still lack the rigid intermolecular distance and orientation that I believe are essential to the functioning photosynthetic unit. This detailed and rigid structure is most likely supplied by the protein component of the reaction center. It is now possible to contemplate the synthesis of such supermolecules by modern organic methods of synthesis, e.g., by the use of the Merrifield solid-phase peptide synthesis [35]. Of necessity, these first attempts at synthesis would be restricted to a fraction of even the small reaction center ($\sim 10^5$ daltons). Although the three-dimensional structure of a reaction center is not known, the theoretical considerations outlined above do suggest a form for this structure. In turn, these structural experiments could be used to test the theory. Norman Kagan in our laboratory has outlined such a synthesis and has built a specific face-to-face dimer of a porphyrin [36]. A second approach to the modeling of a photosynthetic unit is by the Langmuir–Blodget method of monomolecular films [37]. This method has been highly developed and refined in the laboratories of H. Kuhn [38]. They have done exquisite experiments showing organization and environmental effects on the optical properties of dyes. Some experiments on electron transfer in this ordered system have been interpreted with electron tunneling. Whitten and co-workers have shown some aggregation effects of concentrating the pigments in molecular films [39], and Ballard will be studying in our laboratory the photoeffects in pigmented molecular films. The thorough and carefully analyzed photoeffects of chlorophyll films on metal electrodes studied by

Tang and Albrecht [40], although not of the kind discussed here, have an inherent interest. These systems hold the promise of being able to quantitatively test the theoretical framework advanced above.

I hope that the challenge posed by the synthesis of such a macrostructure is sufficient to encourage active research in this field. Although one begins by the usual pastiche of attempting to copy nature, one is sure to end in a new world of possibilities.

NOTE ADDED IN PROOF

The book by E. Broda, "Evolution of the Bioenergetic Process," Pergamon, New York, 1975, contains an excellent summary of our knowledge of bioenergetics and presents an evolutionary argument similar to that given here.

ACKNOWLEDGMENT

The ideas on the evolutionary development of photosynthesis are the product of probing discussions with Dr. S. Granick. The experimental work was made possible by support from the NIH, NSF, and the Rockefeller University.

REFERENCES

1. I. Prigogine and G. Nicolis, *Q. Rev. Biophys.* **4**, 107 (1971).
2. S. L. Miller and L. E. Orgel, "The Origins of Life on the Earth," pp. 33–52. Prentice-Hall, Englewood Cliffs, New Jersey, 1974.
3. S. L. Miller and L. E. Orgel, "The Origins of Life on the Earth," pp. 175–190, and references cited therein. Prentice-Hall, Englewood Cliffs, New Jersey, 1974.
4. S. Granick, *Ann. N.Y. Acad. Sci.* **69**, 292 (1957).
5. A. A. Krasnovskii, *in* "The Origin of Life and Evolutionary Biochemistry" (K. Dose *et al.*, eds.), pp. 233–244. Plenum, New York, 1974.
6. A. R. Battersby and E. McDonald, *in* "Porphyrins and Metalloporphyrins" (K. M. Smith, ed.), pp. 61–122. Elsevier, Amsterdam. 1975.
7. G. S. Marks, "Heme and Chlorophyll," pp. 121–162. Van Nostrand-Reinhold, Princeton, New Jersey, 1969.
8. S. Granick, *in* "The Biochemistry of Chloroplasts" (T. W. Goodwin, ed.), Vol. 2, pp. 373–410. Academic Press, New York, 1967.
9. D. Mauzerall, *Ann. N.Y. Acad. Sci.* **206**, 483 (1973).
9a. D. Mauzerall and F. T. Hong, *in* "Porphyrins and Metalloporphyrins" (K. M. Smith, ed.), pp. 701–725. Elsevier, Amsterdam, 1975.
9b. D. Mauzerall, *Philos. Trans. R. Soc. London, Ser. B* **273**, 287 (1976).
9c. D. Mauzerall, *in* "Encyclopedia of Plant Physiology," New Series, Vol. 5, Photosynthesis (M. Arron and A. Triebs, eds.), pp. 117–124. Springer-Verlag, Berlin and New York (1977).

10. D. Mauzerall, *J. Am. Chem. Soc.* **84**, 2437 (1962).
11. D. Mauzerall, *J. Am. Chem. Soc.* **82**, 2601 (1960).
12. P. George, *Ann. N. Y. Acad. Sci.* **206**, 84 (1973).
13. M. J. Murphy, L. M. Siegel, H. Kamin, and D. Rosenthal, *J. Biol. Chem.* **248**, 2801 (1973).
14. D. Mauzerall, *in* "The Photosynthetic Bacteria" (R. K. Clayton and W. R. Sistrom, eds.). Plenum, New York (in press).
15. R. K. Clayton, *Annu. Rev. Biophys. Bioeng.* **2**, 131 (1973); W. W. Parson and R. J. Cogdell, *Biochim. Biophys, Acta* **416**, 105 (1975); K. Sauer, *in* "Bioenergetics of Photosynthesis" (Govindjee, ed.), pp. 115–181. Academic Press, New York, 1975.
16. J. D. McElroy, D. Mauzerall, and G. Feher, *Biochim Biophys. Acta* **333**, 261 (1974).
17. D. Mauzerall, *in* "The Porphyrins" (D. Dolphin, ed.). Academic Press, New York (in press); D. Mauzerall, Brookhaven Symp. Biol. No. 28, pp. 64–73 (1976).
18. R. E. Dale and J. Eisinger, *Biopolymers* **13**, 1573 (1974).
19. J. J. Hopfield, *Proc. Natl. Acad. Sci. U.S.A.* **71**, 3640 (1974).
20. J. Fuhrhop, *in* "Porphyrins and Metalloporphyrins" (K. M. Smith, ed.), pp. 593–623. Elsevier, Amsterdam, 1975; D. Dolphin and R. H. Felton, *Acc. Chem. Res.* **7**, 26 (1974); K. M. Kadish and D. G. Davis, *Ann. N. Y. Acad. Sci.* **206**, 495 (1973).
21. M. Zerner and M. Gouterman, *Theor. Chim. Acta* **4**, 44 (1966).
22. J. H. Fuhrhop and D. Mauzerall, *J. Am. Chem. Soc.* **91**, 4174 (1969).
23. D. Mauzerall, *J. Am. Chem. Soc.* **82**, 1832 (1960).
24. P. Carapellucci and D. Mauzerall, *Ann. N. Y. Acad. Sci.* **244**, 214 (1975).
25. H. Gaffron and J. Rubin, *J. Gen. Physiol.* **26**, 219 (1942).
26. S. C. Straley, W. W. Parson, D. Mauzerall, and R. K. Clayton, *Biochim. Biophys. Acta* **305**, 597 (1973).
27. J. R. Norris, M. E. Druyan, and J. J. Katz, *J. Am. Chem. Soc.* **95**, 1680 (1973).
28. G. Feher, A. J. Hoff, R. A. Isaacson, and L. C. Ackerson, *Ann. N. Y. Acad. Sci.* **244**, 239 (1975).
29. J. Fajer, D. C. Brune, M. S. Davis, A. Forman, and L. D. Spaulding, *Proc. Natl. Acad. Sci. U.S.A.* **72**, 4956 (1975).
30. R. A. Cellarius and G. A. Peters, *Biochim. Biophys. Acta* **189**, 234 (1969).
31. Govindjee, ed., "Bioenergetics of Photosynthesis." Academic Press, New York, 1975.
32. R. A. Cellarius and D. Mauzerall, *Biochim. Biophys. Acta* **112**, 235 (1966).
33. H. T. Tien, "Bilayer Lipid Membranes." Dekker, New York, 1974.
34. F. T. Hong and D. Mauzerall, *Proc. Natl. Acad. Sci. U.S.A.* **71**, 1564 (1974); *J. Electrochem. Soc.* **123**, 1317 (1976).
35. R. B. Merrifield and R. S. Hodges, *in* "Proceedings of the International Symposium on Macromolecules" (E. B. Maro, ed.), pp. 417–431. Elsevier, Amsterdam, 1975.
36. N. Kagan, Ph.D. Thesis, Rockefeller University, New York; N. E. Kagan, D. Mauzerall, and R. B. Merrifield, *J. Am. Chem. Soc.* **99**, 5484 (1977).
37. C. G. Suits and H. E. Way, eds., "The Collected Works of Irving Langmuir," Vol. 9, pp. 317–347. Pergamon, Oxford, 1961.
38. H. Kuhn, D. Möbius, and H. Bücker, *Tech. Chem. (N.Y.)* **1**, Part IIIB, p. 577–702 (1972).
39. F. H. Quina and D. G. Whitten, *J. Am. Chem. Soc.* **97**, 1602 (1975); F. R. Hopf, D. Möbius, and D. G. Whitten, *ibid.* **98**, 1584 (1976).
40. C. W. Tang and A. C. Albrecht, *J. Chem. Phys.* **62**, 2139 (1975).

12

Mechanisms of Enzymelike Reactions Involving Human Hemoglobin

John J. Mieyal

INTRODUCTION

Hemoglobin has been termed an inert oxygen-carrier hemoprotein [1]. Indeed, the chief physiological function of hemoglobin is to bind oxygen in the lungs and release it to the tissues, where it serves as the terminal electron acceptor in oxidative metabolic reactions. The oxygen-binding capacity of hemoglobin is modulated by several factors including pH, pO_2, oxidation state and conformation of hemoglobin, and concentration of the endogenous modifier 2,3-diphosphoglycerate. Because of the central importance of hemoglobin in physiology, it is probably the most extensively and intensely studied human protein. A vast literature has been amassed over the past century of scientific investigations concerning the physical and chemical properties, molecular structure and function, and comparative biochemistry of hemoglobins from various species. Review articles and reports from symposia on these topics have appeared regularly, and some of the more recent publications are cited in the reference list [2–7]. In contrast to those reviews, this chapter deals primarily with unusual reactions involving hemoglobin whose importance to its normal physiological function has yet to be fully evaluated. The reactions to be considered mimic those that are normally catalyzed by enzymes. They are of special interest to the author because elucidation of their mechanisms may provide new insights with regard to the reaction schemes of the normal enzymes; moreover, the characterization of so-called abnormal reactions often uncovers new aspects of normal functions and may also be predictive of possible pathophysiological mechanisms involving hemoglobin.

Although it has often been used as a model for cooperativity in enzyme-catalyzed reactions and has even been called an "honorary enzyme" for inclusion on the agendas of conferences on enzymes, hemoglobin itself is not considered to be an enzyme. Instead, hemoglobin and myoglobin are considered unique among the hemoproteins in their ability to bind oxygen reversibly. The reversibility of oxygen binding by oxygen-carrier hemoproteins, however, is a relative rather than an absolute quality. Human oxyhemoglobin undergoes slow autoxidation to ferrihemoglobin, even inside the erythrocytes. It is reconverted to the oxyferro form by the action of specific methemoglobin reductase enzymes [8–10]. The closely related hemoprotein myoglobin is more readily autoxidizable than hemoglobin, and shark hemoglobin is very rapidly autoxidized relative to mammalian hemoglobin [2,11,12]. In all cases the autoxidation process results in the conversion of oxygen to reduced forms (e.g., O_2^-, $\cdot OH$, H_2O_2), which may react further.

The dictum that hemoglobin is not an enzyme may also not be an absolute. Many enzymelike reactions have been reported to be mediated by hemoglobin. The oldest and most familiar of these are the peroxidaselike reactions that are catalyzed by hemoglobin when H_2O_2 and an oxidizable substrate are added [13]. This reactivity is enhanced when hemoglobin is complexed with haptoglobin [14] (see the section on peroxidaselike activity, below). In addition to oxidizing the usual peroxidase substrates like guaiacol, hemoglobin has also been reported to catalyze the H_2O_2-dependent oxidation of glutathione [15] and bilirubin [16]. Like catalase, hemoglobin can carry out the decomposition of H_2O_2 [13], and it can serve as a catalyst of lipid peroxidation [17]. Recently, it was reported [18] that hemoglobin, like cytochrome P-450, can utilize cumene hydroperoxide to catalyze the oxidation of NADPH. Hemoglobin has also been reported to accelerate various other reactions including the conversion of Tetralin hydroperoxide to tetralol [19], the decarboxylation of dihydroxyphenylalanine [20,21] (see the section on redox reactions involving hemoglobin, below), the dealkylation of certain aromatic N,N-dimethylamine N-oxides [22], the reduction of p-nitrobenzoic acid [23], and the hydroxylation of aniline [24–28]. The detailed molecular mechanism of none of these reactions has been fully elucidated, and, to the best of my knowledge, only in the last case has hemoglobin been demonstrated to fulfill all of the criteria for classification as an enzyme [27]. That reaction is discussed in detail in a later section (on, mixed-function oxidaselike activity). Nevertheless, the above listing of reactions precludes the classification of hemoglobin as "inert."

The hemoproteins hemoglobin, peroxidase, cytochrome P-450, and catalase all contain the same prosthetic group, protoporphyrin IX, and all of them can carry out peroxidatic reactions to some extent, yet their apparent physiological functions are quite different. Hemoglobin is an oxygen-carrier protein; P-450 and peroxidase are oxidative enzymes; and catalase decomposes H_2O_2 and may also function as an oxidase. The relative effectiveness of these hemoproteins in their designated functions may depend on a balance among various factors including the ease of electron transfer to and from the heme iron atom (or other moieties of the heme protein), the ability to bind O_2 reversibly, the ability to interact with H_2O_2 with minimum damage to the protein or heme prosthetic group, and the ability to bind substrates other than O_2 or H_2O_2. More detailed comparative discussions of these hemoproteins have been published previously [29,30]. It would appear that the categorization of the functions of these hemoproteins may also not be an absolute. For example, it has been suggested that the abundance of P-450 in certain tissues like liver may mean that it could also serve as a tissue oxygen-carrier protein like myoglobin [31]. In the following sections of this chapter, specific examples of hemoglobin-mediated reactions are discussed with regard to their probable mechanisms.

REDOX REACTIONS INVOLVING HEMOGLOBIN

General Considerations

It is conceivable that hemoglobin with its central iron atom might participate in various electron transfer reactions provided that the relative redox potentials of the reactants were appropriate. Whether such reactions could occur *in vivo* would depend on the accessibility of the oxidants or reductants to the hemoglobin that is contained within the erythrocytes. A common example of reversible oxidation and reduction of hemoglobin by an exogenous agent is illustrated by methylene blue, which when administered in high concentrations causes oxidation of ferrohemoglobin to ferrihemoglobin; this reaction has been employed as an antidote to cyanide poisoning [32]. At lower concentrations, methylene blue is reduced to leukomethylene blue by an NADPH-dependent electron transport chain in erythrocytes. The leukomethylene blue in turn can be reoxidized by reducing ferrihemoglobin to ferrohemoglobin. The latter reaction has been used to advantage in the treatment of acute cases of methemoglobinemia [32]. The interactions of various other exogenous redox agents with hemoglobin and its associated NADH- and NADPH-dependent electron transport systems in erythrocytes have been discussed in a recent comprehensive review by Kiese [11] along

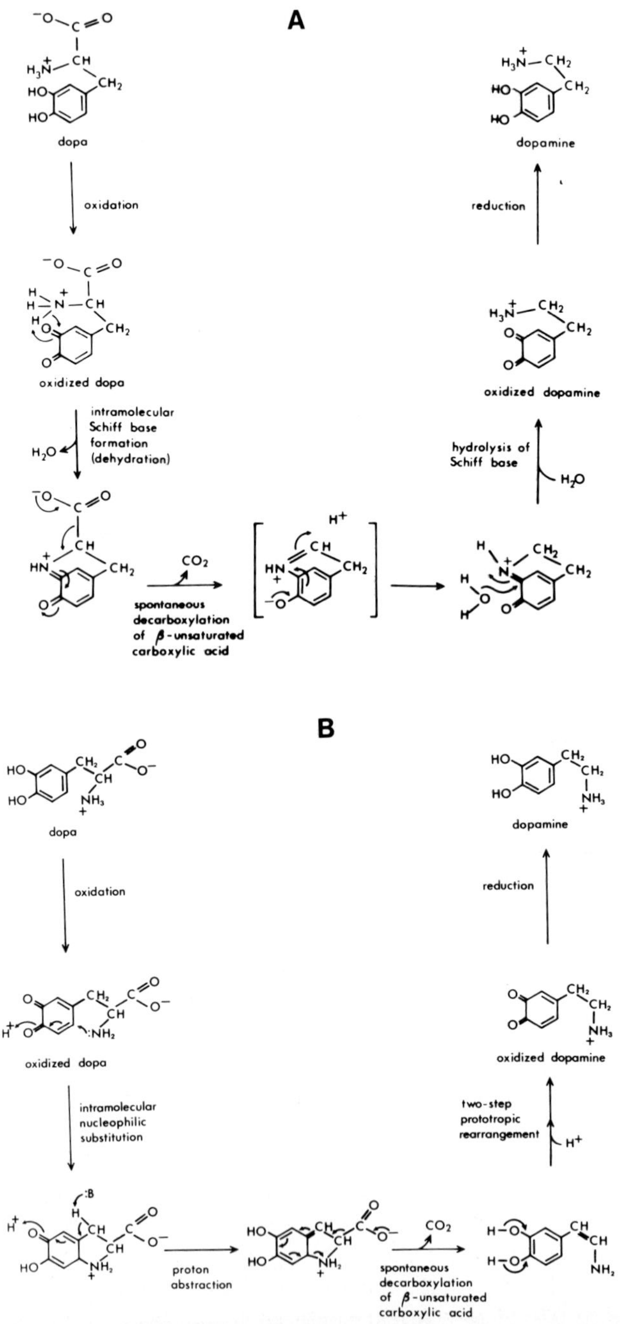

A

dopa

oxidation

oxidized dopa

intramolecular
Schiff base
formation
(dehydration)

spontaneous
decarboxylation
of β-unsaturated
carboxylic acid

dopamine

reduction

oxidized dopamine

hydrolysis of
Schiff base

B

dopa

oxidation

oxidized dopa

intramolecular
nucleophilic
substitution

proton
abstraction

spontaneous
decarboxylation
of β-unsaturated
carboxylic acid

dopamine

reduction

oxidized dopamine

two-step
prototropic
rearrangement

with the detailed mechanisms of ferrihemoglobin formation *in vivo* and *in vitro*. Therefore, this topic in general is not further discussed here. One example of this type of reaction, however, has generated particular interest because of its relationship to the therapy of a relatively common disease, Parkinson's disease (see next section).

Decarboxylation of Dopa

L-Dopa (L-3,4-dihydroxyphenylalanine), which is an endogenous metabolite of the amino acid tyrosine and a precursor of the neurotransmitter dopamine, has become an important drug in the treatment of Parkinson's disease [32]. This drug is normally converted to dopamine by the pyridoxal phosphate-dependent enzyme aromatic L-amino-acid decarboxylase (AADC). Several groups of investigators [20,21,32–34], however, have reported that dopa is also decarboxylated to some extent by lysates of human erythrocytes, and that this reaction cannot be attributed to the action of AADC. Thus, in contrast to the reaction catalyzed by AADC, the reaction in lysed erythrocytes is not stereospecific (i.e., both L- and D-dopa are substrates), it is not inhibited by antibodies directed against AADC, and it is apparently much less sensitive to heat inactivation. As the result of separate investigations, all of these research groups concluded that hemoglobin was involved in the reaction, but Tate *et al.* [20] implicated ferrihemoglobin, while Dairman and Christenson [34] and Yamabe and Lovenberg [21] concluded that oxyhemoglobin must be the form of hemoglobin involved in the reaction. Comparative data from their studies and hypothetical mechanisms for the decarboxylation reaction follow.

Decarboxylation of an amino acid is a reaction typically ascribed to pyridoxal phosphate enzymes wherein the pyridoxal moiety facilitates elimination of CO_2 by creating unsaturation in the form of a Schiff base moiety β to the carboxyl group [35]. In Fig. 1, two hypothetical mechanisms are proposed whereby oxidation of dopa to its *o*-quinone form provides the opportunity for development of unsaturation in the β position, which in turn may facilitate spontaneous decarboxylation of oxidized dopa. Although both schemes, for the sake of simplicity, are depicted as intramolecular reactions with respect to dopa, intermolecular reactions may be more

Fig. 1 Hypothetical reaction mechanisms for the hemoglobin-accelerated decarboxylation of 3,4-dihydroxyphenylalanine (dopa). (A) Scheme involving Schiff base formation. Note that, even though the reaction is depicted for simplicity as intramolecular with respect to dopa (oxidized dopa), it may actually occur more readily as an intermolecular reaction involving two molecules derived from dopa (see text). (B) Scheme involving ring closure at C-6. Note that development of β-unsaturation in this case requires proton abstraction from carbon (see text).

favorable, especially for scheme A (Fig. 1). Whether the proximate decarboxylated products are converted ultimately to dopamine would depend on the relative importance of various side reactions that may occur in the system and on the ease of reduction of oxidized dopamine (*vide infra*).

As alluded to above, there is a controversy concerning the form of hemoglobin that may be involved in accelerating the decarboxylation reaction (Table 1). Both Dairman and Christenson [34] and Yamabe and Lovenberg [21] cited the inhibition of the reaction by carbon monoxide as evidence that oxyhemoglobin (HbO_2) is responsible for the reactivity. It is also possible, however, that CO may inhibit autoxidative conversion of HbO_2 to ferrihemoglobin (Hb^{3+}). In addition, the latter investigators reported (Table 1) [21] that Hb^{3+} was virtually inactive. In contrast, Tate *et al.* [20] reported that Hb^{3+} was much more active than HbO_2, and they ascribed the low activity reported by Yamabe and Lovenberg to the presence of ascorbic acid in their test solutions. That conclusion was based on the finding by Tate *et al.* that 0.1 M ascorbate diminished by 10-fold the decarboxylation activity in their system containing ferrimyoglobin [20]. Examination of Table 1, however, shows that the discrepancy between the specific activities reported by the two groups of Hb^{3+} is almost 2000-fold. On the other hand, there is reasonably good agreement among the values reported by all three groups for systems containing HbO_2. Clearly, additional experimentation is required to resolve the disagreement.

TABLE 1

Decarboxylation of ^{14}C-Carboxyl-Labeled Dopa[a]

Form of hemoglobin	Specific activity (pmole CO_2/min/mg Hb)	Reference
Diluted intact erythrocytes (HbO_2)	~ 0.2[b]	34
Diluted lysed erythrocytes (HbO_2)	~ 6[b]	34
Diluted lysed erythrocytes + CO (Hb–CO)	0	34
"Purified" HbO_2	~ 8–19 (range of values reported)	21
"Purified" Hb^{3+}	< 0.1	21
"Purified" Hb–CO	< 0.4	21
"Purified" Hb^{2+}	< 0.3	21
Diluted lysed erythrocytes (HbO_2)	4.4	20
Diluted lysed erythrocytes treated with $K_3Fe(CN)_6$ (Hb^{3+})	187	20
Commercial Hb^{3+}	260, 373	20

[a] Measured as $^{14}CO_2$ released at 37°C, pH 7–7.2, in the presence of various forms of human hemoglobin.

[b] Values calculated from the data of Dairman and Christenson [34] which were reported in the form: nmoles CO_2/0.217 ml whole blood per 15 minutes.

Since both Hb^{3+} and HbO_2 can serve as electron acceptors, it is conceivable that both forms could mediate the oxidation of dopa. Two heme equivalents of Hb^{3+} would be converted to Hb^{2+} for each molecule of oxidized dopa formed. With HbO_2, electron transfer to O_2 could occur both from the dopa and from the heme Fe^{2+} atoms to give Hb^{3+}, H_2O_2, and oxidized dopa $[Hb^{2+}—(O_2)_4 + 2\ dopa \rightarrow Hb^{3+} + 4\ H_2O_2 + 2\ oxidized\ dopa]$.

Whether either of these reactions might involve a two-step oxidation of dopa via the semiquinone [36,37] remains to be elucidated. The latter reaction (transfer of electrons from Fe^{2+} of hemoglobin to O_2) actually describes autoxidation of HbO_2 to Hb^{3+}; if that were the case, one would expect an autocatalytic acceleration of CO_2 formation (decarboxylation of dopa) due to an increasing concentration of Hb^{3+}. Since none of the current studies [20,21,33,34] reported time courses for the release of CO_2 from dopa, the possibility of autocatalysis with HbO_2 cannot be assessed from presently available data.

Figure 2 shows the simplest scheme by which hemoglobin might mediate dopa decarboxylation. The net reaction is depicted as the conversion of dopa to dopamine and CO_2, with hemoglobin cycling between Hb^{3+} and Hb^{2+} to effect the conversion. Hence, hemoglobin is depicted as a true catalyst in the overall reaction, and only catalytic amounts should be required. In fact, Tate et al. reported that, when relatively high concentrations of Hb^{3+} were combined with dopa in air, dopa was decarboxylated but HbO_2 accumulated in essentially a stoichiometric way, while very little dopamine was formed [20,38]. This means that if either of the schemes in Fig. 1 approximates the mechanism by which dopa is decarboxylated, then at some point following CO_2 release, reactions other than those shown must predominate so that little dopamine is formed. It is not likely that the presence of O_2 (which ties up Hb^{2+} as HbO_2) is the only inhibiting factor, because when cytochrome c^{3+} was substituted for Hb^{3+}, cytochrome c^{2+} (which does *not* bind O_2) accumulated rather than serving to reduce oxidized dopamine [20]. Furthermore, it is known that quinones oxidize both HbO_2 and Hb^{2+} readily [39], so that the formation of HbO_2 should not be inhibitory to the last step in either of the proposed schemes.

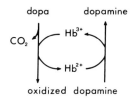

Fig. 2 Simplified representation of the role of hemoglobin as a redox catalyst in dopamine formation from dopa.

Neither of the proposed schemes in Fig. 1 is especially favorable for the complete cycle of dopamine formation. For scheme A, it can be learned from manipulation of either stick or space-filling molecular models that the Schiff base intermediates shown cannot assume a conformation in which all of the sp^2 carbon atoms lie in the same plane. Hence, if a Schiff base mechanism were operating, it would probably involve the coupling of the amine nitrogen of one molecule of dopa (or oxidized dopa) to the ring carbonyl carbon of a second molecule of oxidized dopa. Nevertheless, intermediates analogous to those drawn in Fig. 1A (albeit dimeric in nature) would be expected. The place in the proposed pathway where an abortive reaction would most likely occur is at the stage (shown in brackets) where a prototropic shift must take place in order to give the appropriate Schiff base to yield oxidized dopamine upon hydrolysis. If hydrolysis were to occur at that point (i.e., prior to the prototropic shift), products containing an amino group on the ring, rather than on the side chain, would accumulate.

In the case of scheme B (Fig. 1), the development of β-unsaturation requires the removal of a proton from carbon. That step would be more dependent on general base catalysis than formation of the Schiff base in scheme A. The crucial steps for dopamine formation in scheme B are also prototropic rearrangements indicated after the decarboxylation. The statement by Tate et al. [20] that melaninlike pigments may accumulate in their system could be taken as some support for scheme B, since the intermediate shown after decarboxylation can also serve as a precursor of 5,6-dihydroxyindole, which is normally incorporated into the melanin polymer in vivo [40]. However, before any conclusions about mechanism can be drawn, it must be established whether any dopamine at all is formed in the absence of pyridoxal phosphate. Only Yamabe and Lovenberg [21] omitted pyridoxal from their assay system, but they did not measure dopamine. Also, the method for assay of dopa employed by Tate et al. may not be sufficiently specific to distinguish dopamine from related compounds that may form in these mixtures. Specific experiments, especially those in which the major products other than CO_2 are identified, are required to distinguish among the hypothetical schemes that may be proposed for the hemoglobin-mediated decarboxylation of dopa.

In the above schemes, hemoglobin is shown acting only as a redox catalyst; if that is its only role, the effectiveness of cytochrome c^{3+}, $Fe(CN)_6^{3-}$, etc., in replacing it should relate only to their relative redox potentials. It is conceivable, however, that certain of the acidic and basic residues of the hemoglobin molecule could also participate in catalyzing the proton transfer steps shown above, and the heme metal iron could serve a second function by coordinating with the Schiff base nitrogen atom and thereby increasing its electron affinity. In any event, perhaps the most important distinction between this reaction involving hemoglobin and an enzyme-catalyzed reaction is that hemoglobin may not bind dopa in a specific fashion, or, if it does, the same molecule of dopa does not remain bound throughout a sequence of oxidation, decarboxylation, and rereduction to yield dopamine and regenerate Hb^{3+}; i.e., Hb^{3+} does not "turn over" under these conditions. Experimentation is currently being conducted in our laboratory to test these various possibilities, to evaluate the mechanisms of Fig. 1, and to reassess the relative role of Hb^{3+} and HbO_2 as catalysts of the proposed reaction scheme. Elucidation of this oxidative mechanism of decarboxylation of dopa may also be pertinent to understanding the mechanism of formation of the tissue pigment melanin, for which dopa is also a precursor [40].

PEROXIDASELIKE ACTIVITY

General Characteristics of Peroxidaselike Reactions Catalyzed by Hemoglobin

As alluded to above, the most commonly known enzymelike reactivity of hemoglobin is its ability to catalyze H_2O_2-requiring peroxidaselike reactions. In standard nomenclature this reactivity is designated donor:H_2O_2 oxido-reductase (EC $1 \cdot 11 \cdot 1 \cdot 7$). Although the detailed mechanism for this type of reaction has not been fully elucidated [41–43], it appears that the reaction carried out by methemoglobin may be analogous to the typical peroxidase-catalyzed reactions. As a result of numerous earlier studies [13,44–47], it has been learned that ferrihemoglobin (and myoglobin) can interact directly with H_2O_2 to form complexes that appear to be analogous to those observed for the so-called true peroxidase enzymes, i.e., Compounds I, II, and III [13,41–49]; further reaction with H_2O_2 may lead to decomposition of the H_2O_2 (catalaselike reactivity of hemoglobin [13]) and to the eventual degradation of the hemoglobin [13,48]. When typical peroxidase substrates are present (e.g., guaiacol, pyrogallol, phenols, and aromatic amines) along with H_2O_2, methemoglobin catalyzes their oxidation in a reaction that is dependent on the Hb^{3+} concentration and on the concentrations of both of the substrates, H_2O_2 and the oxidizable donor. Hemoglobin has also been studied as the catalyst of lipid peroxidation reactions [17,50–52], but Tappel [17,50] has suggested that the function of hemoglobin in these cases may be to accelerate homolytic cleavage of the lipid peroxides (or H_2O_2) to yield reactive free radicals and thereby initiate chain reactions. In a more recent study in which the reactivity of hemoglobin and soybean lipoxidase (which apparently is a nonmetallic protein) were directly compared [53], it was found that catalytic activity was retained in the hemoglobin-containing samples at pH 2 and pH 12 and after 10 min at 100° C, whereas the soybean lipoxidase activity was abolished under these conditions. This, the hemoglobin activity could be easily distinguished from the "true" enzymic lipoxidase activity. The remainder of this section is devoted to a discussion of how well hemoglobin mimics the peroxidases, and how the peroxidaselike activity of hemoglobin is altered by factors that modify its other properties.

Comparison of Hemoglobin and True Peroxidases

A recent study [54] has systematically tested hemoglobin as an enzyme in the catalysis of the H_2O_2-dependent oxidation of thiocyanate. It was found that the reaction was linearly dependent on the hemoglobin concentration up to ~ 2 μM (~ 8 nmoles of heme per milliliter). In most studies of hemoglobin-catalyzed peroxidaselike reactions, much higher concentrations have been employed which would be well beyond the region of hemoglobin concentration shown in Fig. 3, where the reactivity has already reached a plateau. Figures 4 and 5 show that typical Lineweaver–Burk plots can be obtained for the dependence of the reaction velocity on the concentrations of thiocyanate and H_2O_2, and the apparent K_m values are relatively low for both substrates; i.e., ~ 1.5 mM and ~ 3 mM, respectively. The reaction was also reported to be inhibited by the typical peroxidase inhibitors sodium azide and 3-amino-1,2,4-triazole. A temperature optimum between $40°$ and $45°C$ was reported, but whether some reactivity of hemoglobin was retained after heating at $\sim 100°C$ was not indicated.

Thermostability has typically been used as a criterion to distinguish between "true" peroxidase reactivity and pseudo peroxidase activity. It

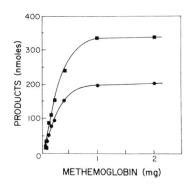

Fig. 3 Effect of methemoglobin concentration of peroxidase activity (■—■, sulfate production; ●—●, cyanide production). The incubation mixture contained in 3 ml of sodium acetate buffer, pH 5.0, 2 μmoles of KSCN, 2 μmoles of H_2O_2, and methemoglobin as shown. The incubation time was 1 min at 37°C. (From Chung and Wood [54]. Reproduced with permission from the *J. Biol. Chem.*)

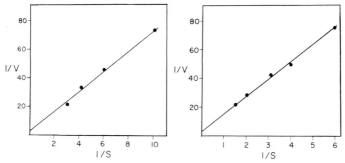

Fig. 4 Thiocyanate peroxidase activity of methemoglobin expressed as a double-reciprocal plot of velocity against concentration of thiocyanate. Thiocyanate concentrations are millimolar, and velocity is expressed as nanomoles of thiocyanate converted to sulfate per minute by 67 μg of methemoglobin at 37°C. The concentration of H_2O_2 was 0.67 mM, and the medium was 0.1 M sodium acetate, pH 5.0. (From Chung and Wood [54]. Reproduced with permission from the *J. Biol. Chem.*)

Fig. 5 Thiocyanate activity of methemoglobin expressed as a double-reciprocal plot of velocity against concentration of hydrogen peroxide. Hydrogen peroxide concentrations are millimolar, and velocity is expressed as nanomoles of thiocyanate converted to sulfate per minute by 67 μg of methemoglobin at 37°C. The concentration of thiocyanate was 0.67 mM, and the medium was 0.1 M sodium acetate, pH 5.0. (From Chung and Wood [54]. Reproduced with permission from the *J. Biol. Chem.*)

may turn out that the apparent thermostability reported for the reactivity of the nonperoxidase hemoproteins like hemoglobin may be due to the relatively high concentrations that are normally employed with these proteins. It would not be surprising, for example, if total loss of activity did occur at ~ 100°C for hemoglobin in the case under consideration (Fig. 3), provided that the test were carried out with a hemoglobin concentration within the region where the reaction velocity is linearly dependent on it. That concentration range is more akin to the range employed for the true peroxidases than it is to the much higher concentrations usually employed for nonprotein catalysts such as chelates of ferric iron.

In another study, the methemoglobin–H_2O_2 system was compared directly to the horseradish peroxidase–H_2O_2 system for the ability to produce free-radical intermediates from a variety of substrates. For all 17 substrates tested, the methemoglobin reactivity (+ or −) paralleled the peroxidase reactivity [55].

It seems from all of the foregoing information that hemoglobin should qualify as a peroxidase enzyme, at least qualitatively. On the other hand, its quantitative effectiveness as a catalyst of this type of reaction falls quite short of the "true" peroxidases. For example, the specific activity of hemoglobin for oxidation of thiocyanate was ~ 10^{-4} times that of lactoperoxidase [54].

TABLE 2

Relative Peroxidase Activities

Hemoprotein	Substrate[b]	Relative effectiveness[a]	Reference
Hemoglobin	Leuco TIP	1	56
Myoglobin	Leuco TIP	33	56
Uterine peroxidase	Leuco TIP	23	56
Lactoperoxidase	Leuco TIP	4.6×10^3	56
Horseradish peroxidase	Leuco TIP	4.6×10^6	56
Myoglobin	Guaiacol	1	13
Horseradish peroxidase	Guaiacol	2×10^3	13
Myoglobin	Pyrogallol	1	13
Horseradish peroxidase	Pyrogallol	3.2×10^4	13
Myoglobin	p-Cresol	1	13
Horseradish peroxidase	p-Cresol	100	13
Myoglobin	Adrenaline	1	13
Horseradish peroxidase	Adrenaline	400	13

[a] The relative effectiveness as peroxidase catalysts was expressed differently in the two references cited. In Nickel and Cunningham [56], the activities were expressed as "units of peroxidase activity per 5 μg iron," where a unit of activity was defined as the amount of catalyst that caused a $\Delta A_{675\,nm}$ of 1.0 per minute at 30°C; the activity for hemoglobin was normalized to 1.0 for this table. In Kielin and Hartree [13], the hemoproteins were compared according to their values of pC, i.e., "the negative logarithm of the haematin molarity necessary to cause 50% of H_2O_2 generated under standard conditions [by glucose plus glucose oxidase] to be utilized in the coupled oxidation." For this table, the antilog of the difference between the pC values for myoglobin and horseradish peroxidase is given relative to 1.0 for myglobin.

[b] Leuco TIP is leuco-2,3',6-trichloroindophenol.

Comparisons of hemoglobin with other peroxidases for various reactions are given in Table 2. In most cases, the effectiveness of hemoglobin or myoglobin as a peroxidaselike catalyst is at least two orders of magnitude less than the peroxidase enzymes. The only exception is the pig uterine peroxidase [56], which appears to be a relatively poor catalyst. In contrast to these results, we [27] and others [18,57] have found that the activity of hemoglobin in catalyzing certain O_2-requiring and organic peroxide-utilizing reactions is comparable to that of the oxidase enzyme cytochrome P-450 (see the section on mixed-function oxidaselike activity).

Enhancement of the Peroxidatic Activity of Hemoglobin by Haptoglobin

Shortly after the peroxidaselike activity of hemoglobin was described, it was reported that complexation with haptoglobin enhanced that activity [58]. The enhanced activity was the basis for the discovery of haptoglobin [58], and it has also become a convenient assay for that serum protein [14]. The magnitude of the enhancement (if any) is quite dependent on the conditions of assay, including pH, concentration and nature of the oxidizable substrate, and concentration of H_2O_2 [14,59]. Although numerous studies of the haptoglobin–hemoglobin interaction have fairly well established the nature of the complex as a "sandwich," i.e., $\beta\alpha$–Hp–$\alpha\beta$, whose formation is apparently dependent on the dissociation of liganded tetrameric hemoglobin into its α,β-dimers (see the recent reviews of Putman [60] and Sutton [61]), the basis for the enhanced peroxidase activity of this complex has not been finally resolved. There may be several possible explanations. For example, (a) the dissociation of the tetrameric structure into dimers might be a sufficient reason for the enhanced activity; (b) haptoglobin may simply protect the hemoprotein from degradation under the usual conditions of the peroxidase assay (i.e., pH 4, H_2O_2, guaiacol); (c) binding of the haptoglobin might elicit conformational changes in the globin moiety which would alter the local environment of the heme iron atom. There is available evidence pertinent to each of these postulates.

With regard to the dissociation phenomenon (a), it is noteworthy that the monomeric myoglobin has a substantially greater peroxidase activity than hemoglobin (e.g., Table 2). It should also be pointed out, however, that Keilin and Hartree [13] specifically chose myoglobin for comparative studies of peroxidase activity, because they said it was less susceptible to degradation by H_2O_2 [see discussion of postulate (b), below]. Shibata *et al.* [62,63] suggested that the peroxidase like activity of horse methemoglobin and catalase seems to be generated in parallel with the dissociation of these tetrameric hemoproteins; i.e., hemoglobin dissociates in acid and alkali and shows a parallel biphasic pattern of increases in peroxidase activity, whereas catalase, which dissociates only in alkali, shows an increased peroxidase activity only in alkali. Also, consistent with this hypothesis, they [63] found that guanidine and formamide, which dissociate hemoglobin into dimers, also markedly enhance its peroxidaselike activity. It is probable, however, that these denaturants and extremes of pH also alter the conformational integrity of the globins [see discussion of postulate (c), below]. Evidence against this hypothesis was reported by Smith and Beck [64], who showed that the isolated α and β subunits of hemoglobin (at protein concentrations equivalent to intact tetrameric hemoglobin) possessed only 10–24% of the peroxidase activity of tetrameric hemoglobin in the absence of serum (haptoglobin). A final conclusion, however, on this matter would depend on whether in the absence of haptoglobin, the separate α,β-dimers of hemoglobin which mimic the separate monomeric subunits in ligand binding and other properties [3] also display peroxidase activity that is less than or equal to that of the tetramer.

The postulate (b) of a protective role for haptoglobin at low pH is perhaps best supported by the studies of Sasazuki *et al.* [65]. They compared the properties of human and bovine hemoglobin and their haptoglobin complexes and studied the relative properties of the human hemoglobin–haptoglobin and hemoglobin–antibody complexes. They found that the pH dependence of the peroxidaselike activity of free bovine hemoglobin was remarkably parallel to that of the human hemoglobin–haptoglobin complex. Thus, the bovine hemoglobin showed relatively high activity at pH 4 in contrast to the near absence of activity of free human hemoglobin at that pH [14,65]. Complexation of the bovine hemoglobin with haptoglobin had no qualitative effect on the pH dependence of its peroxidase activity and actually diminished it somewhat throughout the pH range [65]. Their data on apparent acid denaturation showed that 50% of free human hemoglobin was modified in 5 sec at pH 4, whereas the same change in human hemoglobin complexed with haptoglobin did not occur that rapidly until the pH was lowered to 2. The human hemoglobin–antibody complex, which displayed only a slightly increased peroxidase activity at low pH, was similarly denatured to 50% at pH 3.5. (In light of possibility (a), it is interesting that the antibody apparently binds human hemoglobin as the tetramer [66].) It was also reported that free bovine hemoglobin displayed an acid resistance similar to the human hemoglobin–haptoglobin complex, but the data were not shown. These results indicate that protection by haptoglobin at low pH is probably an important factor in determining enhanced peroxidatic activity of hemoglobin. The fact that the activity of the human hemoglobin–haptoglobin complex is still significantly higher than that of free hemoglobin near pH 5.3, where the activity curve shows a maximum for the free hemoglobin [14,65], would suggest, however, that factors other than protection from acid denaturation may also contribute.

With regard to postulate (c), several groups of investigators [67–73] have reported evidence that would seem to support the concept that haptoglobin-induced conformational alterations of hemoglobin may contribute to its enhanced peroxidase activity. For example, there were two cases in which modification of haptoglobin either by heating [67] or by trinitrophenylation [68] apparently abolished its ability to enhance the peroxidase activity of hemoglobin more rapidly than its ability to bind hemoglobin. These results suggest that specific structural features of haptoglobin are necessary to elicit the peroxidase effect, but one would like to know whether the relative acid stability was retained in those hemoglobin–haptoglobin complexes that had lost the enhanced peroxidatic activity. Direct evidence for haptoglobin-induced conformational changes in hemoglobin affecting the heme environment has been obtained via the application of various spectroscopic techniques including optical rotatory dispersion [69], electric birefringence [69], and electron paramagnetic resonance spectroscopy of spin-labeled hemoglobin [70,71].

On the basis of the strict species specificity of antihuman hemoglobin antibodies relative to the broad specificity of human haptoglobin, Cohen-Dix *et al.* [73] suggested that the contact region between the two α,β-dimers of hemoglobin might contain the haptoglobin binding site, because this region of the molecule has been relatively conserved in evolution. This is also the region of binding of the modifiers that alter the O_2 affinity of hemoglobin [74], namely, 2,3-diphosphoglycerate and inositol hexaphosphate. It is therefore quite interesting that inositol hexaphosphate apparently enhances the peroxidaselike activity of human hemoglobin but does not affect the intrinsic peroxidase activity of cat hemoglobin, which is insensitive (with regard to O_2 affinity) to organic phosphates [75]. One would like to know in this regard whether cat hemoglobin (which has a higher intrinsic peroxidase activity than the human variety [75]) binds human haptoglobin, since this was not one of the species tested by Cohen-Dix *et al.* [73].

A unified theory for the basis of the peroxidaselike activity of hemoglobin is not obvious from present knowledge, but there does seem to be a correlate of the enhancement of hemoglobin peroxidase activity which traverses most of the available data; i.e., factors that lower the oxygen affinity of hemoglobin enhance its peroxidaselike activity. A partial listing follows: (a) Changes in pH between 7 and 5 lower the O_2 affinity by virtue of the Bohr effect, and peroxidase activity is increased in this region; (b) as cited above, inositol hexaphosphate lowers O_2 affinity and enhances peroxidase activity; (c) cat hemoglobin, which has a lower affinity for O_2 than human hemoglobin, also has a higher intrinsic peroxidase activity [75]; (d) tetrameric hemoglobin has a lower O_2 affinity than its separate subunits and has a higher intrinsic peroxidase activity [64]. Factors that decrease the O_2 affinity of hemoglobin generally also accelerate its rate of autoxidation [76].

The most notable exception to the trend developed above is the hemoglobin–haptoglobin complex, which is reported to have a much greater peroxidase activity (at pH 4) but a higher affinity for O_2 (measured at values of pH above 7 [77]) than uncomplexed hemoglobin. However, available data indicate that the peroxidase activities of hemoglobin alone and the hemo-globin–haptoglobin complex would not be much different from each other or much different from zero [14,62,65] at pH 7. It is therefore difficult to assess whether a correlation with O_2 affinity exists in this case without informa-tion about O_2 affinity at pH 4. On the other hand, there may be a correlation with autoxidizability. Upon addition of haptoglobin to a solution containing carbon monoxyhemoglobin, spectral changes occurred which were interpreted to reflect a haptoglobin-induced change in spinstate of the hemoglobin toward high spin [67]. That interpretation may certainly be valid, but it is curious that the new spectral maxima corresponded to those of ferrihemo-globin. Hence, an alternative explanation (if the system under study were not *strictly* anaerobic) would be that the process of haptoglobin binding interfered with ligation and actually might have increased the susceptibility of the ferrohemoglobin to oxidation—thus, the appearance of ferrihemoglobin.

Although further study is certainly necessary, it is attractive to consider changes in the autoxidizability of hemoglobin as the ultimate correlate of its pseudoenzymatic activity, because the ability to transfer electrons (in this case to molecular oxygen) appears to be an intrinsic property of those enzymes to which hemoglobin can be compared, e.g., peroxidase and cytochrome *P*-450 (see the section on mixed-function oxidaselike activity). In this light it is interesting that myoglobin, which binds oxygen more avidly than tetrameric hemoglobin, is nevertheless more autoxidizable and apparently has greater intrinsic peroxidase activity.

Whether peroxidaselike activity of hemoglobin inside normal erythrocytes may have physiological importance is questionable for several reasons. First, the concentration of ferrihemoglobin is usually $\sim 0.1\%$ of the total hemoglobin (i.e., $\sim 2 \ \mu M$); second, the steady-state concentration of H_2O_2 in erythrocytes has been estimated to be only $\sim 0.1 \ \mu M$ [78]; third, there are much better scavengers of H_2O_2 in the red blood cell than hemoglobin, namely, glutathione peroxidase and catalase. In certain diseased states accompanied by extensive hemolysis, however, the peroxidatic activity of hemoglobin may be substantially enhanced because it will be released to the serum and complexed by haptoglobin while at the same time the effective activities of catalase, glutathione peroxidase, and the reductase enzymes responsible for maintaining hemoglobin in the ferrous state will be diminished by dilution. The implications of the latter sequence of events, if it occurs as described, have yet to be explored.

MIXED-FUNCTION OXIDASELIKE ACTIVITY

Artificial Aniline Hydroxylase Systems

The hydroxylation of aromatic compounds like aniline (depicted in the following reaction equation) is a typical O_2-requiring reaction catalyzed by the mixed-function oxidase (monooxygenase) enzyme cytochrome *P*-450 [79]:

$$C_6H_5—NH_2 + NADPH + H^+ + O_2 \longrightarrow HO—C_6H_4—NH_2 + NADP^+ + H_2O$$

Various O_2-utilizing model systems that also catalyze the aniline (or acetanil-ide) hydroxylation reaction [80–84] have been studied in detail and compared with the cytochrome P-450 system. The compositions of such systems have covered a broad spectrum, ranging from simple metal ion systems (e.g., Sn^{2+}–phosphate [82] or Fe^{2+}–EDTA–ascorbate [80]) to hemoprotein systems (e.g., peroxidase–dihydroxyfumarate [84] or cytochrome b_5–cytochrome b_5 reductase–NADH [81]). Reviews of the data obtained from the model studies have appeared elsewhere [85–87]. From them it can be learned that two features of the reaction catalyzed by the cytochrome P-450 system generally distinguish it from the various model systems. First, the predominant product of the P-450-mediated reaction is p-aminophenol, whereas the other systems yield substantial amounts of the ortho and meta isomers in addition to the para compound. Second, the P-450-catalyzed reaction is accompanied to a much greater extent by the so-called NIH shift (the migration of a substituent from the point of substitution by oxygen to an adjacent position in the product [87]) than are most of the other reactions. In none of the cases, however, has the intimate molecular mechanism of the reaction been elucidated.

More recently, in a study designed to characterize the aniline hydroxylase system of human placental tissue, Juchau and Symms [24] reported that the activity was primarily due to the catalytic action of contaminating hemoglobin rather than to cytochrome P-450. Tredger et al. [88] reported, however, that they were unable to detect any p-aminophenol when their rabbit neonatal microsomal fractions were replaced with either neonatal or adult rabbit hemoglobin under their conditions of assay. Nevertheless, in studies detailed below, we have shown [27] that both human and bovine hemoglobin as well as myoglobin are effective catalysts, and Jonen et al. [89] also reported that hemoglobin from ox catalyzes aniline hydroxylation effectively.

In another report, Symms and Juchau [25] tested the relative effectiveness of various metal chelates and various hemoproteins including hemoglobin to catalyze p-aminophenol formation under a specific set of conditions (i.e., metal ion or heme concentration 10 μM, NADPH or NADH concentration 1 mM, and $\pm$ FMN at 0.33 mM at 37°C under 80:20 $N_2:O_2$). The most efficient of these non-P-450 systems was cytochrome b_5 + NADH + FMN, ~ 0.25 nmole of p-aminophenol per minute per nmole cytochrome b_5. The analogous system, in which ferrihemoglobin was substituted for cytochrome b_5, was about 60% as active. This result among others [25] revealed that the hemoproteins that are known to bind O_2 and CO (i.e., hemoglobin and myoglobin) could not be readily distinguished from those that normally do not (cytochrome b_5 and cytochrome c). Thus, the activity with cytochrome b_5 was ~ 1.5 times that with hemoglobin, but ~ 7 times that with cytochrome c; the activity of cytochrome b_5 was not affected by changing the atmosphere to 80:20 $CO:O_2$, but both the hemoglobin and the cytochrome c-mediated reactions were inhibited $> 90\%$. Such results suggest that direct ligation of O_2 with the ferrohemoprotein need not be an important step in the mechanism of catalysis of aniline hydroxylation by the hemoproteins under the specific conditions employed in that case (i.e., relatively high concentrations of ferrihemoprotein, reduced pyridine nucleotide, and flavin). In contrast, under conditions that are more typically employed to reconstitute the solubilized components of the liver microsomal cytochrome *P*-450 system (i.e., flavin present at very low concentration only as an integral component of the *P*-450 reductase enzyme [90,91] see below), we (J. J. Mieyal and R. S. Ackerman, unpublished observations) found that cytochrome *c* could not serve as a catalyst of the aniline hydroxylation reaction, whereas the activities of the O_2-carrier hemoproteins hemoglobin and myoglobin were comparable to cytochrome *P*-450 [27] (see Table 4).

Intracellular Environments of Hemoglobin and Cytochrome P-450

The hemoprotein enzyme cytochrome *P*-450 is predominantly localized *in vivo* in the endoplasmic reticulum of hepatocytes along with two electron transport systems (one NADPH-dependent pathway and one NADH-dependent pathway, each involving a flavoprotein reductase enzyme and the latter also involving cytochrome b_5). The roles of these various components in the overall mechanism of reactions catalyzed by the cytochrome *P*-450 system are a matter of active investigation [79]. Lu *et al.* [90] accomplished solubilization of the liver *P*-450 system, separated it into its component parts, and reconstituted it. The reconstituted system [consisting of cytochrome *P*-450, NADPH-dependent cytochrome *c* (*P*-450) reductase, and a lipid cofactor] has been shown to catalyze most of the reactions of the intact system including hydroxylation of aniline [90,91]. Hemoglobin is localized in the erythrocytes along with two electron transport systems that can mediate its reduction (one NADPH-dependent pathway and one NADH-dependent pathway, each involving a flavoprotein reductase enzyme and the latter probably also involving cytochrome b_5 [8–10,92]). Thus, the *in vivo* environment of hemoglobin is quite similar to that of *P*-450 (although *P*-450 is membrane bound); and it has been reported that cytochrome *c* reductase of liver, which is equated with *P*-450 reductase, can also catalyze reduction of human methemoglobin [93]. The following discussion pertains to how well hemoglobin can substitute one-for-one for cytochrome *P*-450 as the terminal oxidase enzyme in a reconstituted hydroxylase system.

Properties of the Hemoglobin-Cytochrome P-450 Reductase Aniline Hydroxylase System

The requirements for the catalytic hydroxylation of aniline by such a system are depicted in Table 3. It is clear that hemoglobin (either ferri- or

TABLE 3

**Dependence of p-Aminophenol Formation on the Components of the
Reconstituted Hemoglobin System**

System	Percent Activity[a] (as p-aminophenol formed in 15 min)
A. Ferrihemoglobin[b]	
Complete	100% (1.8 nmoles p-aminophenol)
Complete (heated hemoglobin)[c]	nd
Minus hemoglobin	nd
Minus reductase	38
Minus NADPH[d]	nd
Minus O_2[e]	nd
Complete (bubbled with O_2)[f]	174
Complete (bubbled with O_2/N_2: 50/50)[g]	127
Complete (bubbled with O_2/CO: 50/50)[h]	55
Complete (plus 100 μg dilaurolyl phosphatidylcholine)	100
B. Oxyferrohemoglobin[i]	
Complete	67
Minus NADPH	nd

[a] The assay for p-aminophenol is described under *Assays* in Mieyal *et al.* [27]. The activity for the complete system with normal dissolved oxygen from the atmosphere was arbitrarily set to 100%. All other activities are expressed relative to the standard. All values represent the average of at least two determinations which agreed within < 10%. The designation nd means none detectable (< 0.001 $A_{630\,nm}$ relative to the zero time control).

[b] The complete catalytic system for formation of p-aminophenol consisted of human hemoglobin, 1 μM, added to the reaction mixture in the oxidized state (Hb^{3+}), NADPH added in excess (1 mM), solubilized liver microsomal cytochrome P-450 reductase [12 μg (0.006 unit)], aniline (40 mM), and atmospheric O_2, all in 20mM potassium phosphate, pH 6.8, 38°C. The reaction was initiated usually by addition of NADPH to a premixed, preincubated (38°C) solution of the other components; the reaction was allowed to proceed for 15 min. Details of the selection of these specific conditions are given in Mieyal *et al.* [27].

[c] The stock solution was preheated to 90°C, cooled, and the appropriate aliquot was transferred to give 1 μM heated Hb^{3+} in the reaction mixture in place of unheated Hb^{3+}.

[d] The reaction was initiated by addition of aniline.

[e] The premixture was placed in the central portion of a Warburg flask and the aliquot of NADPH in the side arm. Both solutions and the flask were purged with N_2; the flask was then evacuated and repurged with N_2. The NADPH solution was tipped in to initiate the reaction.

[f] The premixture was placed in a special vessel (similar to a Warburg flask but total volume ~ 2 ml); the aliquot of NADPH was placed in the side receptacle. Both solutions were bubbled with pure O_2 for ~ 15 sec; then the atmosphere of the flask was flushed with O_2 and the flask closed off. The NADPH solution was tipped in to initiate the reaction.

[g] Same as "[f]," except that a 50/50 gas mixture of nitrogen and oxygen was substituted for the pure O_2.

[h] Same as "[f]," except that a 50/50 gas mixture of carbon monoxide and oxygen was substituted for pure O_2.

[i] The appropriate aliquot of a stock solution of purified human oxyferrohemoglobin (see Mieyal and Blumer [28] for preparation) was added to the reaction mixture to give 1 μM HbO_2 rather than Hb^{3+} at the outset of the reaction. The concentration of aniline in this case was 0.02 M.

oxyferro hemoglobin) in its native state (i.e., not denatured by heating) is required for the reaction along with molecular O_2 and a source of electrons (NADPH or NADPH plus reductase). These are the typical requirements of a mixed-function oxidase system [79]. It remains to be demonstrated for the hemoglobin-catalyzed reaction, however, via experiments using $^{18}O_2$ and $H_2^{18}O$, that one atom of each O_2 molecule appears in the product p-amino-phenol and one in H_2O. It was confirmed by combination gas chroma-tographic–mass spectroscopic analysis* that p-aminophenol was the predominant product of the reaction; hence, this system mimics the P-450 system in that respect also. Moreover, this system (like P-450) is inhibited by CO. We have not tested the hemoglobin system with regard to the NIH shift.

Besides being thermally labile, hemoglobin also fulfills the other criteria to be classified as an enzyme in this system. First, the rate of aniline hydroxy-lation is linearly dependent on hemoglobin concentration (Fig. 6). The same

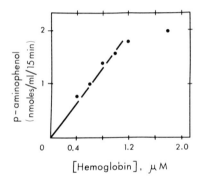

Fig. 6 Dependence of rate of aniline hydroxylation on hemoglobin concentration. Reaction mixtures were set up as described in the footnotes to Table 3, except that the concentration of hemoglobin was varied as shown. Each point on the graph represents the average of at least two determinations (From Mieyal *et al.* [27]. Reproduced with permission from the *J. Biol. Chem.*)

linear concentration range is observed (i.e., up to ~ 1.2 nmoles Hb/ml) whether or not the reductase enzyme is present. A similar linear range of hemoenzyme concentration is generally observed for P-450-catalyzed reactions. Second, the concentration of hemoglobin in this range is $\sim 10^4$ times lower than the concentration of substrate aniline, which elicits half-maximal velocity ($K_m = 8$ mM, Fig. 7B). Third, the dependence of reaction

* We thank Dr. John M. Strong of the Clinical Pharmacology Unit, Northwestern University Medical School, for the analyses and interpretation of mass spectra. (Details are reported in Mieyal *et al.* [27].)

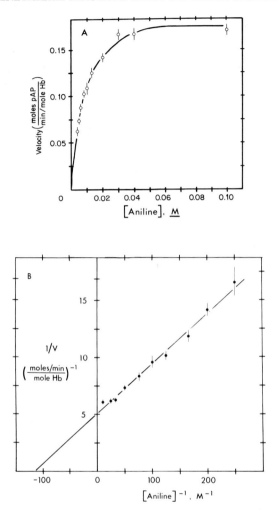

Fig. 7 (A) Dependence of rate of aniline hydroxylation on aniline concentration. Reaction mixtures were set up as described in the footnotes to Table 3, except that the concentration of aniline was varied as shown. Each point represents the average (± standard deviation of the mean) of at least six determinations (pAP, p-aminophenol). (B) Double-reciprocal plot of velocity vs. aniline concentration. Data from Fig. 7A. (From Mieyal et al. [27]. Reproduced with permission from the J. Biol. Chem.)

velocity on aniline concentration displays typical Michaelis–Menten saturation kinetics (Fig. 7A), thereby reflecting an interaction between aniline and hemoglobin, i.e., an enzyme–substrate complex. The K_m for aniline (8 mM) is also comparable to the range of K_m values reported for aniline hydroxylation by liver microsomes from various sources (0.1–6 mM) [94].

Spectral Determination of Aniline–Hemoglobin Interactions

In deriving a mechanism for the hemoglobin-catalyzed reaction, therefore, it was necessary to determine which form(s) of hemoglobin showed an affinity for aniline. Direct combination of aniline with deoxyferrohemoglobin (Hb^{2+}) gave no consistent spectral evidence of an interaction [28]. Similar experiments with Hb^{3+} [26] demonstrated that aniline interacted only weakly ($K_s \sim 105$ mM) and in a cooperative manner (Hill coefficient $n = 2 \cdot 2$); thus, the aniline–Hb^{3+} complex also could not be catalytically important. Figure 8 shows the time-dependent spectral changes elicited when aniline

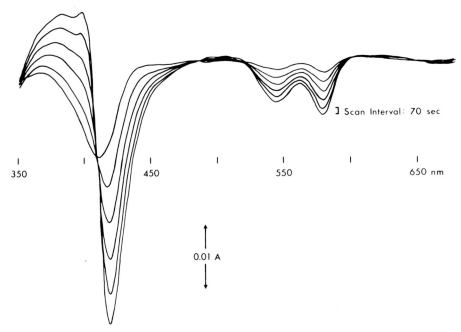

Fig. 8 Aniline interaction with oxyhemoglobin: time course of spectral changes. An Aminco DW-2 dual-beam spectrophotometer operating at 38°C in the split-beam and baseline correction modes was employed for these experiments. Mixing cuvettes (Pyrocell) having two chambers of 4.5 mm path length were used as the sample and reference cuvettes. One milliliter of 2 μM HbO_2 in 20 mM potassium phosphate buffer, pH 6.8, was added to one chamber of both the sample and reference cuvettes. One milliliter of aniline solution of appropriate concentration (see Fig. 9), in this case 80 mM, was added to the other chamber of each cuvette. The cuvettes were placed in the light beam, and a flat baseline was set after the solutions had equilibrated to 38°C. The sample cuvette was inverted at zero time to allow mixing of the HbO_2 and aniline solutions, and it was replaced in the spectrophotometer. (From Mieyal and Blumer [28]. Reproduced with permission from the *J. Biol. Chem.*)

was combined with oxyferrohemoglobin (HbO_2). Diminution in absorbance occurred at the wavelengths associated with the spectral maxima for HbO_2, while increases occurred at spectral regions associated with the maxima for Hb^{3+}, indicating that aniline accelerates the autoxidation of HbO_2 (see

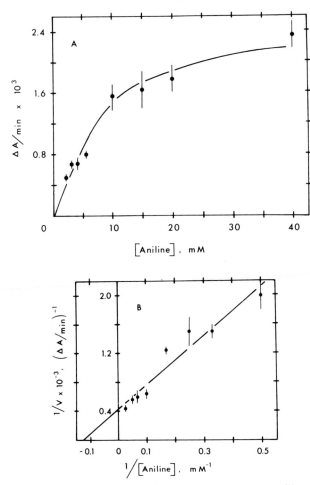

Fig. 9 Dependence of rate of oxyhemoglobin spectral change on aniline concentration. (A) Experiments were set up as described under Fig. 8, except that the aniline concentration was varied as indicated. The absorbance difference ($A_{400\,nm} - A_{420\,nm}$, Fig. 8) was measured at various times and then plotted as a function of time. The rate of change of absorbance, i.e., the tangent to the initial linear portion (~20 min) of each such curve, was determined for each aniline concentration; these rates were then replotted as a function of aniline concentration. (B) Double-reciprocal replot of rate of spectral change vs. aniline concentration. Data from Fig. 9A. (From Mieyal and Blumer [28]. Reproduced with permission from *J. Biol. Chem.*)

discussion of the possible relationship between autoxidation of hemoglobin and its enzymelike behavior in the previous section and below). The initial rate of change of absorbance and the projected extent of change were related to the amount of aniline added. The relationship between initial rates and aniline concentration was indicative of aniline binding (Fig. 9A). The double-reciprocal plot (Fig. 9B) yielded an interaction constant for aniline, $K = 8$ mM. This value is identical to K_m for aniline determined from the kinetics of the hemoglobin-catalyzed formation of p-aminophenol (Fig. 7B). Thus, an HbO_2–aniline interaction is implicated in the mechanism of the overall aniline hydroxylation reaction (see below).

Relative Efficiencies of Hemoprotein-Mediated Aniline Hydroxylase Systems

Table 4 lists for comparison the rates (turnover numbers) of aniline hydroxylation catalyzed by the various reconstituted systems that we have characterized [27] and by the reconstituted P-450 systems studied by Lu *et al.* [91]. All systems including induced P-450 are comparable within a factor of 5, and most agree more closely. These results contrast markedly

TABLE 4

Rates of Hydroxylation of Aniline by Various Reconstituted Enzyme Systems

System	Turnover number $\left(\dfrac{\text{moles } p\text{-aminophenol/min}}{\text{mole hemoprotein}}\right)$	Reference
1a. Rat cytochrome P-450 (uninduced) + P-450 reductase (0.050 unit)[a] + lipid	0.22	91
1b. Methylcholanthrene-induced	0.37	91
1c. Phenobarbital-induced	0.26	91
2. Human hemoglobin + P-450 reductase (0.005 unit)[a]	0.20	27
3. Bovine hemoglobin + P-450 reductase (0.010 unit)[a]	0.23	27
4. Whale myoglobin + P-450 reductase (0.009 unit)[a]	0.08	27

[a] Units of reductase activity are defined as μmoles cytochrome c^{3+} reduced per minute.

with the comparison between methemoglobin or metmyoglobin as pseudo-peroxidases and the true peroxidases (previous section, Table 2), where the O_2 carriers were much poorer catalysts. The difference between the two cases, however, may be in the fact that P-450 itself is much less efficient (turnover number) than are the peroxidases; hence, hemoglobin has a better chance to "catch up." Unlike cytochrome P-450 [90,91], none of the "oxygen-carrier" hemoprotein systems (Mb or Hb) required a lipid cofactor. Also in contrast to the P-450 system, the Hb and Mb systems apparently required much less

TABLE 5

Inhibition of Hemoglobin-Catalyzed Aniline Hydroxylation by Scavengers of Activated Oxygen

Scavenging agent	Concentration	Apparent inhibition[a] (%)
Ethanol	0.0 M	0 (15)
	0.2 M	18 ± 16 (10)
	1.0 M	20 ± 19 (9)
Mannitol	0.0 M	0 (5)
	0.2 M	11 ± 32 (5)
Superoxide dismutase	0 μg/ml	0 (15)
	2 μg/ml	0 ± 15 (14)
	5 μg/ml	0 ± 14 (10)
	10 μg/ml	15 ± 11 (10)
	16 μg/ml	17 ± 9 (12)
	40 μg/ml	12 ± 11 (10)
Catalase	0 μg/ml	0 (8)
	0.1 μg/ml	8 ± 5 (6)
	0.3 μg/ml	52 ± 9 (4)
	0.5 μg/ml	51 ± 9 (4)
	1.0 μg/ml	66 ± 5 (6)
	1.5 μg/ml	80 ± 13 (4)
	20.0 μg/ml	93 ± 6 (4)

[a] All of the numbers were calculated from data reported in reference Mieyal et al. [27]. The values for percent inhibition were calculated according to the relationship

$$\% \text{ Inhibition} = \frac{\text{control value} - \text{sample}}{\text{control}} \times 100$$

Each value represents the mean ±SE, where the SE reflects the variation in both the control values and the "inhibited" values. The number in parentheses represents the number of determinations. As shown in Table 3, some p-aminophenol is formed in the absence of the reductase enzyme. The various potential inhibitors were tested under those conditions also [27], but this table has been limited to the results for the "reductase-mediated" reaction (i.e., the effects of the scavengers on the difference in rates of p-aminophenol formation in the presence or absence of the reductase).

of the reductase enzyme, even though the reductase was prepared by the same procedure and had essentially the same specific activity in all cases. This finding would suggest that electron transport in the Hb and Mb systems may be more efficient than it is in the reconstituted P-450 system.

Oxygen Activation in the Hemoglobin-Catalyzed Reaction

It is known that NADPH alone [95] and NADPH plus the rat liver reductases [93] can reduce human Hb^{3+} and thus effect formation of HbO_2 when O_2 is present. This same combination of NADPH and the reductase can generate superoxide ($O_2^{\bar{\cdot}}$) from O_2 [96]. The superoxide so formed can disproportionate to give H_2O_2 and O_2 [97]. In addition, the very reactive hydroxylating agent $\cdot OH$ can be formed from H_2O_2 and $O_2^{\bar{\cdot}}$ or Hb^{2+} [98,99]. Therefore, since multiple forms of hemoglobin and of oxygen could exist in our reaction mixtures, we investigated which might participate significantly in the catalytic process. The effects of various scavengers of activated oxygen on the activity of the catalytic system are depicted in Table 5. Ethanol and mannitol at relatively low concentrations have been reported to act as scavengers of radicals such as $\cdot OH$ [100], but these agents apparently do not destroy $O_2^{\bar{\cdot}}$ [97]; superoxide dismutase does destroy $O_2^{\bar{\cdot}}$ ($2H^+ + 2O_2^{\bar{\cdot}} \rightarrow H_2O_2 + O_2$) [97], and catalase converts H_2O_2 to O_2 and H_2O. Only catalase caused significant inhibition, and it essentially completely halted the reaction.

The inhibitory effect of catalase (Table 5) suggested either that substrate amounts of H_2O_2 may accumulate during the reaction or that a catalytic complex involving bound H_2O_2 (formed *in situ*) may participate. The data of Fig. 10 rendered the former postulate unlikely. The accumulation of H_2O_2 in a catalytic system can be conveniently monitored by following H_2O_2-dependent, peroxidase-catalyzed oxidation of guaiacol. This is shown for the enzyme glucose oxidase, which converts glucose and O_2 to gluconic acid and H_2O_2 (Fig. 10, curve A; under those conditions, ~ 25 nmoles of H_2O_2 were formed in 10 min). Curve B (Fig. 10) shows that it is possible to monitor H_2O_2 formation in the presence of the complete hemoglobin system, if it is formed. When glucose oxidase is omitted, apparently no detectable H_2O_2 is formed by the hemoglobin system which hydroxylates aniline (curve C, Fig. 10). We also found that the combination of Hb^{3+} with H_2O_2 is effective in hydroxylation of aniline only when the concentrations of these two components are relatively high [27]. The results were comparable to those reported previously for the H_2O_2-dependent, metmyoglobin-catalyzed oxidation of various phenol derivatives, where $\sim 30 \ \mu M$ Hb^{3+} was required to give half-maximal activity [101]. Since the normal catalytic system (Table 3) contains only 1 μM Hb^{3+} and little or no H_2O_2 is accumulated in the system (Fig.

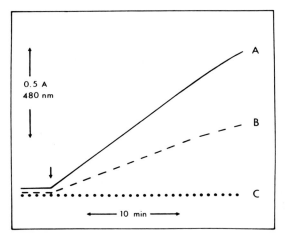

Fig. 10 Analysis for hydrogen peroxide formation. A standard assay for H_2O_2 was employed, whereby the H_2O_2-dependent oxidation of guaiacol by horseradish peroxidase (Sigma) was monitored at the absorption maximum of the product (480 nm) by means of a Gilford recording spectrophotometer. Both the peroxidase (10 μg) and guaiacol (0.4 mM) were added in excess so that only the rate of formation of H_2O_2 would be limiting. The dependence of $A_{480 nm}$ on $[H_2O_2]$ was determined separately (standard curve) and was found to be linear at least up to 100 μM H_2O_2. Thus, in 0.02 M potassium phosphate, pH 6.8, 38°C, 1 nmole of H_2O_2 in 1 ml of reaction mixture gives rise to $\Delta A_{480 nm} = 0.016$. Curve A: Direct tracing from the Gilford chart showing the rate of formation of H_2O_2 by 0.5 μg of glucose oxidase (Sigma) at 38°C; 0.1 ml of 1 M glucose was added to 0.9 ml of reaction mixture (containing the peroxidase and guaiacol in the potassium phosphate buffer) at the time indicated by the arrow. Curve B: Same conditions as curve A, but the reaction mixture also contained 1 nmole of Hb^{3+}, 10 μg of the reductase, and 40 μmoles of aniline; the reaction was initiated by addition of 0.1 ml of 1.0 M glucose, 1 mM NADPH. Curve C: Same curve as B, but glucose and glucose oxidase omitted; 0.1 ml of 1 mM NADPH added (at arrow). (From Mieyal et al. [27]. Reproduced with permission from the *J. Biol. Chem.*)

10), it seems unlikely that the long-known "peroxidatic activity" of methemoglobin (see previous section) is operating in this O_2-requiring system (see reaction scheme, Fig. 11).

Hypothetical Mechanism of the Reaction

Figure 11 represents our working hypothesis for the mechanism of the O_2-requiring, hemoglobin-catalyzed hydroxylation of aniline. The central, enblocked portion represents those reactions that are proposed to be on the direct path to product formation. Either the various other reactions and equilibria, shown on the sides, occur in the system but have been shown to be unimportant, or there is evidence that their occurrence is unlikely (see further

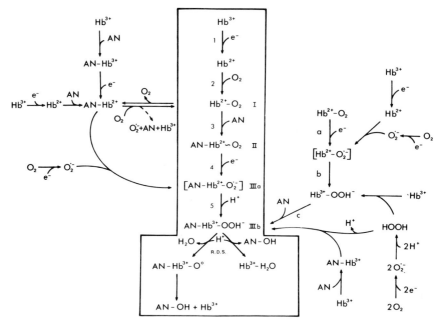

Fig. 11 Hypothetical mechanism for the hemoglobin-catalyzed hydroxylation of aniline. The central, enblocked portion represents those reactions that are proposed to be on the direct path to product formation (see text).

discussion). The only alternative path that cannot be eliminated on the basis of presently available evidence is that designated by steps a, b, and c. The enblocked mechanism would fulfill the basic criteria for a mixed-function oxidase system; i.e., two electrons are utilized, and the oxygen molecule is disproportionated into the products p-aminophenol (AN-OH) and H_2O.

In contrast to most mechanisms proposed for P-450 reactions wherein formation of a substrate–ferrihemoprotein complex is the first step [79], this scheme shows aniline binding to HbO_2 rather than to Hb^{3+}. That postulate was based on the data from spectral studies of the direct combination of aniline with the various forms of hemoglobin [26,28], as discussed above. It seems most likely that aniline may interact at a site other than the heme iron atom with HbO_2, thus diminishing the affinity of hemoglobin for O_2 and concomitantly increasing its susceptibility to autoxidation (Fig. 8) [28]. We found that, whereas oxygen inhibits the aniline-accelerated autoxidation of hemoglobin [28], it enhances the overall rate of Hb-catalyzed hydroxylation of aniline [27]. This relationship also suggests that in the overall catalytic system addition of aniline and O_2 to Hb^{2+} results in a ternary $AN \sim Hb^{2+} \sim (O_2)_{1-4}$ complex(es). This possibility could be tested via a detailed two-substrate kinetic study. However, since we are unable to demonstrate aniline–Hb^{2+} complex formation at present, we have proposed the reaction scheme in an ordered fashion:

$$Hb^{2+} + O_2 \longrightarrow HbO_2 \xrightarrow{\text{AN}} AN\text{–}Hb\sim O_2$$

where the $\sim$ represents an altered form of O_2 binding to Hb. According to the Franck–Condon principle, the more closely related the geometry before and after electron transfer, the faster the transfer will occur [102]. This principle may explain how aniline may convert the inert oxygen carrier HbO_2 to an oxygen-activating enzyme. Thus, aniline binding to HbO_2 may sufficiently distort the $Hb\sim O_2$ bond so that total electron transfer from Fe^{2+} to O_2 would be facilitated.

The above order of addition might also pertain to the *in vivo* situation, in which hemoglobin is maintained as HbO_2 [10]. By analogy we speculate that liver cytochrome *P*-450 might also be maintained in the *P*-450^{2+}–O_2 state *in vivo*. Thus, a drug entering the liver via the portal circulation might convert an inert form of *P*-450–O_2 to an activated substrate–*P*-450$\sim O_2$ complex, which would generate the hydroxylated metabolite. It has been shown both for *P*-450$_{cam}$ [79] and for adrenal mitochondrial *P*-450 [103] that ternary substrate–*P*-450–O_2 complexes are formed which do not yield products unless the specific electron donor (putidaredoxin or adrenodoxin, respectively) is added. Analogously, combination of aniline with HbO_2 does not yield *p*-aminophenol unless a source of electrons such as NADPH plus *P*-450 reductase is also added (Table 3). Since the reductase is a flavoenzyme, it could accept two electrons from NADPH and transfer these in one-electron steps [104]; hence, the reduced reductase could donate the electrons as shown at both step 1 and step 4 (Fig. 11).

The reductase enzyme is known to catalyze reduction of methemoglobin [93], and it does transfer electrons directly to O_2 to form O_2^{-} [96]. We confirmed that O_2^{-} is generated by our preparation of *P*-450 reductase and that the coupled oxidation of epinephrine (assay for O_2^{-} [97]) can be *completely* inhibited by ~ 1 μg of superoxide dismutase. That amount is 40 times less than the highest concentration tested for inhibition of aniline hydroxylation, which had very little effect (if any) on the reductase-mediated reaction (Table 5). Therefore, we concluded that the reductase can donate electrons directly to Hb-bound O_2 and that the following alternatives for step 4 of the scheme are unlikely: (a) that free O_2^{-} generated by the reductase could either bind directly to Hb^{2+}, replace Hb-bound O_2, or donate its electron to the Hb-bound O_2 and (b) that free O_2^{-} could produce enough free H_2O_2 by dismutation to serve as substrate for "peroxidatic" hydroxylation of aniline. It is also unlikely that species IIIa (Fig. 11) is a long-lived intermediate;

otherwise, the action of superoxide dismutase on any $O_2^{\overline{\cdot}}$ dissociated from that complex would drive the equilibrium toward dissociation and should cause complete inhibition by drawing off the $O_2^{\overline{\cdot}}$. By analogy, we conclude that complex IIIb (Fig. 11) is a relatively stable intermediate (i.e., the step following its formation is rate limiting), since complete inhibition of the reductase-mediated reaction can be effected by catalase (Table 5) even though substrate amounts of H_2O_2 for "peroxidatic" catalysis by Hb^{3+} do not accumulate (Fig. 10). It is noteworthy that a peroxide–methemoglobin complex can be detected after addition of excess H_2O_2 to Hb^{3+} [47], and the combination of relatively high concentrations of methemoglobin and H_2O_2 with aniline does generate p-aminophenol [27].

Complex IIIb may be formally equivalent to substrate-bound "peroxidase compound I" [105]. Its breakdown in the rate-determining step is drawn in two ways because the intimate molecular mechanism is not known. In one case, the incipient peroxide ligand is shown to disproportionate in one step to give p-aminophenol and leave H_2O as the ligand. The sequence of steps from complex IIIb might also be written according to an "oxenoid" mechanism [106,86], showing atomic oxygen as a transient intermediate. The postulated proton donation in either case is consistent with the effect of pH on the reaction; i.e., decrease of pH enhances the rate of reaction [27]. It is therefore our hypothesis that the apparent difference between this O_2-requiring catalysis by hemoglobin and the long-known H_2O_2-requiring "peroxidatic" action of methemoglobin is the *in situ* activation of ferrohemoglobin-bound O_2 to *bound* peroxide, which is utilized for hydroxylation of the bound substrate.

The characterization of hemoglobin as an enzyme in this system does not suggest that it might so operate as part of its normal physiological function. However, it does suggest that certain encounters of drugs, environmental pollutants, and even dietary substances with HbO_2 may interfere with its normal function. The remarkable similarities between various cytochrome P-450 oxygenase systems and the catalytic system utilizing hemoglobin as the enzyme indicate that a careful examination of the extent to which such reactions may be catalyzed by the hemoglobin system *in vivo* is warranted.

CONCLUSIONS

Hemoglobin, although often termed the catalyst in various reactions, has heretofore not been considered an enzyme. This reticence has probably been due in part to the knowledge that many of the reactions catalyzed by hemoglobin could also be catalyzed by iron chelates, that the hemoglobin concentrations required were generally high, and that in many cases boiled hemoglobin could catalyze the reactions about as well as the native protein. In this chapter, two notable exceptions to these general statements have been discussed in which hemoglobin has apparently fulfilled the criteria to be classified as an enzyme, namely, in the H_2O_2-dependent catalysis of thiocyanate oxidation [54] and in the O_2-requiring hydroxylation of aniline [27]. One criterion for enzymes that has not been discussed in these two cases is specificity. In a survey of substrate specificity with our reconstituted hemoglobin system [27], we have so far shown that besides aniline, xylidine is also hydroxylated and benzphetamine is dealkylated. The liver microsomal cytochrome P-450's and the peroxidase enzymes themselves, however, are not very specific with respect to substrates. It was mentioned above that quantitative efficiency rather than the nature of the catalyzed reaction may be a distinguishing factor between "pseudo" and "true" enzymes; for example, the relative activities of catalase, peroxidase, and myoglobin in disproportionating H_2O_2 are 500,000, 100, and 1, respectively [13].

Although under certain circumstances hemoglobin possesses the fundamental characteristics of enzymes and may even catalyze certain reactions as efficiently as the titled enzyme to which it is compared, it has not been shown to catalyze a known reaction more efficiently, nor does it carry out any reactions that are unique to it. Hence, it seems that hemoglobin for the present must retain the title of "honorary enzyme." Nevertheless, its high concentration in the blood (~ 2 mM) and the many reactions in which it can participate make it imperative that attention be given to the erythrocyte membrane permeability of exogenous agents such as drugs and environmental toxins which may interact directly with hemoglobin and either alter its normal physiological function or be altered by it.

REFERENCES

1. L. L. Ingraham, *Compr. Biochem.* **14**, 424 (1966).
2. E. Antonini and M. Brunori, *Annu. Rev. Biochem.* **39**, 977 (1970).
3. E. Antonini and M. Brunori, "Hemoglobin and Myoglobin in Their Reactions with Ligands." North-Holland Publ., Amsterdam, 1971.
4. J. V. Kilmartin and L. Rossi-Bernardi, *Physiol. Rev.* **53**, 836 (1973).
5. H. Kitchen and S. H. Boyer, *Ann. N.Y. Acad. Sci.* **241**, 1 (1974).
6. S. J. Edelstein, *Annu. Rev. Biochem.* **44**, 209 (1975).
7. J. M. Baldwin, *Prog. Biophys. Mol. Biol.* **29**, 225 (1975).
8. D. Niethammer and F. M. Huennekens, *Arch. Biochem. Biophys.* **146**, 564 (1971).
9. D. E. Hultquist and P. G. Passon, *Nature (London), New Biol.* **229**, 252 (1971).
10. F. Kuma and H. Inomata, *J. Biol. Chem.* **247**, 566 (1972).
11. M. Kiese, "Methemoglobinemia: A Comprehensive Treatise." CRC Press, Cleveland, Ohio, 1974.
12. H. P. Misra and I. Fridovich, *J. Biol. Chem.* **247**, 6960 (1972).

13. D. Kielin and E. F. Hartree, *Nature (London)* **166**, 513 (1950).
14. G. E. Connell and O. Smithies, *Biochem. J.* **72**, 115 (1959).
15. A. L. Pawlak, *Klin. Wochenschr.* **52**, 645 (1974).
16. R. Brodersen and P. Bartels, *Eur. J. Biochem.* **10**, 468 (1969).
17. A. L. Tappel, *Arch. Biochem. Biophys.* **44**, 378 (1953).
18. E. G. Hrycay, H. G. Jonen, A. Y. H. Lu, and W. Levin, *Arch. Biochem. Biophys.* **166**, 145 (1975).
19. C. Chen and C. C. Lin, *Biochim. Biophys. Acta* **170**, 366 (1968).
20. S. S. Tate, J. Orlando, and A. Meister, *Proc. Natl. Acad. Sci. U.S.A.* **69**, 2505 (1972).
21. H. Yamabe and W. Lovenberg, *Biochem. Biophys. Res. Commun.* **47**, 733 (1972).
22. M. Kiese, G. Renner, and R. Schlaeger, *Naunyn-Schmiederberg's Arch. Pharmakol.* **268**, 247 (1971).
23. K. G. Symms and M. R. Juchau, *Biochem. Pharmacol.* **21**, 2519 (1972).
24. M. R. Juchau and K. G. Symms, *Biochem. Pharmacol.* **21**, 2053 (1972).
25. K. G. Symms and M. R. Juchau, *Drug. Metab. Dispos.* **2**, 194 (1974).
26. J. J. Mieyal and L. S. Freeman, *Biochem. Biophys. Res. Commun.* **69**, 143 (1976).
27. J. J. Mieyal, R. S. Ackerman, J. L. Blumer, and L. S. Freeman, *J. Biol. Chem.* **251**, 3436 (1976).
28. J. J. Mieyal and J. L. Blumer, *J. Biol. Chem.* **251**, 3442 (1976).
29. P. Nicholls, *J. Gen. Physiol.* **49**, Suppl., 131 (1965).
30. C. E. Castro. *J. Theor. Biol.* **33**, 475 (1971).
31. I. S. Longmuir, S. Sun, and W. Soucie, *Oxidases Relat. Redox Syst.*, *2nd, 1971* Vol. 2, p. 451 (1973).
32. L. S. Goodman and A. Gilman, eds., "The Pharmacological Basis of Therapeutics," 5th ed. Macmillan, New York, 1975.
33. S. S. Tate, R. Sweet, F. H. McDowell, and A. Meister, *Proc. Natl. Acad. Sci. U.S.A.* **68**, 2121 (1971).
34. W. Dairman and G. Christenson, *Eur. J. Pharmacol.* **22**, 135 (1973).
35. T. C. Bruice and S. Benkovic, "Bioorganic Mechanisms," Vol. 2. Benjamin, New York, 1966.
36. G. Cohen and R. E. Heikkila, *J. Biol. Chem.* **249**, 2447 (1974).
37. O. Augusto and G. Cilento, *Arch. Biochem. Biophys.* **168**, 549 (1975).
38. H. Hinterberger. *Biochem. Med.* **5**, 412 (1971).
39. T. Kakizaki, M. Sato, H. Tsuruta, and H. Hasegawa. *Ind. Health* **7**, 13 (1969).
40. R. H. Thomson, *Comp. Biochem.* **3**, 727 (1962).
41. A. S. Brill, *Compr. Biochem.* **14**, 447 (1966).
42. Y. Yonetani, *Adv. Enzymol.* **33**, 309 (1970).
43. J. E. Critchlow and H. B. Dunford, *Oxidases Relat. Redox Syst. Proc. Int. Symp.*, *2nd, 1971* Vol. 1, p. 355, 1973.
44. D. Keilin and E. F. Hartree, *Proc. R. Soc. London, Ser. B* **117**, 1 (1935).
45. D. Keilin and E. F. Hartree, *Nature (London)* **164**, 254 (1949).
46. D. Keilin and E. F. Hartree, *Biochem. J.* **49**, 88 (1951).
47. K. Dalziel and J. R. P. O'Brien, *Biochem. J.* **56**, 648 (1954).
48. K. Dalziel and J. R. P. O'Brien, *Biochem. J.* **56**, 660 (1954).
49. B. Chance, *in* "The Enzymes" (J. B. Sumner and K. Myrbäck, eds.), 1st ed., Vol. 2, Part 1, p. 444. Academic Press, New York, 1951.
50. A. L. Tappel, *in* "The Enzymes" (P. D. Boyer, H. Lardy, and K. Myrbäck, eds.), 2nd ed., Vol. 8, p. 275. Academic Press, New York, 1963.
51. E. D. Wills, *Biochem. J.* **99**, 667 (1966).
52. F. Haurowitz, M. Groh, and G. Gansinger, *J. Biol. Chem.* **248**, 3810 (1973).

53. A. M. Siddiqi, *Acta Biol. Adad. Sci. Hung.* **22**, 275 (1971).
54. J. Chung and J. L. Wood, *J. Biol. Chem.* **246**, 555 (1971).
55. T. Shiga and K. Imaizumi, *Arch. Biochem. Biophys.* **167**, 469 (1975).
56. K. S. Nickel and B. A. Cunningham, *Anal. Biochem.* **27**, 292 (1969).
57. E. G. Hrycay and P. J. O'Brien, *Arch. Biochem. Biophys.* **147**, 28 (1971).
58. M. Polonovski and M. F. Jayle, *C. R. Seances Soc. Biol. Ses Fil.* **129**, 457 (1938).
59. H. Mattenheimer and E. C. Adams, Jr., *Z. Klin. Chem. Klin. Biochem.* **6**, 10 (1968).
60. F. W. Putman, ed., "The Plasma Proteins," Vol. 2, p. 2. Academic Press, New York, 1975.
61. H. E. Sutton, *Prog. Med. Genet.* **7**, 163 (1970).
62. Y. Inada, T. Kurozumi, and K. Shibata, *Arch. Biochem. Biophys.* **93**, 30 (1961).
63. T. Kurozumi, Y. Inada, and K. Shibata, *Arch. Biochem. Biophys.* **94**, 464 (1961).
64. M. J. Smith and W. S. Beck, *Biochim. Biophys. Acta* **147**, 324 (1967).
65. T. Sasazuki, H. Tsunoo, H. Nakajima, and K. Imai, *J. Biol. Chem.* **249**, 2441 (1974).
66. T. Sasazuki, *Immunochemistry* **8**, 695 (1971).
67. M. Waks and A. Alfsen, *Biochem. Biophys. Res. Commun.* **23**, 62 (1966).
68. T. Shinoda, *J. Biochem. (Tokyo)* **57**, 100 (1965).
69. M. Makinen, J. B. Milstien, and H. Kon, *Biochemistry* **11**, 3851 (1972).
70. M. Makinen and H. Kon, *Biochemistry* **10**, 43 (1971).
71. B. Malchy, H. Dugas, F. Ofosu, and I. C. P. Smith, *Biochemistry* **11**, 1669 (1972).
72. M. Waks, A. Alfsen, S. Schwaiger, and A. Mayer, *Arch. Biochem. Biophys.* **132**, 268 (1969).
73. P. Cohen-Dix, R. W. Noble, and M. Reichlin, *Biochemistry* **12**, 3744 (1973).
74. A. Arnone, *Nature (London)* **237**, 148 (1972).
75. A. G. Mauk, M. R. Mauk, and F. Taketa, *Nature (London), New Biol.* **246**, 189 (1973).
76. A. Mansouri and K. H. Winterhalter, *Biochemistry* **13**, 3311 (1974).
77. R. L. Nagel, J. B. Wittenberg, and H. M. Ranney, *Biochim. Biophys. Acta* **100**, 286 (1965).
78. A. Muller, S. Rapaport, and R. Knofel, *Folia Haematol. (Leipzig)* **89**, 228 (1968).
79. R. W. Estabrook, J. R. Gillette, and K. C. Liebman, eds. "Microsomes and Drug Oxidations." Williams & Wilkins, Baltimore, Maryland, 1972.
80. S. Udenfriend, C. T. Clark, J. Axelrod, and B. B. Brodie, *J. Biol. Chem.* **208**, 731 (1954).
81. Hj. Staudinger and V. Ullrich, *Biochem. Z.* **339**, 491 (1964).
82. V. Ullrich and Hj. Staudinger, *in* "Microsomes and Drug Oxidations" (J. R. Gillette *et al.*, eds.), p. 199. Academic Press, New York, 1969.
83. G. A. Hamilton, R. J. Workman, and L. Woo, *J. Am. Chem. Soc.* **86**, 3390 (1964).
84. J. W. Daly and D. M. Jerina, *Biochim. Biophys. Acta* **208**, 340 (1970).
85. V. Ullrich and Hj. Staudinger, *Handb. Exp. Pharmakol.* **28**, part 2, 251 (1971).
86. G. A. Hamilton, *in* "Molecular Mechanisms of Oxygen Activation" (O. Hayaishi, ed.), p. 405. Academic Press, New York, 1974.
87. D. M. Jerina and J. W. Daly, *Oxidases Relat. Redox Syst. Proc. Int. Symp., 2nd, 1971* Vol. 1, p. 143 (1973).
88. J. M. Tredger, R. S. Chhabra, and J. R. Fouts, *Drug Metab. Dispos.* **4**, 17 (1976).
89. H. G. Jonen, R. Kahl, and G. F. Kahl, *Xenobiotica* **6**, 307 (1976).
90. A. Y. H. Lu, K. W. Junk, and M. J. Coon, *J. Biol. Chem.* **224**, 3714 (1969).
91. A. Y. H. Lu, M. Jacobson, W. Levin, S. B. West, and R. Kuntzman, *Arch. Biochem. Biophys.* **153**, 294 (1972).

92. F. Kuma, R. A. Prough, and B. S. S. Masters, *Arch. Biochem. Biophys.* **172**, 600 (1976).
93. B. Mondovi, A. S. Benerecetti, and A. Rossi-Fanelli, *Arch. Sci. Biol.* (*Bologna*) **46**, 340 (1962).
94. B. N. LaDu, H. G. Mandel, and E. L. Way, eds., "Fundamentals of Drug Metabolism and Drug Disposition," p. 230. Williams & Wilkins, Baltimore, Maryland, 1970.
95. W. D. Brown and H. E. Synder, *J. Biol. Chem.* **244**, 6702 (1969).
96. M. J. Coon, A. P. Autor, R. F. Boyer, E. T. Lode, and H. W. Strobel, *Oxidases Relat. Redox Syst., Proc. Int. Symp., 2nd, 1971* Vol. 2, p. 529 (1973).
97. I. Fridovich, *Acc. Chem. Res.* **5**, 321 (1972).
98. F. Haber and J. Weiss, *Proc. R. Soc. London, Ser. A* **147**, 332 (1934).
99. C. Walling, *Acc. Chem. Res.* **8**, 125 (1975).
100. P. Neta and L. M. Dorfman, *Adv. Chem. Ser.* **81**, 222 (1968).
101. D. Keilin and E. F. Hartree, *Biochem. J.* **60**, 310 (1955).
102. M. N. Hughes, "The Inorganic Chemistry of Biological Processes," p. 142. Wiley, New York, 1974.
103. H. Schleyer, D. Y. Cooper, and O. Rosenthal, *Oxidases Relat. Redox Syst., Proc. Int. Symp., 2nd, 1971* Vol. 2, p. 469 (1973).
104. B. S. S. Masters, R. A. Prough, and H. Kamlin, *Biochemistry* **14**, 607 (1975).
105. B. Chance, *Adv. Enzymol.* **12**, 153 (1951).
106. P. George, *Oxidases Relat. Redox Syst., Proc. Symp., 1974* Vol. 1, p. 3 (1965).

13

Interaction of Transition-Metal Ions with Amino Acids, Oligopeptides, and Related Compounds

Akitsugu Nakahara, Osamu Yamauchi,
and Yasuo Nakao

INTRODUCTION

Recent interest in the role of metal ions in biological systems has led to a large number of investigations aimed at elucidation of the mechanisms of biological reactions requiring metal ions. Concurrent with progress in sophisticated experimental techniques that are applicable to systems with metal ions and biological macromolecules, more and more chemists have become involved in the scientific field shared by biochemistry and coordination chemistry and thus acquired valuable information on this subject.

Metal ions essential for living organisms play a variety of roles according to their own chemical properties. Non-transition-metal ions such as Na^+, K^+, Mg^{2+}, and Ca^{2+} perform their functions in biological control or trigger mechanisms and act as positive-charge carriers that stabilize the higher-order structures of macromolecules by neutralizing negatively charged groups. Of the transition-metal ions, zinc at the active center of carboxypeptidase and carbonic anhydrase exerts its effects as a strong Lewis acid binding with the substrates regarded as Lewis bases. Other transition-metal ions such as iron, cobalt, copper, and molybdenum play vital roles as biological redox catalysts in electron transfer processes.

Owing to the complexity of biological systems, full understanding of the functions of these metal ions is almost always achieved with difficulty except in very few cases, and little has been disclosed about the interactions between metal ions and biological molecules *in situ*. However, in view of the demand for information on the interactions between transition-metal ions and proteins, a great deal of information has been accumulated for more simplified systems containing small molecules such as amino acids and oligopeptides, enabling us to acquire insight into reactions in biological systems. Since the reactivities exhibited by proteins toward metal ions could be governed largely by those of the component amino acids or oligopeptides, proper knowledge of the fundamental systems would serve as a basis for clearer understanding of the biological reaction mechanisms.

This brief review is not intended to be a thorough survey of the literature but rather aims to summarize some important aspects of the model systems reported so far. In the following sections, we describe, first, complex formation between some transition-metal ions and amino acids or peptides and, second, the metal-ion-induced reactivities of these molecules as the results of complex formation.

COMPLEX FORMATION BETWEEN SOME TRANSITION-METAL IONS AND AMINO ACIDS, OLIGOPEPTIDES OR RELATED COMPOUNDS

As fundamental constituents of living organisms, amino acids and peptides have attracted wide attention, and a number of studies have been reported on their modes of interaction with transition-metal ions [1].

Metal Complexes Containing Only One Kind of Amino Acid

Some 20 L-α-amino acids known as the constituents of proteins are effective bi- and/or terdentate ligands, and their typical mode of reaction with transition-metal ions is the formation of five-membered metal chelates [(1) in Eq. (1)], in which the amino acidate zwitterion is bound to the metal ion M^{n+} through the amino nitrogen and the carboxylate oxygen with the concomitant deprotonation of the $-NH_3^+$ group.

The amino acids with additional donor groups in the side chain R may

$$M^{n+} + H_3\overset{+}{N}-\underset{R}{CH}-COO^- \rightleftharpoons M^{n+}\underset{O^--CO}{\overset{NH_2-CH-R}{\diagup\diagdown}} + H^+ \qquad (1)$$

1

form metal chelates having different structures. For example, the imidazole nitrogen of histidine can take part in coordination under certain conditions, in which histidine reacts as a terdentate ligand with Co(II), Ni(II), and Cd(II) to form the complexes of type [M(II)(L-His)₂] (2) [2–4]. In the 1:2 zinc(II)–

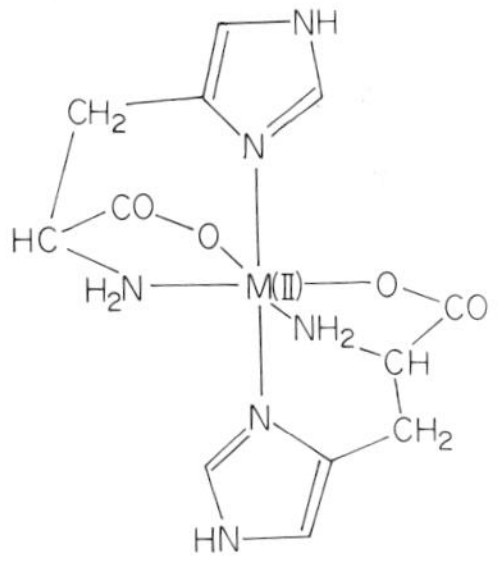

M = Co or Ni

2

histidine complex with a distorted tetrahedral structure [5], histidine acts as a bidentate ligand, linked to the central Zn(II) ion through the amino and imidazole nitrogens while the carboxylate group remains uncoordinated. A similar mode of chelation by histidine is found in glycyl-L-histidinatocopper-(II) (3) [6], where the carboxylate group of the histidyl residue binds not to the Cu(II) within the same molecule but to the adjacent one.

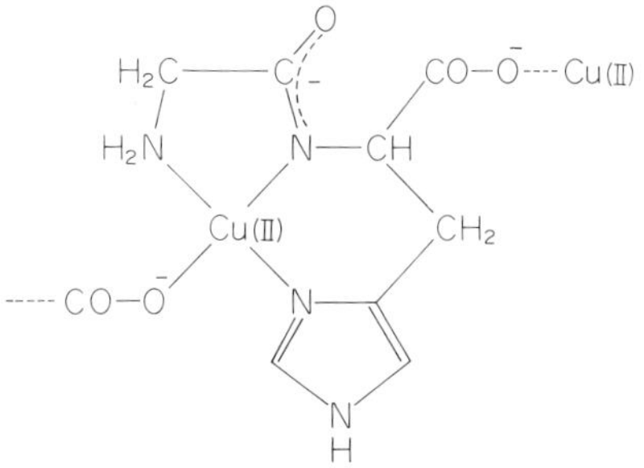

3

Various types of metal complexes of sulfur-containing amino acids such as methionine and cysteine have also been synthesized [7–11]. Hidaka et al.

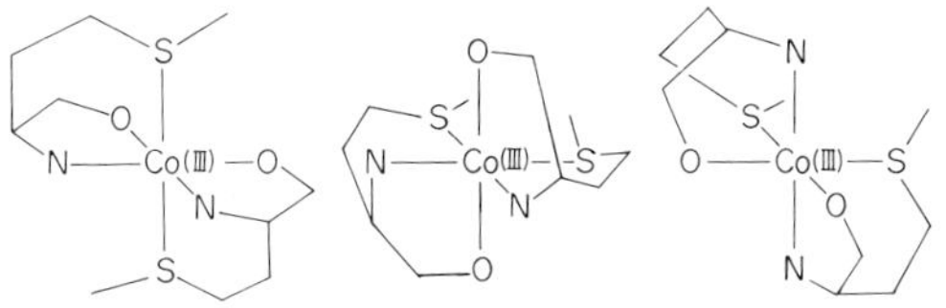

4a   4b   4c

[12] obtained three stereoisomers of 1:2 cobalt(III)–methionine, which were described by the structures (4a)–(4c), involving methionine as a terdentate ligand.

Amino acids are often coordinated to metal ions as unidentate ligands, binding only through the carboxylate oxygen (5) or the amino nitrogen (6), as expressed by Eqs. (2) and (3), respectively [13–16].

$$M^{n+} + {}^-OOC\!-\!\underset{\underset{R}{|}}{CH}\!-\!NH_3{}^+ \;\rightleftharpoons\; M^{n+}\!-\!\overset{-}{O}OC\!-\!\underset{\underset{R}{|}}{CH}\!-\!NH_3{}^+ \qquad (2)$$

(5)

$$M^{n+} + \overset{+}{N}H_3\!-\!\underset{\underset{R}{|}}{CH}\!-\!COO^- \;\rightleftharpoons\; M^{n+}\!-\!NH_2\!-\!\underset{\underset{R}{|}}{CH}\!-\!COO^- + H^+ \qquad (3)$$

(6)

Some inert cobalt(III) complexes have been reported to have amino acids bridging the two neighboring cobalt(III) ions through the amino and the carboxylate groups to form polynuclear structures such as (7a) and (7b) [17,18].

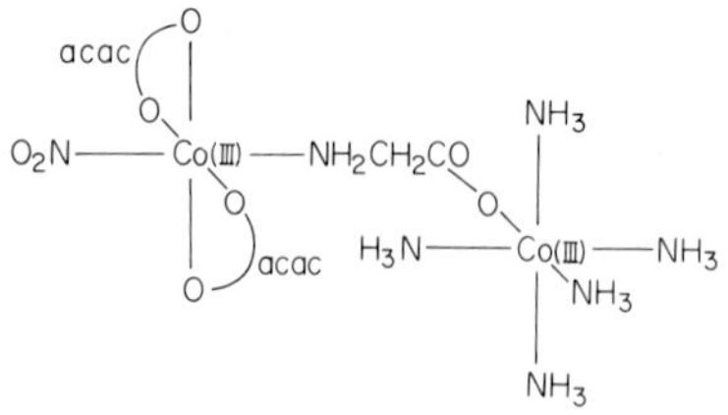

acac : acetylacetone

7a

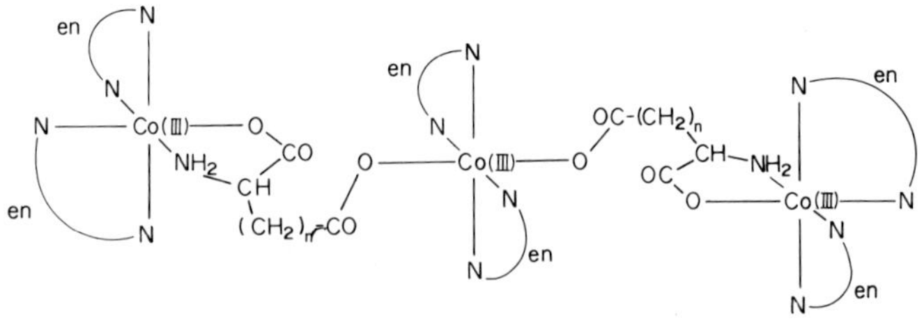

(n= 1 or 2),   en: ethylenediamine

7b

The number of possible modes of coordination may increase with the increasing ratio of ligand to metal ion. Exchange-inert cobalt(III) complexes can be separated chromatographically into their stereoisomers and determined by various spectroscopic methods [19,20]. Square-planar structures give rise to geometric isomerism in inert complexes such as bis(glycinato)platinum(II), the crystal structure of which has been determined for the *trans*-isomer by X-ray analysis [21]. Even for labile 1:2 copper(II)–amino acid complexes, some *cis*- and *trans*-isomers have been isolated as crystals under specific conditions, and their infrared (ir) spectra have been compared [22]. However, ' the characterization of the geometric isomers of labile complexes in solution is usually unsuccessful because rapid equilibration precludes isolation of only one of the two isomers.

In addition to the structural aspects of coordination, stability of metal complexes is another important factor, which is closely related to the structures and reactivities of complexes. The stability of metal–amino acid complexes depends mainly on the sizes of the chelate rings, five-membered rings being the most stable [23]. Thus, the stability of the copper(II)–amino acid complexes decreases in the order glycinate > β-alaninate > γ-aminobutyrate [24]. The central metal ions also affect the stability, and, as far as the typical mode of coordination shown by (**1**) is concerned, the stability constants for complexes of the bivalent ions of the first transition metals follow the Irving–Williams stability series [25]:

$$Mn(II) < Fe(II) < Co(II) < Ni(II) < Cu(II) > Zn(II)$$

Amino acids are known to have a tendency to form Schiff bases with carbonyl compounds, although the azomethine bond $>C=N-$ is often subject to hydrolysis in aqueous solution. Interestingly, the presence of transition-metal ions stabilizes the linkage when it is involved in chelate ring formation [26–29]. In this context, a number of metal chelates of Schiff bases derived from amino acids and salicylaldehyde or pyruvic acid have been prepared [30–34], and not a few of them have been shown to be of interest from synthetic and biological viewpoints.

Metal Complexes Containing Two Kinds of Amino Acids

Recently, mixed ligand complexes have attracted a great deal of attention, particularly because of their relevance to the enzyme–metal–substrate complexes formed as intermediates in many reactions performed by metalloenzymes [35]. In the presence of a 1000-fold excess of amino acid, transition-metal ions in biological fluids most probably form mixed ligand complexes, the simplest of which is a ternary complex, M(A)(B), where A and B refer to different amino acids. Sarkar and Kruck [36] actually isolated a mixed ligand copper(II) complex from human serum and identified it as [Cu(II)(L-His)-(L-Thr)], whose molecular structure was later revealed as (**8**) through X-ray analysis by Freeman *et al.* [37]. This and other histidine-containing ternary

8

copper(II) complexes have been inferred to be the transport forms of copper-
(II) in blood [36,38–40]. Various other ternary copper(II) complexes involving
amino acids were also detected in human serum [38].

Ternary complexes often have higher formation constants [41–45] than the
statistically expected value, which, for example, is calculated to be 4 for
the equilibrium shown in Eq. (4) [46,47]. Of fundamental importance are the

$$M(A)_2 + M(B)_2 \underset{}{\overset{X}{\rightleftharpoons}} 2M(A)(B) \qquad X = \frac{[M(A)(B)]^2}{[M(A)_2][M(B)_2]} \qquad (4)$$

driving forces leading to the formation of ternary complexes, because they
may be basically related to the factors governing the specific formation of the
enzyme–metal–substrate complexes and make it possible to interpret the
biological mechanisms in terms of molecular structures.

There have been proposed for model ternary systems several factors
including electronic effects such as π-back donation in the presence of π-
acceptors, steric effects, and electrostatic effects such as neutralization of
charges [44,45,48]. Sigel *et al.* [41,49,50] have demonstrated that in ternary
complex formation the copper(II)–2,2'-bipyridine complex prefers bonding
with ligands containing negatively charged oxygens to that with ethylenediamine or other ligands containing amino nitrogens. This finding points to
the ligand selectivity due to the electronic effect (π-back donation), which
finds applicability to experimental coordination chemistry. When accumulated, this kind of information may help in the determination of the
coordinating groups involved in metalloenzyme–substrate bondings or give
an explanation for the specificity of enzymatic reactions.

Brookes and Pettit [51] recently showed by potentiometric studies that the
ternary systems composed of copper(II), L-histidine, and an amino acid with
a positively charged side chain, such as L-arginine and L-lysine, exhibited
appreciably higher stability constants than those of the systems with D-
histidine in place of L-histidine. They ascribed the stereoselectivity to the
electrostatic interaction between the positive group and the carboxylate
oxygen of histidine, which would be more effective for two L-ligands (9) than

9   10

for an L- and a D-ligand (10), probably owing to the shorter distance between the charged groups in (9).

Synthetic and spectroscopic studies were made on the ternary copper(II) complexes containing two optically active α-amino acids with oppositely charged groups in their side chains [52]. For all combinations of ligands A and B, where A refers to aspartic or glutamic acid and B to arginine, lysine, or ornithine, the mixed ligand complexes [Cu(II)(L-A)(L-B)] and [Cu(II)(D-A)-(L-B)], except [Cu(II)(L-Asp)(L-Orn)], were isolated as crystals. By comparing the observed circular dichroism (CD) spectra in the visible region of the ternary systems with the spectra estimated from those of the corresponding binary systems, the cd magnitude enhancements were detected for all the [Cu(II)(L-A)(L-B)] systems, whereas no such enhancement was observed when one of the ligands was L-alanine or L-valine with no charged side chain [52a]. Similar observations were also made for the ternary copper(II) systems involving an α-amino acid and ethylenediamine-N-monoacetate [52b]. This CD spectral behavior and the facile isolation of the solid complexes have been interpreted as evidence supporting intramolecular electrostatic bonding between oppositely charged groups in the side chains of the ligands, as illustrated for [Cu(II)(L-Glu)(L-Arg)] (11). Such ligand–ligand interactions

11

probably serve as an effective driving force leading to ternary complex formation. Because of the steric requirements for the interactions, they gave rise to geometric isomerism among [Cu(II)(A)(B)], and molecular models suggested a *trans*-structure for [Cu(II)(L-A)(L-B)] and a *cis*-structure for [Cu(II)(D-A)(L-B)] or [Cu(II)(L-A)(D-B)] [52a]. In fact, stereoselectivity due to stability and/or solubility differences arising from geometric isomerism has led to successful optical resolution of DL-A and DL-B, respectively, *via* formation of ternary complexes [Cu(II)(A)(L-B)] and [Cu(II)(L-A)(B)], both of which preferentially incorporated D-enantiomers [53].

Metal Complexes of Oligopeptides

As described in the foregoing sections, many structures have been reported for the metal complexes of single amino acids. Hence, a far greater number of structures or modes of coordination should be possible for metal complexes of peptides composed of different amino acids. It is beyond the scope of this review to make a detailed survey of the structures of such complexes, so that the description in this section is confined to the structures and chemical properties of copper(II) chelates of small glycine peptides having no side chain.

Two copper(II) complexes of glycylglycine (abbreviated as HGly·Gly, where H indicates a dissociable peptide hydrogen), Cu(HGly·Gly)$^+$ [54] and Cu(Gly·Gly) [55], were isolated from acid and neutral or weakly alkaline solution, respectively. The molecular structures in the crystalline state have been revealed by X-ray analysis as (**12**) [56] and (**13**) [57] for the positive and

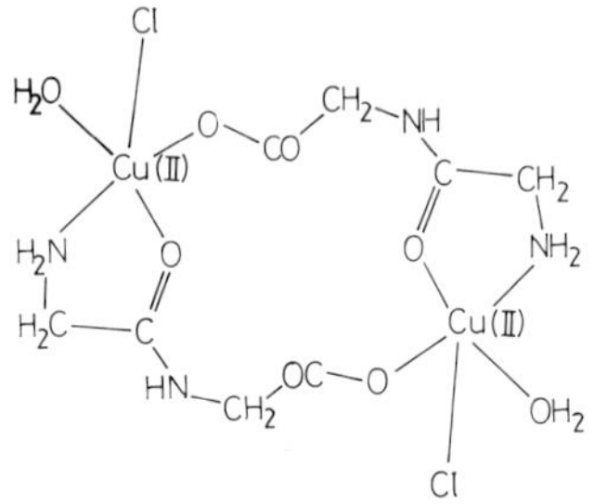

12

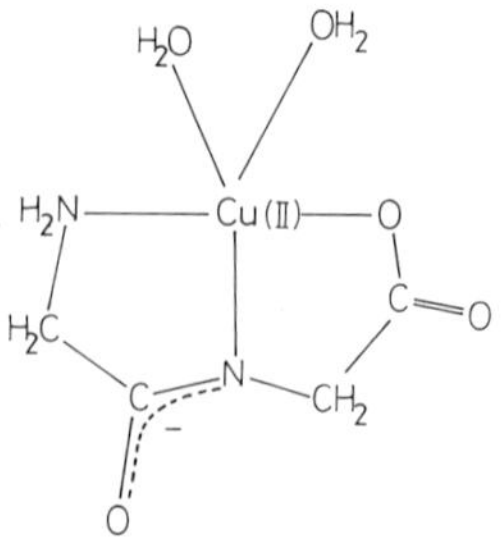

13

the neutral complex, respectively. The former **(12)** possesses a dimeric structure of C_2-like symmetry, where glycylglycine is bound to the copper(II) ion through the amino nitrogen and the peptide oxygen without deprotonation at the peptide linkage and to the adjacent copper(II) ion through the carboxyl oxygen; each of these copper(II) ions binds Cl^- and H_2O, forming a five-coordinate distorted square-pyramidal configuration. The structure **(13)** for the neutral complex shows that deprotonation of the peptide group gives double-bond character to the peptide N—C bond.

In addition to these 1:1 complexes, deprotonated 1:2 complexes were reported to be formed in the presence of an excessive amount of glycylglycine in rather strongly alkaline solution [58–60]. One of them was isolated as the potassium salt [61], which was shown to have the molecular structure **(14)** [62].

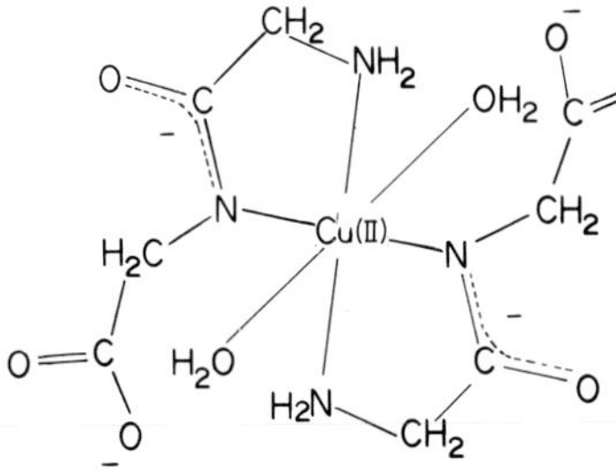

14

Triglycine ($H_2Gly \cdot Gly \cdot Gly$) reacts with copper(II) in a manner similar to that of glycylglycine, and the preparation and X-ray analyses of $Cu(H_2Gly \cdot Gly \cdot Gly)^+$ **(15)** [63] and $Cu(Gly \cdot Gly \cdot Gly)^-$ **(16)** [64] have been performed.

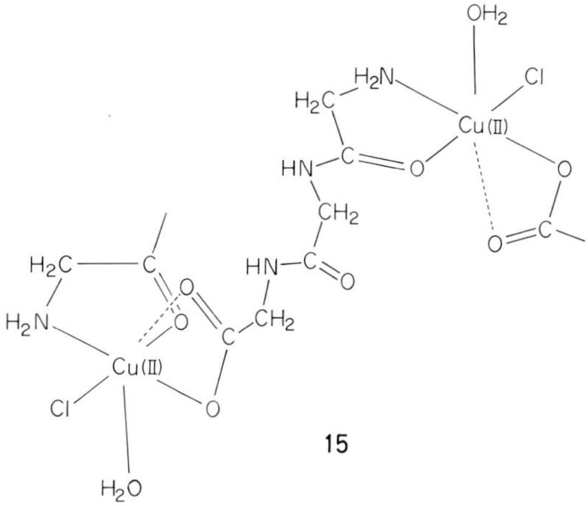

15

Structure (15) indicates that triglycine molecules bridge the neighboring copper(II) ions to form polymeric chains covering the whole crystal structure. On deprotonation, triglycine forms a copper(II) complex with a dimeric structure (16), in which the glycine residue at the C-terminus is coordinated

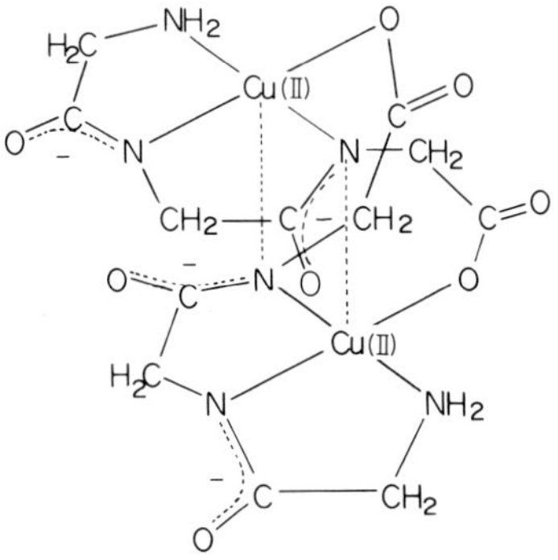

16

to another copper(II) ion above or below the plane of coordination. The distance between the copper(II) and the nitrogen interacting apically was found to be 0.257 nm, which indicates that the apical Cu—N bond is weaker than the bonds involved in the tetragonal plane.

With regard to the copper(II) chelates of tetraglycine ($H_3Gly \cdot Gly \cdot Gly \cdot Gly$) and pentaglycine ($H_4Gly \cdot Gly \cdot Gly \cdot Gly \cdot Gly$), the deprotonated complexes $Cu(Gly \cdot Gly \cdot Gly \cdot Gly)^{2-}$ and $Cu(HGly \cdot Gly \cdot Gly \cdot Gly \cdot Gly)^{2-}$ have been given, through X-ray studies, structures (17) [65] and (18) [66], respectively. In the pentaglycine complex, the carboxyl oxygen is located 0.308 nm from the

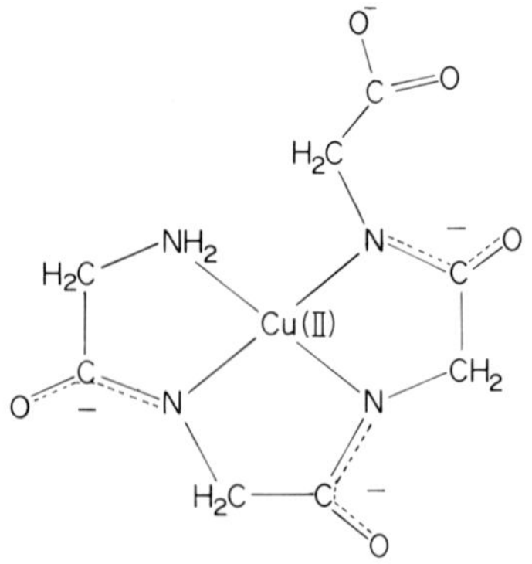

17

18

coordinated amino nitrogen to form an intramolecular hydrogen bond $>N—H\cdots\ ^-O—$.

Although many X-ray analysis data on metal–peptide complexes [1,67,68] have constituted unambiguous information on their complicated structures or modes of interactions in the crystalline state and also valuable evidence suggestive of those in solution, molecular structures in dilute aqueous solution may not necessarily be the same as in the solid state, mainly because of hydration, and polymeric forms such as (12), (15), and (16) are most probably split into monomers. Accordingly, the metal–peptide binding modes can be different from those established in the crystalline state.

Potentiometric studies have shown that the reaction of glycylglycine with copper(II) proceeds in acid to weakly alkaline solution according to Eqs. (5)

$$Cu^{2+} + HGly\cdot Gly^- \underset{}{\overset{K_1}{\rightleftharpoons}} Cu(HGly\cdot Gly)^+ \qquad K_1 = \frac{[Cu(HGly\cdot Gly)^+]}{[Cu^{2+}][HGly\cdot Gly^-]} \qquad (5)$$

$$Cu(HGly\cdot Gly)^+ \overset{K_{c_1}}{\rightleftharpoons} Cu(Gly\cdot Gly) + H^+ \qquad K_{c_1} = \frac{[H^+][Cu(Gly\cdot Gly)]}{[Cu(HGly\cdot Gly)^+]} \qquad (6a)$$

and (6a) [58,69–72]. Equation (6a) represents the deprotonation step of the peptide hydrogen. Additional deprotonation reactions such as shown for triglycine are known to occur for longer peptides with two or more peptide bonds [(6b) and (7)] [60,70,73–76].

$$Cu(H_2Gly\cdot Gly\cdot Gly)^+ \overset{K_{c_1}}{\rightleftharpoons} Cu(HGly\cdot Gly\cdot Gly) + H^+ \qquad (6b)$$

$$Cu(HGly\cdot Gly\cdot Gly) \overset{K_{c_2}}{\rightleftharpoons} Cu(Gly\cdot Gly\cdot Gly)^- + H^+$$

$$K_{c_2} = \frac{[H^+][Cu(Gly\cdot Gly\cdot Gly)^-]}{[Cu(HGly\cdot Gly\cdot Gly)]} \qquad (7)$$

TABLE 1

Acid Dissociation Constants (pK_a) of Oligopeptides and Equilibrium Constants of 1:1 Copper(II)–Oligopeptide Complexes (25°C) [a]

Ligand	pK_a (COOH)	pK_a (NH_3^+)	$\log K_1$	$-\log K_{c1}$	$-\log K_{c2}$	$\log K_1 K_{c1}$	$\log K_1 K_{c1} K_{c2}$	Ionic strength	Reference [b]
Glycylglycine	3.14	8.09	5.50	4.10		1.40		0.1 (KNO₃)	77
	3.23	8.16	5.50	4.30		1.20		0.16 (KCl)	(1)
	3.08	8.09	5.56	4.06		1.50		0.1 (NaClO₄)	78
	3.25	8.30	5.60	4.29		1.31		1 (NaClO₄)	60
β-Alanylglycine	3.22	9.45	5.45	4.09		1.36		0.1 (KNO₃)	77
Glycyl-β-alanine	3.98	8.16	5.70	4.64		1.06		0.1 (KNO₃)	72
	4.02	8.19	5.69	4.57		1.12		0.1 (KNO₃)	(2)
β-Alanyl-β-alanine	4.02	9.37	5.5[c]	6.8[c]		−1.3[c]		0.1 (KNO₃)	72
Triglycine	3.26	7.93	5.25	5.23	6.73	0.02	−6.71	0.1 (KNO₃)	76
	3.18	7.87	5.04	5.06	6.78	−0.02	−6.80	0.1 (NaClO₄)	78
	3.712	8.547	5.66	5.79	6.73	−0.13	−6.86	3.0 (NaClO₄)	75
	3.12	7.88	5.12	5.11	6.68	0.01	−6.67	0.1 (NaClO₄)	(3)
Glycylglycyl-β-alanine	4.08	7.93	5.25	5.27	6.08	−0.02	−6.10	0.1 (KNO₃)	76
Glycyl-β-alanylglycine	3.34	8.09	5.60	5.36	5.74	0.24	−5.50	0.1 (KNO₃)	76
β-Alanylglycylglycine	3.23	9.29	5.28	5.23	5.54	−0.04	−5.58	0.1 (KNO₃)	76
Glycylglycinamide		7.81	4.80	5.05	7.96	−0.25	−8.21	0.1 (KNO₃)	79
		7.78	4.88	5.07	8.01	−0.19	−8.20	0.1 (KNO₃)	(4)
Glycyl-β-alaninamide		7.87	5.22	5.42	8.99	−0.20	−9.19	0.1 (KNO₃)	79
β-Alanylglycinamide		9.18	5.16	5.39	d	−0.23	d	0.1 (KNO₃)	79
β-Alanyl-β-alaninamide		9.25	d	d	d	d	d	0.1 (KNO₃)	79

[a] Variances (3σ values), which are usually within ±0.05 log unit, are omitted for simplicity.

[b] References in parentheses are as follows (other references can be found in reference list): (1) G. F. Bryce, J. M. H. Pinkerton, L. K. Steinrauf, and F. R. N. Gurd, *J. Biol. Chem.* **240**, 3829 (1965). (2) G. Brookes and L. D. Pettit, *J. Chem. Soc., Dalton Trans.* p. 2106 (1975). (3) H. Hauer, E. J. Billo, and D. W. Margerum, *J. Am. Chem. Soc.* **93**, 4173 (1971). (4) T. F. Dorigatti and E. J. Billo, *J. Inorg. Nucl. Chem.* **37**, 1515 (1975).

[c] Not accurate because of precipitation.

[d] Not calculated because of precipitation.

Table 1 [60,72,75–79] summarizes the equilibrium constants for the copper-(II) complexes of various di- and tripeptides and dipeptide amides, and Figs. 1 and 2, respectively, indicate the abundances of the coordinated species in the 1:1 systems copper(II)–HGly·Gly [77] and copper(II)–H$_2$Gly·Gly·Gly [76] at pH < 8.

The degrees of formation of the deprotonated species at a constant pH can be estimated from the relationships shown in Eqs. (8a) and (8b). Figures 1

$$\frac{[\text{Cu(Gly·Gly)}]}{[\text{Cu}^{2+}][\text{HGly·Gly}]} = \frac{K_1 K_{c_1}}{[\text{H}^+]} \tag{8a}$$

$$\frac{[\text{Cu(Gly·Gly·Gly)}^-]}{[\text{Cu}^{2+}][\text{H}_2\text{Gly·Gly·Gly}]} = \frac{K_1 K_{c_1} K_{c2}}{[\text{H}^+]^2} \tag{8b}$$

and 2 indicate that Cu(Gly·Gly) and Cu(Gly·Gly·Gly)$^-$ are nearly completely formed at pH > 6 and at pH > 7.5, respectively. As we see from (13) and (16), these deprotonated complexes have fused-ring structures containing two consecutive five-membered rings (5–5-membered system). In dilute aqueous solution, Cu(Gly·Gly·Gly)$^-$ is inferred to have three consecutive five-membered rings (5-5-5-membered system), as depicted in (19), instead of the dimeric 5-5-membered rings (16). Comparison of the equilibrium constants for the copper(II) complexes of di- and tripeptides composed of glycine and/or β-alanine revealed that the degree of formation or the apparent stabilities of the complexes as estimated from the $K_1 K_{c_1}$ or $K_1 K_{c_1} K_{c_2}$ values [Eqs. (8a) and (8b)] were affected by the fused-ring structures. Thus, the stabilities for the dipeptide series decreased in the order Cu(Gly·Gly) (5–5) > Cu(β-Ala·Gly) (6-5) > Cu(Gly·β-Ala) (5-6) > Cu(β-Ala·β-Ala) (6-6), and those for the tripeptide series decreased in the order Cu(Gly·β-Ala·Gly)$^-$ (5-6-5) > Cu(β-Ala·Gly·Gly)$^-$ (6-5-5) > Cu(Gly·Gly·β-Ala)$^-$ (5-5-6)

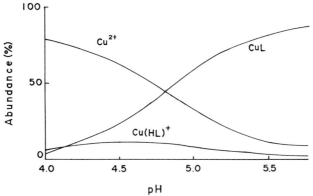

Fig. 1 Abundances of coordinated species in the 1:1 copper(II)–glycylglycine system [77]. Calculated from the equilibrium constants listed in Table 1. Glycylglycine is expressed as follows: HL$^-$ (HGly·Gly$^-$): L^{2-} (Gly·Gly^{2-}).

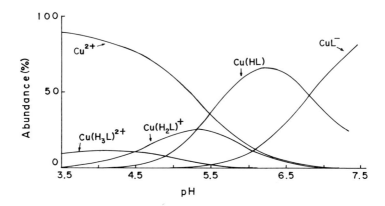

Fig. 2 Abundances of coordinated species in the 1:1 copper(II)–triglycine system [76]. Calculated from the equilibrium constants listed in Table 1. Triglycine is expressed as follows: H_3L ($H_2Gly \cdot Gly \cdot Gly$, zwitterionic form); H_2L^- ($H_2Gly \cdot Gly \cdot Gly^-$); HL^{2-} ($HGly \cdot Gly \cdot Gly^{2-}$); L^{3-} ($Gly \cdot Gly \cdot Gly^{3-}$).

> $Cu(Gly \cdot Gly \cdot Gly)^-$ (5-5-5) (Table 1). The observed stability sequences can be explained by the steric strains due to the fused-ring structures [72,76].

There has been considerable debate concerning the structure of the $Cu(HGly \cdot Gly)^+$ complex formed in acid solution. Two modes of coordination, (**20**) and (**21**), have been proposed for this species. The former assumes

19

bidentate chelation of glycylglycine through the amino nitrogen and the peptide oxygen, and the latter assumes terdentate chelation through the amino nitrogen, the peptide nitrogen, and the carboxyl oxygen. Recent kinetic [80,81], thermodynamic [78,82], and spectroscopic studies [84,85] are more in line with structure (**20**), postulated by Rabin [86], than with (**21**). In the case of glycylglycine, however, there is a possibility that the favorable locations of the terminal amino and carboxyl groups around the central copper(II) ion may enable glycylglycine to behave as a terdentate ligand, for which the bonding through the peptide nitrogen prior to deprotonation (**21**) is a necessary consequence. Terdentate chelation has been suggested on the

20   21

basis of comparative studies of the stability constants for various copper(II)–dipeptide chelates [72,77]. Interestingly, a 1:2 cobalt(III)–glycylglycine complex, $[Co(III)(HGly \cdot Gly)_2]^+$, prepared in acid solution, was shown by X-ray analysis to contain two terdentate glycylglycines with the peptide hydrogen bonded to the carbonyl oxygen [87].

Optical properties and kinetic aspects of metal–peptide complexes in solution have been reviewed recently by Martin [88] and Margerum and Dukes [89], respectively.

Some dipeptides as well as amino acids were reported to form Schiff bases with salicylaldehyde and other carbonyl compounds in the presence of copper(II) [30,90–94]. The corresponding Schiff base complexes were isolated as crystals, and structure–stability relationships were interpreted in terms of the fused-ring systems [91–94].

REACTIVITY OF AMINO ACIDS AND OLIGOPEPTIDES IN THE COORDINATION SPHERE OF TRANSITION METALS

Coordinated ligands acquire remarkable reactivity due to the electron-withdrawing effect of positively charged metal ions. Amino acids and oligopeptides bonded to transition-metal ions are no exceptions, and, aside from the interesting bonding modes involved in their metal complexes outlined in the section on complex formations between transition-metal ions and amino acids, oligopeptides, on related compounds, they exhibit a variety of reactivities upon coordination to metal ions [95,96]. The following sections are concerned primarily with the selective activations of coordinated amino acids and peptides.

Deprotonation of Coordinated Amino Groups

The amino group of glycine coordinated to nickel(II) was found to be deprotonated with potassium amide in liquid ammonia to give (22) and (23)

$$[\text{Ni(II)}(\text{NH}_2\text{CH}_2\text{COO}^-)_2] + \text{NH}_2^- \xrightarrow[(-33.5°C)]{\text{in liquid NH}_3}$$

$$[\text{Ni(II)}(\bar{\text{N}}\text{HCH}_2\text{COO}^-)(\text{NH}_2\text{CH}_2\text{COO}^-)]^- + \text{NH}_3$$

$$\textbf{22} \tag{9}$$

$$[\text{Ni(II)}(\text{NH}_2\text{CH}_2\text{COO}^-)_2] + 2\text{NH}_2^- \xrightarrow[(-33.5°C)]{\text{in liquid NH}_3}$$

$$[\text{Ni(II)}(\bar{\text{N}}\text{HCH}_2\text{COO}^-)_2]^{2-} + 2\text{NH}_3$$

$$\textbf{23}$$

[97,98] [Eqs. (9)]. Similar reactions were observed for tris(glycinato)cobalt(III) and bis(iminodiacetato)cobaltate(III), and the respective deprotonated complexes were isolated [99]. All these complexes were stable under helium atmosphere but decomposed in water to give the corresponding original complexes. These reactions reflect electron attraction by metal ions from the coordinated amino nitrogen.

Reactivity Enhancement of the Methylene Group of Coordinated Glycine

On the basis of a proton magnetic resonance (pmr) study of cobalt(III)–amino acid chelates in deuterium oxide, Williams and Busch found that the methine or methylene protons of α-amino acids are activated by cobalt(III) and become exchangeable with deuterons [100]. This finding suggested that the methylene group of coordinated glycine might be readily deprotonated in the presence of a base and make a subsequent attack by an electrophile feasible.

In fact, threonine was successfully synthesized by the reaction of the copper(II)–glycine complex with acetaldehyde in alkaline solution [101] [Eq. (10)]. The resulting threonine could be isolated from the complex either

$$\text{Eq. (10)}$$

by treating with concentrated aqueous ammonia and removing copper(II)–ammine complexes with a cation-exchange resin or by passing hydrogen sulfide through the aqueous solution of the complex. The intermediate in this reaction was isolated and was shown by X-ray analysis to have structure (**24**), involving oxazolidine rings [102].

Scheme 1 shows that the activated methylene group of *N*-salicylidene-glycinatocopper(II) (**25**) reacts with alkyl halides in the presence of a strong

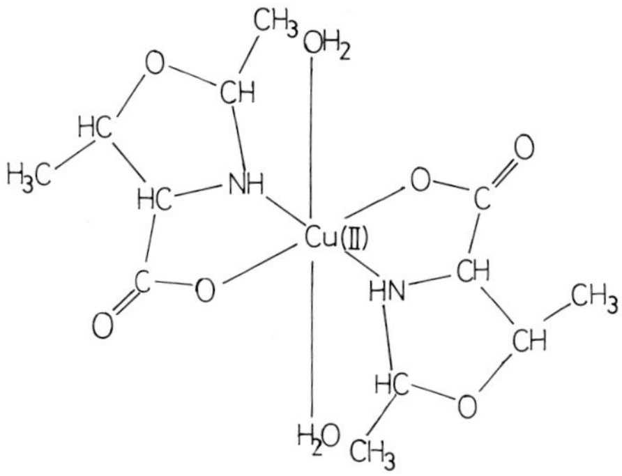

24

alkali to give the Schiff base complexes of α-alanine, valine, leucine, phenyl-alanine, and aspartic acid (26) [103]. The yields of the final products (26) were not high enough to be useful for practical synthetic purposes, but the reaction itself suggests possible natural processes involving conversion of glycine to various other amino acids and might shed light on the buildup of the "fore-protein" in primitive ages. The reactivity of the methylene group is considered

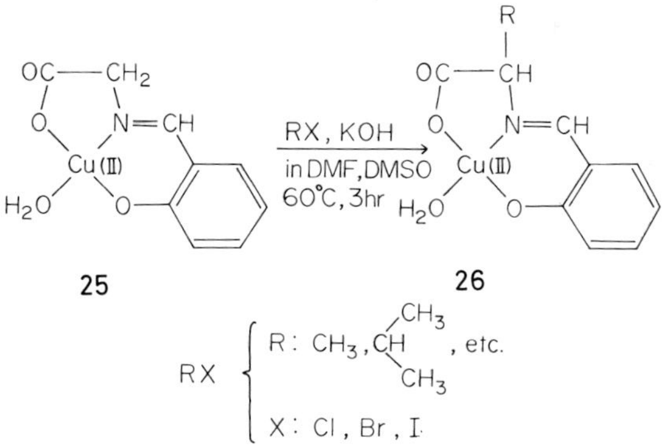

Scheme 1

to be greater when it is contained in a Schiff base chelate than when it is in a simple glycinate complex, because the electron-attracting power of the central copper(II) ion in a Schiff base complex can be transmitted to the methylene group through both the azomethine nitrogen and the carboxyl oxygen [104]. Enhanced reactivity of such a complex and high solubility in organic solvents such as dimethylformamide and dimethyl sulfoxide favor the above reaction and make it an interesting synthetic route to amino acids. In this connection, Harada and Oh-hashi [105] and Ichikawa et al. [106], respectively, reported successful synthesis of β-hydroxy-α-amino acids from N-salicylidene- and N-pyruvylideneglycinatocopper(II).

It is noteworthy that partially asymmetric syntheses of threonine and *allo*-threonine have been performed by using acetaldehyde and optically active glycinatobis(ethylenediamine)cobalt(III) [107,108], bis(glycinato)ethylene-diaminecobalt(III) [108], or bis(N-salicylideneglycinato)cobalt(III) [109].

Selective Activation of the Methylene Groups of Glycylglycine

Deuteration of the cobalt(III)–glycylglycine complex (**27**) revealed that the

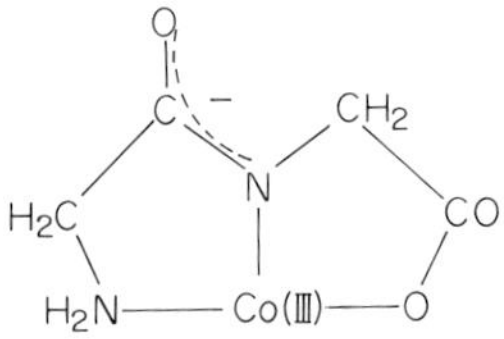

27

protons of the C-terminal methylene group exchanged readily with deuterons in alkaline solution (pD > 11) owing to activation by cobalt(III), whereas the protons of the N-terminal methylene group did not under the same conditions or when treated with alkali for 4 weeks [110]. No evidence of exchange of any methylene protons in alkaline solution was obtained for uncomplexed glycylglycine, which indicates that coordination around cobalt(III) is a prerequisite for selective activation of the C-terminal methylene group [110].

On the other hand, the reaction of glycylglycine with formaldehyde in the presence of an equimolar amount of copper(II) in alkaline solution gave serylglycine but neither serylserine nor glycylserine [111]. This finding suggests that the N-terminal methylene group was activated by chelation with copper-(II), which appears to be contradictory to the observations about the cobalt-(III)–glycylglycine system [110]. To arrive at a solution to this seeming inconsistency, Uyama *et al.* [112] carried out a pmr study of the nickel(II) chelates of the Schiff bases derived from salicylaldehyde and dipeptides

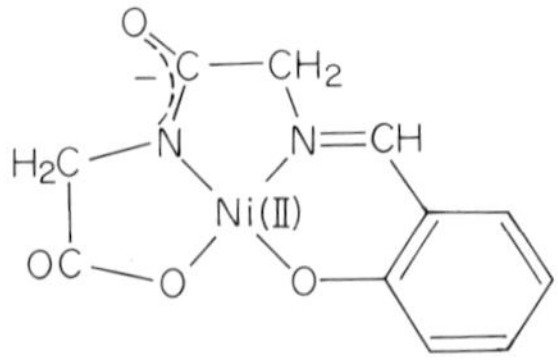

28

containing glycine and/or α-alanine and found that in deuterium oxide (pD > 11) the methylene protons of the N-terminal glycyl residue of the complex (28) were much more readily exchanged with deuterons than were those of the C-terminal residue. That the corresponding cobalt(III) chelates gave the same results [112] indicates that the activation of the N-terminal methylene group is a consequence of Schiff base complex formation, irrespective of the metals used. Therefore, the formation of serylglycine from glycylglycine and formaldehyde is probably explained by a mechanism that assumes initial formation of the intermediate Schiff base complex (29) and

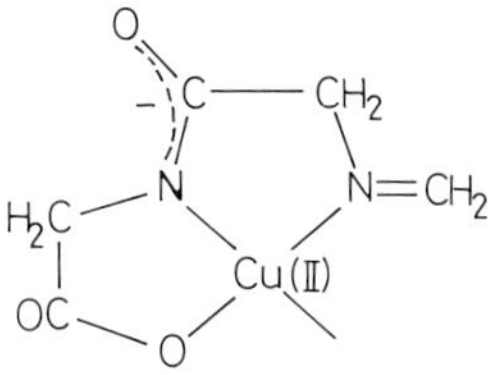

29

subsequent electrophilic attack of another formaldehyde molecule at the N-terminal methylene group.

Selective Activation of the Methylene or Methine Groups of Tripeptides

Studies on the selective activation in glycylglycine have been extended to tripeptides containing glycine and/or α-alanine (30) [113]. In accordance with

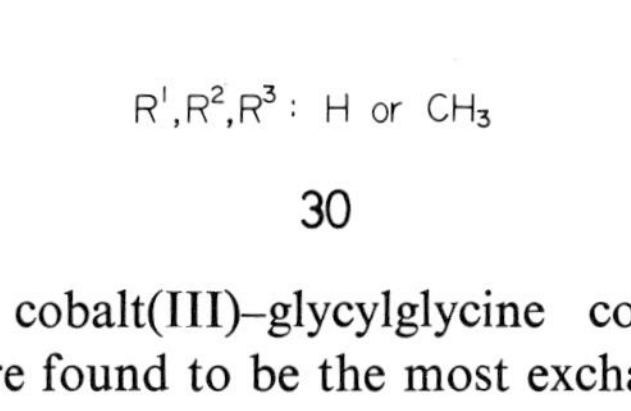

R^1,R^2,R^3 : H or CH$_3$

30

observations of the cobalt(III)–glycylglycine complex, the C-terminal methylene protons were found to be the most exchangeable with deuterons. Schiff base formation with salicylaldehyde (31) shifted the site of activation from the C-terminal glycine residue to the N-terminal glycine residue [114], and the reaction of the copper(II)–triglycine complex with acetaldehyde in

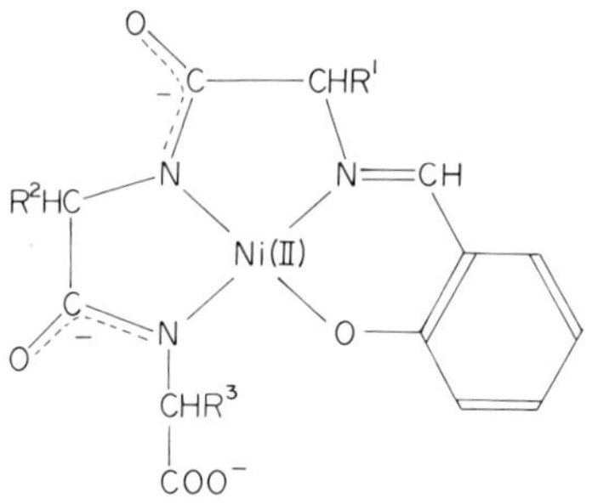

R^1, R^2, R^3 : H or CH_3

31

alkaline solution was shown to give threonylglycylglycine, which proves that the N-terminal methylene group preferentially reacts with an electrophile [115].

Particularly significant is the fact that, whereas a tripeptide is hardly susceptible to the proton exchange reaction at any of the methylene groups except the central one, which shows slight activity [116], it becomes considerably activated upon coordination to a metal ion, being deuterated at the C-terminal methylene group when bound in a fashion represented by (30) and at the N-terminal group when incorporated into a Schiff base complex such as (31). Thus, the selective activation of a certain methylene group of peptides can be attained by suitable modification of the peptide molecules through chelation with transition metals [114].

Hydrolysis of Amino Acid Esters and Amides and Peptides

In addition to the activation of the amino group and the methylene or methine group of amino acids and oligopeptides, metal ions promote hydrolytic cleavage of amino acid esters and peptides by virtue of their polarizing effect on the ester and peptide groups involved in complex formations.

Although the hydrolysis reactions of amino acid esters usually proceed very slowly in neutral solution at room temperature, the presence of transition metals such as copper(II), cobalt(II), and manganese(II) accelerates them considerably. Kroll [117] explained the catalytic role of metal ions in the hydrolysis reactions by a mechanism involving the intermediate (32) or (33) as illustrated in Scheme 2. Formation of (32) was confirmed later by several investigations [118–124].

The diesters of aspartic and glutamic acids suffer selective hydrolysis by copper(II) at the α-ester group because of its involvement in chelate ring formation [125,126]. Amino acid amides have also been reported to be

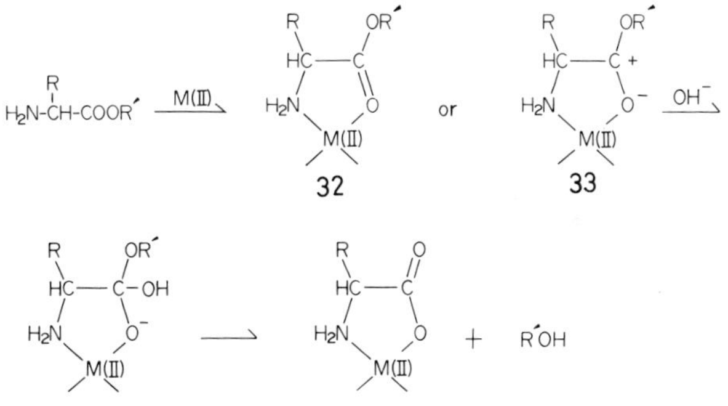

Scheme 2

hydrolyzed in the presence of metal ions, and the reaction mechanism has been postulated [127–130].

Selective cleavage of peptide bonds takes place in the presence of cobalt(III) chelates having two readily exchangeable ligands in the *cis* positions. Very useful examples of peptide hydrolysis were presented by Collman and Buckingham [131] and Buckingham *et al.* [132], who made use of *cis*-β-hydroxyaquatriethylenetetraminecobalt(III), $[Co(III)(OH)(H_2O)trien]^{2+}$, where trien refers to triethylenetetramine [Eqs. (11)]. The reactions were

$$
[Co(III)(OH)(H_2O)\,trien]^{2+}
\begin{cases}
\xrightarrow{HGly\cdot Gly} [Co(III)(Gly)trien]^{2+} + Gly \\
\xrightarrow{H_2Gly\cdot Gly\cdot Gly} [Co(III)(Gly)trien]^{2+} + HGly\cdot Gly \\
\xrightarrow{HGly\cdot Phe} [Co(III)(Gly)trien]^{2+} + Phe \\
\xrightarrow{HPhe\cdot Gly} [Co(III)(Phe)trien)^{2+} + Gly
\end{cases}
\qquad (11)
$$

carried out at pH 7.5 (65°C), and the crystalline products were identified by elemental analysis, ir and pmr studies, and paper chromatography. The reactions have been considered to proceed via either of the two paths shown in Scheme 3. The molecular structure of the reaction intermediate, isolated under suitable conditions, was determined by X-ray analysis [133]. A similar complex, *cis*-$[Co(III)(OH)(H_2O)en_2]^{2+}$, which has two ethylenediamine (en) molecules instead of a triethylenetetramine molecule, also catalyzed peptide hydrolysis, although the reaction was accompanied by complicated by-products [134]. In an analogous fashion 2,2′,2″-triaminotriethylaminecobalt(III) exhibited hydrolytic activity [135].

It is understood from the four reactions shown in Eq. (11) that the hydrolysis reaction is specific for the first peptide linkage from the N-terminus.

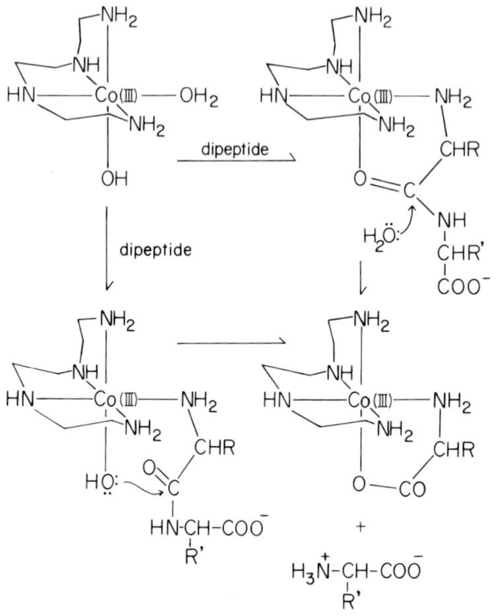

Scheme 3

Another type of peptide hydrolysis, which was specific for the second peptide bond from the N-terminus, was effected by using bis(salicylaldehydato)-copper(II) (**34**), which reacted with an amino acid and a dipeptide to form the corresponding Schiff base complexes according to Schemes 4 and 5, respectively [30]. With a longer peptide at pH 4.5 (70°C), it produced *N*-salicylidenedipeptidatocuprate(II), containing the hydrolyzed peptide (**35**)

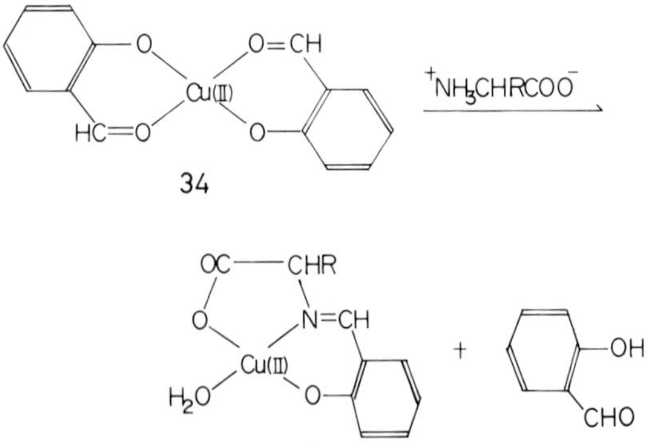

Scheme 4

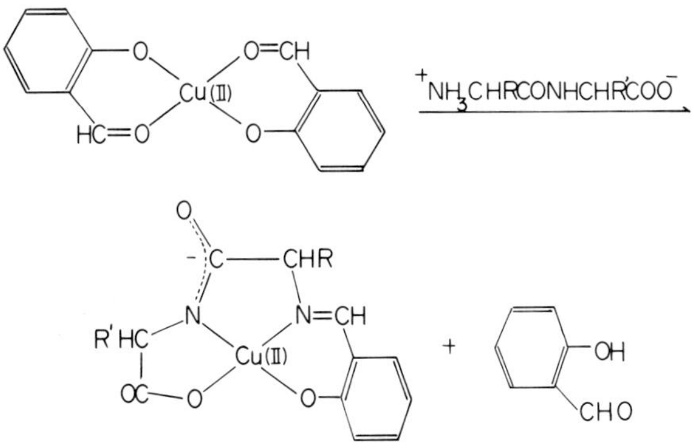

Scheme 5

as indicated in Eqs. (12) and Scheme 6 [90,136], where Sal refers to salicylaldehyde and the double bond between Sal and the peptide denotes the
azomethine bonding formed between them.

$$[Cu(II)(Sal)_2]$$

$H_2Gly \cdot Gly \cdot Gly$ in aqueous EtOH	$[Cu(II)(Sal=Gly \cdot Gly)]^- + Gly + Sal$
$H_2Ala \cdot Gly \cdot Gly$ in aqueous EtOH	$[Cu(II)(Sal=Ala \cdot Gly)]^- + Gly + Sal$
$H_2Gly \cdot Gly \cdot Ala$ in aqueous EtOH	$[Cu(II)(Sal=Gly \cdot Gly)]^- + Ala + Sal$
$H_3Gly \cdot Gly \cdot Gly \cdot Gly$ in aqueous EtOH	$[Cu(II)(Sal=Gly \cdot Gly)]^- + HGly \cdot Gly + Sal$
$H_4Gly \cdot Gly \cdot Gly \cdot Gly \cdot Gly$ in aqueous EtOH	$[Cu(II)(Sal=Gly \cdot Gly)]^- + H_2Gly \cdot Gly \cdot Gly + Sal$

$$(12)$$

The reaction mechanism (Scheme 6) involves the intermediate complexes
(**36**) and (**37**), which were isolated as crystals and identified by elemental
analyses and ir spectra. The driving force of the hydrolysis reactions may be
attributed to the electronegativity of copper(II). It lowers the electron density
at the carbon atom of the coordinated peptide carbonyl group (**36**), which
then suffers nucleophilic attack at the carbon atom by a water molecule to
accomplish the hydrolytic cleavage of the N—C bond.

The situation in the doubly deprotonated complex (**37**) distinctly differs
from that in (**36**), since the electron density of the relevant carbonyl carbon
atom cannot be lowered effectively because of the negative charge resulting
from deprotonation.

The optimal pH of 4.5, as concluded from the pH–reactivity profile for this
reaction, probably corresponds to the pH region where species (**36**) is most

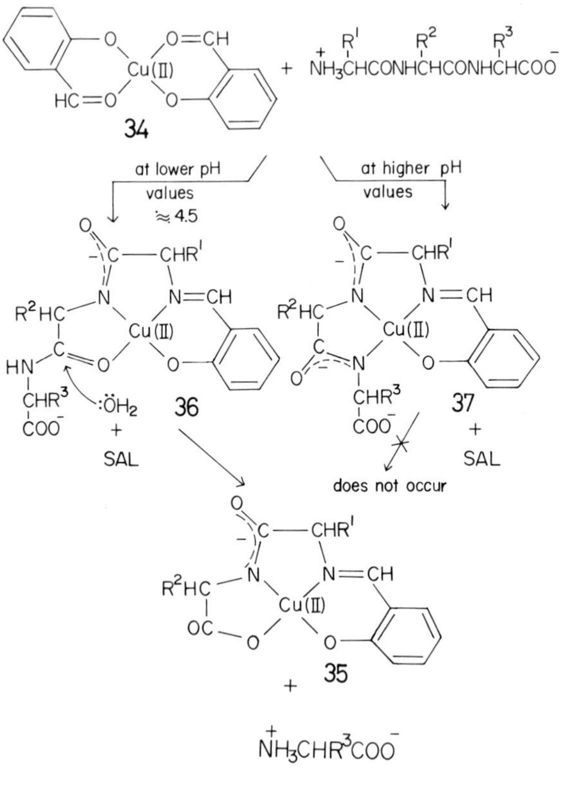

Scheme 6

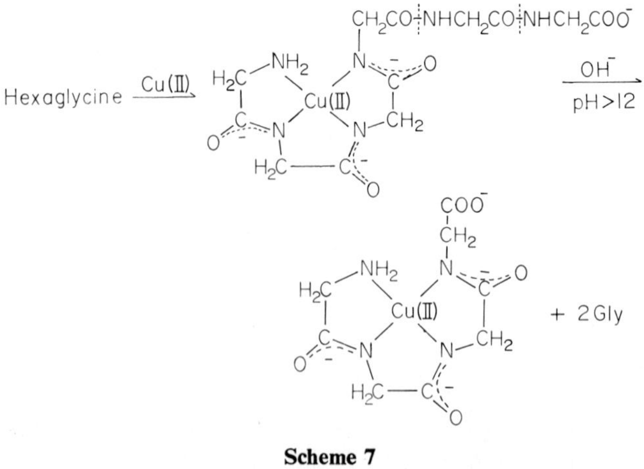

Scheme 7

abundant. The cobalt(II) and nickel(II) chelates of salicylaldehyde exhibited the same reactivity as the copper(II) chelate, but the optimal pH values of the reactions were considerably higher as compared with the reaction involving copper(II), suggesting the difference in species distribution [136].

Selective peptide hydrolysis was reported to occur in alkaline solution in the presence of copper(II) according to Scheme 7 [137], in which the role of

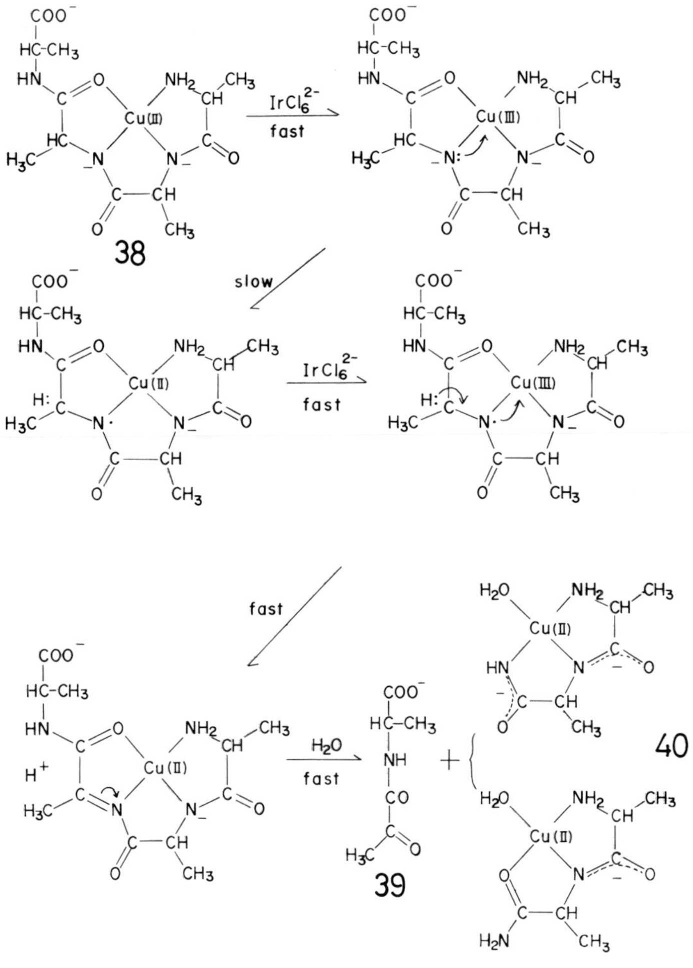

Scheme 8   The structural formulas described in the original scheme [138] were somewhat modified in this scheme, since under the experimental conditions (pH 8) the third amide group from the N-terminus is expected to be bonded to copper(II) through the carbonyl oxygen atom. Also, the final product (40) is probably a mixture of these two species, as judged from the experimental conditions [79].

the metal ion was supposed to be protection of the sequence of four residues of the bound peptide against alkaline hydrolysis. This seems reasonable because the conditions employed were such that all the peptide bonds might have been hydrolyzed in the absence of metal ions. The function of the metal ion may be explained by the same mechanism as described for the Schiff base complex (36).

Levitzki *et al.* [138] reported the interesting finding that the copper(II) chelate of tetra-L-alanine (38) was oxidized by 2 equivalents of hexachloroiridate(IV) at pH 6–9 and subsequently hydrolyzed. Scheme 8 shows the mechanism proposed for this reaction. Copper(II) is oxidized by hexachloroiridate(IV) to copper(III), which then is reduced to give a free-radical intermediate; copper(III) regenerated by a second molecule of hexachloroiridate(IV) oxidizes the peptide group in a subsequent step, leading ultimately to a carbonyl derivative (39).

Peptide Formation

The catalytic function of the cobalt(III)–trien complex in the hydrolysis of some oligopeptides described in the preceding section can also be applied to the formation of peptide bonds by employing suitable conditions. Thus, Buckingham *et al.* [139] succeeded in obtaining β-[Co(III)(HGly·GlyOEt)-trien]$^{3+}$ (41), where HGly·GlyOEt is ethyl glycylglycinate, from ethyl

$$\alpha\text{-[Co(III)(TBP)}_2\text{trien]}^{3+} + 2\text{GlyOEt} \xrightarrow[25°C]{\text{TBP}}$$

$$\beta\text{-[Co(III)(HGly·GlyOEt)trien]}^{3+} + \text{EtOH} + 2\text{TBP}$$

$$\textbf{(41)}$$

$$\beta\text{-[Co(III)Cl(GlyOEt)trien]}^{2+} + \text{GlyOEt} \xrightarrow[25°C]{\text{sulfolane}} \textbf{(41)} + \text{EtOH} + \text{Cl}^- \tag{13}$$

$$\beta\text{-[Co(III)(TBP)}_2\text{trien]}^{3+} + 2\text{GlyOEt} \xrightarrow[25°C]{\text{TBP}} \textbf{(41)} + \text{EtOH} + 2\text{TBP}$$

glycinate (GlyOEt) and α- or β-[Co(III)(TBP)$_2$trien]$^{3+}$ or β-[Co(III) Cl-(GlyOEt)trien]$^{2+}$ (42), where TBP refers to tri-*n*-butyl phosphate, in tri-*n*-butyl phosphate or dimethylformamide as a solvent [Eqs. (13)]. The final product (41) could be obtained in a yield as high as 80% by either of the three

41

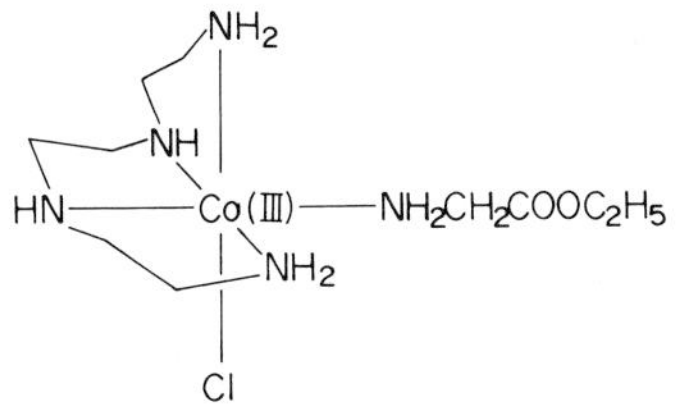

42

methods in Eq. (13) and was identified by elemental analysis and comparison of the ir, pmr, and visible and ultraviolet absorption spectra with those of an authentic sample prepared by a known method.

Similar reactions were carried out by using $[Co(III)(GlyOMe)en_2]^{3+}$, where GlyOMe is methyl glycinate, instead of (**42**) [140] [Eq. (14)], where

$$[Co(III)(GlyOMe)en_2]^{3+} + AAOR \xrightarrow[20°C]{acetone}$$

$$[Co(III)(HGly \cdot AAOR)en_2]^{3+} + MeOH \qquad (14)$$

AAOR denotes an amino acid ester. An analogous complex, cis-$[Co(III)-Cl_2trien]^+$, has also been used as a starting material [141].

A completely different type of reaction of ethyl glycinate in the coordination sphere of copper(II) was found to give the ethyl esters of glycylglycine, triglycine, and tetraglycine at room temperature in nonaqueous solvents [142,143]. The proposed mechanism involves an intramolecular nucleophilic attack of the deprotonated amino nitrogen on the neighboring carbonyl carbon (Scheme 9) [144]. In contrast to ethyl glycinate, some optically active amino acid esters yielded only the corresponding dipeptide esters [145].

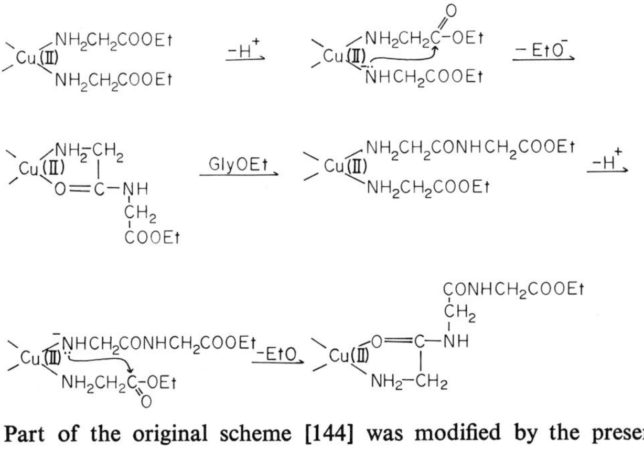

Scheme 9   Part of the original scheme [144] was modified by the present authors.

Apart from the peptide formation reactions, oligopeptides have been reported to undergo metal-ion-catalyzed oxidations, in which the nickel(II) [146] and copper(II) [147] complexes of tetra- and pentapeptides consumed oxygen in neutral solution to give a number of products including amino acid amides, peptide amides, oxo acids, and carbon dioxide.

Transamination Reactions [34]

Various transaminases are widely distributed in biological systems, and their enzymatic activity is believed to require pyridoxal or pyridoxamine phosphate as a coenzyme [148]. It is well established that pyridoxal catalyzes the conversion of α-amino acids to keto acids by nonenzymatic transamination, especially in the presence of metal ions such as copper(II), iron(II), iron(III), and aluminum(III) [34,149].

The catalytic behavior of pyridoxal was explained by Metzler *et al.* [150] by invoking the metal chelate of the Schiff base produced from pyridoxal and an amino acid (Scheme 10). Although the chelate ring structure for the pyridoxal–amino acid system (43) is exactly the same as that for the salicylaldehyde–

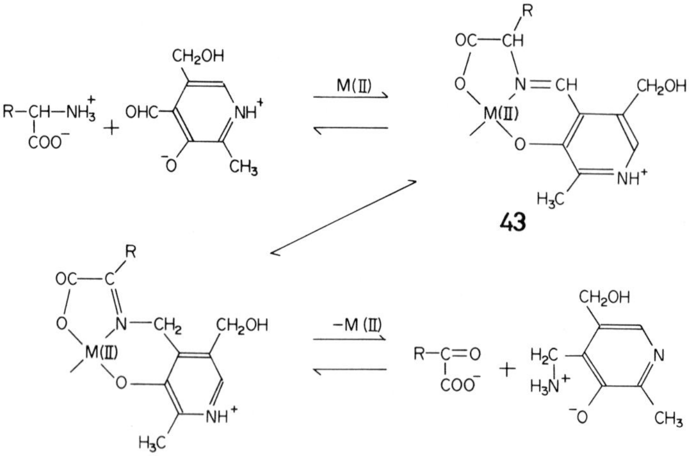

Scheme 10

amino acid system (44), the latter did not show any tendency to undergo the structural change toward (45) except a slight reverse change, and both chelates (44) and (45) could be isolated as stable crystals [93,151]. The finding suggests that the nitrogen atom of the pyridine ring of pyridoxal or pyridoxamine plays an essential role in the coenzyme function in transamination.

Very rapid and effective transamination reactions have been reported for

44   45

some complexes with fused chelate rings without the pyridoxal or pyridoxa-mine moiety [152–157]. The copper(II) complex isolated from a reaction mixture containing equimolar amounts of copper(II), glyoxylic acid, and α-alanine in water at 40°C turned out to be the complex of the Schiff base that could be derived directly from pyruvic acid and glycine (47) and not the

46   47

expected complex (46) [156,157]. The yield of the reaction starting from glyoxylic acid and α-alanine was roughly the same as that attained by direct synthesis from pyruvic acid and glycine, both in the presence of copper(II). The same result was obtained for the system with palladium(II) in place of copper(II) [104]. These observations clearly show that the transamination reaction occurs very effectively from (46) to (47), although it does not exclude the possibility of the transient existence of (46) in the reaction mixture.

An important driving force of this kind of transamination reaction is considered to be the steric strains associated with formation of fused-ring metal chelates. Thus, the strain produced in the Schiff base chelates composed of two consecutive five-membered chelate rings such as (46) and (47) is estimated to be so great [31,93] that the double bond of the azomethine group may be delocalized to decrease the strain as a whole. Because the $C=N$ double bond might be completely delocalized in the transition state of the reaction and consequently the two C—N bond lengths around the nitrogen might be the same, the activation energy for the rearrangement of (46) to (47) would be small, and some factors such as solubility could affect the equilibrium between the two species.

Evidence supporting this argument can be obtained by comparison of the differences between the two C—N distances in several metal chelates. Interestingly, Table 2 shows that such differences are greater for 5-6- or 6-5-membered chelates, where the fused-ring strains may be smaller owing to cancellation of the strains than in the 5-5-membered ones, which retain the

TABLE 2

Azomethine C—N Bond Lengths in Some Schiff Base–Copper(II) Complexes

Complex[a]	Fused-ring system[b]	Bond length (nm)		Difference (nm)
[Cu(II)(Sal=Gly)(H$_2$O)]·$\frac{1}{2}$H$_2$O[c]	6–5	0.128	0.145	0.017
[Cu(II)(Sal=Gly)(H$_2$O)]·4H$_2$O[d]	6–5	0.130	0.146	0.016
[Cu(II)(Pyv=Gly)(H$_2$O)]·2H$_2$O[e]	5–5	0.131	0.143	0.012
[Cu(II)(Pyv=β-Ala)(H$_2$O)]·2H$_2$O[f]	5–6	0.125	0.148	0.023

[a] Sal=Gly, Pyv=Gly, and Pyv=β-Ala are Schiff bases derived from amino acids and salicylaldehyde (Sal) or pyruvic acid (Pyv).

[b] The numbers denote the sizes of the fused-ring system with the ring formed by the Sal or Pyv moiety designated first.

[c] T. Ueki, T. Ashida, Y. Sasada, and M. Kakudo, *Acta Crystallogr.* 22, 870 (1967).

[d] T. Ueki, T. Ashida, Y. Sasada, and M. Kakudo, *Acta Crystallogr., Sect. B* 25, 328 (1969).

[e] A. Torii, H. Tamura, K. Ogawa, and T. Watanabe, *Z. Kristallogr.* 133, 179 (1971).

[f] T. Ueki, T. Ashida, Y. Sasada, and M. Kakudo, *Acta Crystallogr., Sect. B* 24, 1361 (1968).

strains [31]. Therefore, the degree of delocalization of the C=N bond as is estimated from the C—N bond lengths may reflect the accumulated strains and point to the possibility of transamination.

CONCLUDING REMARKS

The above-described interactions between transition-metal ions and biological molecules can hardly explain by themselves the highly efficient and specific catalytic functions exhibited by metalloenzymes and metal-ion-activated enzymes, and innumerable reactions proceeding in cooperation with each other in biological systems are far beyond our present recognition. However, the principles underlying intricate biological processes should be compatible with what we understand from model systems involving small molecules that serve as constituents of biological macromolecules. Specifically, the structures and functions peculiar to the metal binding sites in biological systems may be related to those of the model systems in some way, and it should be possible to approximate the active sites of enzymes more precisely by proper choice of metal complexes.

In view of the enzyme–metal–substrate complexes formed in enzymatic reactions, chemical and biological properties of mixed ligand complexes containing peptides and analogous compounds are attractive and important subjects to be explored and may enable us to simulate the hydrophobic environment, stereospecificity, and other characteristics that are the exclusive properties of enzymes.

REFERENCES

1. J. P. Greenstein and M. Winitz, "Chemistry of the Amino Acids," Vol. 1, pp. 569–682. Wiley, New York, 1961; H. C. Freeman, *Inorg. Biochem.* **1**, 121–166 (1973), R. Österberg, *Coord. Chem. Rev.* **12**, 309 (1974).
2. K. A. Fraser and M. M. Harding, *J. Chem. Soc. A* p. 415 (1967).
3. M. M. Harding and H. A. Long, *J. Chem. Soc. A* p. 2554 (1968).
4. R. Candlin and M. M. Harding, *J. Chem. Soc. A* p. 421 (1967).
5. M. M. Harding and S. J. Cole, *Acta Crystallogr.* **16**, 643 (1963).
6. J. F. Blount, K. A. Fraser, H. C. Freeman, J. T. Szymanski, and C. H. Wang, *Acta Crystallogr.* **22**, 396 (1967).
7. C. A. McAuliffe, J. V. Quagliano, and L. M. Vallarino, *Inorg. Chem.* **5**, 1996 (1966).
8. S. E. Livingstone and J. D. Nolan, *Inorg. Chem.* **7**, 1447 (1968).
9. H. Shindo and T. L. Brown, *J. Am. Chem. Soc.* **87**, 1904 (1965).
10. M. V. Veidis and G. J. Palenik, *Chem. Commun.* p. 1277 (1969).
11. C. A. McAuliffe and S. G. Murray, *Inorg. Chim. Acta, Rev.* **6**, 103 (1972).
12. J. Hidaka, S. Yamada, and Y. Shimura, *Chem. Lett.* p. 1487 (1974).
13. I. Lindqvist and R. Rosenstein, *Acta Chem. Scand.* **14**, 1228 (1960).
14. A. A. Grinberg, A. I. Stetsenko, and E. N. In'kova, *Dokl. Akad. Nauk SSSR* **136**, 821 (1961).
15. L. M. Volshtein and I. O. Volodina, *Russ. J. Inorg. Chem. (Engl. Transl.)* **5**, 840 (1960); L. M. Volshtein and G. G. Motyagina, *ibid.* p. 949.
16. L. E. Erickson, J. W. McDonald, J. K. Howie, and R. P. Clow, *J. Am. Chem. Soc.* **90**, 6371 (1968).
17. T. Yasui, T. Ama, H. Morio, M. Okabayashi, and Y. Shimura, *Bull. Chem. Soc. Jpn.* **47**, 2801 (1974).
18. T. Yasui, H. Kawaguchi, Z. Kanda, and T. Ama, *Bull. Chem. Soc. Jpn.* **47**, 2393 (1974).
19. B. E. Douglas and S. Yamada, *Inorg. Chem.* **4**, 1561 (1965).
20. J. H. Dunlop and R. D. Gillard, *J. Chem. Soc.* p. 6531 (1965).
21. H. C. Freeman and M. L. Golomb, *Acta Crystallogr., Sect. B* **25**, 1203 (1969).
22. R. A. Condrate and K. Nakamoto, *J. Chem. Phys.* **42**, 2590 (1965).
23. A. E. Martell and M. Calvin, "Chemistry of the Metal Chelate Compounds," pp. 134–180. Prentice-Hall, Englewood Cliffs, New Jersey, 1952.
24. A. Nakahara, J. Hidaka, and R. Tsuchida, *Bull. Chem. Soc. Jpn.* **29**, 925 (1956).
25. H. Irving and R. J. P. Williams, *Nature (London)* **162**, 746 (1948).
26. G. L. Eichhorn and N. D. Marchand, *J. Am. Chem. Soc.* **78**, 2688 (1956).
27. D. L. Leussing and D. C. Schultz, *J. Am. Chem. Soc.* **86**, 4846 (1964).
28. D. L. Leussing and E. M. Hanna, *J. Am. Chem. Soc.* **88**, 693 and 696 (1966).
29. D. L. Leussing and N. Huq, *Anal. Chem.* **38**, 1388 (1966).
30. A. Nakahara, *Bull. Chem. Soc. Jpn.* **32**, 1195 (1959).
31. A. Nakahara, H. Yamamoto, and H. Matsumoto, *Bull. Chem. Soc. Jpn.* **37**, 1137 (1964).

32. Y. Nakao, K. Sakurai, and A. Nakahara, *Bull. Chem. Soc. Jpn.* **40**, 1536 (1967).
33. R. C. Burrows and J. C. Bailar, Jr., *J. Am. Chem. Soc.* **88**, 4150 (1966).
34. R. H. Holm, *Inorg. Biochem.* **2**, 1137–1167 (1973).
35. A. S. Mildvan, in "The Enzymes" (P. D. Boyer, ed.), 3rd ed., Vol. 2, pp. 445–536. Academic Press, New York, 1970.
36. B. Sarkar and T. P. A. Kruck, in "Biochemistry of Copper" (J. Peisach, P. Aisen, and W. E. Blumberg, eds.), pp. 183–209. Academic Press, New York, 1966.
37. H. C. Freeman, J. M. Guss, M. J. Healy, R.-P. Martin, C. E. Nockolds, and B. Sarkar, *Chem. Commun.* p. 225 (1969).
38. P. Z. Neumann and A. Sass-Kortsak, *J. Clin. Invest.* **46**, 646 (1967).
39. S.-J. Lau and B. Sarkar, *J. Biol. Chem.* **246**, 5938 (1971).
40. T. P. A. Kruck and B. Sarkar, *Inorg. Chem.* **14**, 2383 (1975).
41. R. Griesser and H. Sigel, *Inorg. Chem.* **9**, 1238 (1970).
42. P. R. Huber, R. Griesser, and H. Sigel, *Inorg. Chem.* **10**, 945 (1971).
43. R. Griesser and H. Sigel, *Inorg. Chem.* **10**, 2229 (1971).
44. H. Sigel, *Angew. Chem., Int. Ed. Engl.* **14**, 394 (1975).
45. R.-P. Martin, M. M. Petit-Ramel, and J. P. Scharff, in "Metal Ions in Biological Systems" (H. Sigel, ed.), Vol. 2, pp. 1–61. Dekker, New York, 1973.
46. R. DeWitt and J. I. Watters, *J. Am. Chem. Soc.* **76**, 3810 (1954).
47. S. Kida, *Bull. Chem. Soc. Jpn.* **29**, 805 (1956).
48. H. Sigel and D. B. McCormick, *Acc. Chem. Res.* **3**, 201 (1970).
49. H. Sigel, P. R. Huber, R. Griesser, and B. Prijs, *Inorg. Chem.* **12**, 1198 (1973).
50. H. Sigel, in "Metal Ions in Biological Systems" (H. Sigel, ed.), Vol. 2, pp. 63–125. Dekker, New York, 1973.
51. G. Brookes and L. D. Pettit, *Chem. Commun.* p. 385 (1975).
52(a). T. Sakurai, O. Yamauchi, and A. Nakahara, *Bull. Chem. Soc. Jpn.* **49**, 169 (1976).
52(b). O. Yamauchi, Y. Nakao, and A. Nakahara, *Bull. Chem. Soc. Jpn.* **48**, 2572 (1975).
53. T. Sakurai, O. Yamauchi, and A. Nakahara, *Chem. Commun.* p. 553 (1976).
54. M. L. Bair and E. M. Larsen, *J. Am. Chem. Soc.* **93**, 1140 (1971).
55. A. R. Manyak, C. B. Murphy, and A. E. Martell, *Arch. Biochem. Biophys.* **59**, 373 (1955).
56. M. Shiro, Y. Nakao, O. Yamauchi, and A. Nakahara, *Chem. Lett.* p. 123 (1972).
57. B. Strandberg, I. Lindqvist, and R. Rosenstein, *Z. Kristallogr.* **116**, 266 (1961).
58. H. Dobbie and W. O. Kermack, *Biochem. J.* **59**, 246 (1955).
59. W. L. Koltun, R. H. Roth, and F. R. N. Gurd, *J. Biol. Chem.* **238**, 124 (1963).
60. R.-P. Martin, L. Mosoni, and B. Sarkar, *J. Biol. Chem.* **246**, 5944 (1971).
61. Y. Nakao, K. Sakurai, and A. Nakahara, *Bull. Chem. Soc. Jpn.* **39**, 1608 (1966).
62. A. Sugihara, T. Ashida, Y. Sasada, and M. Kakudo, *Acta Crystallogr., Sect. B* **24**, 203 (1968).
63. H. C. Freeman, G. Robinson, and J. C. Schoone, *Acta Crystallogr.* **17**, 719 (1964).
64. H. C. Freeman, J. C. Schoone, and J. G. Sime, *Acta Crystallogr.* **18**, 381 (1965).
65. H. C. Freeman and M. R. Taylor, *Acta Crystallogr.* **18**, 939 (1965).
66. J. F. Blount, H. C. Freeman, R. V. Holland, and G. H. W. Milburn, *J. Biol. Chem.* **245**, 5177 (1970).
67. H. C. Freeman, *Adv. Protein Chem.* **22**, 257 (1967).
68. H. C. Freeman, in "Biochemistry of Copper" (J. Peisach, P. Aisen, and W. E. Blumberg, eds.), pp. 77–113. Academic Press, New York, 1966.
69. S. P. Datta and B. R. Rabin, *Trans. Faraday Soc.* **52**, 1123 (1956).
70. C. B. Murphy and A. E. Martell, *J. Biol. Chem.* **226**, 37 (1957).

71. W. L. Koltun, M. Fried, and F. R. N. Gurd, *J. Am. Chem. Soc.* **82**, 233 (1960).
72. O. Yamauchi, Y. Hirano, Y. Nakao, and A. Nakahara, *Can. J. Chem.* **47**, 3441 (1969).
73. H. Dobbie and W. O. Kermack, *Biochem. J.* **59**, 257 (1955).
74. M. K. Kim and A. E. Martell, *J. Am. Chem. Soc.* **88**, 914 (1966).
75. R. Österberg and B. Sjöberg, *J. Biol. Chem.* **243**, 3038 (1968).
76. O. Yamauchi, Y. Nakao, and A. Nakahara, *Bull. Chem. Soc. Jpn.* **46**, 2119 (1973).
77. O. Yamauchi, H. Miyata, and A. Nakahara, *Bull. Chem. Soc. Jpn.* **44**, 2716 (1971).
78. A. P. Brunetti, M. C. Lim, and G. H. Nancollas, *J. Am. Chem. Soc.* **90**, 5120 (1968).
79. O. Yamauchi, Y. Nakao, and A. Nakahara, *Bull. Chem. Soc. Jpn.* **46**, 3749 (1973); **47**, 514 (1974).
80. G. K. Pagenkopf and D. W. Margerum, *J. Am. Chem. Soc.* **90**, 6963 (1968).
81. R. F. Pasternack, M. Angwin, and E. Gibbs, *J. Am. Chem. Soc.* **92**, 5878 (1970).
82. G. H. Nancollas and D. J. Poulton, *Inorg. Chem.* **8**, 680 (1969).
83. G. H. Nancollas, *Coord. Chem. Rev.* **5**, 407 (1970).
84. M. K. Kim and A. E. Martell, *J. Am. Chem. Soc.* **91**, 872 (1969).
85. M. Tasumi, S. Takahashi, T. Nakata, and T. Miyazawa, *Bull. Chem. Soc. Jpn.* **48**, 1595 (1975).
86. B. R. Rabin, *Trans. Faraday Soc.* **52**, 1130 (1956).
87. M. T. Barnet, H. C. Freeman, D. A. Buckingham, I.-N. Hsu, and D. van der Helm, *Chem. Commun.* p. 367 (1970).
88. R. B. Martin, *in* "Metal Ions in Biological Systems" (H. Sigel, ed.), Vol. 1, pp. 129–156. Dekker, New York, 1974.
89. D. W. Margerum and G. R. Dukes, *in* "Metal Ions in Biological Systems" (H. Sigel, ed.), Vol. 1, pp. 157–212. Dekker, New York, 1974.
90. A. Nakahara, K. Hamada, I. Miyachi, and K. Sakurai, *Bull. Chem. Soc. Jpn.* **40**, 2826 (1967).
91. Y. Nakao, N. Nonagase, and A. Nakahara, *Bull. Chem. Soc. Jpn.* **42**, 452 (1969).
92. Y. Nakao, H. Ishibashi, and A. Nakahara, *Bull. Chem. Soc. Jpn.* **43**, 3457 (1970).
93. Y. Nakao, *Nippon Kagaku Zasshi* **92**, 399 (1971).
94. Y. Nakao and A. Nakahara, *Bull. Chem. Soc. Jpn.* **46**, 187 (1973).
95. M. M. Jones, "Ligand Reactivity and Catalysis," Academic Press, New York, 1968.
96. A. E. Martell, *in* "Metal Ions in Biological Systems" (H. Sigel, ed.), Vol. 2, pp. 207–268. Dekker, New York, 1973.
97. G. W. Watt and J. F. Knifton, *Inorg. Chem.* **6**, 1010 (1967).
98. G. W. Watt and D. G. Upchurch, *Adv. Chem. Ser.* **62**, 253–271 (1967).
99. G. W. Watt and J. F. Knifton, *Inorg. Chem.* **7**, 1159 (1968).
100. D. H. Williams and D. H. Busch, *J. Am. Chem. Soc.* **87**, 4644 (1965).
101. M. Sato, K. Okawa, and S. Akabori, *Bull. Chem. Soc. Jpn.* **30**, 937 (1957).
102. J. A. Aune, P. Maldonado, G. Larcheres, and M. Pierrot, *Chem. Commun.* p. 1351 (1970).
103. A. Nakahara, S. Nishikawa, and J. Mitani, *Bull .Chem. Soc. Jpn.* **40**, 2212 (1967).
104. H. Yoneda, Y. Morimoto, Y. Nakao, and A. Nakahara, *Bull. Chem. Soc. Jpn.* **41**, 255 (1968).
105. K. Harada and J. Oh-hashi, *J. Org. Chem.* **32**, 1103 (1967).
106. T. Ichikawa, S. Maeda, T. Okamoto, Y. Araki, and Y. Ishido, *Bull. Chem. Soc. Jpn.* **44**, 2779 (1971).
107. M. Murakami and K. Takahashi, *Bull. Chem. Soc. Jpn.* **32**, 308 (1959).
108. J. C. Dabrowiak and D. W. Cooke, *Inorg. Chem.* **14**, 1305 (1975).

109. Y. N. Belokon', V. M. Belikov, S. V. Vitt, M. M. Dolgaya, and T. F. Savel'eva, *Chem. Commun.* p. 86 (1975).
110. R. D. Gillard, P. R. Mitchell, and N. C. Payne, *Chem. Commun.* p. 1150 (1968).
111. K. Noda, M. Bessho, T. Kato, and N. Izumiya, *Bull. Chem. Soc. Jpn.* **43**, 1834 (1970).
112. O. Uyama, Y. Nakao, and A. Nakahara, *Bull. Chem. Soc. Jpn.* **46**, 496 (1973).
113. Y. Nakao, O. Uyama, and A. Nakahara, *J. Inorg. Nucl. Chem.* **36**, 685 (1974).
114. M. Fujioka, Y. Nakao, and A. Nakahara, *Bull. Chem. Soc. Jpn.* **49**, 477 (1976).
115. M. Fujioka, Y. Nakao, and A. Nakahara, *J. Inorg. Nucl. Chem.* (in press).
116. R. Mathur and R. B. Martin, *J. Phys. Chem.* **69**, 668 (1965).
117. H. Kroll, *J. Am. Chem. Soc.* **74**, 2036 (1952).
118. M. L. Bender and B. W. Turnquest, *J. Am. Chem. Soc.* **79**, 1889 (1957).
119. W. A. Conner, M. M. Jones, and D. L. Tuleen, *Inorg. Chem.* **4**, 1129 (1965).
120. M. D. Alexander and D. H. Busch, *J. Am. Chem. Soc.* **88**, 1130 (1966).
121. D. A. Buckingham, D. M. Foster, and A. M. Sargeson, *J. Am. Chem. Soc.* **90**, 6032 (1968).
122. D. A. Buckingham, D. M. Foster, and A. M. Sargeson, *J. Am. Chem. Soc.* **91**, 4102 (1969); **92**, 5701 (1970).
123. D. A. Buckingham, D. M. Foster, L. G. Marzilli, and A. M. Sargeson, *Inorg. Chem.* **9**, 11 (1970).
124. Y. Wu and D. H. Busch, *J. Am. Chem. Soc.* **92**, 3326 (1970).
125. R. L. Prestidge, D. R. K. Harding, J. E. Battersby, and W. S. Hancock, *J. Org. Chem.* **40**, 3287 (1975).
126. Y. Nakao and A. Nakahara, *Proc. 32nd Annu. Meet. Chem. Soc. Jpn., 1975* Vol. 1, p. 519 (1975).
127. L. Meriwether and F. H. Westheimer, *J. Am. Chem. Soc.* **78**, 5119 (1956).
128. D. A. Buckingham, C. E. Davis, D. M. Foster, and A. M. Sargeson, *J. Am. Chem. Soc.* **92**, 5571 (1970).
129. D. A. Buckingham, D. M. Foster, and A. M. Sargeson, *J. Am. Chem. Soc.* **92**, 6151 (1970).
130. C.-G. Regardh, *Acta Pharm. Suec.* **4**, 335 (1967).
131. J. P. Collman and D. A. Buckingham, *J. Am. Chem. Soc.* **85**, 3039 (1963).
132. D. A. Buckingham, J. P. Collman, D. A. R. Happer, and L. G. Marzilli, *J. Am. Chem. Soc.* **89**, 1082 (1967).
133. D. A. Buckingham, P. A. Marzilli, I. E. Maxwell, and A. M. Sargeson, *Chem. Commun.* p. 488 (1968).
134. D. A. Buckingham and J. P. Collman, *Inorg. Chem.* **6**, 1803 (1967).
135. E. Kimura, S. Young, and J. P. Collman, *Inorg. Chem.* **9**, 1183 (1970).
136. A. Nakahara, K. Hamada, Y. Nakao, and T. Higashiyama, *Coord. Chem. Rev.* **3**, 207 (1968).
137. R. H. Andreatta, H. C. Freeman, A. V. Robertson, and R. L. Sinclair, *Chem. Commun.* p. 203 (1967).
138. A. Levitzki, M. Anbar, and A. Berger, *Biochemistry* **6**, 3757 (1967).
139. D. A. Buckingham, L. G. Marzilli, and A. M. Sargeson, *J. Am. Chem. Soc.* **89**, 2772 (1967).
140. D. A. Buckingham, L. G. Marzilli, and A. M. Sargeson, *J. Am. Chem. Soc.* **89**, 4539 (1967).
141. J. P. Collman and E. Kimura, *J. Am. Chem. Soc.* **89**, 6096 (1967).
142. S. Yamada, S. Terashima, and M. Wagatsuma, *Tetrahedron Lett.* p. 1501 (1970).
143. S. Yamada, M. Wagatsuma, Y. Takeuchi, and S. Terashima, *Chem. Pharm. Bull.* **19**, 2380 (1971).

144. M. Wagatsuma, S. Terashima, and S. Yamada, *Tetrahedron* **29**, 1497 (1973).
145. S. Terashima, M. Wagatsuma, and S. Yamada, *Tetrahedron* **29**, 1487 (1973).
146. E. B. Paniago, D. C. Weatherburn, and D. W. Margerum, *Chem. Commun.* p. 1427 (1971).
147. G. L. Burce, E. B. Paniago, and D. W. Margerum, *Chem. Commun.* p. 261 (1975).
148. E. E. Snell, A. E. Braunstein, E. S. Severin, and Y. M. Torchinsky, eds., "Pyridoxal Catalysis: Enzymes and Model Systems." Wiley, New York, 1968.
149. D. E. Metzler and E. E. Snell, *J. Am. Chem. Soc.* **74**, 979 (1952).
150. D. E. Metzler, M. Ikawa, and E. E. Snell, *J. Am. Chem. Soc.* **76**, 648 (1954).
151. Y. Nakao, K. Sakurai, S. Sasaki, and A. Nakahara, *Bull. Chem. Soc. Jpn.* **40**, 241 (1967).
152. D. E. Metzler, J. Olivard, and E. E. Snell, *J. Am. Chem. Soc.* **76**, 644 (1954).
153. H. Mix, *Hoppe-Seyler's Z. Physiol. Chem.* **315**, 1 (1959).
154. L. W. Fleming and G. W. Crosbie, *Biochim. Biophys. Acta* **43**, 139 (1960).
155. H. Mix, *Hoppe-Seyler's Z. Physiol. Chem.* **325**, 106 (1961).
156. Y. Nakao, K. Sakurai, and A. Nakahara, *Bull. Chem. Soc. Jpn.* **38**, 687 (1965).
157. Y. Nakao, K. Sakurai, and A. Nakahara, *Bull. Chem. Soc. Jpn.* **39**, 1471 (1966).

14

Nonenzymatic Dihydronicotinamide Reductions as Probes for the Mechanism of NAD+-Dependent Dehydrogenases

David S. Sigman, Joseph Hajdu, and
Donald J. Creighton

INTRODUCTION

Nonenzymatic dihydronicotinamide reactions can provide important insights and chemical precedents crucial for understanding the direct and stereospecific hydrogen transfer between coenzyme and substrate catalyzed by NAD+-dependent dehydrogenases. The simplest dehydrogenases such as lactate dehydrogenase and alcohol dehydrogenase catalyze only this process, while enzymes such as glyceraldehyde-3-phosphate dehydrogenase and isocitrate dehydrogenase catalyze hydrogen transfer as part of a more complex sequence of reactions.

For these enzymes, no single set of active-site functional groups is uniquely required for hydrogen transfer. This conclusion was first suggested by early observations that alcohol dehydrogenase contained zinc ion and that lactate dehydrogenase did not [1–4]. X-Ray crystallographic and primary sequence analysis have unambiguously confirmed it [5]. The catalytic groups at the nicotinamide binding site of the dehydrogenases, whose structures have been solved, are quite different, although it is likely that different amino acid residues can perform analogous catalytic functions.

One use of nonenzymatic dihydronicotinamide reactions is to identify the types of noncovalent interactions between enzyme and substrate or enzyme and coenzyme that could be useful in accelerating hydrogen transfer. Results of this nature can permit the assignment of a catalytic function to an amino acid residue or metal ion, known to be at the active site from structural work, by providing a useful chemical precedent. Another purpose is to determine whether hydrogen transfer is a single-step or multistep process. Steady-state or transient kinetic studies of the enzymatic reactions themselves are often ambiguous about the details of hydrogen transfer because coenzyme or substrate association and dissociation or protein conformational changes can be the slow steps in the catalytic sequence [6]. A thorough investigation of nonenzymatic reactions should provide a useful conceptual framework to approach these questions. All investigations discussed here involve oxidation of the reduced form of the coenzyme analogs. The reverse process has not yet been observed in nonenzymatic systems because it is thermodynamically unfavorable at neutral pH.

The early studies of nonenzymatic dihydronicotinamide reductions, following the formulative work of Westheimer, Vennesland, and their colleagues

(1)

in 1951 demonstrating direct hydrogen transfer in alcohol dehydrogenase [7–9], focused on the search for oxidants in which the hydrogen transferred was incorporated into a nonexchangeable position. Before this time various dyes, such as methylene blue, were known to be reduced nonenzymatically by dihydronicotinamides [10]. However, since the transferred hydrogen exchanged with the media, it was not possible to determine if these reactions were sufficiently similar to the enzymatic reaction to be considered valid models.

Malachite green was the first nonenzymatic oxidant of N-alkyldihydronicotinamide discovered in which the reaction product incorporated the transferred hydrogen into a nonexchangeable position [Eq. (1)]. Its reduction by N-benzyldihydronicotinamide proceeds by direct hydrogen transfer from the

TABLE 1

Rate Constants for Oxidants Reduced by Direct Hydrogen Transfer

Dihydro-nicotin-amide	Oxidant	k (M^{-1} sec^{-1})	Solvent	Temp (°C)	Reference
N-Benzyl	Malachite green	Complete after 25 hr	90% EtOH	30	22
N-Benzyl	Thiobenzophenone	0.023	70% EtOH	30	12
N-Benzyl	Hexachloroacetone	0.03[a]	EtOH	25	14
N-Benzyl	α-Chlorophenyl-(2-methyl-3H)indolyidene methane hydrochloride	26	EtOH	25	16
N-Benzyl	Trifluoroacetophenone	0.0014	H$_2$O	25	17
N-Propyl	Trifluoroacetophenone	0.004	H$_2$O	25	17
NADH	3,10-Dimethyl-5-deazaiso-alloxazine	0.03	H$_2$O	30	19
N-Propyl	3,10-Dimethyl-5-diazaiso-alloxazine	6.17	H$_2$O	30	19
NADH	N-Benzyl-3-acetylpyri-dinium chloride	0.0055	H$_2$O	30	15
N-Propyl	3-Hydroxypyridine-4-carboxaldehyde	0.27	MeOH	64	20
N-Propyl	Pyridoxal phosphate	0.45	H$_2$O	30	20
N-Propyl	1,10-Phenanthroline-2-carboxaldehyde–zinc complex	0.32	MeCN	25	18
N-Propyl	N-Methylacridinium ion	2040	H$_2$O	25	21
N-Benzyl	N-Methylacridinium ion	550	H$_2$O	25	23

[a] Approximate value.

4 position of the pyridine ring in complete analogy with the enzymatic process. An isotope effect of 4.5 was obtained by deuterium analysis of the products when 4-monodeutero-*N*-benzyldihydronicotinamide, obtained by reduction of *N*-benzylnicotinamide with dithionite in D_2O [11], was used as a reductant.

Since the report of the reduction of malachite green 20 years ago, a number of oxidants have been discovered in which the transferred hydrogen is incorporated into a nonexchangeable position by reduction with an *N*-alkyl-dihydronicotinamide. They include thiobenzophenone and its derivatives [12,13], hexachloroacetone [14], *N*-benzylacetylpyridine [15], indolenine salts [16], trifluoroacetophenone [17], the zinc complex of 1,10-phenanthroline-2-carboxaldehyde [18], deazaflavin [19], pyridoxal [20], and *N*-methylacridinium ion [21]. The second-order rate constants for the reduction of these oxidants are reported in Table 1 [12,14–23]. All these reactions are of intrinsic interest and provide information as to the mechanism of hydrogen transfer between a dihydronicotinamide and an oxidant. However, those reactions that exhibit the greatest congruence to enzymatic processes will be stressed.

ACTIVATION OF SUBSTRATES

A key question in enzymatic reactions is how substrates are activated for their reversible oxidation. For example, lactate dehydrogenase reduces pyruvate which possesses a carbonyl group that is substantially less electron deficient than hexachloroacetone and trifluoroacetophenone. Nonenzymatic reactions are useful in identifying the possible modes of activation available to the enzyme. The first study that provided the chemical precedent of the predictable importance of general acid catalysis in dihydronicotinamide reactions involved the reduction of a series of thiobenzophenone derivatives by

$$(2)$$

N-benzyldihydronicotinamide [12] [Eq. (2)]. The reduction of thiobenzophenone exhibits no pH dependence, but the second-order rate constants increase with ionic strength in 70% ethanol–water mixtures and decrease when ethanol is partially replaced by the less polar solvents chloroform and tetrahydrofuran. The activated complex must therefore be more polar than the reactants.

Most significantly, substituents on thiobenzophenone markedly affect the rate of reduction. Electron-donating substituents decrease the rate, whereas electron-withdrawing substituents enhance it (Table 2). Clearly the reactivity

TABLE 2

Substituent Effects on Thiobenzophenone Reduction by N-Benzyldihydronicotin-amide[a]

Thiobenzophenone substituent	k (M^{-1} sec^{-1})
p-PH	0.00195
p-OCH$_3$	0.00447
o-OCH$_3$	0.0154
H	0.023
o-OH	0.0539
p-Cl	0.0603

[a] 70°C EtOH; temp, 30°C; Abeles et al. [12].

of the thiocarbonyl group is enhanced by an increase in the partial positive charge on the carbon. An o-hydroxyl group is an important exception to this trend of reactivity. Instead of decreasing the rate constant through resonance effects like the p-hydroxy and o-methoxy groups, the o-hydroxy group enhances it. The reason for the acceleratory effect of the o-hydroxy group is that it can serve as an intramolecular general acid catalyst and stabilize the incipient mercaptide ion in the transition state by hydrogen bonding. Since general acid catalysis can accelerate the rate of his dihydronicotinamide reduction, it follows from microscopic reversibility that the reverse reaction would be susceptible to general base catalysis.

It might be argued, however, that the demonstration of general acid catalysis in this reaction is an unconvincing precedent for this mode of catalysis in the reduction of a carbonyl group in an enzymatic reaction. Thiocarbonyl groups are substantially more reactive than carbonyl groups. For example, the rate of formation of benzophenone phenylhydrazone from thiobenzophenone is 6000-fold faster than from benzophenone at pH 6 [24].

The importance of general acid catalysis in the reduction of carbonyl groups was first noted in the *N*-propyldihydronicotinamide reduction of 3-hydroxy-4-pyridine aldehyde. In refluxing methanol, the second-order rate constant is $0.27 \, M^{-1} \, sec^{-1}$ (Table 1). The reaction proceeds by direct hydrogen transfer, is insensitive to free-radical quenching agents, and is faster in solutions of high dielectric constant [20]. The parent compound, 4-pyridine aldehyde, is not reduced in either methanol or aqueous solution. Assuming that the reactive form of 3-hydroxy-4-pyridine aldehyde is the neutral (**1a**) and not the dipolar form (**1b**), general acid catalysis has a more

pronounced effect on the reduction of carbonyl groups than on the thio-carbonyl. Its rate enhancement is very large given the failure of 4-pyridine aldehyde to react at all.

The pH dependence of the second-order rate constant for the reduction of pyridoxal phosphate by *N*-propyl-1,4-dihydronicotinamide suggests that the neutral form of 3-hydroxy-4-pyridine aldehyde is the reactive one. The rate of the reaction increases with increasing acidity in the pH range 9.22–7.15. If the ionization state of the phosphate does not affect the reduction rate, the rate constants for the three ionic forms (**2a**), (**2b**), and (**2c**) are 0.28, 0.045, and $0.001 \, M^{-1} \, sec^{-1}$, respectively [20].

Recent X-ray crystallographic measurements on an abortive $E \cdot NAD^+-$ pyruvate ternary complex of lactate dehydrogenase suggest a general acid–general base role for an active-site histidine on the basis of its proximity to bound pyruvate [25]. The nonenzymatic reactions discussed above support this suggestion by providing a chemical precedent for the susceptibility of dihydronicotinamide reductions to general acid catalysis. Presumably, potential general acids like tyrosine, serine, and cysteine play roles comparable to that of the histidine of lactate dehydrogenase in other NAD^+-dependent dehydrogenases.

Metal ion catalysis provides another mechanism for the activation of substrates for reduction. The potential importance of metal ions in dehydrogenases was underscored by the discovery of zinc ion in alcohol dehydrogenase and the demonstration of competitive inhibition of this enzyme, with respect to the coenzyme, by the chelating agent 1,10-phenanthroline [26]. Although the zinc ion was initially considered to be required for coenzyme binding, spectroscopic and kinetic studies with 1,10-phenanthroline [27] and 2,2-bipyridine [28] clearly indicated that the zinc ion coordinated by these chelating agents constituted a central feature of the binding site for alcohol and aldehyde substrates. X-Ray crystallographic studies [29] and work with chromogenic substrates [30] have provided firm support for these conclusions. A sensible suggestion for the role of zinc ion, given its location within the active site, is that it catalyzes hydrogen transfer between coenzyme and substrate by polarizing the carbonyl group of the aldehyde substrates and by facilitating deprotonation of the hydroxyl group of the alcohol substrates [12,31].

The first nonenzymatic reaction that provided a chemical precedent for the catalytic function of a metal ion in a dihydronicotinamide reduction involved the zinc ion-catalyzed reduction of 1,10-phenanthroline-2-carboxaldehyde by N-propyldihydronicotinamide [Eq. (3)] [21]. The zinc complex of 1,10-phenanthroline-2-carboxaldehyde was chosen for study since 1,10-phenanthroline derivatives generally have a high affinity for zinc ion, and construction of molecular models of the complex indicated that the carbonyl oxygen could coordinate to the metal ion. Such an interaction in the inner coordination sphere of the zinc ion is possibly analogous to the suggested coordination of the aldehyde to the enzymatic zinc ion in the reactive enzyme–NADH–aldehyde ternary complex. The model reaction was carried out in acetonitrile at 25°C. Aqueous solutions were avoided because the metal ion efficiently catalyzes the hydration of 1,10-phenanthroline-2-carboxaldehyde to the nonreducible aldehydrol form. Unhydrated aldehydes and not their aldehyde forms are the true substrates for alcohol dehydrogenase [32].

The reaction proceeds by direct hydrogen transfer via a mechanism that is insensitive to free-radical quenching agents. The relatively small but significant kinetic isotope effect (1.74) obtained using the dihydro and dideutero forms of N-propyldihydronicotinamide indicates that hydrogen transfer is at least partially rate limiting in the overall reaction. The rate enhancement attributable to the zinc ion catalysis must be very large since no reaction could be

$$(3)$$

detected in the absence of the metal ion. Pyridoxal is the only other aldehyde that can be reduced nonenzymatically and, as noted above, this is most likely the result of intramolecular general acid catalysis. Further support for the efficiency of metal ion catalysis in dihydropyridine reductions is the demonstration that a series of metal ions, including Ni^{2+}, Co^{2+}, Zn^{2+}, Mn^{2+}, and Mg^{2+}, catalyze the rate of reduction of pyridoxal phosphate by the Hantzsch ester, 2,6-dimethyl-3,5-dicarboethoxy-1,4-dihydropropyridine [20].

The relevance of these model reactions to the catalytic mechanism of horse liver alcohol dehydrogenase has recently been brought into question by the results of Sloan *et al.* [33]. On the basis of their paramagnetic line-broadening experiments on the catalytically inactive $E \cdot Co^{2+} \cdot NAD^+ \cdot$ isobutyramide and $E \cdot Co^{2+} \cdot NADH \cdot$ ethanol complexes, isobutyramide and ethanol appear on the average to be in the second coordination sphere of the metal ion, too distant for direct coordination. Assuming a close resemblance to the catalytically competent complex, these authors suggest that a water molecule or hydroxide ion coordinated to the metal ion intervenes between substrate and metal, possibly playing a general acid–general base role during catalysis. On the other hand, the large spectral changes obtained in reactive ternary complexes involving the chromogenic substrate *p*-dimethylamino-cinnamaldehyde indicate direct coordination of substrate to metal ion and suggest that such outer-sphere complexes may not be kinetically competent even if they form at the active site. Certainly, a metal-bound water molecule is not an absolute requirement for the model reaction summarized in Eq. (3), which proceeds readily only under scrupulously anhydrous conditions. The presence of water strongly inhibits the reduction because the metal ion efficiently catalyzes the formation of the nonreducible aldehydrol.

Moreover, mechanistic studies on the metal-ion-catalyzed reduction of chelating aldehydes by borohydride strongly support the contention that inner-sphere complexes are catalytically competent. The reduction of 2-pyridine aldehyde catalyzed by zinc ion is 6000 times faster than the reduction of 4-pyridine aldehyde catalyzed by zinc ion in anhydrous acetonitrile at equal concentrations of borohydride. Proximity and coordination of the metal ion to the zinc ion, and not inductive effects exerted through the aromatic ring, are therefore central features of the metal ion catalysis in this reaction as well as the dihydronicotinamide reduction summarized in Eq. (3). The reduction of both pyridine aldehyde isomers depends linearly on the zinc ion concentrations below the independently measured dissociation constants. However, a surprising and interesting result relevant to the potential importance of inner-sphere complexes to the alcohol dehydrogenase mechanism is that the reduction of the zinc–pyridine-2-aldehyde complex is independent of borohydride concentration. The reduction of the zinc–pyridine-4-aldehyde complex, on the other hand, exhibits the expected first-order dependence on borohydride concentration.

A kinetic scheme consistent with the zero-order dependence on borohydride concentration is summarized in Eq. (4). The initial complex (3) is thermodynamically stable, and its dissociation constant can be readily measured by conventional spectrophotometry. The dissociation constant of this complex is 5-fold greater than that for 4-pyridine aldehyde, suggesting that the carbonyl group of the 2-isomer in complex (3) does not reside in the first coordination sphere of the metal ion. In order for the scheme to be consistent with the observed kinetics, a slow step must precede that involving borohydride reduction. The slow step indicated in Eq. (4) involves the rate-limiting dissociation of an acetonitrile molecule from the inner coordination sphere of the metal ion followed by the direct coordination of the carbonyl group to yield an inner-sphere complex extremely susceptible to reduction by borohydride. The analogous process cannot occur for the 4-pyridine aldehyde complex and, as a result, the rate of its reduction exhibits a first-order dependence on borohydride concentration [34].

While the model dehydrogenase reactions described here emphasize the competency of inner-sphere complexes in metal-catalyzed reductions of aldehydes in nonenzymatic systems, this cannot in itself be construed to eliminate an outer-sphere mechanism of the type suggested by Sloan *et al.* for the enzyme. Ultimately, the value of model dehydrogenase reactions for testing suggested catalytic mechanisms of alcohol dehydrogenase depends on the accuracy with which they duplicate known or suggested interactions in the active site of the enzyme. Examination of the catalytic properties of aquo complexes of Ni, Co(III), or Cr(III) in nonenzymatic systems containing kinetically stable inner-sphere water molecules may be an important future avenue for exploring the potential importance of outer-sphere mechanisms in alcohol dehydrogenase. In addition, the active-site zinc ion of alcohol dehydrogenase appears to be coordinated by two cysteine residues and one histidine residue from the protein. The extent to which the physicochemical properties of the active-site zinc ion, coordinated by two negatively charged, highly polarizable sulfur atoms, resembles those of metal ions not coordinated by such ligands, as in the model dehydrogenase complexes, is a question worthy of future exploration.

NONCOVALENT INTERACTIONS THAT ENHANCE DIHYDRONICOTINAMIDE REDUCTIONS

Specific noncovalent interactions between the dihydronicotinamide moiety of NAD^+ and dehydrogenases as well as those between substrate and enzyme should facilitate hydrogen transfer. The first studies providing a nonenzymatic example of an intramolecular noncovalent interaction that could enhance dihydronicotinamide reactivity involved the demonstration that negatively charged groups could accelerate dihydronicotinamide reductions by electrostatic stabilization of the positive charge that must develop on the nicotinamide moiety during its oxidation. These studies were initially motivated by the report, since retracted, that the carboxylate group of a glutamyl residue was in close proximity to the nicotinamide moiety of the coenzyme at the active site of lactate dehydrogenase [25].

Specifically, it was shown that dihydronicotinamides containing neighboring carboxylate groups could reduce the oxidant N-methylacridinium ion in nonaqueous solution [Eq. (5)] orders of magnitude more rapidly than

$$(5)$$

homologous derivatives lacking the free carboxylate group [23]. For example, 2'-carboxylbenzyldihydronicotinamide exhibits a 100-fold rate acceleration relative to its methyl ester in acetonitrile, while the more conformationally inflexible derivative N-cis-2-carboxylcyclopentyldihydronicotinamide (4)

(4)

TABLE 3

Neighboring-Group Effects of Carboxylate Groups in *N*-Alkyldihydro-
nicotinamide. Reductions of *N*-Methylacridinium Ion

		k (M^{-1} sec^{-1})[a]		
Dihydronicotinamide	X	0.05 M Tris, pH 8.0	Methanol	Acetonitrile
	H	550	180	65
	2-COOCH$_3$	460	220	70
	2-COO$^-$Na$^+$	1,450	4,000	10,000
	H	2,300	350	150
	-COOCH$_3$	200	—	36
	-COO$^-$Na$^-$	1,950	22,000	30,000

[a] Temp, 25°C; Hajdu and Sigman [23]. When X is 4-COO$^-$Na$^+$, adduct production rather than reduction is observed.

exhibits a larger rate enhancement (1000-fold) relative to its reference compound in the same solvent (Table 3) [36]. In order to ensure that these results reflected the electrostatic stabilization of net positive charge on the nicotinamide moiety in the activated complex, it was essential to demonstrate that the transition state was more polar than the initial state. This was accomplished by measuring the rate of reduction of *N*-methylacridinium ion by a series of neutral *N*-benzyldihydronicotinamides in solvents of different polarity (Table 4). Since the reaction rates for all the neutral dihydronicotinamides were 10-fold faster in water than in acetonitrile, a significant amount of positive charge must develop in the nicotinamide ring in the transition state at the expense of the larger acridinium cation. The suggested explanation for the acceleration by the carboxylate groups is therefore plausible. The insensitivity of the observed reaction rate to substituents in the benzyl group excludes inductive effects from responsibility for the kinetic effects of the carboxylate groups.

It is important to stress that neutral dihydronicotinamides and those with

TABLE 4

**Rate Constants for Reduction of *N*-Methylacridinium
Ion by Neutral *N*-Benzyldihydronicotinamides**

Substituent on benzyl group	k (M^{-1} sec^{-1})[a]		
	H$_2$O	MeOH	MeCN
H	550	180	65
4'-CN	180	—	26
4'-COOCH$_3$	270	—	42
4'-CH$_3$	520	—	80
4'-OCH$_3$	530	—	92
2'-CH$_3$	540	—	70
2'-COOCH$_3$	460	220	70

[a] Temp, 25°C; Hajdu and Sigman [23].

carboxylate groups show opposite solvent effects in the reduction of *N*-methyl-acridinium ion. With the negatively charged derivatives there is an absolute increase in rate in going from water to acetonitrile, whereas for the neutral derivatives there is a decrease. There are at least two reasons for this solvent effect and the related finding that *N*-2'-carboxybenzyldihydronicotinamide reduces *N*-methylacridinium ion only twice as fast as the 4'-carboxy derivative in aqueous solution. The first is that a carboxylate is effectively hydrated in aqueous solution. Its approach to the dihydronicotinamide is therefore sterically hindered. Less effective solvation and hence closer proximity would be expected in acetonitrile. Second, electrostatic interactions vary inversely with dielectric constant and therefore would be expected to be least effective in aqueous solution.

Another manifestation of the more intimate interaction of the carboxylate groups with the dihydronicotinamide moiety in acetonitrile comes from the solvent dependence of the characteristic long-wavelength absorption maximum of the dihydronicotinamides [37]. In this transition, there is greater charge separation in the electronically excited state than in the ground state as is evident from the blue shift observed in less polar solvents for the uncharged dihydronicotinamides (Table 5). Of particular interest is the unusual red shift observed in the absorption maximum of the negatively charged 2'-carboxy-*N*-benzylnicotinamide in acetonitrile but not in water. Apparently, the negatively charged carboxylate group can preferentially stabilize the excited state in acetonitrile but not in water, a finding that is strikingly congruent with the kinetic effect of the carboxylate group.

The kinetic effects summarized above relate to the reduction of *N*-methyl-acridinium ion. This oxidant was employed primarily because its reduction

TABLE 5

Effect on Substituents on Absorption Maxima of
a Series of N-Benzyldihydronicotinamides

Substituent on benzyl group	λ_{max} (nm)		
	H_2O	MeOH	MeCN
H	358	353	347.5
$2'$-CH_3	358	353.5	348
$2'$-$COOCH_3$	358	354	348
$2'$-COO^-	358	356	352
$4'$-COO^-	358	353.5	348

can be conveniently measured spectrophotometrically or fluorimetrically and is sufficiently rapid so that the instability of the dihydronicotinamides does not influence the measured kinetics. In addition, it is soluble and stable in aqueous solution and nonaqueous solution, and the products of its reduction are stable in the presence of oxygen. Finally, its reduction proceeds by direct hydrogen transfer, and the overall reaction rate is insensitive to free-radical quenching agents. However, a question that can be reasonably asked is whether the acceleratory effects of the carboxylate group are observable in the reduction of other oxidants. At present, only an incomplete answer is available, partially for technical reasons, but the results obtained to date are of interest.

TABLE 6

Rates of Reduction of Tetrachlorobenzoquinone (Chloranil) by a Series
of N-Benzyldihydronicotinamides

		k (M^{-1} sec^{-1})[a]	
Dihydronicotinamide	X	Methanol	Acetonitrile
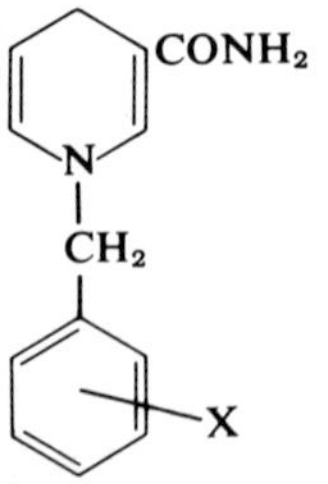	H	1,100	1,900
	2-COO^-K^+	3,000	22,500
	4-COO^-K^+	2,800	3,500

[a] Temp, 25°C; Hajdu and Sigman [23].

The reduction of chloranil by N-2'-carboxy- and N-4'-carboxy-N-benzyl-dihydronicotinamide has been studied in methanol and acetonitrile. The 2'-carboxyl derivative is more reactive than either the 4'-carboxyl- or unsubstituted N-benzyldihydronicotinamide (Table 6) [23]. However, the rate enhancement is only 10 and not 100. These results are explicable if one recognizes that chloranil is reduced 29 times faster than N-methylacridinium ion by N-benzyldihydronicotinamide and that the transition state of the chloranil reaction is therefore probably achieved sooner when a common reaction coordinate involving the nicotinamide moiety is defined for the two reactions. Less positive charge resides on the nicotinamide moiety in the activated complex. As a result, the chloranil reduction would be correspondingly less susceptible to stabilization by an electrostatic interaction involving a neighboring carboxylate group.

Carboxylate residues are not the only potential functional group of amino acids capable of accelerating dihydronicotinamide reductions. For example, neighboring hydroxyl groups enhance dihydronicotinamide reductions in both aqueous and nonaqueous solutions [36]. β-NADH and its unnatural epimer α-NADH, in which the dihydronicotinamide is subject to comparable inductive effects, are a readily accessible set of derivatives that illustrate the ability of hydroxyl groups to enhance dihydronicotinamide reductions via noncovalent interactions in aqueous solution. The data reported in Table 7

TABLE 7[a]

Nonenzymatic Reduction of Epimers of NADH and NMNH[b]

Oxidant	α-NADH	β-NADH	α-NMNH	β-NMNH
N-Methylacridinium ion	1300	101	900	42
Diamide	84	5.7		
10-Methyl-5-deazaisoalloxazine	0.5	0.048		

[a] Hajdu and Sigman [36].
[b] 0.1 M phosphate buffer at pH 8.0, $T = 25°C$.

indicate that α-NADH is a more effective reducing agent than β-NADH for three distinct oxidizing agents. Interactions between the adenine and dihydronicotinamide moiety are probably unimportant because of the greater reactivity of the α-nicotinamide mononucleotide relative to the β-epimer.

Synthetic dihydronicotinamides with neighboring hydroxyl groups indicate that the acceleratory effect of the hydroxyl groups is not restricted to aqueous solution. Modest rate accelerations for the reduction of N-methylacridinium

TABLE 8

**Neighboring-Group Effects in the Reduction of
N-Methylacridinium Ion**

Dihydronicotinamide[a]	k (M^{-1} sec^{-1})		
	0.05 M Tris, pH 8.0	MeOH	MeCN
H—R (H)	2300	350	150
HO—R (H)	1000	770	650
H—R (HO)	410	230	100
R / OH / H	2000	—	1500

[a] R is dihydronicotinamide moiety.

ion are observable in acetonitrile (Table 8). Curiously, these derivatives exhibit smaller rate enhancements in aqueous solution than the epimeric forms of NADH even though the stereochemical relation of the hydroxyl group to the nicotinamide moiety is probably comparable. The pH independence and the absence of a solvent isotope effect on the rate of reduction of *N*-methylacridinium ion by α-NADH suggests that the hydroxyl group is exerting its kinetic effect in the protonated form.

The catalytic effects of the hydroxyl groups are modest in relation to those observed with carboxylate groups, at least in the reduction of *N*-methylacridinium ion. However, they are sufficiently large to indicate that dihydronicotinamide reductions are sensitive to "microscopic solvent effects" induced by neighboring groups. It is interesting that the only polar residues in the hydrophobic nicotinamide binding site of alcohol dehydrogenase are serine and threonine residues [35]. These nonenzymatic reactions provide a chemical precedent, but not proof, for a catalytic role for a hydroxyl group in the reaction mechanism, assuming that it is close enough to the nicotinamide moiety in the ternary complex.

INTERMEDIATES

The demonstration of direct hydrogen transfer in enzymatic and nonenzymatic reactions does not necessarily mean that dihydronicotinamide reductions proceed in a single step via a hydride ion. It is entirely possible that these reactions proceed by multistep pathways in which radical intermediates are produced if hydrogen atom transfer and electron transfer occur in discrete steps. The only restriction in these complex mechanisms is that the hydrogen transferred must never be exchangeable with the media. Indeed, the difference between the simple and complex schemes may be more semantic than real since there is a very low probability that two electrons and a hydrogen ion are transferred simultaneously [38]. Hydride transfer mechanisms may involve formation of a radical pair at low steady-state concentrations.

The first suggestion that multistep reaction pathways might occur in dihydronicotinamide reductions that proceed by direct hydrogen transfer arose from isotope effect studies on the reduction of trifluoroacetophenone by N-propyldihydronicotinamide. The analysis depends on measuring the isotope effects in two distinct ways [17]. The first is by assessing the effect on the observed rate of substituting deuterium for hydrogen in the 4' position of the dihydronicotinamide moiety. This kinetic isotope effect can be measured either with 4'-monodeuterodihydronicotinamide or 4,4'-dideuterodihydronicotinamides. The second involves determining the deuterium content in the reduced oxidant. The use of product analysis to assess the isotope partitioning ratio can be performed only with monodeuterodihydronicotinamides in which some isotopic discrimination is possible. The stereospecificity of enzymatic reactions prohibits the use of product analysis as a means of determining isotope effects for dehydrogenase reactions.

If the nonenzymatic reaction proceeds by a single-step bimolecular reaction mechanism, the isotope effects measured in both ways should coincide after appropriate correction for statistical factors. The relevant expressions can be derived with reference to Eq. (6), where RH_2 is the dihydronicotinamide, A is the oxidant, k_H represents the rate constant for the transfer of a hydrogen from the C-4 of a dihydronicotinamide containing another hydrogen at the C-4 position, $k_{H'}$ is the rate constant for the transfer of a hydrogen from the C-4 position possessing a deuterium at the C-4 position, and k_D and $k_{D'}$ are defined for the transfer of deuterium in an analogous way. The observed second-order rate constants for Eq. (6a,b,c) are $k_{(6a)}$, $k_{(6b)}$, and $k_{(6c)}$, respectively [Eq. (7)]. Three different kinetic isotope

$$
\begin{array}{lll}
\text{(a)} & R\!\!\begin{array}{c}\nearrow H \\ \searrow H\end{array}\!\! + A & \begin{array}{c}\xrightarrow{k_H} \\ \\ \xrightarrow[k_H]{}\end{array} \quad \begin{array}{l}RH + \dot{A}H \\ \\ RH + AH\end{array} \\[2em]
\text{(b)} & R\!\!\begin{array}{c}\nearrow H \\ \searrow D\end{array}\!\! + A & \begin{array}{c}\xrightarrow{k_{H'}} \\ \\ \xrightarrow[k_D]{}\end{array} \quad \begin{array}{l}RD + AH \\ \\ RH + AD\end{array} & \quad (6) \\[2em]
\text{(c)} & R\!\!\begin{array}{c}\nearrow D \\ \searrow D\end{array}\!\! + A & \begin{array}{c}\xrightarrow{k_{D'}} \\ \\ \xrightarrow[k_{D'}]{}\end{array} \quad \begin{array}{l}RD + AD \\ \\ RD + AD\end{array}
\end{array}
$$

$$
\begin{aligned}
k_{(6a)} &= 2k_H \\
k_{(6b)} &= k_{H'} + k_D \\
k_{(6c)} &= 2k_{D'}
\end{aligned} \qquad (7)
$$

effects can be measured, but that involving Eqs. (6a) and (6b) ($k_{(6a)}/k_{(6b)}$) has been most widely used because fewer corrections for complete isotopic substitution are required [Eq. (8)]. The isotope partitioning ratio, PR,

$$
\frac{k_{(6a)}}{k_{(6b)}} = \frac{2k_H}{k_{H'} + k_D} = \frac{2k_H/k_{H'}}{1 + (k_D/k_{H'})} \qquad (8)
$$

which can be measured only with monodeutero derivatives [Eq. (6b)] is simply defined by Eq. (9).

$$
PR = \frac{k_{H'}}{k_D} \qquad (9)
$$

    Table 9 records the measured kinetic isotope effect, the isotope partitioning ratio, and the calculated secondary isotope effect for several reaction systems.

    These data raise the following significant difficulty. Secondary isotope effects of less than 1 have never been previously observed for processes involving the conversion of a carbon atom from an sp^3 to an sp^2 hybridization [39]—exactly the change that occurs in dihydronicotinamide reduction. One is faced with either accepting an anomalous secondary isotope effect of about 0.7 or concluding that a strict bimolecular reaction mechanism is inadequate to account for this reaction scheme because it leads to an unacceptable secondary isotope effect. It must be noted that the reduction of thiobenzophenone by N-benzyldihydronicotinamide is the only reported reaction in which the published kinetic isotope effect and isotope partitioning ratio yield a secondary isotope effect of unity [12].

    Multistep mechanisms provide kinetic expressions that are capable of

TABLE 9

Isotope Effects for Reductions by Monodeuterodihydronicotinamide[a]

Dihydronicotin-amide	Oxidant	$k_{(6a)}/k_{(6b)}$	PR	$k_H/k_{H'}$
N-Propyl	Trifluoroacetophenone	1.16	3.8	0.71
N-Benzyl	Trifluoroaectophenone	1.47	3.8	0.93
N-Propyl	N-Methylacridinium ion	1.36	5.4	0.75
N-Benzyl	N-Methylacridinium ion	1.17	4.0	0.73
(4)	N-Methylacridinium ion	1.11	3.8	0.70

[a] Other experimental data demand a secondary isotope effect of 0.59 for the reduction of the 1,10-phenanthroline-2-carboxaldehyde–zinc complex if a bimolecular reaction mixture is assumed [31].

explaining the observed isotope effects without requiring unusual secondary effects to explain the data. Although some positive evidence for these more complex schemes is accumulating, there is no compelling evidence yet available that implicates analogous complex mechanisms at the active site of enzymes. It is instructive, however, to determine which multistep mechanisms provide acceptable explanations for the isotope effects observed in certain nonenzymatic reactions.

The formation of a covalent intermediate (W_1, W_2), which permits the transfer of either a hydrogen or deuterium, does not provide a satisfactory solution for the anomalous secondary isotope effects (Scheme 1). This scheme requires secondary isotope effects identical to those obtained for the single-step bimolecular mechanism. The funneling of the reaction through a noncovalent intermediate (X) in which either hydrogen may be

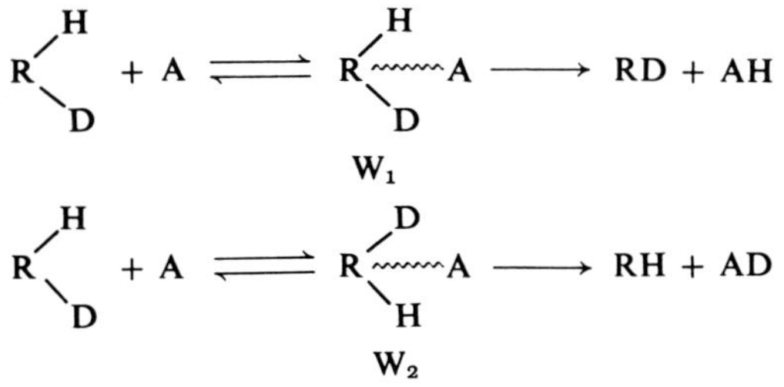

Scheme 1   R, nicotinamide; A, oxidizing agent; ⁓ implies covalent bond.

transferred (Scheme 2) can explain the isotope effect as long as the formation of this species is rate limiting. This scheme, however, is unlikely for two reasons. First, it requires that the rate of formation of this noncovalent complex be approximately the same order of magnitude as the overall

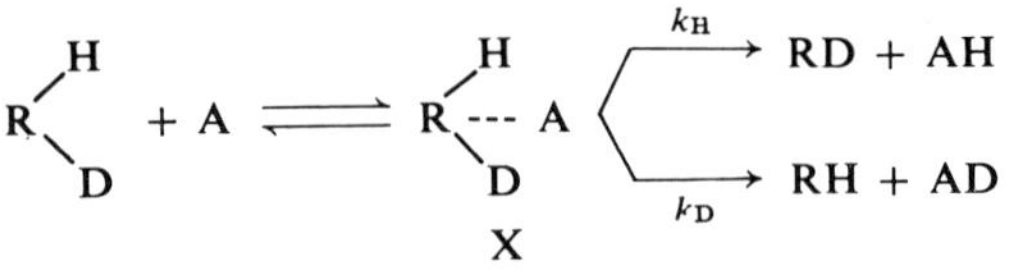

$$R\underset{D}{\overset{H}{\diagup}} + A \rightleftharpoons R\underset{D}{\overset{H}{\diagdown}} {---} A \underset{X}{\overset{k_H}{\diagup}} \begin{array}{c} RD + AH \\ \\ RH + AD \end{array}$$

Scheme 2   --- implies noncovalent complex.

reaction rate. However, the reactions summarized in Table 9 have second-order rate constants which at their fastest are still five orders of magnitude slower than the diffusion-controlled rate of formation of aromatic complexes [40]. Second, a kinetically competent noncovalent complex of change-transfer character has been observed in the N-benzyldihydronicotinamide reduction of N-methylacridinium ion [41]. The rate of disappearance of this complex can be conveniently measured in concentration ranges where saturation kinetics, characteristic of schemes involving complexes, are obtained. The extrapolated first-order rate constants at an infinite concentration of di-hydronicotinamide are 55.5, 47.6, and 40.0 sec^{-1} when the dihydro, mono-deutero, and dideutero forms of N-benzyldihydronicotinamide, respectively, are used. For all three derivatives, the dihydronicotinamide concentration at which half the maximal rate is obtained is 0.32 M.

The isotope effect in the extrapolated first-order rate constant, of the same magnitude as that obtained when the reaction takes place under second-order conditions, eliminates Scheme 2 for this reaction since it requires a large isotope effect in the extrapolated first-order constants. Coupled with the first difficulty associated with this scheme, these results argue for the existence of an additional intermediate. As long as this second intermediate forms from an initial noncovalent complex, it can be covalent (Y_1 or Y_2 in Scheme 3) or noncovalent (Z in Scheme 4). The central difference between the covalent

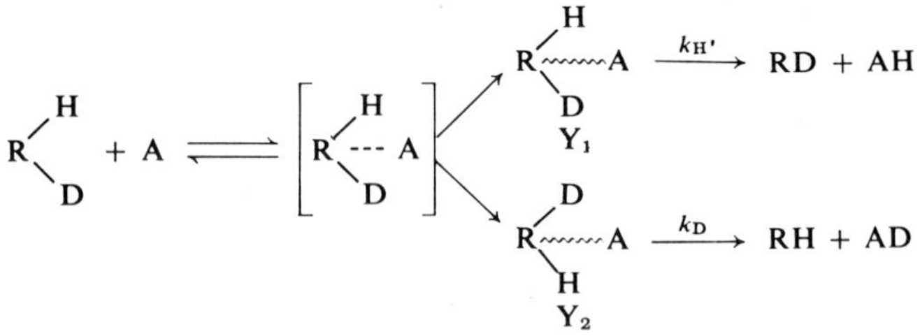

$$R\underset{D}{\overset{H}{\diagup}} + A \rightleftharpoons \left[R\underset{D}{\overset{H}{\diagdown}} {---} A \right] \begin{array}{c} R\underset{D}{\overset{H}{\diagdown}}\wwwA \xrightarrow{k_{H'}} RD + AH \\ Y_1 \\ \\ R\underset{H}{\overset{D}{\diagdown}}\wwwA \xrightarrow{k_D} RH + AD \\ Y_2 \end{array}$$

Scheme 3

$$R\underset{D}{\overset{H}{\diagup}} + A \left[R\underset{D}{\overset{H}{\diagdown}} {---} A \right] \rightleftharpoons \left[R\underset{D}{\overset{H}{\diagdown}}{\cdot +} + \cdot A \right] \underset{Z}{\overset{k_H}{\diagup}} \begin{array}{c} RD + AH \\ \\ RH + AD \end{array}$$

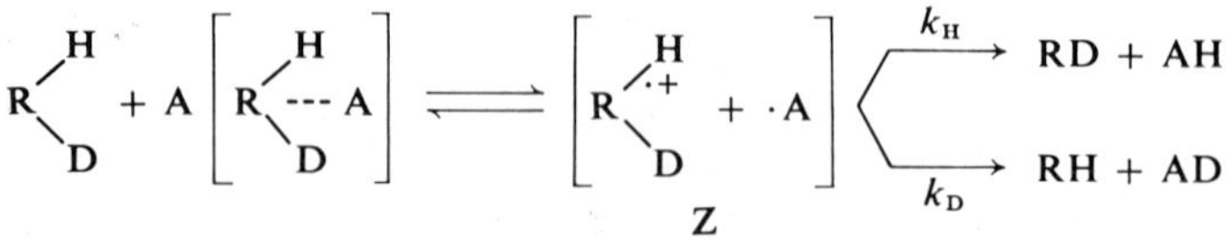

Scheme 4

and noncovalent complex is that the former commits transfer of either a hydrogen or deuterium, while the latter permits transfer of either, subject only to the primary isotope effect of the hydrogen transfer step.

Although the indirect kinetic evidence summarized here argues for the existence of an intermediate subsequent to an initial charge-transfer complex, the chemical nature of the second intermediate is as yet unknown. Calculated energy contours for the dihydroflavin reduction of formaldehyde have supported the possible importance of intimate radical pairs in these reactions. By analogy, comparable intermediates might be expected in dihydronicotin- amide reductions after the initial production of the reversible charge-transfer complexes [42]. A brief report has described the detection of the thiol anion radical (5) by electron paramagnetic resonance in the N-propyldihydro- nicotinamide reduction of thiobenzophenone, but the kinetic competence of

$$
\langle \text{Ph} \rangle - \underset{\cdot}{\overset{\overset{\displaystyle S^-}{|}}{C}} - \langle \text{Ph} \rangle \qquad (13)
$$

(5)

this putative intermediate has not yet been established [13]. It should further be remembered that the reduction of thiobenzophenone is the one reaction in which isotope effect studies have not demanded the existence of any intermediate. Nevertheless, continued search for intermediates via chemical and physical techniques seems justified from the kinetic studies described here.

It is worthwhile to note that kinetic studies on the pre-steady-state produc- tion of NADH by alcohol dehydrogenase and lactate dehydrogenase have exhibited variable isotope effects. Lactate dehydrogenase exhibits no iso- tope effect in its pre-steady-state burst when deuterolactate is used as sub- strate [13]. Since dissociation of coenzyme and substrate cannot be rate limiting in these single-turnover measurements, it is tempting to suggest that the formation of an intimate radical pair, or a conformational change that energizes this process, might be the step masking the isotope effect inherent in hydrogen transfer. The full kinetic isotope effect of 6 observed for alcohol dehydrogenase suggests that hydrogen transfer is the rate-limiting step in the ternary complex composed of enzyme, NAD^+, and dideutero- ethanol [43]. Intimate radical pair formation may occur in this case as well, but its collapse by hydrogen transfer, rather than its formation, would be rate limiting.

The importance of intermediates in nonenzymatic and enzymatic di- hydronicotinamide remains uncertain. Further work is required before the chemical nature or even existence of such species is established. Since non- enzymatic reactions are providing useful leads in this direction, continued work in these systems seems justified. Certainly, nonenzymatic reactions are providing valuable chemical precedents for understanding the catalytic function of amino acid residues and cofactors at the active site of NAD-dependent dehydrogenases.

ACKNOWLEDGMENTS

D.S.S. wishes to acknowledge the generous financial support of the Josiah Macy Foundation and the gracious hospitality of Dr. G. K. Radda during the 1975–1976 academic year while he was on sabbatical leave at the Department of Biochemistry, Oxford University. Research in the authors' laboratory has been supported by USPHS Grant 21199.

REFERENCES

1. H. Theorell, A. P. Nygaard, and R. Bonnichsen, *Acta Chem. Scand.* **9**, 1148 (1955).
2. B. L. Vallee and F. L. Hoch, *J. Biol. Chem.* **225**, 185 (1957).
3. A. Akeson, *Biochem. Biophys. Res. Commun.* **17**, 211 (1964).
4. G. Pleiderer, D. Jeckel, and T. Wieland, *Biochem. Z.* **330**, 296 (1959).
5. M. G. Rossman, A. Liljas, C.-I. Branden, and J. L. Banaszek, *in* "The Enzymes" (P. D. Boyer, ed.), 3rd ed., Vol. 11, p. 62. Academic Press, New York, 1975.
6. J. J. Holbrook and H. Gutfreund, *FEBS Lett.* **31**, 157 (1973).
7. F. H. Westheimer, H. F. Fisher, E. E. Conn, and B. Vennesland, *J. Am. Chem. Soc.* **73**, 2403 (1951).
8. H. F. Fisher, E. E. Conn, B. Vennesland, and F. H. Westheimer, *J. Biol. Chem.* **202**, 687 (1953).
9. B. Vennesland and F. H. Westheimer, *in* "The Mechanism of Enzyme Action" (W. D. McElroy and B. Glass, eds.), p. 357. Johns Hopkins Press, Baltimore, Maryland, 1954.
10. J. G. Dewan and D. E. Green, *Biochem. J.* **32**, 626 (1938).
11. G. W. Rafter and S. P. Colowick, *J. Biol. Chem.* **209**, 773 (1954).
12. R. H. Abeles, R. F. Hutton, and F. H. Westheimer, *J. Am. Chem. Soc.* **79**, 712 (1957).
13. A. Ohno and N. Kito, *Chem. Lett.* p. 369 (1972).
14. D. C. Dittmer and R. Fouty, *J. Am. Chem. Soc.* **86**, 91 (1964).
15. G. Cilento, *Arch. Biochem. Biophys*, **88**, 352 (1960).
16. K. A. Schellenberg, G. W. McLean, H. C. Lipton, and P. S. Lietman, *J. Am. Chem. Soc.* **89**, 1948 (1967).
17. J. J. Steffens and D. M. Chipman, *J. Am. Chem. Soc.* **93**, 6694 (1971).
18. D. J. Creighton and D. S. Sigman, *J. Am. Chem. Soc.* **93**, 6314 (1971).
19. M. Brutslein and T. C. Bruice, *J. Am. Chem. Soc.* **94**, 6548 (1972).
20. S. Shinkai and T. C. Bruice, *Biochemistry* **12**, 1750 (1973).
21. D. J. Creighton, J. Hajdu, G. Mooser, and D. S. Sigman, *J. Am. Chem. Soc.* **95**, 6855 (1973).
22. D. Mauzerall and F. H. Westheimer, *J. Am. Chem. Soc.* **77**, 2261 (1955).
23. J. Hajdu and D. S. Sigman, *J. Am. Chem. Soc.* **97**, 3524 (1975).
24. J. C. Powers and F. H. Westheimer, *J. Am. Chem. Soc.* **82**, 5431 (1960).

25. M. J. Adams, M. Buchner, K. Chandrasekhar, G. C. Ford, M. L. Hackert, A. Liljas, M. G. Rossman, I. E. Smiley, W. S. Allison, J. Everse, N. O. Kaplan, and S. S. Taylor, *Proc. Natl. Acad. Sci. U.S.A.* **70**, 1968 (1973); S. S. Taylor, *J. Biol. Chem.* **252**, 1799 (1977).
26. B. L. Valee, R. J. P. Williams, and F. L. Hoch, *J. Biol. Chem.* **234**, 2621 (1959).
27. T. Yonetani, *Biochem. Z.* **338**, 300 (1963).
28. D. S. Sigman, *J. Biol. Chem.* **242**, 3815 (1967).
29. C. Branden, H. Eklund, B. Nordstrom, T. Bowie, G. Soderland, E. Zeppezauer, I. Ohlsson, and A. Akeson, *Proc. Natl. Acad. Sci. U.S.A.* **70**, 2439 (1973).
30. M. F. Dunn and J. S. Hutchinson, *Biochemistry* **12**, 4882 (1973).
31. H. Sund and H. Theorell, *in* "The Enzymes" (P. D. Boyer, H. Lardy, and K. Myrbäck, eds.), 2nd ed., Vol. 7, p. 26 Academic Press, New York, 1963.
32. J. F. Naylor, III and I. Fridovich, *J. Biol. Chem.* **243**, 341 (1968).
33. D. L. Sloan, J. M. Young, and A. S. Mildvan, *Biochemistry* **14**, 1998 (1975).
34. D. J. Creighton, J. Hajdu, and D. S. Sigman, *J. Am. Chem. Soc.* **98**, 4619 (1976).
35. C.-I. Brändén, H. Jörnvall, H. Eklund, and B. Furugren, *in* "The Enzymes" (P. D. Boyer, ed.), 3rd ed., Vol. 11, p. 104. Academic Press, New York, 1975.
36. J. Hajdu and D. S. Sigman, *Biochemistry* **16**, 2841 (1977).
37. E. M. Kosower, *in* "Molecular Biochemistry." p. 205. McGraw-Hill, New York, 1962.
38. K. Schellenberg, *in* "Pyridine Nucleotide Dependent Dehydrogenases" (H. Sund, ed.), p. 15. Springer-Verlag, Berlin and New York, 1970.
39. W. P. Jencks, "Catalysis in Chemistry and Enzymology," p. 193 McGraw-Hill, New York, 1969.
40. D. H. Turner, G. W. Flynn, S. K. Lundenberg, L. D. Faller, and N. Sutin, *Nature (London)* **239**, 215 (1972).
41. J. Hajdu and D. S. Sigman, *J. Am. Chem. Soc.* **98**, 6060 (1976).
42. T. C. Bruice, *Proc. Natl. Acad. Sci. U.S.A.* **72**, 1763 (1975).
43. J. D. Shore and H. Gutfreund, *Biochemistry* **9**, 4655 (1970).

CHAPTER

15

The Use of Puromycin Analogs and Related Compounds to Probe the Active Center of Peptidyl Transferase on *Escherichia coli* Ribosomes

Robert H. Symons, Raymond J. Harris,
Philip Greenwell, David J. Eckermann, and
Elio F. Vanin

INTRODUCTION*

Peptidyl transferase is an integral part of the 50 S bacterial and 60 S mammalian ribosomal subunit. It catalyzes peptide bond formation by a process involving transfer of a nascent peptide chain from peptidyl-tRNA

* Abbreviations: A-amino acid, $2'(3')$-O-aminoacyladenosine; $2'$-dA-Phe, $3'$-O-phenylalanyl-$2'$-deoxyadenosine; C, cytosine; G, guanosine; I, inosine; U, uridine; N, any nucleoside; Pan, puromycin aminonucleoside; Pan-amino acid, $3'$-N-aminoacyl-Pan; Cys(Bz), S-benzylcysteine; Ser(Bz), O-benzylserine; His(Bz), imbenzylhistidine; pA-Phe, $5'$-O-phosphoryl-A-Phe and similarly for other derivatives; pA-fMet, $5'$-O-phosphoryl-$2'(3')$-O-(N-formylmethionyl)adenosine; CNEt, cyanoethyl; CpPan-Phe, cytidylyl-($3',5'$) derivative of Pan-Phe and similarly for other derivatives of Pan-amino acid and A-amino acid; Cp-$2'$-dA-Phe, cytidylyl-($3',5'$)-($3'$-O-phenylalanyl)-$2'$-deoxyadenosine; Cp-$3'$-dA-Phe, cytidylyl-($3',5'$)-($2'$-O-phenylalanyl)-$3'$-deoxyadenosine; CpA-($2'$-O-methyl)Phe, $2'$-O-methyl, $3'$-O-phenylalanyl derivative of CpA and similarly for related derivatives; Bap-Pan-Phe, $5'$-O-(N-bromoacetyl-p-aminophenylphosphoryl)-$3'$-N-L-phenylalanylpuromycin aminonucleoside; $A_{oxid/red}$, adenosine oxidized with periodate and reduced with borohydride. For clarity the numbering system of the intact ribose is used. All amino acids are the L-isomer.

(donor substrate) on the ribosome to aminoacyl-tRNA (acceptor substrate). The sites to which the peptidyl-tRNA and the aminoacyl-tRNA are bound on the ribosome are known as the *P* site and A site, respectively. We have defined [1,2] the P′ and A′ sites as those parts of peptidyl transferase responsible for binding of the 3′-terminal parts of these molecules (Fig. 1).

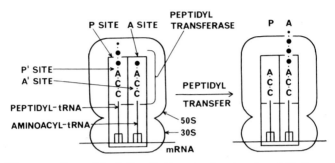

Fig. 1. Diagrammatic model of *E. coli* ribosome showing A and P sites occupied by aminoacyl- and peptidyl-tRNA and the A′ and P′ sites of peptidyl transferase occupied by the 3′-terminal portions, -CpCpA-amino acid and -CpCpA-peptide, respectively. As indicated, peptidyl transferase catalyses the transfer of the peptide to a covalent peptide link with the α-amino group of aminoacyl-tRNA.

The bacterial ribosome is very complex. The 50 S subunit from *Escherichia coli*, for example, contains 33 proteins and one molecule each of 23 S and 5 S RNA. It is only a matter of time before the complete amino acid sequence of all these proteins will be determined [3], while various studies using protein cross-linking reagents and antibodies to ribosomal proteins to allow the electron microsope visualization of their location on the ribosome have encouraged model building of the three-dimensional structure of the ribosome [3].

In spite of the continuing process in these areas, our knowledge of the structure and mechanism of action of peptidyl transferase is much more limited. As is discussed below, the 23 S RNA and several ribosomal proteins are implicated in the active center of this enzyme, but there is really nothing known as to how these molecules may be arranged in three-dimensional form. Even less is known about the molecular mechanism by which peptide bond formation is brought about.

We have been interested for several years in the latter aspects of *E. coli* peptidyl transferase. Our approach has been to make use of low molecular weight substrates and inhibitors of the enzyme in order to probe the active center in terms of the structural requirements needed for substrate activity, the RNA and protein components that make up the active center, and the molecular mechanism of peptide bond formation. By using low molecular weight compounds that are structural analogs of aminoacyl-tRNA and peptidyl-tRNA, we hoped to simplify the system as much as possible by eliminating the complicating effects of tRNA molecules, the necessity for mRNA and the 30 S subunit, etc.

It is the aim of this review to summarize our results and conclusions and to relate them to data in the literature. We first describe the use of a range of analogs of puromycin (Fig. 2) and related aminoacyl nucleoside

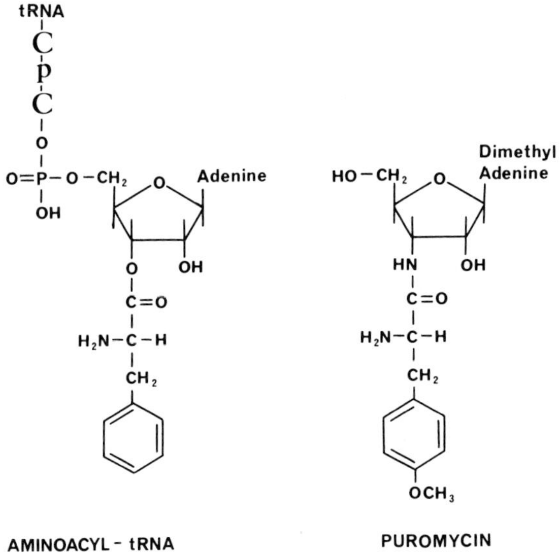

Fig. 2 Structure of puromycin and comparison with the 3′-terminal end of aminoacyl-tRNA.

derivatives in defining the structural requirements for acceptor substrate activity at the A′ site and donor substrate activity at the P′ site of peptidyl transferase and how these indicate the presence of binding sites for various groups in the active center of the enzyme. In addition, we outline a model for the structure of the A′ and P′ sites. We then describe our attempts to investigate the mechanism of peptide bond formation, and finally we look at the use of affinity labeling derivatives of puromycin to probe the active center of peptidyl transferase and the problems inherent in this approach.

For further details of peptide bond formation on the ribosome, the reader is referred to the review of Harris and Pestka [4].

SUBSTRATE SPECIFICITY OF THE A' SITE OF PEPTIDYL TRANSFERASE OF *E. COLI*

As outlined in Fig. 1, the acceptor or A' site is responsible for binding the 3'-terminal end of aminoacyl-tRNA during peptide bond formation. We and several other groups have looked at the acceptor substrate activity of a range of puromycin analogs and aminoacyl nucleoside derivatives in order to define the structural requirements for binding in the A' site. A summary of much of this data is presented in Table 1. We have included data from experiments in which the various substrates acted as acceptors and have omitted data from experiments in which only the inhibition of the overall process of protein synthesis was measured. The classic data of Nathans and Neidle [5,6] on the inhibition of poly(Phe) synthesis by puromycin analogs have therefore not been included nor has later experimental material [7] even though all the data support the trends shown in Table 1.

In the five assay systems of Table 1, all results have been expressed relative to the activity of the well-characterized acceptor molecule puromycin. Since

TABLE 1

Acceptor Substrate Activity Relative to Puromycin of Various Aminoacyl and Nucleotidyl Analogs[a]

		Release assays			
Compound[b]	Fragment reaction [9,17,18]	Ac-Phe-tRNA[c] [9,12,13,17–20]	Poly(Lys)-tRNA [12,18]	Peptidyl-tRNA ribosomes [8]	Peptidyl-tRNA polysomes [14]
Pan-Phe	+ + +	+ + +	—	+ + +	+ + +
A-Phe	+ + +	+ + +	+ + +	—	—
2'-dA-Phe	0/+	0	0	—	—
Pan-Tyr	+	+ +	—	+ +	+ + +
Pan-Cys(Bz)	0	+	—	+ +	—
Pan-Ser(Bz)	0	+	—	+ +	+ +
Pan-His(Bz)	0	+	—	+ +	0
A-His(Bz)	+	+ + +	+	—	—
Pan-Trp	0	0	—	0	0
Pan-Met	0	0	—	—	—
A-Met	+ + +	+ + +	+ +	—	—
Pan-Leu	0	0	—	0	+
A-Leu	0	+ +	+ +	—	—
Pan-Val	0	+	—	—	—
A-Val	0	+	+	—	—

TABLE 1 (*Continued*)

Compound[b]	Fragment reaction [9,17,18]	Release assays			
		Ac-Phe-tRNA[c] [9,12,13,17–20]	Poly(Lys)-tRNA [12,18]	Peptidyl-tRNA ribosomes [8]	Peptidyl-tRNA polysomes [14]
Pan-Ala	0	0	—	0	—
A-Ala	0	+	+ +	—	—
Pan-Gly	0	0	—	0	—
A-Gly	0	0	0	—	—
Pan-Lys	—	—	—	—	+
A-Lys	+	+ + +	0	—	—
Pan-Pro	0	0	—	0	0
A-Pro	0	0	+	—	—
A-Glu	0	0	+	—	—
A-Ser	0	+	+ +	—	—
I-Phe	+ +/+ + +	+ +	+	—	—
C-Phe	+	0	0/+	—	—
G-Phe	—	0	0	—	—
U-Phe	0	0	0	—	—
pPan-Phe	0/+	+	—	—	+ + +
pA-Phe	—	+ + + +	—	—	—
Methyl-pA-Phe	—	+ + + +	—	—	—
pA-Val	—	+ +	—	—	—
pA-Pro	—	+	—	—	—
pPan-Gly	0	0	—	0	—
pA-Gly	—	—	0	—	—
CNEt-pPan-Phe	0/+	+	—	—	—
CpPan-Phe	+ + +	+ + +	—	—	+ + +
CpA-Phe	—	+ + + +	—	—	—
CpPan-Gly	0/+	0	—	+ +	+ +
CpA-Gly	—	—	+/+ +	—	—
ApPan-Gly	0	0	—	0	0
GpPan-Gly	0	0	—	0	0
UpPan-Gly	0	0	—	0	0
UpA-Gly	—	—	0	—	—
UpU-Gly	—	—	0	—	—
Cp-2'-dA-Phe	—	+ + + +	—	—	—
Cp-3'-dA-Phe	—	0	—	—	—

[a] All activities are expressed relative to puromycin and are broadly grouped into five classes: very high activity (+ + + +), appreciably more active than puromycin; high activity (+ + +), about 100 to > 50% of that of puromycin; medium (+ +), 25–50%; low (+), 5 to < 25%; and negligible (0), < 5%. For details of assays, see text.

[b] All amino acids are the L-isomer.

[c] Includes data for five compounds using poly(Phe)-tRNA as donor substrate [19].

the acceptor activity of the compounds listed varies somewhat with the assay system used, activities relative to puromycin have been grouped broadly into five classes [8,9], as indicated in footnote *a* of Table 1. This variation of activity with assay system emphasizes the care that must be taken in the extrapolation of these *in vitro* experiments to the *in vivo* situation. The five assay systems used are as follows:

1. The fragment reaction of Monro and Marcker [10]. This is the most defined assay of peptidyl transferase, which is carried out with washed ribosomes, buffer, salts, ethanol or methanol, puromycin (or analog), and a low molecular weight fragment of the 3′ end of peptidyl-tRNA, e.g., CpAp-CpCpA-(Ac-[³H]Leu). The assay measures the synthesis of *N*-Ac-[³H]Leu-puromycin (or analog).

2. The ribosome-catalyzed transfer of Ac-Phe from Ac-Phe-tRNA to acceptor substrate in the presence of ribosomes and poly(U) [9,11,12].

3. As for assay system 2 but with poly(Lys)-tRNA and poly(A) [12,13].

4. The acceptor-substrate-induced release of peptide from [¹⁴C]pep-tidyl-tRNA formed on ribosomes with natural mRNA [8].

5. The inhibition of the reaction of [³H]puromycin with peptidyl-tRNA on polysomes [14]. The assay measures the inhibition by the added acceptor substrate of the formation of peptidyl[³H]puromycin and is considered to be most representative of the *in vivo* situation [15,16]. It was assumed that inhibition occurred by direct competition of the added acceptor substrate with [³H]puromycin for the A′ site [14].

The data of Table 1 [8,9,12–14,17–20] indicate that few of the many puromycin and aminoacyl nucleoside analogs tested showed a high acceptor activity relative to that of puromycin under the same conditions. Of the nonphosphorylated compounds, Pan-Phe and A-Phe were consistently the best substrates followed by Pan-Tyr. Other analogs that showed some activity were the benzylated derivatives of Pan-Ser, Pan-Cys, and A-His. Of the nonaromatic aminoacyl derivatives, only A-Met and to a lesser extent A-Lys showed any consistent activity. The acceptor activity of I-Phe and C-Phe and the lack of activity of G-Phe and U-Phe are further considered below. Of interest is the negligible activity of the aromatic Pan-Trp and of the cyclic Pan-Pro.

Of the phosphorylated and nucleotidyl analogs, only CpPan-Phe and CpA-Phe showed consistently high activity that was usually higher than that of puromycin. The appreciably lower activity of CpPan-Gly and CpA-Gly emphasizes the important role of the aromatic aminoacyl residue of phenyl-alanine with its single benzene ring. The somewhat variable activity among pPan-Phe, pA-Phe, cyanoethyl-pPan-Phe, and methyl-pA-Phe indicates a variable effect of a 5′-phosphate without the accompanying nucleoside. For detailed discussions of the activities of the analogs of Table 1, the reader is referred to the relevant references given (see also Harris and Symons, [1,2]).

MODEL OF THE A′ SITE OF PEPTIDYL TRANSFERASE
OF *E. coli*

On the basis of the type of data given in Table 1 and a theory on the mechanism of action of a number of inhibitors and substrates of peptidyl transferase [1], we proposed a model for the active center of this enzyme [2]. A slightly modified version of this model is presented in Fig. 3. The A′ site is considered first, and the P′ site is described below. The A′ site is composed of the following binding sites.

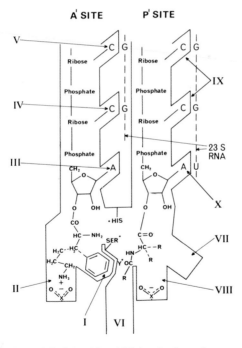

Fig. 3 Diagrammatic model of the A′ and P′ sites in the active center of *E. coli* peptidyl transferase. Details of numbered binding sites are given in text. Attachment is shown to peptidyl transferase of the CpCpA-amino acid (A′ site) and CpCpA-peptide (P′ site), the 3′ termini of aminoacyl- and peptidyl-tRNA, respectively. HIS*, SER*, and Y* represent catalytic or binding residues essential for activity.

Hydrophobic Site (I)

This site is relatively specific for the binding of the aromatic aminoacyl R groups. The high activity of the more hydrophobic derivatives of Table 1 can be understood in terms of binding to this site. The less hydrophobic derivatives show appreciably less activity with a trend to decreasing activity with decreasing hydrophobicity of the R group, i.e., Leu, Val, Ala, Gly. The lack of activity of Pan-Trp is considered to be due to the bulkiness of the indole ring, which does not allow binding to this site. Likewise, the lower activities of the benzylated derivatives of Cys, Ser, and His relative to those of the Phe derivatives have their basis in the greater distance of the benzene ring from the α-amino group of the amino acid. The negligible activity reported by us [9] for Pan-Met may have been due to the presence of the sulfoxide of Pan-Met rather than Pan-Met itself in view of the activity reported for A-Met (Table 1). Support for a hydrophobic binding site is the greater binding to the A' site observed for the Phe-pentanucleotide fragment relative to the Leu- and Val-pentanucleotides and the Met-hexanucleotide [21]. Nathans and Neidle [5,6] and Rychlík *et al.* [18] have suggested similar hydrophobic interactions.

Of considerable interest is the remarkable gap between the high activity in the fragment reaction shown by Pan-Phe, A-Phe, A-Met, and CpPan-Phe relative to the negligible activity shown by essentially all other compounds listed in Table 1. It seems unlikely that this could be explained by some favorable conformation of the aminoacyl-sugar linkage in a way recently suggested by Krayevsky *et al.* [22]. It is not apparent how the aminoacyl R group could uniquely affect the conformation of puromycin and related molecules in view of the extended nature of the puromycin molecule as determined by X-ray analysis [23]. All the evidence indicates that the aminoacyl esters of nucleosides have the same extended structure [23–26]. In addition, this puromycin structure does not support any proposal [7,27,28] in which *intramolecular* stacking of the benzene ring of an amino acid with the adenine ring of the nucleoside is involved. This cannot occur since the atoms that make up the amide and ester linkages are coplanar [23–26]. Overall, the presence of a rather specific hydrophobic binding site appears to be the best explanation of all the data.

Hydrophilic Site (II)

This site was proposed on the basis of the high acceptor activity of A-Lys compared to the lack of activity of A-Gly and related uncharged analogs [18]. It is presumed that the site functions *in vivo* to bind the aminoacyl R groups of Arg-tRNA and Lys-tRNA to a negatively charged group on the ribosome (e.g., an α-, β-, or γ-carboxyl or a phosphoryl group).

Adenine Site (III)

This site is involved in the binding of the terminal adenine of aminoacyl-tRNA. The necessity of the purine base for high acceptor activity is shown by the activity of analogs of the type *N*-Phe, where A $\gg$ I $\gg$ C while the G and U derivatives were inactive (Table 1).

Several alterations of the C-6 amino of the adenine ring have been tested [29]. All analogs had high activity, but it decreased with the decrease in the electron-donating property of the substituent, i.e., NH_2, $N(CH_3)_2 > S-CH_3$, $O-CH_3 > H$. Two possible explanations [29] for this effect are as follows: (1) A decreased electron density in the nebularine ring (unsubstituted purine ring) may decrease hydrogen bond formation between the N-3 nitrogen and either the 2′-hydroxyl group of the ribose (intramolecular) or some group on the ribosome; (2) a decreased electron density may lead to a decreased hydrophobic interaction with some part of the ribosome.

Further, the high activity of these C-6-substituted analogs does not support the possibility of base pairing between the 3′-terminal adenosine of aminoacyl-tRNA and a uridine residue in the 23 S RNA. Hence, our original suggestion [30,31] that the terminal -CpCpA of aminoacyl-tRNA is base paired to a -UpGpGp- sequence of 23 S RNA must be modified so that only the -CpCp- sequence base pairs to -GpGp- (Fig. 2). However, a -UpGpGp-binding sequence is still considered part of the P′ site (see below).

CpCp Site (IV, V)

In view of the large increase in the acceptor activity of Pan-Gly following 5′-substitution with Cp but not with Ap, Gp, or Up [8,9,14], a relatively specific site for the penultimate nucleotide, Cp, of aminoacyl-tRNA was proposed [2,8]. The negligible acceptor activity of A-Gly was similarly specifically increased by substitution with Cp but not with Up [11] (Table 1).

An additional binding site for the third nucleotide, Cp, from the 3′ end of aminoacyl-tRNA appears likely since CpPan-Gly and CpA-Gly had only low activity in releasing peptides from peptidyl-tRNA, while Takanami [32] found that the high puromycinlike activity of various $(Np)_x$CpCpA-amino acids was independent of the nature of the amino acid. The existence of a second Cp-specific site is consistent with the observation of Scolnik *et al.* [33] that CpCpA, possibly by binding to the A′ site, allowed the nucleophilic attack by ethanol on tRNA-fMet to form fMet-*O*-ethyl ester. CpA, ApCpA, GpCpA, and UpCpA were less than 5% as active as CpCpA.

More direct data on this proposed CpCp site are provided by binding assays. Thus, A-Ser was bound very poorly to *E. coli* ribosomes and was just detectable, whereas CpCpA-Ser bound strongly [21]. Likewise, the binding of puromycin was just detectable [34], whereas CpCpA-Phe bound strongly [35]. As suggested above, it is possible that a GpG sequence in 23 S RNA may constitute the CpC binding site.

ACCEPTOR SUBSTRATE ACTIVITY OF DERIVATIVES OF 2′-AMINOACYL-tRNA

Both 2′- and 3′-aminoacyl-tRNA's can act as acceptor substrates, e.g., tRNA-CpCp(3′-deoxy)A-Phe and tRNA-CpCpA$_{oxid-red}$(2′-Phe) [36] as well as tRNA-CpCp(3′-amino,3′-deoxy)A-(3′-Phe) [37]. In the case of the first two derivatives, acceptor activity was appreciably less than with the naturally occurring tRNA-CpCpA-Phe. However, the results indicate that the lone pair of electrons of the α-amino group of phenylalanine is able to attack the carbonyl of the peptidyl ester link whether the amino acid is on the 2′- or the 3′ position. In order to explain this, Chinali *et al.* [36] and Ringer and Chládek [38] have both suggested a model, shown in Fig. 4a, in which the amino group of both the 2′ and 3′ isomers occupies the same position. In order for this to occur, there must be some distortion of the normal angles of the aminoacyl groups with the ribose ring [23–26]. However, an equal or more plausible model is given in Fig. 4b in which no distortion is required and the amino group of the 2′-isomer can attack from one side of the carbonyl group and the amino group of the 3′-isomer from the other.

In contrast to the low but significant acceptor activity of the 2′-aminoacyl-tRNA's, the 2′-aminoacyl derivatives of low molecular weight substrates

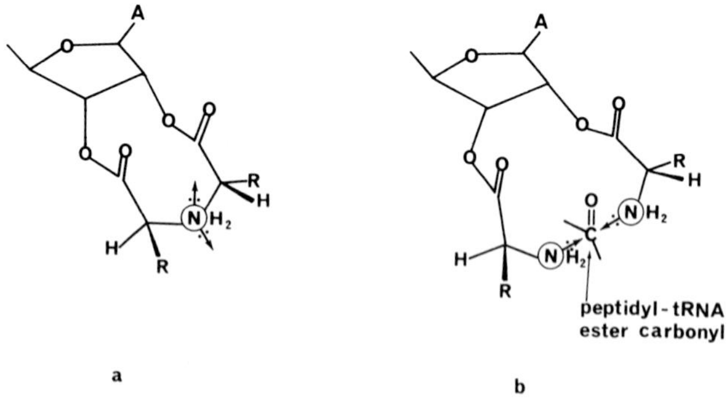

Fig. 4 Diagram of two models for the molecular conformation of 2′- and 3′-*O*-aminoacyl derivatives of adenosine at the 3′ terminus of tRNA. The aminoacyl residues are attached to both the 2′ and 3′ positions in order to show the spatial relationship of the α-amino groups and side chains. (a) Model suggested by Chinali *et al.* [36] and Ringer and Chládek [38]. (b) Model suggested in text in which there is no distortion of the normal angles of the aminoacyl groups with the ribose ring [23–26].

showed negligible acceptor activity. Thus, Chládek *et al.* [39] have looked at the acceptor activity of a number of nonisomerizable 2'- and 3'-*O*-aminoacyl dinucleoside phosphates. The 3'-*O*-aminoacyl derivatives of CpA-(2'-*O*-methyl)Phe, Cp(2'-deoxy)A-Phe, and A-(2'-*O*-methyl)Phe were active acceptors of the Ac-Phe residue in the peptidyl transferase reaction with Ac-Phe-tRNA as the donor substrate (see Table 1). On the other hand, the 2'-*O*-aminoacyl derivatives, CpA-(3'-*O*-methyl)Phe, Cp(3'-deoxy)A-Phe, A-(3'-*O*-methyl)Phe, and 3'-deoxyA-Phe, were essentially inactive. Taken together, these results with both high and low molecular weight compounds indicate that the 3' linkage is either the exclusive or strongly preferred one for peptide bond formation. Consistent with this is the observed inactivity of the 2'-isomer of puromycin in inhibiting protein synthesis [5,6].

Perhaps the role of the 2'-hydroxyl of the terminal adenosine on acceptor activity can be summarized as follows:

1. It is essential for high acceptor activity; e.g., the 2'-deoxyA-Phe has negligible acceptor activity [12,17]. Blocking of the 2'-hydroxyl by methylation decreases but does not remove the acceptor activity, e.g., with A-(2'-*O*-methyl)-Phe [39].

2. It has been suggested that the 2'-*O*-aminoacyl derivative of tRNA is the form in which elongation factor Tu (EF-Tu) presents the aminoacyl-tRNA to the ribosome [38]. Peptidyl transferase would then transfer the amino acid from the 2' to the 3' position, where it can function as an acceptor [38]. Support for this is provided by the low molecular weight 2'-aminoacyl dinucleoside phosphates of Chládek *et al.* [39,40], which bind strongly to the A' site as determined by inhibition of CpApCpCpA-Phe binding. In addition, Ringer and Chládek [38] have shown that it is only these 2'-aminoacyl dinucleoside phosphates, and not the 3'-isomers, that react with the binary complex EF-Tu·GTP in a manner similar to that of aminoacyl-tRNA.

It is of interest that specific aminoacylation of both *E. coli* and yeast tRNA can occur on either the 2'- or 3'-hydroxyl depending on the amino acid, while with some amino acids the attachment is not specific as both hydroxyls can be involved [41,42]. Whatever the biological reason for this difference among amino acids, it does not affect the above discussion in view of the rapid equilibration in solution of the aminoacyl residue between the 2'- and 3'-hydroxyls [43]. The question then is, Why is the amino acid presented to the ribosome as the 2'-*O*-aminoacyl-tRNA even though it is the 3'-*O*-aminoacyl derivative that functions as the acceptor substrate? It seems feasible that this overall process is some fail-safe mechanism. Only the 3'-terminal part of the correctly mRNA programmed 2'-aminoacyl-tRNA·EF-Tu·GTP complex is allowed entry to the A' site of peptidyl transferase by way of the

$2' \rightarrow 3'$-aminoacyl transfer. The 3'-terminal part of any uncoded aminoacyl-tRNA does not undergo this transfer and is thus prevented from acting as an acceptor substrate.

THE P' SITE OF PEPTIDYL TRANSFERASE

This site is responsible for binding the 3'-terminal end of peptidyl-tRNA during peptide bond formation (Fig. 1). Studies of the substrate specificity of this site are more limited than those of the A' site mainly because of the difficulty until recently of obtaining a wide range of low molecular weight donor substrates. Table 2 [22,44–47] lists many of the various α-*N*-acyl-aminoacyl oligonucleotides and mononucleotides that have been tested. A blocked α-amino group is essential for high activity as compared to the unblocked derivative [44].

On the basis of this type of information, a model for the P' site was proposed by Harris and Symons [2]; a slightly modified form is presented in Fig. 3, the details of which are as follows.

pCpC Site (IX)

Sites exist for binding the two 3'-terminal cytidylic acid residues of tRNA since donor activity in the fragment reaction decreases dramatically in the order CpCpA-fMet ≫ CpA-fMet ≫ pA-fMet. Nucleotide sequences 5'-distal to the terminal pCpCpA have little effect on donor activity (Table 2). A GpG sequence in either 23 S RNA or 5 S RNA may constitute this site.

Hydrophobic Site (VII)

This site is involved in the binding of the aminoacyl R groups of the hydrophobic amino acids. The affinity for the R groups of the following amino acids decreases in the order Met > Leu > Phe, while Gly is not bound to this site in view of the negligible donor activity of the Ac-Gly-oligonucleotides (Table 2).

Adenine Site (X)

The activity of donor substrate analogs of the type pN-fMet decreased in the order A > I > G, while the C and U analogs were inactive [47]. These results support the idea that the 3'-terminal adenine of peptidyl-tRNA may be hydrogen bonded to a uracil in 23 S RNA [30,31]. Hydrogen bonding of I and G to U, but not of C and U to U, is allowable in terms of the wobble hypothesis [48].

TABLE 2

P'-Site Donor Activity of *E. coli* Ribosomes

Donor substrate	Relative activity[a]	Comments	References
1. CpApApCpCpA-(Ac-Met)	100		44
CpApApCpCpA-(fMet)	100		44
ApApCpCpA-(fMet)	100		44
ApCpCpA-(fMet)	100		44
CpCpA-(fMet)	100		44
CpA-(fMet)	approx 0.2–0.6	Calculated[b]	45
pA-(fMet)	<0.1 but >0.01	Calculated[b]	45
(Np)$_x$CpCpA-(Ac-Arg)	100	Sequences unknown	44
Cp(or Up)ApCpCpA-(Ac-Leu)	approx 50		44
CpApCpCpA-(Ac-Phe)	approx 25		44
(Np)$_x$CpCpA-(Ac-Gly)	Very low	Actual activities not given	44
(Np)$_x$CpCpA-(Ac-Asp)	Inactive	Sequences unknown	44
CpCpA-(Ac-Leu)	approx 40		44,46
CpCpA-(Ac-Phe)	approx 10		46
2. pA-(fMet)	100		22,47
pA-(Ac-Met)	50		22,47
pI-(fMet)	50		22,47
pG-(fMet)	15		22,47
pC-(fMet)	0		22,47
pU-(fMet)	0		22,47
3. pA-(fMet)		Order of decreasing donor activity; activities not given	22
pA-(fLeu)			22
pA-Phe			22
pA-Gly	0		22
pA-(fGly-Leu)		Active; activity not given	22,47

[a] Approximate activity within each group of compounds expressed relative to activity of 100 for the most active donor substrate.

[b] Approximate relative activities taken from the relative concentrations of donor substrates at which activity could be detected.

Nascent Peptide Site (VI)

The site was proposed [2] on the basis of the protection by the ribosome from proteases of the nascent peptide [49]. Cantor *et al.* [50] have recently affinity labeled several proteins forming this site.

Hydrophilic Site (VIII)

In view of the high donor activity of $(Np)_xCpCpA$-(Ac-Arg) compared to the low activity of the Ac-Gly-oligonucleotide [44], this site was proposed [2]. It is specific for the binding of the basic arginyl R group and possibly also for the lysyl R group. As in the case of a similar binding site in the A′ site, the positively charged amino acid is considered to bind to a negatively charged carboxyl or phosphate group.

Additional Evidence for a Specific Binding Site for the Second pC Residue from the 3′ End of Peptidyl-tRNA

The remarkable drop in the donor activity of CpA-fMet as compared to CpCpA-fMet (Table 2) indicates a major role for this second Cp residue. For the sake of discussion, let us label the binding sites as follows:

$$pC \mid pC \mid pA \mid \text{-fMet}$$
$$3 \mid 2 \mid 1$$

All data can be explained if the binding of pC to site 3 is much greater than that of pC to site 2, which is much greater than pA binding to site 1, and that nucleotide binding sites distal to the terminal CpCpA do not appreciably affect the binding of the terminal CpCpA.

1. The large variation in the binding ability of the three sites explains the decrease in donor activity in order of CpCpA-fMet, CpA-fMet, and pA-fMet [44,45,51]. The presence of a blocked α-amino group is, of course, always required.

2. Addition of 1–5 mM 5′-CMP increased the donor activity of pA-fMet to that of CpA-fMet in the fragment reaction, but it did not affect the activity of CpA-fMet [52,53]. However, it inhibited the donor activity of CpApCpCpA-(Ac-Leu) [52]. These data can be explained by assuming that CpA-fMet successfully competes with 5′-CMP for site 2 but that 5′-CMP binding to site 3 is sufficient to hinder the binding, and hence donor activity, of the Ac-Leu-pentanucleotide. Also, occupation by 5′-CMP of site 2 and/or site 3 must somehow stimulate the donor activity of pA-fMet.

It is possible that entry of pC to site 3 is the initial event in binding of the 3′ terminus of peptidyl-tRNA to the P′ site. It is of interest that none of the tRNA molecules sequences as of 1974 [54] contain free (not hydrogen bonded) cytosine at the fourth position from the 3′ end. Perhaps misalignment of the CpCpA ends of tRNA on peptidyl transferase could occur if -CpCpCpA sequences were present.

The Presence of RNA in the P' Site

There are two lines of evidence indicating that ribosomal RNA is in or near the P' site of peptidyl transferase and may play some role in its action. Černá et al. [55] have shown that digestion of E. coli ribosomes by ribonuclease T1 resulted in the inactivation of peptidyl transferase and substrate binding to the donor site. However, substrate binding to the acceptor site was not affected. Further, a number of affinity labeling derivatives of N-blocked Phe-tRNA have been shown to label 23 S RNA as well as ribosomal proteins (Table 4). Since these tRNA derivatives contained a blocked α-amino group, and some acted as donor substrates subsequent to affinity labeling, they would be expected to bind to the P site on the ribosome.

In the absence of any supporting or contrary evidence, it is suggested that a -UpGpG- sequence of 23 S RNA is located in the P' site and is responsible for the binding of the 3'-terminal CpCpA of peptidyl-tRNA (Fig. 2). A similar -GpG- site has been implicated above in the A' site.

THE MECHANISM OF PEPTIDE BOND FORMATION

Any model of the A' and P' sites of peptidyl transferase and any consideration of the mechanism of action of puromycin, its analogs, and other inhibitors [1,2] must eventually consider the mode of peptidyl transfer on the ribosome. There are two possibilities. In the first, a single-displacement mechanism, the peptidyl group is transferred from peptidyl-tRNA to aminoacyl-tRNA on the surface of the enzyme without covalent interaction with the enzyme. The reactions can be represented as shown in Eq. (1). In

$$\text{Peptidyl-tRNA} + \text{aminoacyl-tRNA} \xrightleftharpoons{\text{ribosome}}$$
$$\text{tRNA} + \text{peptidyl aminoacyl-tRNA} \quad (1)$$

the second possibility, a double-displacement mechanism, the ribosome (peptidyl transferase) transfers the peptidyl chain to aminoacyl-tRNA via a covalent peptidyl ribosome intermediate. The two partial reactions can be represented as shown in Eqs. (2) and (3).

$$\text{Peptidyl-tRNA} + \text{ribosome} \rightleftharpoons \text{tRNA} + \text{peptidyl ribosome} \quad (2)$$
$$\text{Peptidyl ribosome} + \text{aminoacyl-tRNA} \rightleftharpoons$$
$$\text{ribosome} + \text{peptidyl aminoacyl-tRNA} \quad (3)$$

On the basis of the available evidence in a survey of the known transferase reactions, Spector [56] emphasized the substantial and growing body of evidence supporting double-displacement as the usual mechanism. Although very little is known about the actual mechanism of peptidyl transfer by peptidyl transferase, it is considered most probable that the double-displacement mechanism will also hold for this enzyme.

One approach that can be used to detect the presence of a peptidyl ribosome intermediate is to examine an exchange reaction between various P' site compounds and the peptidyl transferase (E) [J. F. B. Mercer, private communication, 1971]. In the unpublished experiments of D. J. Eckermann, the following was attempted [Eqs. (4) and (5)]:

$$\text{UpApCpCpA-(Ac-[}^3\text{H]Leu)} + E \rightleftharpoons \text{UpApCpCpA} + \text{Ac-[}^3\text{H]Leu-E} \quad (4)$$

$$\text{Ac-[}^3\text{H]Leu-E} + \text{CpCpA (or tRNA)} \rightleftharpoons$$
$$\text{CpCpA-(Ac-[}^3\text{H]Leu) (or Ac-[}^3\text{H]Leu-tRNA)} + E \quad (5)$$

The assay involved following the transfer of Ac-[³H]Leu from UpApCpCpA-(Ac-[³H]Leu) to either CpCpA-(Ac-[³H]Leu) or Ac-[³H]Leu-tRNA. However, no exchange product was detected under assay conditions in which pH, Mg^{2+}, K^+, tRNA, and ethanol were varied over wide concentration ranges. Likewise, the addition of poly(U) or *N*-(3-phenylpropionyl)Pan (with the expectation that this compound would bind to the A' site but not act as an acceptor) had no effect.

The attempted isolation of an Ac-[³H]Leu-ribosome intermediate after incubation of ribosomes with UpApCpCpA-(Ac-[³H]Leu) under a variety of conditions was also unsuccessful. Thus, the mechanism of catalysis by peptidyl transferase remains an open question.

POSSIBLE ALLOSTERIC EFFECTS BY VARIOUS COMPOUNDS IN THE A' AND P' SITES

Several observations on the stimulation of binding or substrate activity of A' and P' substrates can be interpreted in terms of an open and closed configuration of peptidyl transferase; thus, access to either site by low molecular weight substrates (e.g., the 3'-terminal aminoacyl and peptidyl fragments of tRNA) is allowed in the open but not in the closed configuration (Table 3) [21,30,31,33,45,52,53,57,58]. The data of Table 3 suggest that at least some of these effects are due to allosterically induced changes in the conformation of peptidyl transferase although an alternative explanation may be that some of the effectors increase the number of active ribosomes [59]. However, any allosteric effect observed *in vitro* may indicate a similar effect *in vivo*. For example:

i. Interaction of the ribosome with some tRNA sequence of mRNA-coded aminoacyl- or peptidyl-tRNA is essential for the binding of a productive acceptor substrate (3'-terminal CpCpA-amino acid of aminoacyl-tRNA) to peptidyl transferase. This interaction may be effected by tRNA or ethanol under certain *in vitro* conditions.

2. An especially appealing idea is that the binding of the acceptor substrate to peptidyl transferase stimulates the binding of the donor substrate; i.e., an ordered reaction sequence operates. The A'-specific antibiotics sparsomycin, gougerotin, and amicetin [58], effectively illustrate this point by their blockage of the A' site and stimulation of substrate binding to the P' site. However, the alternative should be considered that part of the acceptor molecule may supply part of the donor binding site, e.g., by stacking of acceptor and donor 3'-terminal adenosines of aminoacyl- and peptidyl-tRNA [27,29; R. J. Harris, unpublished].

An interesting observation relative to the present concept of the open ⇌ closed configuration of the A' site is that the removal of monovalent cations specifically inactivates the A' site, while the P' site is unaffected [60]. Both heat and monovalent cations are required for reactivation [61]. Similar to the ideas described above, these authors suggest that this active ⇌ inactive (open ⇌ closed?) conversion may be an inherent part of the mechanism of action of peptidyl transferase during various stages of protein synthesis. Possibly this conversion process may mediate the various allosteric phenomena of the A' site listed in Table 3.

LOW MOLECULAR WEIGHT ANALOGS OF PUROMYCIN AS AFFINITY LABELS

A number of groups have sought to exploit the technique of affinity labeling to identify the component(s) of the 50 S ribosomal subunit of *E. coli* that makes up the active center of peptidyl transferase. Most studies have used alkylating, acylating, or photoalkylating groups attached to the α-amino group on aminoacyl-tRNA [50,62–65]. Such analogs would be expected to bind preferentially to the P' site of the enzyme. Other studies have been directed toward the A' site by use of the α-N-iodoacetyl derivative of puromycin [66] and reactive analogs of the A'-site-specific inhibitor chloramphenicol [67,68].

Our approach has been to prepare low molecular weight affinity labeling analogs of puromycin in which the α-amino group is not blocked. The aim here was to ensure that the analog acted as an A'-site-specific compound as shown by its ability to act as an acceptor in the fragment reaction. In addition, it was hoped that covalently bound affinity label would still be able to act as an acceptor substrate, as this would provide very strong evidence that the bound affinity label was correctly positioned in the active center of peptidyl transferase.

TABLE 3

Possible Allosteric Effects on Substrate Binding to the A' and P' Sites and on Enzymatic Activity of *E. coli* Peptidyl Transferase

Additions to assays[a]	A' site		Comments	P' site		Comments
	Open	Closed		Open	Closed	
Nil	—	Yes	Little binding of CpApCpCpA-Phe occurs to 70 S ribosomes [21]	—	Yes	—
Ethanol	Yes		Essential for binding CpApCpCpA-Phe to 50 S ribosomes; stimulates binding to 70 S ribosomes (30×) [21,57]	Yes		Essential for binding of CpApCpCpA-(Ac-Leu) [58]
tRNA	Yes		Greatly stimulates binding of CpApCpCpA-Phe to 50 S ribosomes [57]	Not determined		—
CpCpA	Yes		Permits ethanol access to A' site to act as acceptor substrate [33]	Not determined		—
pC	Not determined		—	Yes		Stimulates donor substrate activity to pA-fMet (3.5 to >10×) [45,52]
Terminating codon plus release factor	Yes		Natural termination process whereby water (normally excluded) enters A' site and acts as acceptor substrate (hydrolysis) [53]	Not determined		—
Sparsomycin	Blocks A' site		—	Yes		Stimulates binding of CpApCpCpA-(Ac-Leu) [58]
Gougerotin	Blocks A' site		—	Yes		Stimulates binding of CpApCpCpA-(Ac-Leu) [58]
Amicetin	Blocks A' site		—	Yes		Stimulates binding of CpApCpCpA-(Ac-Leu) [58]
Erythromycin	A' site unaffected		—	Yes		Stimulates binding of CpApCpCpA-(Ac-Leu) [58]

[a] Binding or enzyme assays contained ribosomes, salts, donor and/or acceptor substrates, and additions as given above. Compounds added were not equally effective; a positive activity indicates a trend or major effect on the ribosome rather than complete activity.

These ideas were realized with 5'-O-(N-bromoacetyl-p-aminophenyl-phosphoryl)-3'-N-L-phenylalanylpuromycin aminonucleoside (Bap-Pan-Phe) [30,31] (Fig. 5). The basis for the selection of this previously unsynthesized compound was as follows. The phenylalanyl analog of puromycin (Pan-Phe) is very accessible chemically and has been shown to be nearly as active as an inhibitor of protein synthesis and as an acceptor substrate as puromycin [7,8,9,14]. Our model of the peptidyl transferase active center predicts a binding site for a 5'-phosphate of the terminal -CpA of transfer RNA [2]. The attachment of the bromoacetylaminophenyl group was based as much on the success of this group to affinity label nucleic acid binding proteins [69,70] as on the basis of our model of the A' site. However, a binding site has been proposed for the cytosine next to the 3'-terminal adenine (Fig. 2) so it was possible that the benzene ring of the alkylating group would bind in this region.

It has been possible to prepare [32P]Bap-Pan-Phe in addition to the non-

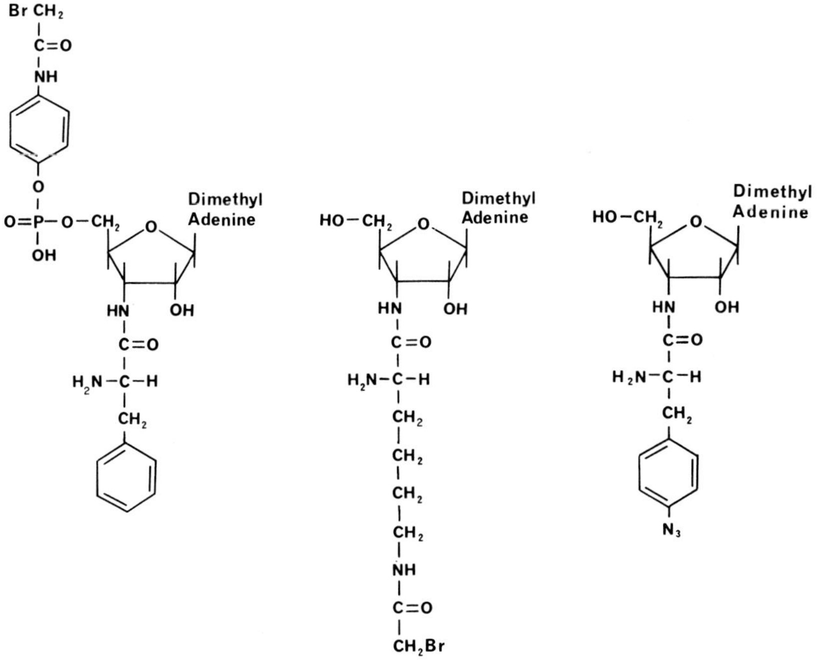

Fig. 5 Structure of three affinity labeling derivatives of puromycin. I, Bap-Pan-Phe, 5'-O-(N-bromoacetyl-p-aminophenylphosphoryl)-3'-N-L-phenylalanylpuromycin amino-nucleoside. II, 3'-N-(ε-N-bromoacetyllysyl)puromycin aminonucleoside. III, 3'-N-(p-azidophenylalanyl)puromycin aminonucleoside.

radioactive compound, and this has allowed an extensive characterization of its reaction with *E. coli* ribosomes [30,31]. The affinity label acted as an acceptor substrate at 0°C in the fragment reaction assay of peptidyl transferase, but is was only about 10% as efficient as puromycin even though both compounds had the same K_m of about 0.1 mM (Table 4). Preincubation of 70 S ribosomes with affinity label at 27°C led to irreversible inactivation of peptidyl transferase activity (pH optimum of 8.8) and extensive covalent attachment of the affinity label to the ribosomes. The inactivation was roughly proportional to the time of incubation or to the concentration of affinity label. At 100% inactivation, up to 25 molecules of affinity label were bound per ribosome. All the inactivation of the enzyme, but only a small proportion of the total covalent labeling (up to 2 molecules per ribosome), could be prevented by the presence of chloramphenicol. The chloramphenicol-insensitive affinity labeling, designated nonspecific, was shown to be distributed over many 50 S and 30 S ribosomal proteins.

On the other hand, the specific, chloramphenicol-sensitive affinity labeling was shown to be attached only to the 23 S RNA. This was most fortunate since it indicated that some part of the 23 S RNA was in or very close to the active center of peptidyl transferase. It has also allowed the ready separation of the specifically labeled 23 S RNA from the complex mixture of nonspecifically labeled ribosomal proteins. Further, a comparison of the stoichiometry

TABLE 4

Comparison of Some Properties of Three Affinity Labeling Analogs of Puromycin with *E. coli* Ribosomes

Property	Puromycin	Bap-Pan-Phe	Bromo-acetyl-Lys-Pan	*p*-Azido-Phe-Pan
K_m, fragment reaction [30,31]	0.1 mM	0.1 mM	0.4 mM	0.15 mM
K_i, polysomes [14]	1 μM	ND[a]	0.3 mM	1.8 μM
Moles affinity label bound/ribosome	—	25[b]	56[c]	17[d]
% Inactivation of peptidyl transferase	—	100[b]	42[c]	0[d]
Protection of inactivation				
By puromycin	—	Yes	Yes	—
By chloramphenicol	—	Yes	No	—
Acceptor substrate activity of bound affinity label	—	Yes	No	—

[a] Not determined.

[b] Ribosomes incubated with 0.5 mM affinity label for 17 hr at 26°C and pH 8.8 prior to washing and assay for peptidyl transferase activity [31].

[c] Ribosomes incubated with 4.2 mM affinity label at 37°C for 16 hr.

[d] Ribosomes were irradiated with ultraviolet light for 5 min in the presence of 1 mM affinity label.

of the RNA labeling with the percent inactivation of peptidyl transferase revealed a linear correlation between the two quantities such that 2.1 ± 0.2 moles of affinity label per 23 S RNA corresponded to complete inactivation. The significance of this figure is discussed below.

It is important to note that this calculation of stoichiometry was based on the assumption that 100% of the ribosomes were affinity labeled. This is a very difficult assumption to check. For example, any measure of the proportion of ribosomes that are active in peptidyl transfer is not relevant since, as mentioned above, ribonuclease T1 digestion of ribosomes inactivated peptidyl transfer and substrate binding to the P′ site but not to the A′ site [55]. However, more serious is the observation by Harris and Pestka [71] that only 11% of their washed 70 S ribosomes were active in binding CpApCpCpA-Phe; such measurements were not done during the affinity label work with Bap-Pan-Phe. Whether such a measurement of a reversible binding reaction would provide an estimate of the proportion of ribosomes that can be covalently and irreversibly alkylated in the A′ site by the affinity label under the prolonged incubation conditions used (see Table 4) is not known. Therefore, the value of 2.1 moles of affinity label per 23 S RNA achieved at 100% inactivation of peptidyl transferase must be taken as a minimum estimate.

Incubation of the covalent ribosome–affinity label complex with the donor substrate, CpApCpCpA-(Ac-[³H]Leu), caused Ac-[³H]Leu to become covalently attached to 23 S RNA; i.e., the covalently bound affinity label could still act as an acceptor substrate. This transfer of Ac-[³H]Leu was linearly correlated both with the extent of ³²P affinity labeling of the 23 S RNA and with the percent inactivation of peptidyl transferase activity. However, a most puzzling feature has been that only about 0.5% of the ribosome–affinity label complexes were able to effect Ac-[³H]Leu transfer and that this extent of transfer was independent of the extent of inactivation.

In order to explain these observations on the action of Bap-Pan-Phe, the following hypothesis was proposed [31]. The affinity label binds to and subsequently reacts with first the A′ site and then the P′ site in an ordered and strongly positive cooperative manner. This explains the linear 2:1 correlation between specific affinity labeling and loss of peptidyl transferase activity and also the small proportion of ribosome–affinity label complexes (0.5%) able to effect peptidyl transfer. The latter is considered to be the population of ribosomes labeled only in the A′ site, because P′ site labeling lags slightly behind that of the A′ site.

However, more recent work has demonstrated that the P′ site is not Up blocked by affinity labeling since the P′ site substrate CpApCpCpA-(Ac-[³H]Leu) could still bind to affinity-labeled ribosomes (D. J. Eckermann and

R. H. Symons, unpublished). In addition, only one affinity-labeled sequence, GpUpC*pCpG (where C* is affinity labeled), has been isolated after ribonuclease T1 digestion of affinity labeled 23 S RNA. It is possible that two affinity label molecules are attached to one cytosine residue or that there are two such sequences in 23 S RNA. We have no hypothesis that explains all the data.

One puzzling observation (R. J. Harris and R. H. Symons, unpublished) made on three separate occasions over a period of two years is that the reaction with ribosomes of the presumptive 2-hydroxyacetyl derivative of Bap-Pan-Phe (prepared by paper chromatography of Bap-Pan-Phe in an aqueous ammonia–isopropanol solvent and shown to contain negligible amounts of reactive bromine) caused up to one-quarter of the inactivation of peptidyl transferase as that found with Bap-Pan-Phe under the same conditions. Further, these partly inactivated ribosomes were better as acceptors for the transfer of Ac-[^{3}H]Leu from CpApCpCpA-(Ac-[^{3}H]Leu) than the Bap-Pan-Phe-inactivated ribosomes; up to 25% of the ribosome–affinity label complexes were active compared to about 0.5%, respectively. However, the mechanism of action of this 2-hydroxyacetyl derivative of Bap-Pan-Phe remains unknown.

In spite of our lack of knowledge of the nature and mechanism of the covalent coupling of the affinity label to 23 S RNA, all the evidence clearly indicates that one or more parts of the 23 S RNA are involved in the active center of peptidyl transferase. Although Bap-Pan-Phe is the only A' site-specific reagent reported so far to label 23 S RNA, a number of α-N-acylated derivatives of Phe-tRNA have been reported as doing so (Table 5) [65,72–76]. One must assume that these have all covalently attached in the region of the P' site. In the case of Bap-Pan-Phe, the reaction with 23 S RNA was dependent on the integrity of the ribosome, as there was little reaction of the affinity label with purified ribosomal RNA [31] or with poly(A), poly(C), poly(G), or poly(U) [D. J. Eckermann, unpublished].

In terms of our model for the active center of *E. coli* peptidyl transferase [2] (see above), we consider that Bap-Pan-Phe alkylates the ribosome in the vicinity of the binding site(s) for the penultimate cytidine residue of tRNA molecules and that 23 S RNA therefore has a direct function in binding the 3' terminus of aminoacyl-tRNA. We have postulated [31] above that this would occur by base pairing of the 3'-terminal -CpCp- sequence of aminoacyl-tRNA to a -GpGp- sequence in 23 S RNA in the A' site and of the 3'-terminal-CpCpA of peptidyl-tRNA to a -UpGpGp- sequence of 23 S RNA in the P' site. However, the recent observation that the 23 S RNA sequence labeled by Bap-Pan-Phe is GpUpC*pCpG (see above) does not support this model. Direct functional, rather than merely structural, roles have been indicated recently for 5 S RNA in tRNA binding [77] and for 16 S rRNA in the binding of mRNA [78,79].

TABLE 5

**Affinity Labeling of 23 S RNA in *E. coli* Ribosomes
by Aminoacyl-tRNA Derivatives [a]**

N-Acyl derivative of Phe-tRNA	Ribosome subunit labeled	Distribution of affinity label (% of total)		RNA labeled	Reactivity of bound affinity label with puromycin or Phe-tRNA	Refer-ence
		Protein	RNA			
1. N-Chlorambucyl	50 S	ND	ND	23 S	Yes	65
2. N-Iodoacetyl	50 S, 30 S?	20	80	23 S	Yes	72
3. N-Bromoacetyl	50 S	< 50	> 50	23 S	ND	73
4. N-Ethyl-2-diazomalonyl	50 S	6	94	23 S	ND	74
5. N-2-Nitro-4-azidobenzoyl	Both	35	65	23 S and/or 5 S	ND	75
6. p-Benzophenone propionyl	ND	ND	ND	23 S	Yes	76

[a] Not determined.

We are also trying to map the position(s) of the bound affinity label(s) on the 23 S RNA. The most promising approach appears to be to attach covalently coupled ferritin and avidin [80] to the biotin–affinity label–23 S RNA complex for visualization under the electron microscope. It is hoped that the position(s) of the affinity label can be oriented relative to the 3' end of the 23 S RNA, which would be visualized by oxidation with periodate followed by reaction with biotin hydrazide [80] and then the coupled ferritin–avidin. One problem here is the nicking of the 23 S RNA that occurs during the prolonged incubation times required for the affinity labeling of the ribosomes [30,31].

OTHER PUROMYCIN ANALOGS AS AFFINITY LABELS

The affinity label Bap-Pan-Phe is considered to bind covalently in the region of the -CpCpA binding site for tRNA on the ribosome (see above). In order to probe another region of the puromycin binding site, two aminoacyl analogs were prepared [E. F. Vanin, unpublished] in which the p-methoxy-phenylalanyl group of puromycin was replaced by ε-N-bromoacetyllysine and by p-azidophenylalanine (Fig. 5). The latter compound is a photoaffinity label in that the azido group is chemically unreactive under normal conditions but is converted by irradiation with ultraviolet light into a highly reactive nitrene, which can covalently couple to carbon or nitrogen atoms [81].

Both analogs were active as acceptor substrates in the fragment reaction with k_m values a little higher than that of puromycin (Table 4). When tested as inhibitors of the puromycin-dependent release of peptides from peptidyl-tRNA on polysomes [14], p-azido-Phe-Pan had a K_i similar to the K_m for puromycin, while that of bromoacetyl-Lys-Pan gave the same high value as that obtained for the fragment reaction. On this basis, p-azido-Phe-Pan was potentially a better affinity label than bromoacetyl-Lys-Pan. However, the reaction of these radioactive affinity labels with ribosomes gave extensive covalent attachment, with only 42% inactivation of peptidyl transferase in the case of the bromoacetyl-Lys analog and no inactivation with the p-azido-Phe analog (Table 4). Inactivation by bromacetyl-Lys-Phe was protected against by puromycin but not by chloramphenicol, while no acceptor activity was found for the bound affinity label. In view of the extensive nonspecific labeling by the bromoacetyl-Lys analog and the great difficulty in selecting out any specific labeling of peptidyl transferase [E. F. Vanin, unpublished], no further work has been attempted.

The relative success of Bap-Pan-Phe as a specific affinity label and the failure of the other two emphasizes the difficulties of using low molecular weight analogs of puromycin for affinity labeling studies on a structure as complex as the ribosome. It was most fortunate that specific labeling with Bap-Pan-Phe occurred on the 23 S RNA and could therefore be readily separated from the nonspecifically labeled ribosomal proteins.

Instead of using a chemically reactive or a photoactivatable group for covalent coupling, Cooperman *et al.* [82] found that irradiation of ribosomes and [³H]puromycin with ultraviolet light led to the coupling of puromycin to *E. coli* ribosomes. However, 36% of the label was associated with the 30 S proteins and RNA, and the remainder with the 50 S proteins and RNA. Although the coupling is obviously very nonspecific, L23 was the major protein labeled. It is possible that an analysis of the acceptor activity of the bound puromycin would determine if any of the coupled puromycin was correctly located in the A′ site, either on RNA or on protein.

Two N-acylated derivatives of puromycin have been used in affinity labeling studies, but again extensive nonspecific labeling occurred with *E. coli* ribosomes. Thus, Pongs *et al.* [66] found that N-iodacetylpuromycin reacted with many 50 S and 30 S ribosomal proteins, but there was no reaction with RNA. The 50% reduction in labeling that occurred in the presence of a 100 molar excess of puromycin provides little evidence that any of the covalent coupling was in either the A′ or the P′ site. In addition, Cooperman *et al.* [82] concluded from their study of the photoactivatable N-(ethyl-2-diazomalonyl)-puromycin that it labeled both 30 S and 50 S proteins and RNA in a nonspecific manner; in fact, it seemed that covalent coupling proceeded through the puromycin residue rather than the photoactivatable side chain.

DERIVATIVES OF AMINOACYL-tRNA AS AFFINITY LABELS FOR 23 S RNA IN *E. coli* RIBOSOMES

A number of derivatives of aminoacyl-tRNA that are known to couple to 23 S RNA on incubation with *E. coli* ribosomes are listed in Table 5. The first three contain an alkylating group and the last three a photoactivatable group. In the three cases tested, the bound affinity label reacted with an acceptor substrate, either puromycin or Phe-tRNA, which indicates that some at least of the bound affinity label was correctly positioned in the P' site. This is an important test for specificity of coupling; it also allows the differentiation between specifically and nonspecifically bound affinity label. In the case of *N*-bromoacetyl-Phe-tRNA, Pellegrini *et al.* [83] showed that affinity label bound to protein was still capable of accepting an amino acid from the Phe-tRNA or puromycin, although this has still to be shown for the 23 S RNA-labeled material of Table 5.

In terms of our model described above (Fig. 3) for the active center of the A' and P' sites of peptidyl transferase, that part of the 23 S RNA involved in this affinity label would be located in the peptide binding site VI and would presumably be separate from the -UpGpG- sequence involved in base pairing to the 3'-terminal -CpCpA of peptidyl-tRNA (Fig. 3).

On the basis of the results presented, it is clear that more consistent success with affinity labeling studies will be obtained with the high molecular weight aminoacyl-tRNA derivatives than with the lower molecular weight analogs of puromycin and possibly other antibiotics. The much lower binding constants of these high molecular weight compounds appear to be necessary for reasonable specificity of the covalent coupling.

MAMMALIAN PEPTIDYL TRANSFERASE

Available information on the active center of peptidyl transferase of ribosomes from mammalian sources is considerably less than that from *E. coli*. However, the limited data do indicate that the mammalian enzyme does have many similarities to the one from *E. coli*. For detailed information, the reader is referred to general reviews [4,84] and to studies on substrate specificity [9] and affinity labeling [85,86].

It is of interest that Bap-Pan-Phe is a poor analog of puromycin and a poor affinity label with rat liver ribosomes [D. J. Eckermann, unpublished]. Thus, its K_m in the fragment reaction was about 4 mM, and its K_i for the inhibition of the puromycin release of nascent peptides from polysomes was 0.14 mM (cf. Table 2). On incubation with rat liver ribosomes, 40–50%

inhibition of peptidyl transferase activity as determined by the fragment reaction corresponded with 40–50 moles of affinity label bound per ribosome, and there was only partial protection of this inactivation by promycin. Less than 7% of the bound affinity label was coupled to RNA.

REFERENCES

1. R. J. Harris and R. H. Symons, *Bioorg, Chem.* **2**, 266 (1973).
2. R. J. Harris and R. H. Symons, *Bioorg. Chem.* **2**, 286 (1973).
3. H.-G. Wittmann, *Eur. J. Biochem.* **61**, 1 (1976).
4. R. J. Harris and S. Pestka, *in* "Molecular Mechanisms of Protein Biosynthesis" (H. Weissbach and S. Pestka, eds.), p. 413. Academic Press, New York, 1977.
5. D. Nathans and A. Neidle, *Nature (London)* **197**, 1076 (1963).
6. D. Nathans, *Fed. Proc., Fed. Am. Soc. Exp. Biol.* **23**, 984 (1964).
7. R. H. Symons, R. J. Harris, L. P. Clarke, J. F. Wheldrake, and W. H. Elliott, *Biochim. Biophys. Acta* **179**, 248 (1969).
8. R. J. Harris, J. E. Hanlon, and R. H. Symons, *Biochim. Biophys. Acta* **240**, 244 (1971).
9. D. J. Eckermann, P. Greenwell, and R. H. Symons, *Eur. J. Biochem.* **41**, 547 (1974).
10. R. E. Monro and K. A. Marcker, *J. Mol. Biol.* **25**, 347 (1967).
11. I. Rychlík, S. Chládek, and J. Žemlička, *Biochim. Biophys. Acta* **138**, 640 (1967).
12. I. Rychlík, J. Černá, S. Chládek, J. Žemlička, and A. Haladová, *J. Mol. Biol.* **43**, 13 (1969).
13. D. Ringer and S. Chládek, *FEBS Lett.* **39**, 75 (1974).
14. E. F. Vanin, P. Greenwell, and R. H. Symons, *FEBS Lett.* **40**, 124 (1974).
15. S. Pestka, *J. Biol. Chem.* **247** 4669 (1972).
16. S. Pestka, *Proc. Natl. Acad. Sci. U.S.A.* **69**, 624 (1972).
17. J. Černá, I. Rychlík, J. Žemlička, and S. Chládek, *Biochim. Biophys. Acta* **204**, 203 (1970).
18. I. Rychlík, J. Černá, S. Chládek, P. Pulkrábek, and J. Žemlička, *Eur. J. Biochem.* **16**, 136 (1970).
19. B. P. Gottikh, L. V. Nikolayeva, A. A. Krayevski, and L. L. Kisselve, *FEBS Lett.* **7**, 112 (1970).
20. D. Ringer and S. Chládek, *Biochem. Biophys. Res. Commun.* **56**, 760 (1974).
21. S. Pestka, T. Hishizawa, and J. L. Lessard, *J. Biol. Chem.* **245**, 6208 (1970).
22. A. A. Krayevsky, M. K. Kukhanova, and B. P. Gottikh, *Nucleic Acids Res.* **2**, 2223 (1975).
23. M. Sundaralingam and S. K. Arora, *J. Mol. Biol.* **71**, 49 (1972).
24. A. McL. Mathieson, *Tetrahedron Lett.* **46**, 4137 (1965).
25. W. Saenger and D. Suck, *Acta Crystallogr., Sect B* **27**, 2105 (1971).
26. N. Yathindra and M. Sundaralingam, *Biochim. Biophys. Acta* **308**, 17 (1973).
27. I. D. Raacke, *Biochem. Biophys. Res. Commun.* **43**, 168 (1971).
28. M. Ariatti and A. O. Hawtrey, *Biochem. J.* **145**, 169 (1975).
29. J. Žemlička, S. Chládek, D. Ringer, and K. Quiggle, *Biochemistry* **14**, 5239 (1975).
30. R. J. Harris, P. Greenwell, and R. H. Symons, *Biochem. Biophys. Res. Commun.* **55**, 117 (1973).
31. P. Greenwell, R. J. Harris, and R. H. Symons, *Eur. J. Biochem.* **49**, 539 (1974).

32. M. Takanami, *Proc. Natl. Acad. Sci. U.S.A.* **52**, 1271 (1964).
33. E. Scolnik, G. Milman, M. Rosman, and T. Caskey, *Nature (London)* **225**, 5228 (1970).
34. R. Fernandez-Muñoz and D. Vasquez, *Mol. Biol. Rep.* **1**, 75 (1973).
35. J. L. Lessard and S. Pestka, *J. Biol. Chem.* **247**, 6901 (1972).
36. G. Chinali, M. Sprinzl, A. Parmeggiani, and F. Cramer, *Biochemistry* **13**, 3001 (1974).
37. T. H. Fraser and A. Rich, *Proc. Natl. Acad. Sci. U.S.A.* **72**, 3004 (1975).
38. D. Ringer and S. Chládek, *Proc. Natl. Acad. Sci. U.S.A.* **72**, 2950 (1975).
39. S. Chládek, D. Ringer, and K. Quiggle, *Biochemistry* **13**, 2727 (1974).
40. D. Ringer, K. Quiggle, and S. Chládek, *Biochemistry* **14**, 514 (1975).
41. M. Sprinzl and F. Cramer, *Proc. Natl. Acad. Sci. U.S.A.* **72**, 3049 (1975).
42. S. M. Hecht and A. C. Chinault, *Proc. Natl. Acad. Sci. U.S.A.* **73**, 405 (1976).
43. R. B. Loftfield, *Prog. Nucleic Acid Res. Mol. Biol.* **12**, 87 (1972).
44. R. E. Monro, J. Černá, and K. A. Marcker, *Proc. Natl. Acad. Sci. U.S.A.* **61**, 1042 (1968).
45. A. A. Krayevsky, L. S. Victorova, V. V. Kotusov, M. K. Kukhanova, A. D. Treboganov, N. B. Tarussova, and B. P. Gottikh, *FEBS Lett.* **62**, 101 (1976).
46. J. F. B. Mercer and R. H. Symons, *Eur. J. Biochem.* **28**, 38 (1972).
47. J. Černá, I. Rychlík, A. A. Krayevsky, and B. P. Gottikh, *Acta Biol. Med. Ger.* **33**, 877 (1974).
48. F. H. C. Crick, *J. Mol. Biol.* **19**, 548 (1966).
49. L. I. Malkin and A. Rich, *J. Mol. Biol.* **26**, 329 (1967).
50. C. R. Cantor, M. Pellegrini, and H. Oen, *in* "Ribosomes" (M. Nomura, A. Tissières, and P. Lengyel, eds.), p. 573. Cold Spring Harbor Lab. Cold Spring Harbor, New York, 1974.
51. J. Černá, I. Rychlík, A. A. Krayevsky, and B. P. Gottikh, *FEBS Lett.* **37**, 188 (1973).
52. J. Černá *FEBS Lett.* **58**, 94 (1975).
53. C. T. Caskey, *Adv. Protein Chem.* **27**, 243 (1973).
54. B. G. Barrell and B. F. C. Clark, *in* "Handbook of Nucleic Acid Sequences." Joynson-Bruvvers Ltd., London, 1974.
55. J. Černá, I. Rychlík, and J. Jonák, *Eur. J. Biochem.* **34**, 551 (1973).
56. L. B. Spector, *Bioorg. Chem.* **2**, 311 (1973).
57. T. Hishizawa and S. Pestka, *Arch. Biochem. Biophys.* **147**, 624 (1971).
58. M. L. Celma, R. E. Monro, and D. Vasquez, *FEBS Lett.* **6**, 273 (1970).
59. S. Pestka, R. Vince, S. Daluge, and R. Harris, *Antimicrob. Agents Chemother.* **4**, 37 (1973).
60. J. Černá, *FEBS Lett.* **15**, 101 (1971).
61. R. Miskin, A. Zamir, and D. Elson, *J. Mol. Biol.* **54**, 355 (1970).
62. O. Pongs, R. Bald, V. A. Erdmann, and E. Reinwald, *in* "Topics in Infectious Diseases" (J. Drews and F. E. Hahn, eds.), Vol. 1, p. 179. Springer-Verlag, Berlin and New York, 1975.
63. E. Kuechler, R. Hauptmann, A. P. Czernilofsky, I. Fiser, A. Barta, H. O. Voorma, G. Stoffler, and K. H. Scheit, *Acta Biol. Med. Ger.* **33**, 633 (1974)
64. A. S. Girshovich, E. S. Bochkareva, and V. A. Pozdnyakov, *Acta Biol. Med. Ger.* **33**, 639 (1974).
65. D. G. Knorre, *Acta Biol. Med. Ger.* **33**, 649 (1974).
66. O. Pongs, R. Bald, T. Wagner, and V. A. Erdmann, *FEBS Lett.* **35**, 137 (1973).
67. N. Sonenberg, M. Wilchek, and A. Zamir, *Proc. Natl. Acad. Sci. U.S.A.* **70**, 1243 (1973).

68. R. Bald, V. A. Erdmann, and O. Pongs, *FEBS Lett.* **28**, 149 (1972).
69. M. B. Sporn, D. M. Berkowitz, R. P. Glinski, A. B. Ash, and C. L. Stevens, *Science* **164**, 1408 (1969).
70. P. Cuatrecasas, M. Wilchek, and C. B. Anfinsen, *J. Biol. Chem.* **244**, 4316 (1969).
71. R. Harris and S. Pestka, *J. Biol. Chem.* **248**, 1168 (1973).
72. M. Yukioka, T. Hatayame, and S. Morisawa, *Biochim. Biophys. Acta* **390**, 192 (1975).
73. J. B. Breitmeyer and H. F. Noller, *J. Mol. Biol.* **101**, 297 (1976).
74. L. Bispink and H. Matthaei, *FEBS Lett.* **37**, 291 (1973).
75. A. S. Girshovich, E. S. Bochkareva, V. M. Kramarov, and Y. A. Ovchinnikov, *FEBS Lett.* **45**, 213 (1974).
76. A. Barta, E. Kuechler, C. Branlant, J. S. Widada, A. Krol, and J. P. Ebel, *FEBS Lett.* **56**, 170 (1975).
77. V. A. Erdmann, M. Sprinzl, and O. Pongs, *Biochem. Biophys. Res. Commun.* **54**, 942 (1973).
78. J. Shine and L. Dalgarno, *Proc. Natl. Acad. Sci. U.S.A.* **71**, 1342 (1974).
79. J. A. Steitz and K. Jakes, *Proc. Natl. Acad. Sci. U.S.A.* **72**, 4734 (1975).
80. H. Heitzmann and F. M. Richards, *Proc. Natl. Acad. Sci. U.S.A.* **71**, 3537 (1974).
81. J. R. Knowles, *Acc. Chem. Res.* **5**, 155 (1972).
82. B. S. Cooperman, E. N. Jaynes, D. J. Brunswick, and M. A. Luddy, *Proc. Natl. Acad. Sci. U.S.A.* **72**, 2974 (1975).
83. M. Pellegrini, H. Oen, D. Eilat, and C. R. Cantor, *J. Mol. Biol.* **88**, 809 (1974).
84. I. G. Wool and G. Stöffler, *in* "Ribosomes" (M. Nomura, A. Tissières, and P. Lengyel, eds.), p. 417. Cold Spring Harbor Lab. Cold Spring Harbor, New York, 1974.
85. J. Stahl, K. Dressler, and H. Bielka, *FEBS Lett.* **47**, 167 (1974).
86. M. A. Minks, M. Ariatti, and A. O. Hawtrey, *Hoppe-Seyler's Z. Physiol Chem.* **356**, 109 (1975).

CHAPTER

16

Hemoprotein Oxygen Transport:
Models and Mechanisms

T. G. Traylor

INTRODUCTION

Hemoproteins, ubiquitous among plants and animals, perform at least four important functions related to oxygen and the production of energy: (1) oxygen transport to tissue [1,2]; (2) catalytic oxidation of organic compounds [3,4]; (3) decomposition of hydrogen peroxide [5]; and (4) electron transfer [2,3]. They have in common a heme (Hm) molecule [6], e.g., (**1a**), (**1b**), and (**1c**).

These heme molecules, in which Fe(II) has four nitrogen ligands in an electron-rich ring system, have the property of binding electron-rich bases such as imidazole (Im) rather poorly (K_1, $K_2 \leq 10^4 \, M^{-1}$) [6–9] and are easily oxidized to Fe^{3+} [Eqs. (1) and (2)] [1, p. 1;5]. Hemoproteins that function as oxygen carriers, e.g., myoglobins and hemoglobins, almost

$$\text{Hm} + \text{Im} \xrightleftharpoons{K_1} \text{Im—Hm} \tag{1}$$

$$\text{Im—Hm} + \text{Im} \xrightleftharpoons{K_2} \text{Im—Hm—Im} \tag{2}$$

invariably contain protoheme (**1a**), an imidazole ring attached to the iron on the *proximal* side, and a rather open but hydrophobic cavity on the sixth, *distal* side [1, p. 1;10–12]. The stereoscopic view of a portion of myoglobin shown in Fig. 1 is rather typical of oxygen-transporting hemoproteins. In

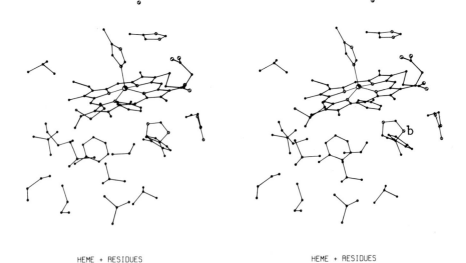

	R_1	R_2	Name
(1a)	—CH=CH$_2$	—CH=CH$_2$	Protoheme
(1b)	—CH=CH$_2$	—C$_{17}$H$_{29}$O	Heme a[a]
(1c)	—CH(CH$_3$)SR′	—CH(CH$_3$)SR′	Heme c[b]
(1d)	—Et	—Et	Mesoheme
(1e)	—Et	—Et	Pyrroheme[c]

[a] Formyl (CHO) in position 8 instead of methyl.
[b] CHO in position 8 instead of methyl.
[c] H in position 6 instead of propionic acid.

HEME + RESIDUES   HEME + RESIDUES

Fig. 1 Stereoscopic view of the active site region of myoglobin.

this figure only those protein side chains closest to the heme have been included. The carboxyl groups, on the surface of the protein, are on the right, and the residue side chains on the distal (lower) side are alkyl or phenyl groups except for one imidazole group (b), which cannot bind to iron. The imidazole on the proximal (upper) side is seen to bind directly to the iron. In the reduced, Fe(II) form, hereafter referred to as (Hm), the sixth position on iron is open, whereas the ferric form (Hm$^+$) binds a water molecule [10]. This contrasts to the electron transport proteins cytochrome b_5 and cytochrome c, in which the sixth position is occupied by an imidazole [13] or a thiomethyl ether group [14], respectively.

In contrast to cytochrome c and b_5, myoglobins and hemoglobins bind oxygen, carbon monoxide, and other ligands reversibly. While simple hemes (1) bind carbon monoxide, cyanide, nitrogenous bases, etc., reversibly in solution, they are oxidized to hemins (Hm$^+$) by molecular oxygen almost instantaneously [1,3]. On the other hand, solutions of myoglobins and hemoglobins are generally stable under 1 atm. oxygen for several days [15,16]. Neither this stability nor the affinity for oxygen in heme proteins is greatly altered by changes in the heme (1) structure [1, p. 286;17] in hemoproteins. These and other observations led to the conclusion that the protein surrounding the heme affords both oxidative stability and variations in oxygen affinity which are so important to oxygen transport [1].

Although much has been learned about the nature of reversible oxygen binding through the study of various normal and abnormal myoglobins and hemoglobins [1, p. 286], these studies have not afforded a detailed understanding of the molecular nature of oxygen binding. This is most vividly seen in comparing the two accurately determined heme protein structures whale myoglobin [11,12] and *Chironomus* hemoglobin [18]. There are some differences in the ligand binding sites of these two proteins, but it is not at all clear from the structures why the kinetics of reactions with ligands are so different [19].

These difficulties have prompted attempts to prepare small molecules with which systematic studies of structural and environmental effects on reversible oxygenation can be made. This review summarizes some of the recent progress made in this effort. Table 1 [1,1, p. 225;19–27] lists some important properties of oxygen transport proteins that the small-molecule model systems are designed to duplicate.

These studies proposed to answer the following questions: (1) How is oxidation to Hm$^+$ to be retarded so that physical measurements such as X-ray crystallography, spectra, kinetics of oxygen combination, etc., can be made? (2) What is the detailed molecular geometry of the heme, heme–CO, and heme–O$_2$ complexes? (3) How are oxygen affinity and oxidation rate controlled in the protein? Before the second and third questions could be

TABI

Protein	MW	Hemes/ molecule	λ_max (nm) Hm α	Hm β	Hm S	HmO₂ α	HmO₂ β	HmO₂ S	HmCO α	HmCO β
Myoglobin (horse)	18,800	1	560		434	580	542	418	579	540
Myoglobin (mesoheme)	17,600	1	546		422	568	533	404	556	528
Hemoglobin (deuteroheme)	17,600	1	544		421	565	532	403	556	528
Hemoglobin A	68,000	4	555		430	577	541	415	569	540
Hemoglobin (*Chironomus*)	16,000	1	559		428				569	538

[a] See Antonini and Brunori [1].

answered, some means of answering the first question had to be provided. Part of this problem was solved by the realization that only X-ray crystallography required more than millisecond stability of the heme. But the long-term stability was also achieved by methods described below.

MODEL PREPARATIONS

Design

Following reports in the late 1950's of reversible oxygen binding to heme–imidazole mixtures as solids [25], in anhydrous pyridine [26], and in polystyrene matrices [27], little was done in the study of model heme–oxygen interaction until recently. Instead, extensive low-temperature studies of the ostensibly more stable cobalt–heme–oxygen complexes were carried out. These studies, more recently related to cobalt–myoglobin and cobalt–hemoglobin studies, have been reviewed elsewhere [28].

The recently revived interest in hemoprotein models has taken the more tactical approach toward the synthesis of heme compounds incorporating in the heme molecule some of the necessary attributes thought to be provided by the protein [1, p. 286]. In our laboratories this approach consists of what we term "active-site" synthesis [29,30], that is, identifying those chemical moieties considered necessary for reversible oxygen binding, excising these from the protein, and synthesizing a simpler organic scaffolding that holds them in their exact natural positions. This would satisfy the structural requirements for the synthetic active sites. But, to serve as a model for perties of Oxygen-Transporting Heme Proteins[a]

ν_{CO} (cm⁻¹)	l' (M⁻¹ sec⁻¹)	l (sec⁻¹)	k' (M⁻¹ sec⁻¹)	k (sec⁻¹)	K (M⁻¹)	L (M⁻¹)	References
44	5×10^5	0.02	1.4×10^7	10	1.3×10^6	3×10^7	1,p.225,20
	3×10^5	0.04	1.0×10^7	10.0	1.0×10^6	2×10^7	1,p.225,20b 1,20,19
51.5 1st			9×10^6	1080			20,22,23
2nd			4×10^6	244			21,22,23
3rd			3.5×10^6	28			20,22,23
4th			4×10^7	48			20,22,23
	2.7×10^7	0.095	3×10^8	218	2×10^6	3×10^8	19,24

myoglobin, such a synthetic site must meet certain behavioral requirements. It should bind oxygen with equilibria, kinetics, and mechanisms similar to those of myoglobin itself and should have similar spectral properties.

Because the proximal imidazole, protoheme, and a hydrophobic covering of the distal side appear to be invariant [1, p. 1], it seemed that three parts were required to meet the above criteria, as shown in Fig. 2a.

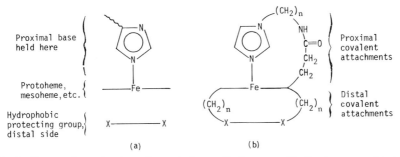

Fig. 2 Proposed structural requirements for synthetic oxygen binding heme compounds.

It was thought that covalent attachment of both the proximal base and the distal protecting group (Fig. 2a) would serve as the scaffolding described above and that, with the addition of some carboxylic acid groups to assist in water solubility, this kind of structure would duplicate myoglobin without protein [29] (Fig. 2b).

One structure, conceived to accomplish these ends, is (2) [29]. While the proximal imidazole structure (3) and the cyclophane porphyrin (4) have been

(2)

(3)

synthesized separately [29,30], they have not been combined into one structure, principally because the structure (3) alone makes possible most of the oxygen binding studies [31–33]. Elegant alternatives to (4), which allow structural studies to be made, are described below [34,35].

(4)

Prevention of Heme Oxidation

Several approaches have accomplished reversible oxygenation without oxidation. All seem designed to prevent one or both of the oxidation mechanisms (3) and (4) [24,25].

$$Hm\text{---}O_2 + H^+ \longrightarrow Hm^+ + H_2O \text{ (or } H_2O_2\text{)} \tag{3}$$

$$Hm\text{---}O_2 + Hm \longrightarrow Hm^+ + H_2O \tag{4}$$

Although all heme–oxygen complexes seem more sensitive to oxidation in hydroxylic solvents and in acid than in aprotic media, no thorough study of reaction (3) has been made [36]. On the other hand, reaction (4) has been strongly implicated by kinetic studies of reactions of protocheme–pyridine mixtures [37] or compounds such as (3) [9a] with oxygen. Reaction (4) requires the kinetics of oxidation to be second order in heme and either first order in oxygen at low oxygen pressure or inverse first order in oxygen at high pressure, all of which have been observed [9a,37]. Therefore, it would seem imperative that reaction (4) be prevented. It can be retarded in the following ways:

1. Hemes can be immobilized in solid matrices. Heme–base mixtures [27] or compounds like (3) [38] have been dissolved in viscous polystyrene or

polysytrene films or attached to silica gel [39], all of which prevent rapid oxidation and allow reversible binding of oxygen.

2. Because reaction (4) is temperature dependent and because the amount of free Hm can be reduced at low temperatures by converting it to HmO$_2$, low temperatures retard this reaction. This technique was first used to study cobalt–porphyrin–oxygen complexes [28] because it was thought that iron porphyrins would oxidize rapidly even at low temperature. Recently, solutions of (5a) [31] or of heme–alkylimidazoles [40–42], heme–*tert*-butyl-

Compound	A	Base
(5a)	—H	—NCH$_2$CH$_2$CH$_2$N (with H on N) imidazole
(5b)	—CH$_2$CH$_2$COCH$_3$	—NCH$_2$CH$_2$CH$_2$N (with H on N) imidazole
(5c)	—CH$_2$CH$_2$CNCH$_2$CH$_2$CH$_2$N imidazole (with OH)	—NCH$_2$CH$_2$CH$_2$N (with H on N) imidazole
(5d)	—CH$_2$CH$_2$COCH$_3$	—OCH$_2$CH$_2$CH$_2$ pyridine
(5e)	—CH$_2$CH$_2$COCH$_2$CH$_2$CH$_2$ pyridine	—OCH$_2$CH$_2$CH$_2$ pyridine

amine [40], or heme–pyridine [43] mixtures have been oxygenated in solution below − 20°C and are in some cases stable for hours under 1 atm oxygen [41]. In certain solvents such as dimethylformamide (DMF), dimethylacetamide, N-methylpyrrolidone, or dimethylsulfoxide, (5)–oxygen complexes are stable for several minutes even at room temperature [41]. Thus, the general impressions of sensitivity of hemes to oxidation has changed drastically in recent years.

3. Steric prevention of Fe—OO—Fe formation by distal side protection has been attempted by several synthetic techniques. The porphyrin cyclophane mentioned above, although synthesized for this purpose, was unfortunately not tested [29]. More recently, a cyclophane-type structure (6), prepared in an elegant one-step synthesis by Baldwin et al. [35], was shown to be much more stable toward oxidation than unprotected tetraphenylheme or other

(6)

simple hemes. The iron complex (7), first reported to be stabilized toward oxidation by steric hindrance [44], is actually much more easily oxidized than are simple hemes under the some conditions. This attempt at steric blocking of the proximal side was therefore unsuccessful, and compound (7) is less of a model for myoglobin than is protoheme itself. In a more successful approach toward hindrance around the periphery of the heme. Collman and his collaborators synthesized the hindered heme (8) [34,45]. This ferroporphyrin, like (6), prevents heme–heme approach provided that a

nitrogeneous base occupies the proximal side (opposite to the tert-BuC groups). The (8)–N-methylimidazole–oxygen complex is stable for long periods at room temperature and can be crystallized as such [34].

4. Carbon monoxide protection. Although there was a widespread impression that heme–base mixtures were oxidized immediately even in the presence of carbon monoxide [1,6,46], we find that such systems are stable under various pressures of oxygen for hours [33]. This results from the strong binding of carbon monoxide to the heme–base complex [Eq. (5)]. This

$$B_2Hm \rightleftharpoons B + HmB \underset{\longleftarrow}{\overset{CO}{\rightleftharpoons}} B-Hm-CO \qquad (5)$$

and the discovery that the HmB complex does not oxidize during flash photolysis kinetic determinations [33,47] make all the flash photolysis kinetic and spectroscopic methods devised for hemoproteins [1, p. 225] directly applicable to simple heme compounds such as (1) and (5).

A summary of stabilities of various heme–base–O_2 mixtures toward oxidation to Fe(III) is given in Table 2. This table shows that stable heme–oxygen complexes can be prepared [especially with (6) or (8)] and that simple heme–base systems can be stabilized to some extent by proper choice of solvent or stabilized greatly by carbon monoxide.

TABLE 2

Stability of Hemes toward Oxidation in 1 atm O_2 [a]

Heme	Proximal base	Solvent	Temp (°C)	Approx $t_{1/2}$ for oxidation	References
Protoheme	N-MeIm	DMF, CH_2Cl_2	−45	several hours	41
Protoheme	N-MeIm	DMF	25	2 min	41
(5a), (5b)		DMF, CH_2Cl_2	−45	Several hours	41
(5a), (5b)		DMF	25	10 min	41
(5a), (5b)	+ >100 torr CO	H_2O, CH_2Cl_2, etc.	25	Several hours	47
(6)	N-MeIm	Benzene	25	1 hr	35
(7)	Pyridine	THF–DMF	−78	Several hours	44
(7)	Pyridine	THF–DMF	−50	Few minutes	44
(8)	N-MeIm	Benzene	25	∼30 hr [b]	45,34
(8)	N-MeIm	Solid	25	Indefinite	34
(5a)	Alkylimidazole	Polystyrene film	25	Several hours	30
Myoglobin	Imidazole	H_2O	25	Several days	15,16

[a] Time for half-oxidation.

[b] This is probably a low estimate because this compound is stable at high heme concentrations, where oxidation is accelerated.

(7)

(8)

PHYSICAL PROPERTIES OF MODEL COMPOUNDS

Structure of Heme–Ligand Complexes

Although X-ray crystallographic studies on myoglobins and hemoglobins clearly defined the structures of the proteins and even indicated that high-spin iron complexes have the iron out of the porphyrin plane [1, p. 1;10–12], they did not allow definitive geometry determinations of the Fe—O_2 bond [48,49]. This awaited the crystal study of the simple heme–oxygen complex (**8**)-*N*-MeIm–O_2, which was reported in 1974 [50]. This study indicated the geometry shown below, in agreement with the conclusions tentatively

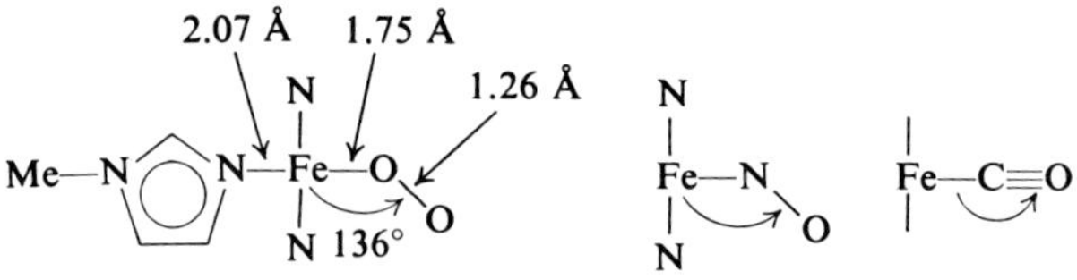

reached by Hobbs and Watson [49] from a study of myoglobin. The Fe—N—O angle in heme —NO adducts is rather similar to the Fe—O—O angle [51]. The Fe—C—O angle in the carbon monoxide complexes is 180° in simple heme complexes [34,52] but ~140° in myoglobins [18,53]. The reduced angle in myoglobin is attributed to steric hindrance on the distal side [22] which has also been suggested as a contributor to the variations in CO binding in hemoproteins [10]. All the crystal studies agree that hexacoordinate heme structures have a planar heme system [54].

Ultraviolet–Visible Spectra

When a heme–base mixture or model compound gives an oxygen complex, its spectrum is essentially identical with that of the corresponding natural or reconstituted myoglobin [29,31,40]. Furthermore, the "synthetic-site" compound (**5b**) has spectra in water or organic solvents that are identical with those of myoglobin (mesoheme reconstituted) as the deoxy, oxy, carbon monoxy, isonitrile complexes or the oxidized form [31]. In those cases in which magnetic moments have been measured, they correspond to the spin state indicated by the spectrum [34,40], i.e., the oxy, CO, and other hexacoordinate compounds have no spin, and the deoxy is high-spin ferrous iron.

Infrared Spectra

The carbonyl stretching frequencies of the carbon monoxide complexes of hemoglobin and myoglobin are observed at about 1950 cm^{-1}, whereas those of heme–base–CO complexes are generally near 1970 cm^{-1} in solution [20, 22]. This is usually taken as evidence for some steric hindrance of the CO moiety in the hemoproteins [20]. The observations that (5a)–CO has a CO frequency of 1965 cm^{-1} in solution and 1951 cm^{-1} as a solid seems to agree with this postulate [30]. Stretching frequencies observed in various hemoprotein and heme complexes with carbon monoxide are shown in Table 3 [20,22,31,55].

Because of the intense interest in the iron–oxygen bonding, several attempts have been made to measure the O—O stretching frequency in the heme–O$_2$ complex. Collman et al. [56] reported a frequency of 1385 cm^{-1} at liquid-nitrogen temperatures but later found this infrared band to be an artifact not related to the heme–O$_2$ bond [57]. Caughey et al. reported a band at 1107 cm^{-1} in oxyhemoglobin, which they attributed to the O—O stretch [20,58]. This superoxidelike stretching frequency is consistent with the polarity deduced for the iron–dioxygen bond by other means described below.

Mössbauer Spectra

The first study of Mössbauer spectra of simple heme–oxygen complexes was carried out using Mössbauer emission spectroscopy from the decay of [^{57}Co](PPIX)PyrO$_2$ in frozen solution [59]. It was found that this complex had Mössbauer parameters similar to those of myoglobin and concluded that the protein had little effect on the electronic structure of the heme–oxygen complex. However, it was also reported that changes in the proximal base (from pyridine to imidazoles, etc.) had little effect on the Mössbauer parameters. This indicates that, except for gross changes such as dissociation of the oxygen, Mössbauer parameters simply do not provide an accurate description of the binding state of the heme–oxygen–base complex. This insensitivity is further documented by the finding that the Mössbauer center shifts in an iron–tetraphenylporphyrin–base complex [34] are very similar to those of hemoglobin, in which protoheme is present.

TABLE 3

C—O Stretching Frequencies in Heme–CO Complexes

Heme	ν_{CO} (cm^{-1})	Medium	References
Myoglobin	1945	H$_2$O	20,22
(5a)–CO	1965	CHCl$_3$ solution	31
(5a)–CO	1951	Amorphous solid	31
n-BuS–Hm–CO	1923	Dimethylacetamide	55

REVERSIBLE BINDING OF O_2 AND OTHER LIGANDS IN SOLUTION

Qualitative observation indicating reversible binding of oxygen in solution are now numerous. The first such observation was actually one of Professor J. Wang's experiments done in 1958, in which a solution of protoheme in 70% phenethylimidazole–30% polystyrene was reversibly oxygenated at room temperature [27]! This mixture is not a polystyrene film, but a viscous solution [9–9b]. More recently, reversible oxygenations have been carried out at $-30°$ to $-80°$ with iron(II) complexes of mesoporphyrin and proto-porphyrin IX and their esters, as well as synthetic iron complexes and the special porphyrins described above. Proximal bases used in these observations are 1-methylimidazole [41,42], 2-methylimidazole [40], 1,2-dimethylimidazole [60], 1-butylimidazole [40], 1-phenethylimidazole [27], pyridine [43], *tert*-butylamine [40], piperidine [28], dimethylformamide (or its dimethylamine impurity) [41], and tetrahydrofuran [34]. Wherever quantitative comparisons have been made, oxygen affinity has been observed to decrease with base structure in the order imidazoles $\geq$ alkylamine > pyridines [28] in both Co(II) and Fe(II) porphyrins.

A comparison of oxygen and carbon monoxide affinities of models with those of myoglobin is given in Table 4. Although quantitative aspects of oxygen binding are discussed in detail in a later section, these data indicate how successfully the oxygen affinities of heme proteins have been duplicated.

There is little question that, except for the high-affinity heme proteins such as legoglobin ($P_{1/2} = 0.04$ torr), the hemoprotein O_2 and CO binding affinities are duplicated by these model systems, even in solution. Further-

TABLE 4

Oxygen and Carbon Monoxide Affinities of Myoglobin Models and Myoglobin at 20°–25°C

Heme	Solvent	K_{O_2} (M^{-1})	(torr)	K_{CO}, M^{-1}	References
(5b)	H$_2$O (CTAB)	$1.7 \times 10^{6\,a}$	0.3^{a}	$>3 \times 10^{8\,b}$	47
PPIX-DME^{c} (0.1 M)	Benzene			$4.8 \times 10^{8\,b}$	61
(1-Me-Im)					
(8)–(1-Me-Im)	Solid		0.31^{b}		57
Myoglobin (whale)	H$_2$O	$1.9 \times 10^{6\,b}$	~ 0.3	$3.5 \times 10^{7\,b}$	1, p. 22

^{a} Obtained by kinetic methods.
^{b} Obtained by ligand titration.
^{c} Protoporphyrin IX dimethyl ester.

more, the synthetic myoglobin site **(5a)** binds oxygen in water as well as does myoglobin itself!

The stoichiometry of oxygen binding to iron porphyrins invariably shows a 1:1 ratio of bound oxygen to iron [30,34,35,43,62] in contrast to cobalt(II) compounds, in which either 1:1 ($Co-O_2$) or 2:1 ($Co-O_2-Co$) ratios have been observed [28]. This 1:1 stoichiometry has been observed in solids [30,57], as well as in solutions at low temperature [45,62] and at room temperature [Eq. (6)] [34].

$$Base-\overset{|}{\underset{|}{Fe}} + O_2 \overset{K_{O_2}}{\rightleftharpoons} Base-\overset{|}{\underset{|}{Fe}}-O_2 \qquad (6)$$

Oxygen binding in aqueous solutions of heme–polyvinylpyridine solutions reported recently [63] could be reversible or irreversible oxidation. The rates of oxygen absorption are orders of magnitude too slow for simple oxygenation, and the systems have not been sufficiently documented by visible, electron spin resonance, or other spectral methods.

An interesting oxyheme formation through the reaction of Fe(III) protoporphyrin IX with superoxide ion in DMF at $-50°C$ has been reported [Eq. (7)] [64]. Although the visible spectrum of the observed oxyheme was

$$\overset{|}{\underset{|}{Fe}}(III)(DMF) + n\text{-}Bu_4\overset{+}{N}O_2^{-} \text{ (1 equiv.)} \longrightarrow (DMF)\overset{|}{\underset{|}{Fe}}-O_2 \qquad (7)$$

identical to that found by reaction of oxygen with the ferroheme in the same solvent, the latter reaction was said not to occur in these experiments. This system deserves further study.

Although spectral properties and general affinities for carbon monoxide and oxygen of these heme–base complexes discussed so far rather resemble those of heme proteins, there are some striking differences. Other than the sterically hindered hemes, all the model systems are much more sensitive to oxidation than is e.g., myoglobin. Furthermore, cyanide ion binds strongly to compound **(5b)** in water ($K \cong 10^4 \text{ M}^{-1}$) [60] and very poorly to myoglobin ($K \cong 2 \text{ M}^{-1}$) [1, p. 227], probably due to the hydrophobic pocket in myoglobin. Additionally, whereas **(5a)** or **(5b)** (Fe(II)) shows the high-spin Fe(II) spectral behavior of myoglobin (broad single band at ~ 550 nm rather than sharp α, β bands), the spectra change to the low-spin spectra as the temperature of aqueous alcoholic or DMF solutions is lowered [9–9b,30,40], indicating definite but weak water binding to the sixth position on iron [7]. Myoglobin shows no such change. Nevertheless, it is estimated that only 10% of complexes such as **(5b)** have water so bound at 20°C [7] ,where most comparisons are made in the following sections. The point is that at ambient temperatures water need *not* be kept out of the myoglobin pocket [65] by the hydrophobic environment as is often suggested. The heme–imidazole system simply does not bind water at room temperature even in aqueous solvents.

EQUILIBRIA AND MECHANISMS OF LIGATION OF HEMOPROTEIN MODELS

Now that it is possible to observe reversible oxygen binding by both static and kinetic methods in a variety of model systems, we can begin to determine what aspects of structure, solvent polarity, viscosity, etc., are responsible for the widely varying ability of heme proteins to bind oxygen, be oxidized and reduced, or, e.g., act as oxygenation catalysts. But first it is necessary to establish that the model system resemble heme proteins such as myoglobin, not only their spectral and oxygen binding properties but in detailed structure and mechanisms of reaction. Because some model systems react with carbon monoxide (and perhaps with oxygen) by mechanisms that differ from those of myoglobin [66], the mechanisms of reaction and co-ordination behavior in solution will be discussed first.

Coordination of Heme–Base–CO Mixtures

Recently, a thorough study by Brault and Rougee [61] has afforded a rather clear picture of coordination of strong field ligands (e.g., imidazoles and other nitrogenous bases) and weak field ligands (e.g., ethers, water, and alcohols) to hemes and the subsequent coordination of carbon monoxide to these heme–base complexes. The summary of their results given in Table 5 for reactions

TABLE 5

Equilibrium Constants for Bases and CO with Deuteroheme Dimethyl Ester in Benzene at 25°C [a]

Base	K_8 (M^{-1})	K_8 (M^{-1})	K_{11} (M^{-1})	K_{12} (M^{-1})	Reference
H$_2$O	~0.1	Small	5 × 10^6	~10	61
EtOH	1.5	Small	8 × 10^6	250	61
4-methylimidazole (N–H, CH$_3$)	1.3 × 10^4	Small	2.5 × 10^6	6.5 × 10^5	61
imidazole (N–H, H)	4.5 × 10^3	6.8 × 10^4	4.8 × 10^8	4.3 × 10^7	61
N-methylimidazole (N–CH$_3$, H)[b]	~30[b]	~30[b]			60

[a] $K_{CO} = 5 \times 10^4$ M^{-1} (0.51 torr^{-1}).
[b] Mesoheme dimethyl ester in aqueous CTAB at 25°C.

(8)–(12) (in which B denotes base and Hm denotes heme) will help in understanding the mechanisms discussed below. This table indicates rather large

$$B + Hm \; \overset{K_8}{\rightleftharpoons} \; B\!-\!Hm \tag{8}$$

$$B\!-\!Hm + B \; \overset{K_9}{\rightleftharpoons} \; B\!-\!Hm\!-\!B \tag{9}$$

$$Hm + CO \; \overset{K_{CO}}{\rightleftharpoons} \; Hm\!-\!CO \tag{10}$$

$$B\!-\!Hm + CO \; \overset{K_{11}}{\rightleftharpoons} \; B\!-\!Hm\!-\!CO \tag{11}$$

$$HmCO + B \; \overset{K_{12}}{\rightleftharpoons} \; B\!-\!Hm\!-\!CO \tag{12}$$

$(10^3–10^4)$ values of both K_8 and K_9 in benzene. Furthermore, pyridines bind to iron(II) heme better than do imidazoles, although the order is reversed with iron(III) [9–9b]. These facts help to explain the poor oxygen binding and slow oxidation of pyridine hemochromes.

A further important point is the effect of solvent on K_8 and K_9. In 50% glycerol:water, K_8 for 2-methylimidazole is about 30 M^{-1} [65], and K_8 and K_9 for 1-methylimidazole can be estimated to be between 30 and 50 M^{-1} in water [9,67]. Thus, the values of K_1 and K_2 for imidazoles appear to increase by factors of 20–200 in going from water to hydrocarbon solvents, probably as a result of the hydrogen bonding of the imidazoles to water or alcohols.

Finally, changing from weak-field proximal bases (H_2O, etc.) to strong-field proximal bases (imidazole, pyridine) results in a 200-fold increase in K_{CO} [$K_{13} \simeq 200K_{14}$; Eqs. (13) and (14)]. This effect is striking and has much to do with mechanism changes discussed below.

$$Im\!-\!Hm + CO \; \overset{K_{13}}{\rightleftharpoons} \; Im\!-\!Hm\!-\!CO \tag{13}$$

$$(H_2O)\!-\!Hm + CO \; \overset{K_{14}}{\rightleftharpoons} \; (H_2O)\!-\!Hm\!-\!CO \tag{14}$$

Comparison of Deoxymyoglobins with Deoxyheme Models. The Effect of Covalent Base Attachment

In our original design [29] of a "myoglobin active site" to be synthesized, we took advantage of the known effect of covalent binding of the proximal imidazole in cytochrome c and the proximity of the carboxyl side chain and the proximal imidazole in myoglobin (Fig. 1 and 3). The encircled part of Fig. 3 is not a part of the myoglobin, but represents a connecting linkage which fits into this structure and which has been used ostensibly to hold the imidazole in place in the "synthetic sites" [(5a) and (5b)] that have been prepared by coupling the proper alcohols or amines to mesoporphyrin or pyrroporphyrin [53,54]. Whether this attachment of a flexible side chain achieves this [40,42,43] has been probed by estimating the change in the

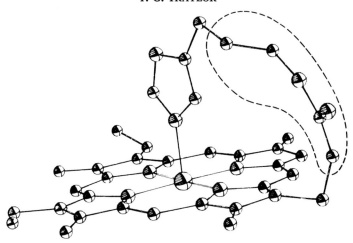

Fig. 3 Proposed covalent connection to insure five-coordination of iron as in myoglobin.

basicity of the proximal base B [Eq. (15)] engendered by the presence of the covalently attached heme [68]. Although alkylimidazoles have pK_a values of 6.3–7.5 (pK_{15}), the pK_a of [(**5b**)–H]$^+$ (pK_{17}) is about 3.6 [as measured either by pH–CO rate profile or by change in the Soret band of (**5b**) with pH].

The change of >2 pK units with covalent binding to heme means that the pK for closure (pK_{16}) is >2 and the $K_{closure}$ that is >100. A pH-rate profile for the pyridine compound (**5d**) indicates a $K_{closure}$ that is even higher (Fig. 4).

Because the binding of imidazoles to iron(II) increases as solvent polarity decreases, the $K_{closure}$ for imidazole compounds like (**5a**), (**5b**), and (**5d**) should be greater in other solvents than it is in water. Therefore, the covalent attachment has the effect of maintaining five-coordinate structures like myoglobin in all solvents. The failure of water to bind strongly to the sixth position [65,68], taken with the coordination behavior just discussed, suggests that the synthetic myoglobin active site [67,68] behaves like deoxy-myoglobin. Model compounds lacking this connection [34,40–43] do not usually behave in this way, as we see below.

$$\begin{array}{ccc}
\overset{+}{HB}\diagdown & & B\diagdown \\
& \xrightarrow{\quad K_{15} \quad} & \\
\text{——Fe——} & \rightleftharpoons & \text{——Fe——} \\
{}_{17}\diagdown\! K_{17} & & {}^{16}\diagup\! K_{16} \\
& B\diagdown & \\
& | & \\
& \text{——Fe——} &
\end{array} \qquad (15\text{–}17)$$

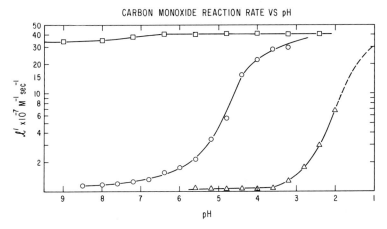

Fig. 4 Plots of rate constants for reaction of carbon monoxide with heme compounds against pH. □, Mesoheme dimethyl ester (**1d**)-DME. ○, Mesoheme monomethyl ester mono[3-(1-imidazolyl)propyl]amide (**5b**). △, Mesoheme monomethyl ester mono[3-(3-pyridyl)propyl]ester (**5d**). [Reprinted with permission from *J. Am. Chem. Soc.* **97**, 20 (1975). Copyright by the American Chemical Society.]

Mechanisms of Heme Model Reactions with CO

The reversible oxygenation of (**5a**) in solution has been qualitatively duplicated with heme–base mixtures without covalent attachment of the proximal imidazole [34,35,40–43]. This raises some doubts concerning the necessity of covalent attachment of bases to achieve complete modeling of myoglobin behavior [28,40,42,43]. By comparing the kinetic behavior of covalently bound heme–base compounds with that of external heme–base mixtures, it has been shown that the covalent attachment is nesessary to achieve the same mechanism of reaction with carbon monoxide as observed with myoglobin [67]. An alternative mechanism for reaction of heme–base mixtures with carbon monoxide was also discovered [67].

Although the reaction of carbon monoxide with heme–base complexes has invariably been written as a single-step process [34,40,69] with a single rate constant [69], it can (and does) take place by two distinct pathways, which have been called the *association mechanism* and the *base elimination mechanism* [67]. These processes are described in the general scheme below, in which B denotes an external or covalently attached base that attaches to Fe(II).

Flash photolysis of B—Hm—CO occurs with a quantum yield near 1.0 to produce B—Hm, which can do one of the three things shown. Let us consider first the system (**5b**) in which only one B is covalently attached and no B_2Hm is formed. In aqueous solution at pH 7, K_1 highly favors B—Hm (see above), and therefore the B—Hm simply captures CO with a rate constant

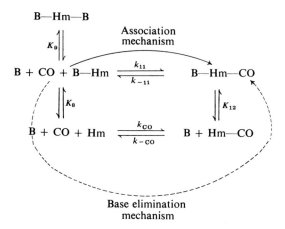

$$B—Hm—B$$

Association mechanism

$$B + CO + B—Hm \underset{k_{-11}}{\overset{k_{11}}{\rightleftharpoons}} B—Hm—CO$$

$$B + CO + Hm \underset{k_{-CO}}{\overset{k_{CO}}{\rightleftharpoons}} B + Hm—CO$$

Base elimination mechanism

k_{11} of 1×10^7 M^{-1} sec^{-1}, the association mechanism. However, at pH 2.5 the B rapidly ($\sim 10^4$ sec^{-1}) dissociates and becomes protonated [see Eqs. (15)–(17)]. This allows the heme to react by the faster process $k_{CO} = 4 \times 10^8$ M^{-1} sec^{-1} typical of hemes in the absence of any base (see mesoheme dimethyl ester kinetics in Fig. 4). The Hm—CO, having more affinity for B than does free Hm (see Table 4), then recaptures B to yield B—Hm—CO (base elimination mechanism). Flash spectroscopy reveals as the intermediate Hm—B (λ_{max} 417) at pH 7.3 and Hm—CO (λ_{max} 403) at pH 2.5. If $k_{CO} \simeq 40k_{11}$ the reaction proceeds 50% by each mechanism when HmB/Hm $\simeq 40$ or at pH $\simeq 5.2$. At physiological pH levels (7–8.5) or in organic solvents the reaction proceeds predominantly through the association mechanism, the same mechanism used by myoglobin. The pyridine compound, **5d**, having a larger $K_{closure}$ (K_{16}), proceeds more by the association mechanism than does **5b**. While there is still some uncertainty about the size of $K_{closure}$ in water, it seems clear that the association mechanism is operative in all solvents.

Now consider two kinds of external base systems that have been proposed as myoglobin models: the N-alkylimidazole–heme [34,35,43] and the hindered 2-methylimidazole–heme [40] mixtures. In the hindered base systems there is strong evidence that little or no B—Hm—B is formed [65,67], although B—Hm formation is as facile as in the case of unhindered imidazoles [61]. For both kinds of external bases

$$K_8 = \frac{(B—Hm)}{(B)(Hm)} \quad \text{and} \quad K_9 = \frac{(B—HmB)}{(B—Hm)(B)}$$

and the observed rate of return to B—Hm—CO after flash photolysis (l'_{obs}) can be expressed in terms of the equation

$$\frac{1}{l'_{obs}} = \frac{K_8 K_9 B^2 + K_8 B + 1}{k_{CO} + k_{11} K_8 B} \tag{18}$$

For 1,2-dimethylimidazole–mesoheme dimethyl ester in toluene, where $K_1 \cong 10^4$ M^{-1} and $k_1 = 4 \times 10^8$ M^{-1} sec^{-1} (see above), the l/l'_{obs} vs. B plot has two slopes and is best fit to Eq. (18) with $K_9 < 1$ M^{-1}, $k_{11} \cong 2 \times 10^5$ M^{-1} sec^{-1}. This means that the base elimination mechanism is followed at the low base concentration (10^{-1}–10^{-5} M) [34,40,43] usually employed in model studies. Because the fraction proceeding through the base elimination mechanism is

$$\text{Fraction base elimination} = \frac{k_{CO}}{k_{11}BK_8 + k_{CO}} \tag{19}$$

this fraction becomes 0.5 in toluene at a concentration of 1,2-dimethylimidazole of $4 \times 10^8 / 2 \times 10^5 \times 1.5 \times 10^4 \cong 0.12$ M. In aqueous or alcoholic solutions the concentration at which the mechanism becomes 50% associative is much higher because k_{CO}/k_{11} remains about the same but K_8 becomes ten times smaller. Therefore, exclusive association mechanisms cannot be achieved in water with hindered imidazoles. Even in the case of the unhindered 1-methylimidazole, flash spectroscopy shows that the reaction proceeds by the base elimination mechanism below about 1 M concentration of the imidazole in water suspension [60].

These results show that mixtures of hemes and imidazoles, although having the ability to reversibly bind oxygen and carbon monoxide [40–43], nevertheless react with carbon monoxide by mechanisms that differ from those usually written for myoglobin reactions except at high concentration.

However, the base elimination mechanism that these mixtures employ could have important implications in heme protein chemistry. The best fit of Eq. (18) to the kinetics of reaction of mesoheme dimethyl ester, 2-methylimidazole, with carbon monoxide suggests that the $k_{20} \cong 2 \times 10^5$ M^{-1} sec^{-1} even though this mechanism is difficult to observe [Eq. (20)]. Thus, the

$$\tag{20}$$

hindered imidazole slows the association reaction with carbon monoxide by about 50-fold compared to those with unhindered imidazoles. This agrees with the suggestion of Perutz [10] that pulling on the proximal imidazole affects the reaction with ligands. This same pull also shifts the mechanism more toward base elimination [Eq. (21)]. Then why should the base elimination not occur in heme proteins? Indeed, Maxwell and Caughey [70] and

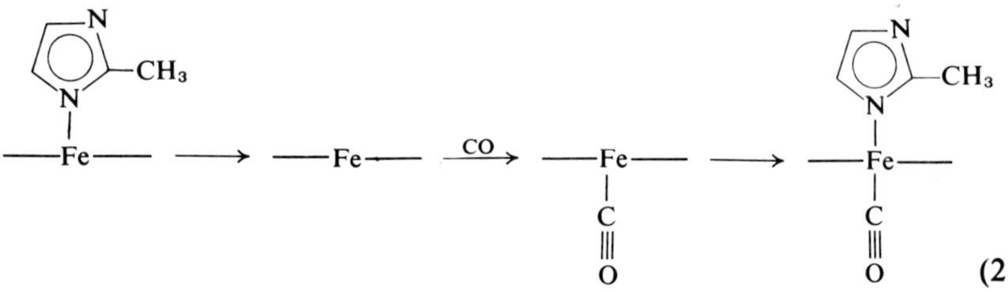

$$\text{(21)}$$

Perutz *et al.* [71] recently presented evidence for a five-coordinate Hm–NO complex in HbA reactions with NO, in the presence of inositol hexaphosphate, analogous to the base elimination mechanism shown above. Furthermore, Chance *et al.* [72] had previously observed a first-order carbon monoxide reaction with hemoglobin crystals with a rate constant $k = 2 \times 10^4 \text{ sec}^{-1}$. This rate is very similar to the estimation of the base dissociation rate [Eq. (22)] [60,62]. These results suggest that four-coordinate heme reactions

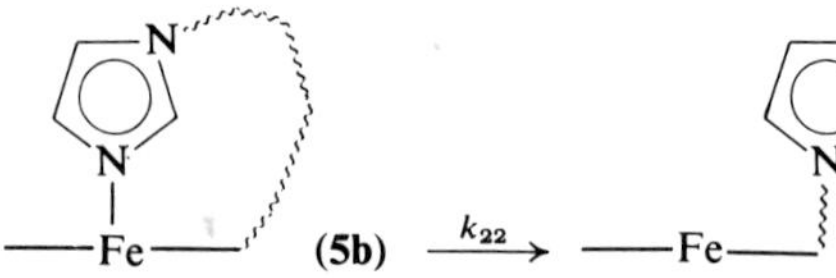

$$\text{(22)}$$

(base elimination mechanism) might represent an important additional means of controlling heme protein reactions. Where either pH changes [as in cytochrome *P*-450, Eq. (23)] or steric changes [as in hemoglobin, Eq. (24)] would bring about a change in proximal base binding, the mechanism could shift from one of association to base elimination.

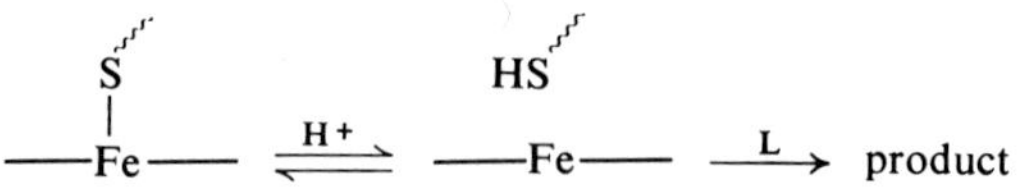

$$\text{(23)}$$

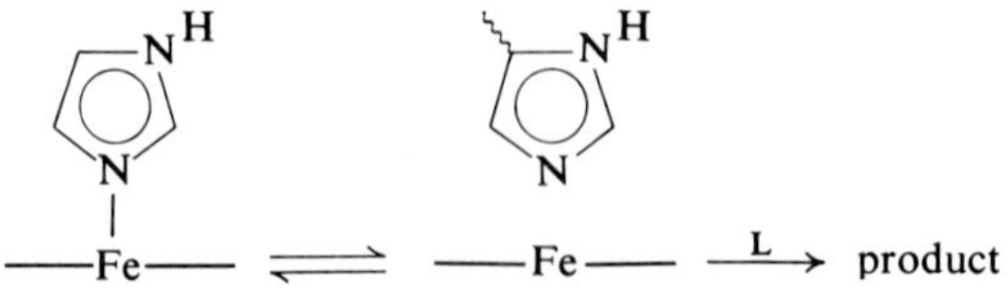

$$\text{(24)}$$

STRUCTURAL AND SOLVENT EFFECTS ON OXYGEN BINDING

While changes in structure or solvent can bring about changes in ligation mechanisms, the problems arising from such mechanism changes can be minimized by either making equilibrium measurements without regard to mechanisms or making kinetic measurements under conditions in which only the association mechanism occurs. By using covalently attached proximal base compounds such as (5b) the association mechanism can be ensured. Both systems are reviewed here.

Equilibrium Studies: Proximal Base Effects

Equilibrium measurements of oxygenation of both Fe(II) and Co(II) mesoporphyrin derivatives have been made at −45°C [28,41,47]. Because Co porphyrin derivatives do not readily form hexacoordinate bisimidazole complexes, the measurements are simple. However, since iron porphyrin–(base)$_2$ complexes can form it is necessary to use a monoliganded heme compound such as (5b) or (5d). A comparison of oxygen binding to (5a), (5b), (5d), (9), and Co(II) protoporphyrin IX is shown in Table 6 [96,41,62,73].

(9)

The first striking fact seen in Table 6 is that Co(II) and Fe(II) porphyrins show the same trends in oxygen binding with structural and solvent changes even though (1) the Fe(II) porphyrins bind oxygen much better than do Co(II) porphyrins; (2) the Fe(II) but not Co(II) porphyrins bind CO; (3) the CO binding by Fe(II) porphyrins is not very solvent dependent. The second important conclusion is that oxygen binding is improved by increasing the basicity of the proximal base. Additionally, that imidazoles improve this

TABLE 6

Oxygenation of Hemes (5b), (5d), (9), and Co(II) Protoporphyrin IX Dimethyl Ester at $-45°C$ [a]

Heme	Proximal base	pK_a (proximal base)	$P_{0.5}^{O_2}$ (torr) DMF	Toluene	References
(5b)	Alkylimidazole	~6.6	0.2^b		9b,62
(5d)	Alkylpyridine	~5.1	5^b	~400	9b,41
(9)	Alkylimidazole	~6.6	$1.8 (17)^c$		9b
Co(PPIX-DME)	4-Cyanopyridine	2.0		6300	73
	Pyridine	5.7		690	73
	4-Aminopyridine	9.2		112	73
	N-Me-imidazole	7.0	$(12.5)^d$	$50(417)^d$	73
	Piperidine	11		224	73

[a] Expressed as $P_{1/2}$ (pressure for half-saturation).
[b] These values are probably high because DMF binds to the sixth position in (5b) and (5d) at $-45°C$, interfering with oxygen binding. A better comparison of (5b) and (5d) at 20°C, where such interference is absent, is seen in Table 7. The ratio of K_{O_2} values is about 25 at both temperatures.
[c] At $-25°C$.
[d] At $-23°C$.

oxygen binding more than their basicity would suggest leads to the conclusion that not only σ basicity (basicity toward protons) but π basicity is important [28,41]. Finally, increased solvent polarity increases oxygen affinity by either metalloporphyrin.

Kinetics of Oxygen and Carbon Monoxide Reactions

A more definitive study of structural effects on binding can be obtained by the flash photolysis methods of Gibson [74]. Because carbon monoxide protects most model systems against oxidation and yet can be flashed off easily with most light frequencies, the Gibson equation (25) can be used to estimate both on and off rates for oxygen binding (R denotes rate of return from oxyheme B—HmO_2 to carbon monoxyheme B—HmCO). A summary

$$\frac{1}{R} = \frac{1}{k_{off}^{O_2}} + \frac{k_{on}^{O_2}[O_2]}{k_{off}^{O_2} k_{on}^{CO}[CO]} \tag{25}$$

of rates obtained by this method is given in Table 7 [47]. This table and the equilibrium data given above allow some conclusions to be drawn concerning the various structural effects on hemoprotein oxygen binding.

TABLE 7

Kinetic Constants for Reactions of Hemes with Carbon Monoxide and Oxygen at 22°C and Hemoproteins at 20°C[a]

Heme	Solvent	l' $(M^{-1}\,sec^{-1}$ $\times 10^{-6})$	k' $(M^{-1}\,sec^{-1}$ $\times 10^{-6})$	k (sec^{-1})	$k'/k = K_{O_2}$ $(M^{-1} \times$ $10^{-6})$	K_{O_2} $(torr^{-1})$
(5a)	CH_2Cl_2	10	90	600	0.15	1.8
	DMF	8	90	160	0.5	2.9
	DMF/H_2O (2/1)	10	85	90	1.0	4.9
	H_2O (CTAB, pH 7.3)[b]	12	30	35	1.7	2.3
(5c)	H_2O (CTAB, pH 7.3)[b]	2				
(5d)	Toluene	7.5	35	2500	0.001	0.01
	DMF	14	96	450	0.02	0.12
	DMF/H_2O (2/1)	1.5	20	400	0.05	0.24
	H_2O (CTAB, pH 7.3)[b]	10	14	150	0.09	0.12
(5e)	H_2O (CTAB, pH 7.3)[b]	0.1				
Isolated hemoglobin α chains[c]	H_2O		55	31	1.8	2.4
Myoglobin (whale)	H_2O	0.5	19	10	1.9	2.6
Myoglobin (Aplysia)[c]	H_2O		15	70	0.22	0.3
Hemoglobin (Hb + O_2 ⇌ HbO_2)[d]	H_2O	0.5	9	1080	0.009	0.012
Hemoglobin [Hb$(O_2)_3$ + O_2 ⇌ Hb$(O_2)_4$][d]	H_2O		40	48	0.8	1.1

[a] From Chang and Traylor [47].

[b] Aqueous 0.1 M sodium phosphate, pH 7.3, buffer containing 2.2% cetyltrimethylammonium bromide.

[c] Antonini and Brunori [1, p. 225].

[d] See Table 1.

EFFECT OF CHANGING THE METAL

Replacement of iron by cobalt in myoglobin reduces oxygen affinity by about 100, and we find a reduction in oxygen binding at room temperature upon such metal replacement in the synthetic site (**5a**) → (**9**) of about 300. Furthermore, recent studies indicate that neither model manganese porphyrins nor manganese heme proteins bind oxygen [28,74].

PROXIMAL BASE ELECTRONIC EFFECT

The kinetic data in Table 7 agree with equilibrium data that increased basicity increases oxygen affinity. These changes seem to affect both oxygen on and off rates, whereas the on rate for carbon monoxide is little affected. These results seem to suggest an importance of oxidation potential of iron on oxygen-binding equilibrium and rates. Perhaps a transition state having charge-transfer properties is involved for oxygen but not for carbon monoxide [Eq. (26)]. This would explain both the faster rates with oxygen and their dependence on proximal basicity.

$$B-\overset{|}{\underset{|}{Fe}} + O_2 \longrightarrow \left[B-\overset{|}{\underset{|}{Fe}}(III)\cdots O_2^{\overline{\cdot}} \right]^{\ddagger} \longrightarrow B-\overset{|}{\underset{|}{Fe}}-O_{\diagdown O} \qquad (26)$$

SOLVENT EFFECTS

The increased oxygen affinity with increased solvent polarity can be seen in Table 7 to be due almost entirely to the off rates. This is a clear indication of the Fe—OO stabilizing influence of solvent polarity and, together with the observation that imidazoles are especially good at increasing oxygen affinity, suggests a strong dipolar or ionic iron–oxygen bond. The imidazole is then seen to act in a synergistic way with the electronegative oxygen by both σ and π back donation [28,41]:

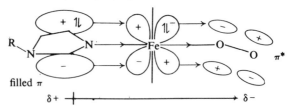

According to this view anything that alters the σ or π basicity of the proximal imidazole or any dipole that interacts with the strong $^+$Fe—OO$^-$ dipole would affect oxygen affinity.

STERIC EFFECTS ON PROXIMAL IMIDAZOLE

In the dual-mechanism studies discussed above, two effects of steric pull on the proximal imidazole were noted. First, the introduction of a methyl group in the 2 position slows the rate of carbon monoxide association by a factor of about 50 $[k_H/k_{CH_3} \cong 50$ in Eq. (27)]. The k_{CH_3} is about 2×10^5 M^{-1} sec^{-1},

$$\tag{27}$$

which is very similar to the rate of reaction of myoglobin or hemoglobin. Therefore, pulling on the proximal imidazole can greatly reduce the reaction rates with external ligands, provided that the proximal base does not dissociate. However, the second effect of pulling on the proximal base is to cause a change from the association mechanism to the base elimination mechanism, and this change results in faster reaction rate. Therefore, hemes and heme proteins have a delicately balanced way of changing reaction rates by either altering the kinetics within one mechanism or switching mechanisms, all accomplished by small changes in proximal base steric effects.

$$\tag{28}$$

STERIC EFFECT IN THE HEME POCKET

It has been suggested that the hydrophobic heme pocket, being rather rigid, is more accurately modeled by a solid heme complex such as (8)–Im crystals than by a simple heme in solution [34,57]. This is based on an observation that either a polymer-bound cobalt complex [28] or an (8)–Im cobalt complex apparently binds oxygen more strongly than do complexes in solution. There are several problems with this notion. First, the solid complexes absorb oxygen or carbon monoxide six to ten orders of magnitude more slowly than solution or crystalline hemoproteins [depending on the size of the (**8**) crystal] [57]. Second, the cobalt synthetic site (**9**), having none of the distal immobilization effects of (**8**), binds oxygen almost as well as does (**8**) (Co). Furthermore, the synthetic site (**5a**) binds both oxygen and carbon monoxide as well in solution as does myoglobin, and the rates of reaction for (**5a**) are between those of myoglobin and erythrocruorin (Tables 1 and 7). The crystal structures of myoglobin and erythrocruorin do not indicate a large difference in either rigidity or steric hindrance in the heme pocket [19] and, even though the carbon monoxide *on* rate is about 50 times faster that that of myoglobin, the Fe—CO bond angle is similarly bent (by steric effects?) in both [18,53].

The Question of Myoglobin Site Polarity

Another observation that leads to the conclusion that there is some special entropic property of the protein on oxygen binding is that the binding constant of cobalt porphyrins (in toluene) for oxygen [Eq. (29)] is some 300 times less than aqueous Co–myoglobin. But this value depends on the choice

$$
\begin{array}{ccc}
\overset{\displaystyle R}{\underset{|}{\text{Im}}} & & \overset{\displaystyle R}{\underset{|}{\text{Im}}} \\
| & & | \\
\text{—Co—} \quad + \ O_2 & \underset{\longrightarrow}{\overset{K}{\longleftarrow}} & \text{—Co—} \\
& & \underset{O_2}{|}
\end{array}
\qquad (29)
$$

of solutions. Thus, the synthetic site (**5b**) binds oxygen in water *exactly* as well as does myoglobin although, if toluene solutions were used for comparison as was done with the cobalt complexes discussed above the conclusion would be that the protein has a large effect. The fact is that we can change oxygen affinities of hemes greatly by changing solvent, and it is not yet clear which solvent system best represents the protein environment. Rather than postulate two effects in transferring the synthetic site from myoglobin to water, it seems preferable (although not necessarily correct) to simply propose that the myoglobin site resembles a polar environment, perhaps by using the distal imidazole to control site polarity. This possibility has been discussed by Basolo, Hoffman, and Ibers [28]. Some site polarity probe is needed to settle this question.

At least distal side steric restrictions do not seem to be a general property of heme proteins because both isolated hemoglobin α chains and erythrocruorin bind carbon monoxide and oxygen with rates equal to those of the chelated heme (**5b**) in solution [1, p. 225; 47].

EFFECT OF PROXIMAL BASE ROTATION

In myoglobin the proximal imidazole plane rotated only about $10°$ from one of the N–Fe–N lines in the porphyrin ($\phi = 10°$) (the porphyrin is in the paper plane and the imidazole is perpendicular to it):

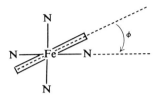

The effect of rotation further from this N–Fe–N line is not known, but comparison of the binding of bases in (5b) vs. (5c) with external bases suggests that it is important. Whereas the K_8 for binding the first external base is somewhat smaller than that for the second (e.g., 4×10^3 vs. 7×10^4 M^{-1} in benzene, Table 5, or $K_8 \simeq K_9 = 30–40$ M^{-1} in water [60]), the opposite is true for the covalently bound bases in (5b), (5c), (5d), and (5c) [Eqs. (30) and (31)].

$$\qquad\qquad (30)$$

$$\qquad\qquad (31)$$

We attribute at least a part of this inability to bind the second base (K_9) to the constraints placed on base rotation by the covalent side chain. This constitutes yet another means of controlling ligand binding in hemoproteins.

EFFECTS OF PORPHYRIN RING STRUCTURE

Although some differences in binding of oxygen to myoglobins and hemoglobins reconstituted with mesoheme, protoheme, and deuteroheme have been observed [1, p. 229], the differences are small and have led to some disagreement concerning the source of the differences. The effect of ring substituents has not been studied in sufficient detail to draw any conclusions, although there is a general impression that electron-donating substituents increase oxygen affinity [75].

SUMMARY

The syntheses of a variety of heme protein model compounds in recent years has made possible detailed formulations of structures and mechanisms of reaction of heme–oxygen complexes. Such studies have indicated an end-on, polar oxygen–iron bond,

$$B\overset{\delta+}{-}Fe-O\diagdown_{O}^{\delta-}$$

and have revealed some of the chemical means by which heme proteins could alter the heme reactivity. But the principal advance has been in the demonstration that simple heme compounds can be made to show the oxygen-binding characteristics of heme proteins. With such model systems in hand many of the mysteries of heme protein functions should soon be solved.

REFERENCES

1. F. Antonini and M. Brunori, "Hemoglobin and Myoglobin and Their Reactions with Ligands." North-Holland Publ., Amsterdam, 1971.
2. J. M. Rifkind, *Inorg. Biochem.* **2**, 832 (1973).
3. O. Hayaishi, ed., "Molecular Mechanisms of Oxygen Activation," p. 1. Academic Press, New York, 1974.
4. R. Lemberg and J. Barrett, "The Cytochromes," p. 1. Academic Press, New York, 1973.
5. P. Nicholls and G. R. Schonbaum, *in* "The Enzymes" (P. D. Boyer, H. Lardy, and K. Myrbäck, eds.), 2nd ed., Vol. 8, p. 147. Academic Press, New York, 1963.
6. J. F. Falk, "Porphyrins and Metalloporphyrins," p. 1. Am. Elsevier, New York, 1964.
7. G. G. Wagner and R. J. Kassner, *Biochim. Biophys. Acta* **392**, 319 (1975).
8. D. Brault and M. Rougee, *Biochemistry* **13**, 4591 (1974); *Biochem. Biophys. Res. Commun.* **57**, 654 (1974).
9. C. K. Chang, Thesis, p. 172. University of California, San Diego (1973).
9a. C. K. Chang, D. Powell, and T. G. Traylor, *Croat. Chem. Acta*, **49**, 295 (1977).
9b. C. K. Chang, unpublished results.
10. M. F. Perutz, H. Muirhead, J. M. Cox, and L. C. G. Goaman, *Nature (London)* **219**, 131 (1968).
11. J. C. Kendrew, *Brookhaven Symp. Biol.* **15**, 216 (1962).
12. H. C. Watson, *Prog. Stereochem.* **4**, 321 (1968).
13. F. F. Mathews, M. Levins, and P. Argos, *J. Mol. Biol.* **64**, 449 (1972).
14. R. E. Dickerson, T. Takano, D. Eisenberg, O. B. Kallai, L. Samson, A. Cooper, and E. Margoliash, *J. Biol. Chem.* **246**, 1511 (1971).
15. W. D. Brown and L. B. Mebine, *J. Biol. Chem.* **244**, 6696 (1969).
16. M. Kiese, *Pharmacol. Rev.* **18**, 1091 (1966).
17. T. Yonetani, H. Yamamoto, and G. V. Woodrow, III, *J. Biol. Chem.* **249**, 682 (1974).
18. R. Huber, O. Epp, W. Steigemann, and H. Formanek, *Eur. J. Biochem.* **19**, 42 (1971).

19a. G. Amiconi, E. Antonini, M. Brunori, H. Formanek and R. Huber, *Eur. J. Biochem.* **31**, 52 (1972).
19b. E. Antonini, M. Brunori, A. Caputo, E. Chiancone, A. Rossi-Fanelli, and J. Wyman, *Biochem. Biophys. Acta* **79**, 284 (1963).
20. I. A. Cohen and W. S. Caughey, *in* "Hemes and Hemoproteins" (B. Chance, R. W. Estabrook and T. Yonetani, eds.), p. 139. Academic Press, New York, 1966.
21a. Y. Sugita and Y. Yonetani, *J. Biol. Chem.* **246**, 389 (1971).
21b. M. Tamura, D. W. Woodrow, III, and Y. Yonetani, *Biochim. Biophys. Acta* **317** 34 (1973).
22. J. O. Alben and W. S. Caughey, *Biochemistry* **7**, 175 (1968).
23. G. Ilgenfritz and T. M. Schuster, *J. Biol. Chem.* **249**, 2959 (1974).
24. W. Scheler and I. Fischbach, *Acta Biol. Med. (Ger.)* **1**, 194 (1958).
25. W. S. Caughey, J. O. Alben, and C. A. Beaudreau, *Oxidases Relat. Redox Syst. Proc. Symp., 1964*, p. 97 (1965).
26. A. H. Corwin and Z. Reyes, *J. Am. Chem. Soc.* **78**, 2437 (1956); A. H. Corwin and S. D. Bruck, *ibid.* **80**, 4736 (1958).
27. J. H. Wang, *J. Am. Chem. Soc.* **80**, 3168 (1958).
28. F. Basolo, B. M. Hoffman, and J. A. Ibers, *Acc. Chem. Res.* **8**, 384 (1975).
29. H. Diekmann, C. K. Chang, and T. G. Traylor, *J. Am. Chem. Soc.* **93**, 4068 (1971).
30. C. K. Chang and T. G. Traylor, *Proc. Natl. Acad. Sci. U.S.A.* **70**, 2647 (1973).
31. C. K. Chang and T. G. Traylor, *J. Am. Chem. Soc.* **95**, 5810 (1973).
32. W. S. Brinigar and C. K. Chang, *J. Am. Chem. Soc.* **96**, 5595 (1974).
33. C. K. Chang and T. G. Traylor, *Biochem. Biophys. Res. Commun.* **62**, 729 (1975).
34. J. P. Collman, R. R. Gagne, C. A. Reed, T. R. Halbert, G. Lang, and W. T. Robinson, *J. Am. Chem. Soc.* **97**, 1427 (1975).
35. J. Almog, J. E. Baldwin, and J. Huff, *J. Am. Chem. Soc.* **97**, 227 (1975).
36. J. H. Wang, *Acc. Chem. Res.* **3**, 90 (1970).
37. I. A. Cohen and W. S. Caughey, *Biochemistry* **7**, 636 (1968); *in* "Hemes and Hemoproteins" (B. Chance, R. W. Estabrook, and T. Yonetani, eds.), p. 577. Academic Press, New York, 1966.
38. C. K. Chang and T. G. Traylor, *J. Am. Chem. Soc.* **95**, 8477 (1973).
39. O. Leal, D. L. Anderson, R. G. Bowman, F. Basolo, and R. L. Burwell, Jr., *J. Am. Chem. Soc.* **97**, 5125 (1975).
40. G. C. Wagner and R. J. Kassner, *J. Am. Chem. Soc.* **96**, 5593 (1974).
41. W. S. Brinigar, C. K. Chang, J. Geibel, and T. G. Traylor, *J. Am. Chem. Soc.* **96**, 5597 (1974).
42. J. Almog, J. E. Baldwin, R. L. Dyer, J. Huff, and C. J. Wilkerson, *J. Am. Chem. Soc.* **96**, 5600 (1974).
43. D. L. Anderson, C. J. Weschler, and F. Basolo, *J. Am. Chem. Soc.* **96**, 5599 (1974).
44. J. E. Baldwin and J. Huff, *J. Am. Chem. Soc.* **95**, 5757 (1973).
45. J. P. Collman, R. R. Gagne, T. R. Halbert, J. C. Marchon, and C. A. Reed, *J. Am. Chem. Soc.* **95** 7868 (1973).
46. J. H. Wang, *in* "Oxygenases" (O. Hayaishi, ed.), p. 469. Academic Press, New York, 1962.
47. C. K. Chang and T. G. Traylor, *Proc. Natl. Acad. Sci. U.S.A.* **72**, 1166 (1975).
48. C. L. Hobbs and H. C. Watson, "Mössbauer Colloquium." Springer-Verlag, Berlin and New York, 1968.
49. C. L. Hobbs and H. C. Watson, *Colloq. Ges. Biol. Chem.* **19**, 37 (1968).
50. J. P. Collman, R. R. Gagne, C. A. Reed, W. T. Robinson, and G. A. Rodley, *Proc. Natl. Acad. Sci. U.S.A.* **71**, 1326 (1974).

51. W. R. Scheidt and M. E. Frisse, *J. Am. Chem. Soc.* **97**, 17 (1975).
52. V. L. Goedken and S. Peng, *J. Am. Chem. Soc.* **96**, 7826 (1974).
53. J. C. Norvell, A. C. Nunes, and B. P. Schoenborn, *Science* **190**, 568 (1975).
54. J. L. Hoard, *Science* **174**, 1295 (1971).
55. C. K. Chang and D. Dolphin, *Proc. Natl. Acad. Sci. U.S.A.* **73**, 3338 (1976).
56. J. P. Collman, R. R. Gagne, H. B. Gray, and J. Hare, *J. Am. Chem. Soc.* **96**, 6522 (1974).
57. J. P. Collman, J. I. Brauman, and K. S. Suslick, *J. Am. Chem. Soc.* **97**, 7186 (1975).
58. C. H. Barlow, J. C. Maxwell, W. J. Wallace, and W. S. Caughey, *Biochem. Biophys. Res. Commun.* **55**, 91 (1973).
59. L. Marchant, M. Sharrock, B. M. Hoffman, and E. Munck, *Proc. Natl. Acad. Sci. U.S.A.* **69**, 2396 (1972).
60. J. Cannon and A. Berginis, unpublished work.
61. M. Rougée and D. Brault, *Biochemistry* **14**, 4100 (1975).
62. J. Geibel, unpublished results.
63. E. Tsuchida, E. Hasegawa, and K. Honda, *Biochem. Biophys. Res. Commun.* **67**, 864 (1975).
64. H. A. Hill and D. R. Turner, *Biochem. Biophys. Res. Commun.* **56**, 739 (1974).
65. C. L. Hobbs, H. C. Watson, and J. C. Kendrew, *Nature (London)* **209**, 339 (1966).
66. J. B. Cannon, J. Geibel, M. Whipple, and T. G. Traylor, *J. Am. Chem. Soc.* **98**, 3395 (1976).
67. J. Geibel, C. K. Chang, and T. G. Traylor, *J. Am. Chem. Soc.* **97**, 5924 (1975).
68. C. J. Weschler, D. L. Anderson, and F. Basolo, *J. Am. Chem. Soc.* **97**, 6707 (1975).
69. J. C. Maxwell and W. S. Caughey, *Biochemistry* **14**, 388 (1976).
70. M. F. Perutz, J. V. Kilmartin, K. Nagai, A. Szabo, and S. R. Simon, *Biochemistry* **15**, 378 (1976).
71. T. Reed, J. Bunkenberg, and B. Chance, *in* "Probes of Structure and Function of Macromolecules and Membranes" (B. Chance, T. Yonetani, and A. S. Mildvan, eds.), Vol. 2, p. 3. Academic Press, New York, 1971.
72. D. V. Stynes, H. C. Stynes, B. R. James, and J. A. Ibers, *J. Am. Chem. Soc.* **95**, 1796 (1973).
73. R. W. Noble and Q. H. Gibson, *J. Biol. Chem.* **244**, 3905 (1969).
74. B. Gonzalez, J. Kouba, S. Yee, C. A. Reed, J. F. Kirner, and W. R. Sheidt, *J. Am. Chem. Soc.* **97**, 3247 (1975).
75. Recent results of Dr. Dwane Campbell in our laboratories show that substitution of acetyl groups for ethyl groups in **5b** greatly decreases its oxygen affinity.

INDEX